Smithsonian Institution

Smithsonian meteorological tables

Smithsonian Institution

Smithsonian meteorological tables

ISBN/EAN: 9783743333352

Manufactured in Europe, USA, Canada, Australia, Japa

Cover: Foto ©berggeist007 / pixelio.de

Manufactured and distributed by brebook publishing software
(www.brebook.com)

Smithsonian Institution

Smithsonian meteorological tables

Smithsonian Miscellaneous Collections
1032

SMITHSONIAN

METEOROLOGICAL TABLES

[BASED ON GUYOT'S METEOROLOGICAL AND PHYSICAL TABLES]

(SECOND REVISED EDITION)

CORRECTED TO OCTOBER, 1897

CITY OF WASHINGTON
PUBLISHED BY THE SMITHSONIAN INSTITUTION
1897

ADVERTISEMENT TO SECOND REVISED EDITION.

The edition of the Smithsonian Meteorological Tables issued in 1893 having become exhausted, a careful examination of the work was made, at my request, by Mr. Alexander McAdie, of the United States Weather Bureau, and a revised edition was published in 1896 with corrections upon the plates, and a few slight changes. The International Meteorological Symbols and an Index were also added.

The demand for the work has been so great that it becomes necessary to print a new edition of the revised work, which is here presented with corrections to date.

S. P. Langley,
Secretary.

Smithsonian Institution,
Washington City, *October 30, 1897.*

PREFACE

In connection with the system of meteorological observations established by the Smithsonian Institution about 1850, a collection of meteorological tables was compiled by Dr. ARNOLD GUYOT, at the request of Secretary HENRY, and published in 1852 as a volume of the *Miscellaneous Collections*.

Five years later, in 1857, a second edition was published after careful revision by the author, and the various series of tables were so enlarged as to extend the work from 212 to over 600 pages.

In 1859 a third edition was published, with further amendments.

Although designed primarily for the meteorological observers reporting to the Smithsonian Institution, the tables obtained a much wider circulation, and were extensively used by meteorologists and physicists in Europe and in the United States.

After twenty-five years of valuable service, the work was again revised by the author; and the fourth edition, containing over 700 pages, was published in 1884. Before finishing the last few tables, Dr. GUYOT died, and the completion of the work was intrusted to his assistant, Prof. WM. LIBBEY, JR., who executed the duties of final editor.

In a few years the demand for the tables exhausted the edition, and thereupon it appeared desirable to recast entirely the work. After very careful consideration, I decided to publish the new tables in three parts: METEOROLOGICAL TABLES, GEOGRAPHICAL TABLES, and PHYSICAL TABLES, each representative of the latest knowledge in its field, and independent of the others; but the three forming a homogeneous series.

Although thus historically related to Dr. Guyot's Tables, the present work is so substantially changed with respect to material, arrangement, and presentation that it is not a fifth edition of the older tables, but essentially a new publication.

In its preparation the advantage of conformity with the recently issued *International Meteorological Tables* has been kept steadily in view, and so far as consistent with other decisions, the constants and methods there employed have been followed. The most important difference in constants is the relation of the yard to the metre. The value provisionally adopted by the Bureau of Weights and Measures of the United States Coast and Geodetic Survey,

$$1 \text{ metre} = 39.3700 \text{ inches},$$

has been used here in the conversion-tables of metric and English linear measures, and in the transformation of all formulæ involving such conversions.

A large number of tables have been newly computed; those taken from the *International Meteorological Tables* and other official sources are credited in the introduction.

To Prof. WM. LIBBEY, JR., especial acknowledgments are due for a large amount of attention given to the present work. Prof. LIBBEY had already completed a revision, involving considerable recomputation, of the meteorological tables contained in the last edition of Guyot's Tables, when it was determined to adopt new values for many of the constants, and to have the present volume set with new type. This involved a large amount of new computation, which was placed under the direction of Mr. GEORGE E. CURTIS, who has also written the text, and has carefully prepared the whole manuscript and carried it through the press. To Mr. Curtis's interest, and to his special experience as a meteorologist, the present volume is therefore largely due.

Prof. LIBBEY has contributed Tables 38, 39, 55, 56, 61, 74, 77, 89, and 90, and has also read the proof-sheets of the entire work.

I desire to express my acknowledgments to Prof. CLEVELAND ABBE, for the manuscript of Tables 32, 81, 82, 83, 84, 85, 86; to Mr. H. A. HAZEN, for Tables 49, 50, 94, 95, 96, which have been taken from his *Hand-book of Meteorological Tables;* and also to the Superintendent of the United States Coast and Geodetic Survey, the Chief Signal Officer of the Army, and the Chief of the Weather Bureau, for much valuable counsel during the progress of the work.

S. P. LANGLEY,
Secretary.

TABLE OF CONTENTS.

INTRODUCTION.

		PAGE
Description and use of the Tables		xi to lix

THERMOMETRICAL TABLES.

TABLE		PAGE
	Conversion of thermometric scales—	
1	Reaumur scale to Fahrenheit and Centigrade	2
2	Fahrenheit scale to Centigrade	3
3	Centigrade scale to Fahrenheit	7
4	Centigrade scale to Fahrenheit, near the boiling point of water	9
5	Differences Fahrenheit to differences Centigrade	9
6	Differences Centigrade to differences Fahrenheit	9
7	Reduction of temperature to sea level—English measures	10
8	Reduction of temperature to sea level—Metric measures	11
9	Correction for the temperature of the mercury in the thermometer stem. For Fahrenheit and Centigrade thermometers	12

BAROMETRICAL TABLES.

	Reduction of the barometer to standard temperature—	
10	English measures	14
11	Metric measures	34
	Reduction of the barometer to standard gravity at latitude 45°—	
12	English measures	58
13	Metric measures	59
	Reduction of the barometer to sea level — English measures.	
14	Values of $2000\,m$	60
15	Correction of $2000\,m$ for latitude	69
16	$B_o - B = B(10^m - 1)$	70
	Reduction of the barometer to sea level — Metric measures.	
17	Values of $2000\,m$	78
18	Correction of $2000\,m$ for latitude	90
19	$B_o - B = B(10^m - 1)$	91

TABLE OF CONTENTS.

BAROMETRICAL TABLES.—Continued.

TABLE		PAGE
	Determination of heights by the barometer — English measures.	
20	Values of $60368 \left[1 + 0.0010195 \times 36\right] \log \dfrac{29.90}{B}$	100
21	Term for temperature	104
22	Correction for latitude and weight of mercury	106
23	Correction for an average degree of humidity	108
24	Correction for the variation of gravity with altitude	109
	Determination of heights by the barometer — Metric measures.	
25	Values of $18400 \log \dfrac{760}{B}$	110
26	Term for temperature	111
27	Correction for humidity	112
28	Correction for latitude and weight of mercury	114
29	Correction for the variation of gravity with altitude	115
30	Difference of height corresponding to a change of 0.1 inch in the barometer — English measures	116
31	Difference of height corresponding to a change of 1 millimetre in the barometer — Metric measures	117
	Determination of heights by the barometer.	
32	Formula of Babinet	118
	Barometric pressures corresponding to the temperature of the boiling point of water —	
33	English measures	119
34	Metric measures	119

HYGROMETRICAL TABLES.

	Pressure of aqueous vapor (*Broch*) —	
35	English measures	122
36 } 43 }	Metric measures	{ 128 142
37	Pressure of aqueous vapor at low temperatures (*C. F. Marvin*) — English and Metric measures	130
38	Weight of aqueous vapor in a cubic foot of saturated air — English measures	132
39	Weight of aqueous vapor in a cubic metre of saturated air — Metric measures	133
	Reduction of psychrometric observations — English measures.	
40	Pressure of aqueous vapor	134
41	Values of $0.000367\, B\,(t - t_1) \left(1 + \dfrac{t - t_1}{1571}\right)$	136
42	Relative humidity — Temperature Fahrenheit	138

HYGROMETRICAL TABLES.—*Continued.*

Reduction of psychrometric observations — Metric measures.

Table		Page
43	Pressure of aqueous vapor	142
44	Values of $0.000660\, B\,(t - t_1)\left(1 + \dfrac{t - t_1}{873}\right)$	143
45	Relative humidity—Temperature Centigrade	144

Reduction of snowfall measurements.

46	Depth of water corresponding to the weight of snow (or rain) collected in an 8-inch gage	146
47	Rate of decrease of vapor pressure with altitude	146

WIND TABLES.

Mean direction of the wind by Lambert's formula—

48	Multiples of cos 45°; form and example of computation	148
49	Values of the mean direction (a) or its complement ($90 - a$)	149
50	Synoptic conversion of velocities	154
51	Miles per hour into feet per second	155
52	Feet per second into miles per hour	155
53	Metres per second into miles per hour	156
54	Miles per hour into metres per second	157
55	Metres per second into kilometres per hour	158
56	Kilometres per hour into metres per second	159
57	Beaufort wind scale and its conversion into velocity	160

GEODETICAL TABLES.

58	Relative acceleration of gravity at different latitudes	162
59	Length of one degree of the meridian at different latitudes	164
60	Length of one degree of the parallel at different latitudes	165
61	Duration of sunshine at different latitudes	166
62	Declination of the sun for the year 1894	177
63	Relative intensity of solar radiation at different latitudes	178

CONVERSION OF LINEAR MEASURES.

64	Inches into millimetres	180
65	Millimetres into inches	187
66	Feet into metres	200
67	Metres into feet	202
68	Miles into kilometres	204
69	Kilometres into miles	206
70	Interconversion of nautical and statute miles	208
71	Continental measures of length with their metric and English equivalents	208

TABLE OF CONTENTS.

CONVERSION OF MEASURES OF TIME AND ANGLE.

Table		Page
72	Arc into time	210
73	Time into arc	211
74	Days into decimals of a year and angle	212
75	Hours, minutes and seconds into decimals of a day	216
76	Decimals of a day into hours, minutes and seconds	216
77	Minutes and seconds into decimals of an hour	217
78	Mean time at apparent noon	217
79	Sidereal time into mean solar time	218
80	Mean solar time into sidereal time	218

MISCELLANEOUS TABLES.

81	Density of air at different temperatures Fahrenheit	220
	Density of air at different humidities and pressures—English measures.	
82	Term for humidity: auxiliary to Table 83	221
83	Values of $\dfrac{h}{29.921} = \dfrac{b - 0.378\,e}{29.921}$	222
84	Density of air at different temperatures Centigrade	224
	Density of air at different humidities and pressures—Metric measures.	
85	Term for humidity: auxiliary to Table 86	225
86	Values of $\dfrac{h}{760} = \dfrac{b - 0.378\,e}{760}$	226
87	Conversion of avoirdupois pounds and ounces into kilogrammes	226
88	Conversion of kilogrammes into avoirdupois pounds and ounces	230
89	Conversion of grains into grammes	230
90	Conversion of grammes into grains	231
91	Conversion of units of magnetic intensity	231
92	Quantity of water corresponding to given depths of rainfall	232
93	Dates of Dove's pentades	232
94	Division by 28 of numbers from 28 to 867 972	233
95	Division by 29 of numbers from 29 to 898 971	234
96	Division by 31 of numbers from 31 to 960 969	235
97	Natural sines and cosines	236
98	Natural tangents and cotangents	238
99	Logarithms of numbers	240
100	LIST OF METEOROLOGICAL STATIONS	243

APPENDIX.

Constants	258
Synoptic conversion of English and metric units	260
Dimensions of physical quantities	262
International Meteorological Symbols	263

INTRODUCTION.

DESCRIPTION AND USE OF THE TABLES.

THERMOMETRICAL TABLES.

COMPARISON OF THERMOMETRIC SCALES.

Conversion of readings of the Reaumur thermometer to readings of the Fahrenheit and Centigrade thermometers.   **TABLE 1.**

The argument is given for every Reaumur degree from + 80° to − 40° Reaumur, and the corresponding readings Fahrenheit and Centigrade are given to hundredths of a degree, permitting the exact values to be expressed. A column of proportional parts gives the values corresponding to tenths of a Reaumur degree. By the help of the column of proportional parts, the table is also conveniently used for converting Fahrenheit to Centigrade and Reaumur, and Centigrade to Fahrenheit and Reaumur throughout the thermometric scale from the boiling point of water to − 60° F. or − 51° C.

The formulæ expressing the relation between the different scales are given at the bottom of the table, where

$F°$ = Temperature Fahrenheit.
$C°$ = Temperature Centigrade.
$R°$ = Temperature Reaumur.

Examples:

To convert 18°.3 Reaumur to Fahrenheit and Centigrade.

From the table,	18°.0 $R.$ =	72°.50 $F.$ =	22°.50 $C.$
From column Prop. Parts,	0.3 =	0.675 =	0.375
	18°.3 $R.$ =	73°.2 $F.$ =	22°.9 $C.$

To convert 147°.7 Fahrenheit to Centigrade and Reaumur.

From the table,	146°.75 $F.$ =	63°.75 $C.$ =	51°.0 $R.$
From column Prop. Parts,	0.95 =	0.53 =	0.4
	147°.7 $F.$ =	64°.3 $C.$ =	51°.4 $R.$

To convert 16°.9 Centigrade to Fahrenheit and Reaumur.

From the table,	16°.25 $C.$ =	61°.25 $F.$ =	13°.0 $R.$
From column Prop. Parts,	0.65 =	1.17 =	0.5
	16°.9 $C.$ =	62°.4 $F.$ =	13°.5 $R.$

TABLE 2. *Conversion of readings of the Fahrenheit thermometer to readings Centigrade.*

The conversion of Fahrenheit temperatures to Centigrade temperatures is given for every tenth of a degree from $+130°.9\,F.$ to $-70°.9\,F.$ The side argument is the whole number of degrees Fahrenheit, and the top argument, tenths of a degree Fahrenheit; interpolation to hundredths of a degree, when desired, is readily effected mentally. The tabular values are given to hundredths of a degree Centigrade.

The formula for conversion is

$$C° = \frac{5}{9}(F° - 32°)$$

where $F°$ is a given temperature Fahrenheit, and $C°$ the corresponding temperature Centigrade.

Example:

To convert $79°.7$ Fahrenheit to Centigrade.
The table gives directly $26°.50\,C.$

For conversions of temperatures above $131°\,F$, use Table 1.

TABLE 3. *Conversion of readings of the Centigrade thermometer to readings Fahrenheit.*

The conversion of Centigrade temperatures to Fahrenheit temperatures is given for every tenth of a degree Centigrade from $+50°.9$ to $-50°.9\,C.$ The tabular values are expressed in hundredths of a degree Fahrenheit.

The formula for conversion is

$$F° = \frac{9}{5}C° + 32°$$

where $C°$ is a given temperature Centigrade, and $F°$ the corresponding temperature Fahrenheit.

For conversions of temperatures above the upper limit of the table, use Tables 1 and 4.

TABLE 4. *Conversion of readings of the Centigrade thermometer near the boiling point to readings Fahrenheit.*

This is an extension of Table 3 from $90°.0$ to $100°.9$ Centigrade.

Example:

To convert $95°.74$ Centigrade to Fahrenheit.
| From the table, | $95°.70\,C.$ | $=$ | $204°.26\,F.$ |
| By interpolation, | 0.04 | $=$ | 0.07 |

$$95°.74\,C. = 204°.33\,F.$$

Conversion of differences Fahrenheit to differences Centigrade. **TABLE 5.**

The table gives for every tenth of a degree from 0° to 20°.9 F. the corresponding lengths of the Centigrade scale.

Conversion of differences Centigrade to differences Fahrenheit. **TABLE 6.**

The table gives for every tenth of a degree from 0° to 9°.9 C. the corresponding lengths of the Fahrenheit scale.

Example:
 To find the equivalent difference in Fahrenheit degrees for a difference of 4°.72 Centigrade.
 From the table,   4°.70 C. = 8°.46 F.
 From the table by moving the decimal point for 0.2,   0.02 = 0.04
                                                                                  4°.72 C. = 8°.50 F.

REDUCTION OF TEMPERATURE TO SEA LEVEL.
 English Measures.   **TABLE 7.**
 Metric Measures.   **TABLE 8.**

These tables give for different altitudes and for different uniform rates of decrease of temperature with altitude, the amount in hundredths of a degree Fahrenheit and Centigrade, which must be added to observed temperatures in order to reduce them to sea level.

The rate of decrease of temperature with altitude varies from one region to another, and in the same region varies according to the season and the meteorological conditions; being in general greater in warm latitudes than in cold ones, greater in summer than in winter, and greater in cyclones than in anti-cyclones. For continental plateau regions, the reduction often becomes fictitious or illusory. The use of the tables therefore requires experience and judgment in selecting the rate of decrease of temperature to be used.

The tables are given in order to facilitate the reduction of temperature either upwards or downwards in special investigations, but the reduction is not ordinarily applied to meteorological observations.

The tables, 7 and 8, are computed for rates of temperature change ranging from 1° Fahrenheit in 200 feet to 1° Fahrenheit in 900 feet, and from 1° Centigrade in 100 metres to 1° Centigrade in 500 metres; and for altitudes up to 5,000 feet and 3,000 metres respectively.

Example, Table 7:
 Observed temperature at an elevation of 2,500 feet,   52°.5 F.
 Reduction to sea level for an assumed decrease in temperature of 1°F. for every 300 feet,   + 8°.3
 Temperature reduced to sea level,   60°.8 F.

Example, Table 8:

Observed temperature at an elevation of 500 metres,	12°.5 C.
Reduction to sea level for an assumed decrease in temperature of 1° C. for every 200 metres,	+ 2°.5
Temperature reduced to sea level,	15°.0 C.

CORRECTION FOR THE TEMPERATURE OF THE MERCURY IN THE THERMOMETER STEM.

TABLE 9. *Fahrenheit thermometers; Centigrade thermometers.*

When the temperature of the thermometer stem is materially different from that of the bulb, a correction needs to be applied to the observed reading in order to correct it for the difference in the length of the mercury column caused by this difference in its temperature. This correction frequently becomes necessary in physical experiments where the bulb only is immersed in a bath whose temperature is to be determined, and in meteorological observations it may become appreciable in wet-bulb, dew point, and solar radiation thermometers, when the temperature of the bulb is considerably above or below the air temperature.

If t' be the average temperature of the mercury column, t the observed reading of the thermometer, n the length of mercury in the stem in scale degrees, and a the apparent expansion of mercury in glass for 1°, the correction is given by the expression

$$- an (t' - t)$$

in which, for Centigrade temperatures, $a = 0.000154$ or 0.000155.

The average temperature of the mercury column can not be directly observed and is difficult to determine, for it differs from the temperature of the glass stem by an amount depending on the conduction of heat between the bulb and the mercury column. Practically however it is possible to use the actually observed temperature of the glass stem as the value of t' by making a small compensating change in the value of a, and this appears to be the simplest method that has been proposed. Mr. T. E. Thorpe (*Journal of the Chemical Society*, vol. 37, 1880, p. 160) has determined by a series of experiments that the proper thermometric corrections will be obtained by this method if 0.000143 be used as a coefficient (for Centigrade temperatures) instead of the value of a given above, and this value has been adopted in the present tables.

The correction formulæ are, then,

$$T = t - 0.0000795\ n\ (t' - t) \quad \text{Temperature Fahrenheit.}$$
$$T = t - 0.000143\ n\ (t' - t) \quad \text{Temperature Centigrade.}$$

in which $T =$ Corrected temperature.
$t =$ Observed temperature.
$t' =$ Mean temperature of the glass stem.
$n =$ Length of mercury in the stem in scale degrees.

When t' is $\begin{Bmatrix} \text{greater} \\ \text{less} \end{Bmatrix}$ than t, the numerical correction is to be $\begin{Bmatrix} \text{subtracted.} \\ \text{added.} \end{Bmatrix}$

Example:

The observed temperature of a black bulb thermometer is 120°.4 $F.$, the temperature of the glass stem is 55°.2 $F.$ and the length of mercury in the stem is 130° $F.$ To find the corrected temperature.

With $n = 130°$ $F.$ and $-t'$ $t = [-]$ 65° $F.$, as arguments, the table gives the correction 0°.7 $F.$, which by the above rule is to be added to the observed temperature. The corrected temperature is therefore 121°.1 $F.$

BAROMETRICAL TABLES.

REDUCTION TO A STANDARD TEMPERATURE OF OBSERVATIONS MADE WITH BAROMETERS HAVING BRASS SCALES.

The indicated height of the mercurial column in a barometer varies not only with changes of atmospheric pressure, but also with variations of the temperature of the mercury and of the scale. It is evident therefore that if the height of the barometric column is to be a true relative measure of atmospheric pressure, the observed readings must be reduced to the values they would have if the mercury and scale were maintained at a constant standard temperature.

This reduction is known as the reduction for temperature, and combines both the correction for the expansion of the mercury and that for the expansion of the scale, on the assumption that the attached thermometer gives the temperature both of the mercury and of the scale.

The freezing point is universally adopted as the standard temperature of the mercury, to which all readings are to be reduced. The temperature to which the scale is reduced is the normal or standard temperature of the adopted standard of length. For English scales, which depend upon the English yard, this is 62° Fahrenheit. For metric scales, which depend upon the metre, it is 0° Centigrade.

As thus reduced, observations made with English and metric barometers become perfectly comparable when converted by the ordinary tables of linear conversion, viz.: millimetres to inches and inches to millimetres (see Tables 64, 65), for these conversions refer to the metre at 0° Centigrade and the English yard at 62° Fahrenheit.

The general formula for reducing barometric readings to a standard temperature is

$$C = -B \frac{m(t-T) - l(t-\theta)}{1 + m(t-T)},$$

in which $C =$ Correction for temperature.
$B =$ Observed height of the barometric column.
$t =$ Temperature of the attached thermometer.
$T =$ Standard temperature of the mercury.
$m =$ Coefficient of expansion of mercury.
$l =$ Coefficient of linear expansion of brass.
$\theta =$ Standard temperature of the scale.

The accepted determination of the coefficient of expansion of mercury is that given by Broch's reduction of Regnault's experiments, viz:

$$m \text{ (for } 1° C.) = 10^{-9}(181792 + 0.175 t + 0.035116 t^2).$$

As a sufficiently accurate approximation, the intermediate value

$$m = 0.0001818$$

has been adopted uniformly for all temperatures in conformity with the usage of the *International Meteorological Tables*.

Various specimens of brass scales made of alloys of different composition show differences in their coefficients of expansion amounting to eight and sometimes ten per cent. of the total amount. The *Smithsonian Tables* prepared by Prof. Guyot were computed with the average value l (for $1° C$) $= 0.0000188$; for the sake of uniformity with the *International Meteorological Tables*, the value

$$l = 0.0000184$$

has been used in the present volume. For any individual scale, either value may easily be in error by four per cent.

A small portion of the tables has been independently computed, but the larger part of the values have been copied from the *International Meteorological Tables*, one inaccuracy having been found and corrected.

TABLE 10. *Reduction of the barometer to standard temperature—English measures.*

For the English barometer the formula for reducing observed readings to a standard temperature becomes

$$C = - B \frac{m (t - 32°) - l (t - 62°)}{1 + m (t - 32°)}$$

in which $B =$ Observed height of the barometer in English inches.
$t =$ Temperature of attached thermometer in degrees Fahrenheit.

$$m = 0.0001818 \times \frac{5}{9} = 0.000101$$

$$l = 0.0000184 \times \frac{5}{9} = 0.0000102$$

The combined reduction of the mercury to the freezing point and of the scale to 62° Fahrenheit brings the point of no correction to approximately

28°.5 Fahrenheit, and this is therefore the standard temperature to which all readings are reduced. For temperatures above 28°.5 Fahrenheit, the correction is subtractive, and for temperatures below 28°.5 Fahrenheit, the correction is additive, as indicated by the signs (+) and (−) inserted throughout the table.

The table gives the corrections for every half degree Fahrenheit from 0° to 100°. The limits of pressure are 19 and 31.6 inches, the corrections being computed for every half inch from 19 to 24 inches, and for every two-tenths of an inch from 24 to 31.6 inches.

Example:

Observed height of barometer	= 29.143
Attached thermometer, 54°.5 F.	
Reduction for temperature	= − 0.068
Barometric reading corrected for temperature	= 29.075

TABLE 11.

TABLE 11. *Reduction of the barometer to standard temperature—Metric measures.*

For the metric barometer the formula for reducing observed readings to the standard temperature, 0° $C.$, becomes

$$C = -B \frac{(m-l)t}{1+mt}$$

in which C and B are expressed in millimetres and t in Centigrade degrees.

$$m = 0.0001818 \ ; \quad l = 0.0000184.$$

In the tables, the limits adopted for the pressure are 440 and 795 millimetres, the intervals being 10 millimetres between 440 and 600 millimetres, and 5 millimetres between 600 and 795 millimetres.

The limits adopted for the temperature are 0° + and + 35°.8, the intervals being 0°.5 and 1°.0 from 440 to 560 millimetres, and 0°.2 from 560 to 795 millimetres.

For temperatures above 0° Centigrade the correction is *negative*, and hence is to be subtracted from the observed readings.

For temperatures below 0° Centigrade the correction is *positive*, and from 0° $C.$ down to − 20° $C.$ the numerical values thereof, for ordinary barometric work, do not materially differ from the values for the corresponding temperatures above 0° C. Thus the correction for − 9° $C.$ is *numerically* the same as for + 9° $C.$ and is taken from the table. In physical work of extreme precision, the numerical values given for positive temperatures may be used for temperatures below 0° $C.$ by applying to them the following corrections:

INTRODUCTION.

Corrections to be applied to the tabular values of Table 11 in order to use them when the temperature of the attached thermometer is below 0° Centigrade.

Temperature.	PRESSURE IN MILLIMETRES.							
	450	500	550	600	650	700	750	800
C.	mm.	mm.	mm.	mm.	mm.	mm.	mm.	mm.
− 1°	0.00	0.00	0.00	0.00	0.00	0.00	0.00	0.00
− 9	.00	.00	.00	.00	.00	.00	.00	.00
− 10	0.00	0.00	0.00	0.00	0.00	+0.01	+0.01	+0.01
11	.00	.00	.00	.00	+0.01	.01	.01	.01
12	.00	.00	.00	+0.01	.01	.01	.01	.01
13	.00	.00	+0.01	.01	.01	.01	.01	.01
− 14	.00	+0.01	.01	.01	.01	.01	.01	.01
− 15	+0.01	+0.01	+0.01	+0.01	+0.01	+0.01	+0.01	+0.01
16	.01	.01	.01	.01	.01	.01	.01	.01
17	.01	.01	.01	.01	.01	.01	.01	.02
18	.01	.01	.01	.01	.01	.01	.01	.02
− 19	.01	.01	.01	.01	.01	.01	.02	.02
− 20	+0.01	+0.01	+0.01	+0.01	+0.01	+0.02	+0.02	+0.02
21	.01	.01	.01	.02	.02	.02	.02	.02
22	.01	.01	.02	.02	.02	.02	.02	.02
23	.01	.02	.02	.02	.02	.02	.02	.02
− 24	.01	.02	.02	.02	.02	.02	.02	.03

Example:

Observed height of barometer, 763.17^{mm}.: Temperature of the attached thermometer, − 12° C.

Numerical value of the reduction for + 12° C.	=	1.50
Correction for temperature below 0° C.	= +	0.01
Reduction for − 12° C.	= +	1.51
Observed height of barometer	=	763.17
Barometer corrected for temperature	=	764.68

REDUCTION OF THE BAROMETER TO STANDARD GRAVITY AT LATITUDE 45°.

The atmospheric pressure is measured by the weight of the mercurial column of the barometer, but by common usage the pressures are expressed in terms of the *height* of the barometric column instead of by its *weight*. The observed height however is not a true measure of the pressure, because it changes with the temperature of the mercury and with the variations in the value of gravity. Therefore to obtain a height that shall be a true relative measure of the atmospheric pressure, the observed height of the mercurial column must be reduced to that which would be measured at a standard temperature and under a uniform standard value of gravity.

BAROMETRICAL TABLES.

The standard value of gravity adopted is that prevailing at latitude 45° and sea level. The reduction, accordingly, consists of two parts — a correction for altitude and a correction for latitude. The gravity correction for altitude is usually combined with the reduction of the barometer to sea level; the gravity correction for latitude, which is here given, is commonly called simply the "gravity correction," or the "reduction to standard gravity."

If B_ϕ and B_{45} represent the barometric heights (corrected for temperature) at latitudes ϕ and 45°, and g_ϕ, g_{45} the acceleration of gravity at these latitudes, we have

$$\frac{B_\phi}{B_{45}} = \frac{g_{45}}{g_\phi},$$

and the correction to the observed height will be

$$C = B_{45} - B_\phi = -B_\phi \left(1 - \frac{g_\phi}{g_{45}}\right).$$

If the earth be an ellipsoid of revolution composed of homogeneous homofocal layers arranged according to any law of density,

$$g_\phi = g_{45} (1 - k \cos 2\phi)$$

in which k is a constant depending on the ellipticity of the earth; and the correction becomes

$$C = -k \cos 2\phi \, B_\phi.$$

The value of k adopted here is that determined by Prof. Harkness,*

$$k = 0.002662.$$

The correction is the same numerically for $\phi = 45° + a$ and $\phi = 45° - a$. It is negative for latitudes below 45° and positive for latitudes above 45°.

TABLES 12, 13.

TABLE 12 (*English measures*) gives the correction in thousandths of an inch for every degree of latitude and for each inch of barometric pressure from 19 to 30 inches.

TABLE 13 (*Metric measures*) gives the correction in hundredths of a millimetre for each 20 millimetres barometric pressure from 520 to 770 millimetres.

Example:

Barometric reading (corrected for temperature) at Dodge City, latitude 37° 45′,	= 27.434
Gravity correction for latitude from Table 12,	= − 0.018
Barometer reduced to latitude 45°,	= 27.416

---

*WM. HARKNESS: *The solar parallax and its related constants.* Washington, 1891, 4°, pp. 169.

REDUCTION OF THE BAROMETER TO SEA LEVEL.

The fundamental formula for reducing the barometer to sea level and for determining heights by the barometer is the original formula of Laplace, amplified into the following form—

$$Z = K(1 + a\theta)\left(\frac{1}{1 - 0.378\frac{e}{b}}\right)(1 + k\cos 2\phi)\left(1 + \frac{h + h_0}{R}\right)\log\frac{p_0}{p}$$

in which h = Height of the upper station.
h_0 = Height of the lower station.
$Z = h - h_0$.
p = Atmospheric pressure at the upper station.
p_0 = Atmospheric pressure at the lower station.
R = Mean radius of the earth.
θ = Mean temperature of the air column between the altitudes h and h_0.
e = Mean pressure of aqueous vapor in the air column.
b = Mean barometric pressure of the air column.
ϕ = Latitude of the stations.
K = Barometric constant.
a = Coefficient of the expansion of air.
k = Constant depending on the figure of the earth.

The pressures p_0 and p are computed from the height of the column of mercury at the two stations; the ratio $\frac{B_0}{B}$ of the barometric heights may be substituted for the ratio $\frac{p_0}{p}$, if B_0 and B are reduced to the values that would be measured at the same temperature and under the same relative value of gravity.

The correction of the observed barometric heights for instrumental temperature is always separately made, but the correction for the variation of gravity with altitude is generally introduced into the formula itself.

If B_0, B represent the barometric heights corrected for temperature only, we have the equation

$$\frac{p_0}{p} = \frac{B_0}{B}\left(1 + \mu\frac{Z}{R}\right),$$

μ being a constant depending on the variation of gravity with altitude.

$$\log\frac{p_0}{p} = \log\frac{B_0}{B} + \log\left(1 + \mu\frac{Z}{R}\right).$$

Since $\frac{\mu Z}{R}$ is a very small fraction, we may write

$$\text{Nap. log}\left(1 + \frac{\mu Z}{R}\right) = \frac{\mu Z}{R}, \text{ and } \log\left(1 + \frac{\mu Z}{R}\right) = \frac{\mu Z}{R}M,$$

M being the modulus of common logarithms.

By substituting for Z its approximate value $Z = K \log \frac{B_0}{B}$, we have

$$\log\left(1 + \frac{\mu Z}{R}\right) = \frac{\mu K}{R} M \log \frac{B_0}{B}.$$

With these substitutions the barometric formula becomes

$$Z = K(1 + a\theta)\left(\frac{1}{1 - 0.378\frac{e}{b}}\right)(1 + k \cos 2\phi)\left(1 + \frac{h + h_0}{R}\right)$$
$$\left(1 + \frac{\mu K}{R} M\right) \log \frac{B_0}{B}.$$

As a further simplification we shall put

$$\beta = 0.378 \frac{e}{b}, \quad \gamma = k \cos 2\phi \text{ and } \eta = \frac{\mu K}{R} M,$$

and write the formula—

$$Z = K(1 + a\theta)\left(\frac{1}{1 - \beta}\right)(1 + \gamma)\left(1 + \frac{h + h_0}{R}\right)(1 + \eta) \log \frac{B_0}{B}.$$

Values of the constants.—The barometric constant K is a complex quantity defined by the equation

$$K = \frac{\Delta \times B_n}{\delta \times M}.$$

B_n is the normal barometric height of Laplace, 760 *mm*.

Δ is the density of mercury at the temperature of melting ice. M. Marek (*Travaux et Mémoires du Bureau international des Poids et Mesures*, t. II, p. D 55) gives the value, $\Delta = 13.5956$, and finds that different specimens of mercury purified by different processes differ from this by several units in the fourth decimal. The International Meteorological Committee have taken the value

$$\Delta = 13.5958,$$

and for the sake of uniformity this value is here adopted.

δ is the density of dry air at $0°$ C. and under the pressure of a column of mercury B_n at the sea level and at latitude $45°$. The value adopted by the International Bureau of Weights and Measures (*Travaux et Mémoires*, t. I, p. A 54) is

$$\delta = 0.001293052.$$

M (the modulus of common logarithms) $= 0.4342945$.

These numbers give for the value of the barometric constant

$$K = 18400 \text{ metres.}$$

INTRODUCTION.

For the remaining constants, the following values have been used:

$a = 0.00367$ for 1° Centigrade. (International Bureau of Weights and Measures: *Travaux et Mémoires*, t. I, p. A 54.)

$\gamma = k \cos 2\phi = 0.002662 \cos 2\phi$. (Harkness: *The solar parallax, etc.*, see p. xix.)

$R = 6367324$ metres. (A. R. Clarke: *Geodesy*, 8°, Oxford, 1880.)

$\eta = \dfrac{\mu K M}{R} = 0.002396$. (Ferrel: *Report Chief Signal Officer*, 1885, pt. 2, p. 393.)

In reducing the barometer to sea-level, $h_o = o$, and the factor $\left(1 + \dfrac{h + h_o}{R}\right)$ becomes $\left(1 + \dfrac{Z}{R}\right)$. Taking the product of this factor and $K(1 + a\theta)$ $(1 + n)$, and neglecting the term in θZ, the formula becomes

in *metric measures*

$$Z \text{ (metres)} = (18444 + 67.53\, \theta^{\circ C.} + 0.003\, Z)\left(\dfrac{1}{1-\beta}\right)(1+\gamma)\log\dfrac{B_o}{B},$$

and in *English measures*

$$Z \text{ (feet)} = (56573 + 123.1\, \theta^{\circ F.} + 0.003\, Z)\left(\dfrac{1}{1-\beta}\right)(1+\gamma)\log\dfrac{B_o}{B}.$$

The form adopted for the tables is that of M. Angot.*

Taking the formula in English measures, let

$$m = \dfrac{Z}{56573 + 123.1\,\theta + 0.003\,Z} \cdot \dfrac{1}{1+\beta}.$$

Then disregarding the small correction for gravity, $m = \log\dfrac{B_o}{B}$ gives an approximate value of B_o, and the correction to be added to the observed pressure to obtain the sea-level pressure is

$$C = B_o - B = B(10^m - 1).$$

If m_1 be the value of m corrected for gravity, we have

$$m_1 = \dfrac{m}{1+\gamma} \text{ or, approximately,} = m - m\gamma.$$

The correction for gravity is therefore made by applying to the approximate value m the small correction $m\gamma$. With this corrected value of m, the reduction to sea-level is given by the expression

$$B(10^m - 1).$$

The above fraction designated m contains the altitude Z, the mean temperature θ, and the humidity factor $\dfrac{1}{1+\beta}$.† In the *Smithsonian tables*,

---

*A. ANGOT: *Annales du Bureau Central Météorologique*. Année 1878, t. I, p. C. 13.

†In the humidity factor, when the temperature is below 75°F, $1 - \beta = \dfrac{1}{1+\beta}$. At higher temperatures, the substitution is incorrect, and the more general form $m = \dfrac{Z(1-\beta)}{56573 + 123.1\theta + .003Z}$ is therefore used on pages 60–68, 78–89.

meteorological and physical, by Dr. A. Guyot, the distinguished author in treating of this humidity factor in connection with hypsometric tables took the following position:

"To introduce a separate correction for the expansion of aqueous vapor "is, in the writer's view, a doubtful improvement. The laws of the distri-"bution and transmission of moisture through the atmosphere are too little "known, and its amount, especially in mountain regions, is too variable, and "depends too much upon local winds and local condensation, to allow a "reasonable hope of obtaining the mean humidity of the layer of air between "the two stations by means of hygrometrical observations made at each of "them. These doubts are confirmed by the experience of the author and "of many other observers, which shows that, on an average, Laplace's "method works not only as well as the other, but more uniformly well. At "any rate the gain, if there be any, is not clear enough to compensate for "the undesirable complication of the formula."

Since this position was taken by Dr. Guyot forty years ago, there has been no such advance in our knowledge as to impair the practical conclusion in conformity with which he constructed his hypsometric table. Accordingly in treating this portion of the formula in the construction of the present tables for the reduction of the barometer to sea level, it has been deemed advantageous to retain the method adopted by Guyot, and to incorporate the humidity factor in the temperature term, thereby assuming the air to contain the average degree of humidity corresponding to the actually prevailing condition of temperature.

In evaluating the humidity factor as a function of the air temperature, the tables given by Prof. Ferrel have been adopted (*Meteorological researches. Part iii.—Barometric hypsometry and reduction of the barometer to sea level.* Report, U. S. Coast Survey, 1881. Appendix 10.) These tables by interpolation, and by extrapolation below 0° F., give the following values for β:

For Fahrenheit temperatures,

θ	β	θ	β	θ	β	θ	β
F.		F.		F.		F.	
−20°	0.00008	10°	0.00104	36°	0.00267	62°	0.00724
−16	.00020	12	.00111	38	.00293	64	.00762
−12	.00032	14	.00118	40	.00322	66	.00801
−8	.00044	16	.00126	42	.00353	68	.00839
		18	.00134	44	.00386	70	.00877
−6	0.00050	20	.00143	46	.00421	72	.00914
−4	.00056	22	.00153	48	.00458		
−2	.00062	24	.00163	50	.00496	76	0.00990
0	.00068	26	.00174	52	.00534	80	.01065
+2	.00075	28	.00187	54	.00572	84	.01141
4	.00082	30	.00203	56	.00610	88	.01217
6	.00089	32	.00222	58	.00648	92	.01293
8	0.00096	34	0.0243	60	.00686	96	.01369

For Centigrade temperatures,

θ	β	θ	β	θ	β
C.		C.		C.	
−18°	0.0007	0°	0.0022	18°	0.0077
−16	.0008	+2	.0026	20	.0084
−14	.0009	4	.0031	22	.0091
−12	.0010 +	6	.0037	24	.0097 +
−10	.0012	8	.0043	26	.0104
−8	.0013	10	.0050	28	.0111
−6	.0015	12	.0056	30	.0118
−4	.0017	14	.0063	32	.0125
−2	.0019	16	.0070	34	.0132
				36	.0139

The practical tables consist essentially of two mutually dependent parts:—the first gives values of 2000 m in a table of double entry of which the altitude of the station and the mean temperature of the air between the station and sea level are the arguments; the second gives the reduction to sea level in a table of double entry of which the arguments are 2000 m and the observed barometric height corrected for temperature. In addition, a subsidiary table gives the small correction for latitude to be applied to the values of 2000 m. This correction, while of theoretical interest, seldom becomes of practical importance, since its effect is in general overshadowed by the relatively large uncertainties incident to the determination of the true mean temperature.

The mean temperature of the air column is to be obtained from the observed temperature at the station by employing some assumption as to the rate of change of temperature with altitude. In the discussion of barometric observations made in the mountain and plateau regions of the United States, it has been found that this rate of change is a climatic factor which needs to be determined for every station for different seasons of the year, and for different atmospheric conditions. When the results of such investigations are embodied in tables for reduction to sea level, the tables and the method of their use may be simplified and the labor of obtaining the reduction greatly abridged; but in the nature of the case, these special methods can not be utilized in the construction of general tables which are to be applicable to all phases of topography and climate.

Whatever method be used for obtaining the mean temperature of the air column (θ) from the observed temperature at the station, the former and hence the latter is subject to the important condition that it shall not contain the diurnal fluctuation. Hence in reducing to sea level any individual observation of the barometer, the simultaneous observation of air temperature used in obtaining θ should be reduced to the daily mean by a correction, or, better, the actual mean temperature of the preceding twenty-four hours should be taken.

BAROMETRICAL TABLES.

TABLES 14, 15, 16.

TABLES 14, 15, 16. *Reduction of the barometer to sea level — English measures.*

Table 14 gives values of $2000 \times m$.

$$m = \frac{Z}{56573 + 123.1\,\theta + 0.003\,Z} \cdot \frac{1}{1+\beta}.$$

The temperature θ varies by intervals of $2°$ from $-20°\,F.$ to $96°\,F.$, except near the extremities of the table where the interval is $4°$. The altitude Z varies by intervals of 100 feet from 100 to 9000 feet. The values of $2000\,m$ are given to one decimal.

In order to facilitate interpolations for fractions of a 100 feet in altitude, the tabular differences for 100 feet have been added on each line.

Table 15 gives a small correction to $2000\,m$ for latitude, computed from the expression

$$2000\,m \times 0.002662 \cos 2\,\phi.$$

The arguments are $2000\,m$, which varies by tens from 10 to 350, and the latitude, which varies by $5°$ from $0°$ to $90°$. The correction is to be subtracted for latitudes below $45°$ and added for latitudes above $45°$. The tabular values are given to one decimal.

Table 16, with the value of $2000\,m$ thus corrected, gives the correction which must be applied to the barometric reading B (corrected for temperature) to reduce it to sea level. The arguments are B, which varies by 0.5 inch from 31.00 inches to 19.5 inches, and values of $2000\,m$, which are given for every unit from 1 to 334.

The reduction values $B_\circ - B$ are given to 0.01 inch.

Example:

Let $B = 26.24$ inches be the barometric reading (corrected for temperature) observed at a station whose altitude is 3572 feet, and latitude $32°$. Suppose the mean temperature of the air column $\theta = 63°.0\,F.$

Table 14 gives (p. 63) with $Z = 3,500$ feet and $\theta = 62°.8\,F.$, $2000\,m = 108.0$
The difference for 72 feet is   2.2

The approximate value of $2000\,m$ is   110.2

Table 15, with $2000\,m = 110$ and latitude $= 32°$, gives the subtractive correction 0.1. Hence the corrected value of $2000\,m$ is 110.1.

With $2000\,m = 110.1$ and $B = 26.24$, Table 16 (p. 72) gives the reduction to sea level, 3.55 inches. Accordingly the barometric pressure reduced to sea level is

$$B_\circ = 26.24 + 3.55 = 29.79 \text{ inches.}$$

TABLES 17, 18, 19. *Reduction of the barometer to sea level—Metric measures.*

For reducing to sea level readings of the metric barometer, the barometric formula in metric measures derived on page xxii is treated in the same manner as the formula in English measures just described in detail, and the method of construction of the tables is the same.

Table 17 gives values of 2000 m.

$$m = \frac{Z}{18444 + 67.53\,\theta + 0.003\,Z} \cdot \frac{1}{1+\beta}.$$

The temperature θ varies by intervals of 2° from $-16°$ C. to $+36°$ C. except near the extremities of the table where the interval is 4°. The altitude Z varies by 10 metres from 10 to 3000 metres. The values of 2000 m are given to one decimal.

Table 18 gives the small correction to 2000 m for latitude. The arguments are 2000 m, which varies by tens from 10 to 350, and the latitude which varies by 5° from 0° to 90°. The correction is to be subtracted for latitudes below 45° and added for latitudes above 45°. The tabular values are given to one decimal. The value of 2000 m thus corrected is then used in entering Table 19.

Table 19 gives the correction which must be applied to the barometric reading B (corrected for temperature) to reduce it to sea level. The arguments are B, which varies by 10 mm. from 790 mm. to 480 mm., and values of 2000 m which vary by units from 1 to 345. The tabular values $B_o - B$ are given to 0.1 mm.

Example:

Let $B = 648.7$ mm. be the barometric reading observed and corrected for temperature at a station whose altitude is 1353 metres and latitude 32°. Suppose the mean temperature of the air column $\theta = 14°.3$ C.

Table 17 gives (p. 83) for $\theta = 14°$ and $Z = 1353$,   2000 $m = 138.6$
The proportional part for 0°.3 is   .15

Hence the approximate value of 2000 m is   138.45

Table 18, with 2000 $m = 138$ and latitude 32°, gives the subtractive correction 0.15. Hence the corrected value of 2000 m is 138.3. With this value and $B = 649$ mm. as arguments, Table 19 gives $B_o - B = 112.0$ mm. Accordingly the barometric reading reduced to sea level is

$$B_o = 648.7 + 112.0 = 760.7 \text{ mm.}$$

THE DETERMINATION OF HEIGHTS BY THE BAROMETER.

TABLES 20, 21, 22, 23, 24. *English Measures.*

The barometric formula developed in the preceding section (see p. xxi) is arranged in the following form for determining heights by the barometer.

$$Z = K(\log B_\circ - \log B) \begin{bmatrix} (1 + a\theta) \\ (1 + \beta) \\ (1 + k\cos 2\phi)(1 + \eta) \\ \left(1 + \dfrac{Z + 2h_\circ}{R}\right) \end{bmatrix}$$

in which $K(\log B_\circ - \log B)$ is an approximate value of Z and the factors in the brackets are correction factors depending respectively on the air temperature, the humidity, the variation of gravity with latitude, the variation of gravity with altitude in its effect on the weight of mercury in the barometer, and the variation of gravity with altitude in its effect on the weight of the air. With the constants already given, the formula becomes in English measures:

$$Z \text{ (feet)} = 60368 (\log B_\circ - \log B) \begin{bmatrix} [1 + 0.002039\,(\theta - 32°)] \\ (1 + \beta) \\ (1 + 0.002662 \cos 2\phi)(1 + 0.00239) \\ \left(1 + \left(\dfrac{Z + 2h_\circ}{R}\right)\right) \end{bmatrix}$$

In order to make the temperature correction as small as possible for average air temperatures, 50° F. will be taken as the temperature at which the correction factor is zero. This is accomplished by the following transformation:

$$1 + 0.002039\,(\theta - 32°) = [1 + 0.002039\,(\theta - 50°)][1 + 0.0010195 \times 36°].$$

The second factor of this expresssion combines with the constant, and gives $60368\,(1 + 0.0010195 \times 36°) = 62583.6$.

The first approximate value of Z is therefore

$$62583.6\,(\log B_\circ - \log B).$$

In order further to increase the utility of the tables, we shall make a further substitution for $\log B_\circ - \log B$, and write

$$62583.6\,(\log B_\circ - \log B) = 62583.6 \log\left(\frac{29.9}{B} - \log \frac{29.9}{B_\circ}\right).$$

Table 20 contains values of the expression

$$62583.6 \log \frac{29.9}{B}$$

for values of B varying by intervals of 0.01 inch from 12.00 inches to 30.90 inches.

The first approximate value of Z is then obtained by subtracting the tabular value corresponding to $B_\circ$ from the tabular value corresponding to B (B and $B_\circ$ being the barometric readings observed and corrected for temperature at the upper and lower stations respectively).

Table 21 gives the temperature correction

$$Z \times 0.002039 \, (\theta - 50°).$$

The side argument is the mean temperature of the air column (θ) given for intervals of 1° from 0° to 100° F. The top argument is the approximate difference of altitude Z obtained from Table 20.

For temperatures above 50° F., the correction is to be added, and for temperatures below 50° F., the correction is to be subtracted. It will be observed that the correction is a linear function of Z, and hence, for example, the value for $Z = 1740$ is the sum of the corrections in the columns headed 1000, 700, and 40.

In general, accurate altitudes can not be obtained unless the temperature used is freed from diurnal variation.

Table 22 gives the correction for latitude, and for the variation of gravity with altitude in its effect on the weight of the mercury. When altitudes are determined with aneroid barometers the second factor does not enter the formula. In this case the effect of the latitude factor can be obtained by taking the difference between the tabular value for the given latitude and the tabular value for latitude 45°. The side argument is the latitude of the station given for intervals of 2°. The top argument is the approximate difference of height Z.

Table 23 gives the correction for the average humidity of the air at different temperatures; the values of the factor $(1 + \beta)$ adopted by Prof. Ferrel and given on page xxiii have been used. This correction could have been incorporated with the temperature factor in Table 21, but it is given separately in order that the magnitude of the correction may be apparent, and in order that, when the actual humidity is observed, the correction may be computed if desired, by the expression

$$Z \left(0.378 \, \frac{e}{b} \right)$$

where e is the mean pressure of vapor in the air column, and b the mean barometric pressure.

The side argument is the mean temperature of the air column, varying by intervals of 2° from $-20°$ F. to 96° F., except near the extremities of the table where the interval is 4°. The top argument is the approximate difference of altitude Z.

Table 24 gives the correction for the variation of gravity with altitude in its effect on the weight of the air. The side argument is the approximate difference of altitude Z, and the top argument is the elevation of the lower station h_o.

The corrections given by Tables 22, 23 and 24 are all additive.

Example:

Let the barometric pressure observed, and corrected for temperature, at the upper and lower stations be, respectively, $B = 23.61$ and $B_0 = 29.97$. Let the mean temperature of the air column be $35°\,F.$, and the latitude $44°\,16'$. To determine the difference of height.

	Feet.
Table 20, argument 23.61, gives	6420
Table 20, " 29.97, "	− 64
Approximate difference of height (Z)	= 6484
Table 21, with $Z = 6484$ and $\theta = 35°\,F.$, gives	− 198
Table 22, with $Z = 6300$ and $\phi = 44°$, gives	+ 16
Table 23, with $Z = 6300$ and $\theta = 35°\,F.$, gives	+ 17
Table 24, with $Z = 6300$ and $h_0 = 0$, gives	+ 2
Final difference of height (Z)	= 6321

If in this example the barometric readings be observed with aneroid barometers, the correction to be obtained from Table 22 will be simply the portion due to the latitude factor, and this will be obtained by subtracting the tabular value for $45°$ from that for $44°$, the top argument being $Z = 6300$. This gives $16 - 15 = 1$.

TABLES 25, 26, 27, 28, 29. *Metric Measures.*

The barometric formula developed on page xxi is, in metric units,

$$Z \text{ (metres)} = 18400 \,(\log B_0 - \log B) \begin{bmatrix} (1 + 0.00367\,\theta\,C.) \\ (1 + 0.378\frac{e}{b}) \\ (1 + 0.00266 \cos 2\phi)(1 + 0.00239) \\ 1 + \frac{(Z + 2h_0)}{6\,367\,323} \end{bmatrix}$$

The approximate value of Z (the difference of height of the upper and lower station) is given by the factor $18400\,(\log B_0 - \log B)$. This expression is computed by means of two entries of a table whose argument is the barometric pressure. In order that the two entries may result at once in an approximate value of the elevation of the upper and lower stations, a transformation is made, which gives the following identity:

$$18400\,(\log B_0 - \log B) = 18400\,\left(\log \frac{760}{B} - \log \frac{760}{B_0}\right).$$

TABLE 25

Table 25 gives values of the expression $18400 \log \frac{760}{B}$ for values of B varying by intervals of 1 mm. from 300 mm. to 779 mm. The first approximate value of Z is then obtained by subtracting the tabular value corresponding to B_0 from the tabular value corresponding to B (B and B_0 being the barometric readings observed and reduced to $0°\,C.$ at the upper and lower stations respectively). The first entry of Table 25 with the argument B gives an approximate value of the elevation of the upper station above sea level, and the second entry with the argument B_0 gives an approximate value of the elevation of the lower station.

Table 26 gives the temperature correction: $0.00367 \, \theta \, C. \times Z$.

The side argument is the approximate difference of elevation Z and the top argument is the mean temperature of the air column. The values of Z vary by intervals of 100 m. from 100 to 4000 metres and the temperature varies by intervals of 1° from 1° C. to 10° C. with additional columns for 20°, 30°, and 40° C. Attention is called to the fact that the formula is linear with respect to θ, and hence that the correction, for example, for 27° equals the correction for 20° plus the correction for 7°. When the table is used for temperatures below 0° C., the tabular correction must be subtracted from, instead of added to, the approximate value of Z.

Table 27 (pp. 112 and 113) gives the correction for humidity resulting from the factor $0.378 \, \frac{e}{b} \times Z = \beta \, Z$.

Page 112 gives the value of $0.378 \, \frac{e}{b}$ multiplied by 10000. The side argument is the mean pressure of aqueous vapor, e, which serves to represent the mean state of humidity of the air between the two stations. $e = \frac{1}{2}(f + f_0)$ (f and f_0 being the vapor pressures observed at the two stations) has been written at the head of the table, but the value to be assigned to e is in reality left to the observer, independently of all hypothesis. The top argument is the mean barometric pressure $\frac{1}{2}(B + B_0)$.

The vapor pressure varies by millimetres from 1 to 40, and the mean barometric pressure varies by intervals of 20 mm. from 500 mm. to 760 mm. The tabular values represent the humidity factor β or $0.378 \, \frac{e}{b}$, multiplied by 10000.

Page 113 gives the correction for humidity, with Z and $10000 \times 0.378 \, \frac{e}{b}$ (derived from page 112) as arguments.

The approximate difference of altitude is given by intervals of 100 metres from 100 to 4000 metres, and the values of 10000β vary by intervals of 25 from 25 to 300. The tabular values are given in tenths of metres to facilitate and increase the accuracy of interpolation.

Table 28 gives the correction for latitude, and for the variation of gravity with altitude in its effect on the weight of the mercurial column. When altitudes are determined with aneroid barometers, the latter factor does not enter the formula. In this case the effect of the latitude factor can be obtained by subtracting the tabular value for latitude 45° from the tabular value for the latitude in question.

The side argument is the approximate difference of elevation Z, varying by intervals of 100 metres from 100 to 4000. The top argument is the latitude varying by intervals of 5° from 0° to 75°.

BAROMETRICAL TABLES.

TABLE 29.

Table 29 gives the correction for the variation of gravity with altitude in its effect on the weight of the air.

The side argument is the same as in Table 28; the top argument is the height of the lower station varying by intervals of 200 metres from 0 to 2000, with additional columns for 2500, 3000 and 4000 metres.

Example:

Let the barometric reading (reduced to 0° C.) at the upper station be 655.7 mm.; at the lower station, 772.4 mm. Let the mean temperature of the air column be $\theta = 12°.3$ C., the mean vapor pressure $e = 9$ mm. and the latitude $\phi = 32°.$

Table 25, with argument 655.7, gives	1179 metres.
Table 25, " " 772.4, "	− 129
Approximate value of Z	= 1308
Table 26, with $Z = 1300$ and $\theta = 12°.3$ C, gives	59
Table 27, with $e = 9$ mm. and $Z = 1370$, gives	7
Table 28, with $Z = 1370$ and $\phi = 32°$, gives	5
Table 29, with $Z = 1370$ and $h_e = 0$, gives	0
Corrected value of Z	= 1379 metres.

TABLE 30.

TABLE 30. *Difference of height corresponding to a change of 0.1 inch in the barometer—English measures.*

If we differentiate the barometric formula, page xxvii, we shall obtain, neglecting insensible quantities,

$$dZ = -26281 \frac{dB}{B} \left(1 + 0.002039 \, (\theta - 32°)\right) (1 + \beta).$$

in which B represents the mean pressure of the air column dZ.

Putting $dB = 0.1$ inch,

$$dZ = -\frac{2628.1}{B} \left(1 + 0.002039 \, (\theta - 32°)\right) (1 + \beta).$$

The second member, taken positively, expresses the height of a column of air in feet corresponding to a tenth of an inch in the barometer on the parallel of 45° latitude. Since the last factor $(1 + \beta)$, as given on page xxiii, is a function of the temperature, the function has only two variables and admits of convenient tabulation.

Table 30, containing values of dZ for short intervals of the arguments B and θ, has been taken from the Report of the U. S. Coast Survey, 1881, Appendix 10,—*Barometric hypsometry and reduction of the barometer to sea level*, by Wm. Ferrel.*

---

* Due to the use of a slightly different value for the coefficient of expansion, Prof. Ferrel's formula, upon which the table is computed, is

$$dZ = -\frac{2628.4}{B} \left(1 + 0.002034 \, (\theta - 32°)\right) (1 + \beta).$$

The temperature argument is given for every 5° from 30° F. to 85° F., and the pressure argument for every 0.2 inch from 22.0 to 30.8 inches.

This table may be used in computing small differences of altitude, and, up to a thousand feet or more, very approximate results may be obtained.

Example:

Mean pressure at Augusta, October, 1891, 29.94; temperature, 60°.8 F.
Mean pressure at Atlanta, October, 1891, 28.97; temperature, 59°.4
Mean pressure of air column, $B = 29.455$; $\theta = 60°.1$

Entering the table with 29.455 and 60°.1 as arguments, we take out 94.95 as the difference of elevation corresponding to a tenth of an inch difference of pressure. Multiplying this value by the number of tenths of inches difference in the observed pressures, viz. 97, we obtain the difference of elevation 921 feet.

TABLE 31. *Difference of height of air corresponding to a change of 1 millimetre in the barometer—Metric measures.*

This table has been computed by converting Table 18 into metric units. The temperature argument is given for every 2° from $-2°$ C. to $+36°$ C.; the pressure argument is given for every millimetre from 760 to 560 mm.

TABLE 32. *Babinet's formula for determining heights by the barometer.*

Babinet's formula for computing differences of altitude[*] represents the formula of Laplace quite accurately for differences of altitude up to 1000 metres, and within one per cent for much greater altitudes. As it has been quite widely disseminated among travellers and engineers, and is of convenient application, the formula is here given in English and metric measures. It might seem desirable to alter the figures given by Babinet so as to conform to the newer values of the barometrical constants now adopted; but this change would increase the resulting altitudes by less than one-half of one per cent without enhancing their reliability to a corresponding degree, on account of the outstanding uncertainty of the assumed mean temperature of the air.

The formula is, in English measures,

$$Z \text{ (feet)} = 52494 \left[1 + \frac{t_o + t - 64°}{900}\right] \frac{B_o - B}{B_o + B};$$

and in metric measures,

$$Z \text{ (metres)} = 16000 \left[1 + \frac{2(t_o + t)}{1000}\right] \frac{B_o - B}{B_o + B},$$

in which Z is the difference of elevation between a lower and upper station at which the barometric pressures corrected for all sources of instrumental error are B_o and B, and the observed air temperatures are t_o and t, respectively.

---

[*] *Comptes Rendus*, Paris, 1850, vol. xxv., page 309.

For ready computation the formula is written

$$Z = C \times \frac{B_0 - B}{B_0 + B},$$

and the factor C, computed both in English and metric measures, has been kindly furnished by Prof. Cleveland Abbe. The argument is $\frac{1}{2}(t_0 + t)$ given for every 5° Fahrenheit between 10° and 100° $F.$, and for every 2° Centigrade between 10° and 40° Centigrade.

In using the table, it should be borne in mind that on account of the uncertainty in the assumed temperature, the last two figures in the value of C are uncertain, and are here given only for the sake of convenience of interpolation. Consequently one should not attach to the resulting altitudes a greater degree of confidence than is warranted by the accuracy of the temperatures and the formula. The table shows that the numerical factor changes by about one per cent of its value for every change of five degrees Fahrenheit in the mean temperature of the stratum of air between the upper and lower stations; therefore the computed difference of altitude will have an uncertainty of one per cent if the assumed temperature of the air is in doubt by 5° $F.$ With these precautions the observer may properly estimate the reliability of his altitudes whether computed by Babinet's formula or by more elaborate tables.

Example:

Let the barometric pressure observed and corrected for temperature at the upper and lower stations be, respectively, $B = 635$ mm. and $B_0 = 730$ mm. Let the temperatures be, respectively, $t = 15°$ $C.$, $t_0 = 20°$ $C.$ To find the approximate difference of height.

With $\frac{1}{2}(t_0 + t) = \frac{20° + 15°}{2} = 17°.5$ $C.$, the table in metric measures gives

$C = 17120$ metres. $\quad \frac{B_0 - B}{B_0 + B} = \frac{95}{1365}.$

The approximate difference of height $= 17120 \times \frac{95}{1365} = 1191.5$ metres.

THERMOMETRICAL MEASUREMENT OF HEIGHTS BY OBSERVATION OF THE TEMPERATURE OF THE BOILING POINT OF WATER.

When water is heated in the open air, the elastic force of its vapor gradually increases, until it becomes equal to the incumbent weight of the atmosphere. Then, the pressure of the atmosphere being overcome, the steam escapes rapidly in large bubbles and the water boils. The temperature at which water boils in the open air thus depends upon the weight of the atmospheric column above it, and under a less barometric pressure the water will boil at a lower temperature than under a greater pressure. Now, as the weight of the atmosphere decreases with the elevation, it is obvious that, in ascending a mountain, the *higher* the station where an observation is made, the *lower* will be the temperature of the boiling point.

The difference of elevation between two places therefore can be deduced from the temperature of boiling water observed at each station. It is only necessary to find the barometric pressures which correspond to those temperatures, and, the atmospheric pressures at both places being known, to compute the difference of height by the tables given herein for computing heights from barometric observations.

From the above, it may be seen that the heights determined by means of the temperature of boiling water are less reliable than those deduced from barometric observations. Both derive the difference of altitude from the difference of atmospheric pressure. But the temperature of boiling water gives only *indirectly* the atmospheric pressure, which is given *directly* by the barometer. This method is thus liable to all the chances of error which may affect the measurements by means of the barometer, besides adding to them new ones peculiar to itself, the principal of which is the difficulty of ascertaining with the necessary accuracy the true temperature of boiling water. In the present state of thermometry it would hardly be safe, indeed, to rely, in the most favorable circumstances, upon quantities so small as hundredths of a degree, even when the thermometer has been constructed with the utmost care; moreover, the quality of the glass of the instrument, the form and substance of the vessel containing the water, the purity of the water itself, the position at which the bulb of the thermometer is placed, whether in the current of the steam or in the water, — all these circumstances cause no inconsiderable variations to take place in the indications of thermometers observed under the same atmospheric pressure. Owing to these various causes, an observation of the boiling point, differing by one-tenth of a degree from the true temperature, ought to be still admitted as a good one. Now, as the tables show, an error of one-tenth of a degree Centigrade in the temperature of boiling water would cause an error of 2 millimetres in the barometric pressure, or of from 70 to 80 feet in the final result, while with a good barometer the error of pressure will hardly ever exceed one-tenth of a millimetre, making a difference of 3 feet in altitude.

Notwithstanding these imperfections, the hypsometric thermometer is of the greatest utility to travellers and explorers in rough countries, on account of its being more conveniently transported and much less liable to accidents than the mercurial barometer. A suitable form for it, designed by Regnault (*Annales de Chimie et de Physique*, Tome xiv, p. 202), consists of an accurate thermometer with long degrees, subdivided into tenths. For observation the bulb is placed, about 2 or 3 centimetres above the surface of the water, in the steam arising from distilled water in a cylindrical vessel, the water being made to boil by a spirit-lamp.

TABLES 33, 34.

Barometric pressures corresponding to the temperature of boiling water.

TABLE 33. *English Measures.*
TABLE 34. *Metric Measures.*

Table 33 is a conversion into English measures of Table 34. The argument is the temperature of boiling water for every tenth of a degree from 185°.0 to 212°.9 Fahrenheit. The tabular values are given to the nearest 0.01 inch.

Table 34 is Regnault's table of barometric pressures corresponding to temperatures of boiling water, revised by A. Moritz (*Acad. Sci. Bull.*, St. Petersburg, xiii., 1855, col. 41-44). To the degree of precision here desired, these values do not differ from the more recent reduction by Broch. The argument is given for every tenth of a degree from 80°.0 to 100°.9 C. The tabular values are given to the nearest 0.1 mm.

HYGROMETRICAL TABLES.

PRESSURE OF AQUEOUS VAPOR IN SATURATED AIR.

Tables 35, 36, and 43, giving the pressure of aqueous vapor in saturated air, are based upon Dr. Broch's reduction of the observations of Regnault (*Travaux et Mémoires du Bureau international des Poids et Mesures*, t. I, p. A 19-39). This reduction assumes that the observations may be represented by the empirical formula

$$F = A \times 10^{\frac{bt + ct^2 + dt^3 + et^4 + ft^5}{1 + at}}$$

in which F is the pressure of aqueous vapor expressed in millimetres of standard mercury, that is at 0° C. and at latitude 45° and sea level, its density being 15.59593.

t, the temperature expressed in normal Centigrade degrees.

$a = 0.003667458$

By using the simultaneous values of F and t given by Regnault's observations, Dr. Broch obtained a series of observation equations whose solution by the method of least squares gave the following values for the coefficients:

$$A = 4.568\ 685\ 9$$
$$b = 10^{-2} \times 3.134\ 366\ 174$$
$$c = -10^{-5} \times 1.416\ 112\ 423$$
$$d = 10^{-7} \times 1.935\ 338\ 308$$
$$e = -10^{-9} \times 2.646\ 535\ 103$$
$$f = 10^{-11} \times 1.139\ 377\ 158$$

From this formula Broch's tables of vapor pressure were computed.

TABLE 35. *Pressure of aqueous vapor—English measures.*

This table is a conversion into English measures of Table 36. It gives the vapor pressure in saturated air for temperatures varying by 0°.2 from − 20°.0 to 214°.0 Fahrenheit.

The tabular values are given in inches to four decimals.

A column of differences for 0°.1 is added for convenience in interpolating.

TABLES 36, 43. *Pressure of aqueous vapor.—Metric measures.*

These tables, taken from Broch, give the pressure of aqueous vapor to hundredths of a millimetre for temperatures varying by 0°.1 C. from − 29°.0 to 100°.9 Centigrade. The values for temperatures between 0° C. and 45° C. are given in Table 43, the remainder in Table 36.

TABLE 37. *Pressure of aqueous vapor at low temperature.—(C. F. Marvin.)*

Broch's vapor pressures at temperatures below 0° C. (32° F.) as given in Tables 35 and 36, when compared with the actual observed values of Regnault are found to be systematically too large. This discrepancy signifies that the empirical formula adopted by Broch fails to represent accurately the law of variation of vapor pressure for temperatures both above and below the freezing point. Moreover, the failure in the application of the formula might be inferred from the laws of diffusion following from the kinetic theory of gases, for these give no reason to suppose that the function expressing the relation between vapor pressure and temperature is continuous between the two states of water and ice.

Under proper conditions water can be cooled far below 0° C. (32° F.) before solidifying, so that at the same temperature we may have it either in the liquid or the solid state, and experiments confirm the theory of diffusion in showing that the pressure of the vapor is different according as it is in contact with its liquid or its solid at the same temperature. The method hitherto employed of combining vapor pressures above and below freezing, and attempting to represent them by a single continuous function, must therefore be considered as radically erroneous.

Recognizing the systematic errors of the vapor pressures given by Broch's formula for temperature below freezing, the Chief Signal Officer lately authorized a new determination by direct observation. This experimental investigation has been carried out by Prof. C. F. Marvin, from the results of which (*Annual Report Chief Signal Officer*, 1891; Appendix No. 10,) Table 37 is reproduced. The interpolation between the observed pressures which were noted at intervals of about 5° F., was effected graphically and not by mathematical formula.

The vapor pressures were determined for the case of the vapor in contact with ice and not a water surface. For the temperature of melting ice (0° C. or 32° F.) all values agree. Below this temperature Marvin's vapor pressures are slightly smaller than Regnault's, but differ from the latter less than any other tabular values.

The argument of the table is given for every two-tenths of a degree Fahrenheit from −60°.0 to 32°.0 Fahrenheit. The tabular values are given in millimetres and inches to three and four decimals respectively.

TABLES 38, 39.

TABLE 38. *Weight of aqueous vapor in a cubic foot of saturated air—English measures.*

TABLE 39. *Weight of aqueous vapor in a cubic metre of saturated air—Metric measures.*

The weight of aqueous vapor in a *cubic metre* of saturated air is given by the expression

$$W = \frac{a\delta}{1+\alpha t} \cdot \frac{F}{760},$$

a is the weight of a cubic metre of dry air (free from carbonic acid) at temperature 0° C., and pressure of 760 millimetres of standard mercury at 45° latitude and sea-level: $a = 1.29278$ kg. (Bureau International des Poids et Mesures: *Travaux et Mémoires*, t. I, p. A 54.)

δ is the density of aqueous vapor: $\delta = 0.6221$

F is the pressure of aqueous vapor in saturated air whose temperature is t; Broch's values are adopted, expressed in millimetres.

α is the coefficient of expansion of air for 1° C.: $\alpha = 0.003667$

t is the temperature in Centigrade degrees.

Whence we have

$$W \text{ (grammes)} = 1.05821 \times \frac{F}{1 + 0.003667\,t}.$$

Table 39 is computed from this formula and gives the weight of vapor in grammes in a cubic metre of saturated air for dew-points from −29° to 40° C., the intervals from 6° to 40° C. being 0°.1 C. The tabular values are given to three decimals.

The weight W' of aqueous vapor in a *cubic foot* of saturated air is obtained by converting the foregoing constants into English measures.

The weight of a cubic foot of dry air at temperature 32° F. and at a pressure of 760 mm. or 29.921 inches is

$$a' \text{ (grains)} = \frac{1292.78 \times 15.43235}{(3.280833)^3} = 564.94.$$

We have therefore,

$$W' \text{ (grains)} = \frac{a'\delta}{29.921} \times \frac{F'}{1 + \alpha'(t'-32°)}$$

$$= 11.7459 \frac{F'}{1 + 0.002037\,(t'-32°)}.$$

The temperature t' is expressed in degrees Fahrenheit; the vapor pressure F', expressed in inches, is obtained from Table 35.

Table 38* gives the weight of aqueous vapor in grains in a cubic foot of saturated air for dew-points given to every 0°.5 from $-19°.5$ to $115°$ $F.$, the values being computed to the thousandth of a grain.

The computation of Tables 38 and 39 has been furnished by Prof. Wm. Libbey, jr.

REDUCTION OF OBSERVATIONS WITH THE PSYCHROMETER AND DETERMINATION OF RELATIVE HUMIDITY.

The psychrometric formula derived by Maxwell, Stefan, August, Regnault and others is, in its simplest form,

$$f = f_1 - A B (t - t_1),$$

in which $t =$ Air temperature.

$t_1 =$ Temperature of the wet-bulb thermometer.
$f =$ Pressure of aqueous vapor in the air.
$f_1 =$ Pressure of aqueous vapor in saturated air at temperature t_1.
$B =$ Barometric pressure.
$A =$ A quantity which, for the same instrument and for certain conditions, is a constant, or a function depending in a small measure on t_1.

The important advance made since the time of Regnault consists in recognizing that the value of A differs materially according to whether the wet-bulb is in quiet or moving air. This was experimentally demonstrated by the distinguished Italian physicist, Belli, in 1830, and was well known to Espy, who always used a whirled psychrometer. The latter describes his practice as follows: "When experimenting to ascertain the dew-point by means of the wet-bulb, I always swung both thermometers moderately in the air, having first ascertained that a moderate movement produced the same depression as a rapid one."

The principles and methods of these two pioneers in accurate psychrometry have now come to be adopted in the standard practice of meteorologists, and psychrometric tables are adapted to the use of a whirled or ventilated instrument.

The factor A depends in theory upon the size and shape of the thermometer bulb, largeness of stem and velocity of ventilation, and different formulæ and tables would accordingly be required for different instruments. But by using a ventilating velocity of three metres or more per second, the differences in the results given by different instruments vanish, and the same tables can be adapted to any kind of a thermometer and to all changes of velocity above that which gives sensibly the greatest depression of the wet-bulb temperature; and with this arrangement there is no necessity to measure or estimate the velocity in each case further than to be certain that it does not fall below the assigned limit.

---

* The table has been computed with the factor 11.7449, which results from Clarke's value for the conversion of the metre, instead of with the value 11.7459 above derived.

HYGROMETRICAL TABLES.

The formula and tables here given for obtaining the vapor pressure and dew-point from observations of the whirled or ventilated psychrometer are those deduced by Prof. Wm. Ferrel (*Annual Report Chief Signal Officer*, 1886, Appendix 24) from a discussion of a large number of observations.

Taking the psychrometric formula in metric units, pressures being expressed in millimetres and temperatures in Centigrade degrees, Prof. Ferrel derived for A the value

$$A = 0.000656 \, (1 + 0.0019 \, t_1)$$

In this expression for A, the factor depending on t_1 arises from a similar term in the expression for the latent heat of water, and the theoretical value of the coefficient of t_1 is 0.00115. Since it would require a very small change in the method of observing to cause the difference between the theoretical value and that obtained from the experiments, Prof. Ferrel adopted the theoretical coefficient 0.00115 and then recomputed the observations, obtaining therefrom the final value

$$A = 0.000660 \, (1 + 0.00115 \, t_1).$$

With this value the psychrometric formula in metric measures becomes

$$f = f_1 - 0.000660 \, B \, (t - t_1) \, (1 + 0.00115 \, t_1)$$

In order to adapt the formula to convenient tabulation, Prof. Ferrel substituted $t - t_1$ for t_1 in the last factor, a modification which produces appreciable error only in extreme cases. The error in the computed vapor pressure will be

$$E = 0.00000076 \, B \, (t - t_1) \, (t - 2t_1).$$

Expressed in English measures, the formula is

$$f = f_1 - 0.000367 \, B \, (t - t_1) \, [1 + 0.00064 \, (t_1 - 32°)]$$

and with the same modification in order to render the formula more convenient for tabulation, we have

$$f = f_1 - 0.000367 \, B \, (t - t_1) \, (1 + 0.00064 \, (t - t_1)),$$

in which f = Vapor pressure in inches.

f_1 = Vapor pressure in saturated air at temperature t_1.
t = Temperature of the air in Fahrenheit degrees.
t_1 = Temperature of the wet-bulb thermometer in Fahrenheit degrees.
B = Barometric pressure in inches.

TABLES 40, 41.

Reduction of Psychrometric Observations—English measures.

TABLE 40. *Pressure of aqueous vapor.*

TABLE 41. Values of $0.000367 \, B \, (t - t_1) \left(1 + \dfrac{t - t_1}{1571}\right)$

These two tables provide for computing the vapor pressure and dew-point from observations of ventilated wet- and dry-bulb Fahrenheit thermometers.

Table 40, with the wet-bulb temperature t_1 as an argument, gives the value of f_1, the first term of the formula for the vapor pressure f, given above. It is simply an abbreviation of Table 35 for temperatures above 32° $F.$, and of Table 37 for temperatures below 32° $F.$, reprinted for convenience.

Table 41, with $t-t_1$ and B as arguments, gives the value of the second term of the formula, viz:

$$0.000\,367\,B\,(t-t_1)\left(1+\frac{t-t_1}{1571}\right)$$

The top argument is given for every half inch from 30.5 to 18.5 inches; the side argument, $t-t_1$, is given for every whole degree up to 40° $F.$ Tabular values are given to thousandths of inches.

With the two tables we then have,

$$f\text{ (vapor pressure)} = \text{Table 40} - \text{Table 41}.$$

The value of t in Table 40, corresponding to the vapor pressure thus obtained, is the *dew-point*.

Examples:

1. Given $t = 84°.3$; $t_1 = 66°.7$, and $B = 30.00$ inches, to find the vapor pressure and dew-point.

   Table 40, with $t_1 = 66°.7$, gives $f_1 = 0.654$ inches.
   Table 41, with $t - t_1 = 84°.3 - 66°.7 = 17°.6$ and $B = 30.00$ inches as arguments, gives 0.196 inch as the value of the last term of the expression above. Hence we have the vapor pressure $f = 0.654 - 0.196 = 0.458$ inch. The temperature (Table 40) corresponding to this value of f is the dew-point, $d = 56°.6\ F.$

2. Given $t = 34°.5$; $t_1 = 29°.4$, and $B = 22.3$ inches, to find the vapor pressure and dew-point.

   Table 40, with $t_1 = 29°.4$, gives $f_1 = 0.162$ inch.
   Table 41, with $t - t_1 = 34°.5 - 29°.4 = 5°.1$ and $B = 22.5$ inches (the nearest value in the table to 22.3 inches) as arguments, gives 0.042 inch as the value of the second term of the expression for f. Hence we have the vapor pressure $f = 0.162 - 0.042 = 0.120$ inch.
   The temperature in Table 40, corresponding to this value of f, is the dew-point, $d = 22°.0$.

Note—In using Table 40, the proportional part for tenths of the argument, $t-t_1$, may be easily obtained by taking one-tenth of the tabular value belonging to the same number of degrees; for instance, in the first example, the tabular value for 17° is 0.189, and the proportional part for 0°.6 is one-tenth the tabular value for 6°.0, viz., one-tenth of .066, or .007. Hence we get $0.189 + 0.007 = 0.196$.

TABLE 42. *Relative humidity—Temperature Fahrenheit.*

Table 42 gives the relative humidity of the air in hundredths, having given the air temperature t and the dew-point d in Fahrenheit degrees.

It is computed by the formula

$$\text{Relative humidity} = \frac{f}{F}.$$

f and F are the maximum pressures of vapor corresponding respectively to the temperatures d and t as given in Table 35 for temperatures above 32° $F.$ and in Table 37 for temperatures below 32° $F.$

The top argument is $t-d$, extending by half degree intervals from 0° to 15° $F.$, and by increasing intervals from 15° to 75° $F.$

The side argument is the air temperature t, given for intervals of four degrees from $-32°$ to 120° $F.$

Example:

Let the air temperature be 62° $F.$ and the dew-point 51° $F.$, to find the relative humidity.

With $t-d=11°$ for the top argument, and $t=62°$ for the side argument, the table gives 67.5 per cent as the relative humidity.

TABLES 43, 44.

Reduction of Psychrometric Observations—Metric measures.

TABLE 43. *Pressure of aqueous vapor.*

TABLE 44. *Values of* $0.000660\ B\ (t-t_1)\left(1+\dfrac{t-t_1}{873}\right).$

These two tables provide for computing the vapor pressure and dew-point from observations of ventilated wet and dry-bulb thermometers Centigrade.

Table 43, with the wet-bulb temperature t_1 as an argument, gives the value of f_1, the first term of the formula for the vapor pressure f, viz:

$$f = f_1 - 0.000660\ B\ (t-t_1)\ [1+0.00115\ (t-t_1)].$$

It gives the vapor pressure to hundredths of a millimetre from $-30°.0\ C.$ to $45°.9\ C.$, the intervals being 1° for temperatures below 0° $C.$ and 0°.1 for temperatures above 0° $C.$

Table 44, with the depression of the wet-bulb $t-t_1$, and the barometric pressure B as arguments, gives the value of the second term of the formula.

The top argument is given for every 10 millimetres from 770 to 460 mm.; the side argument $t-t_1$ is given for every whole degree up to 20. Tabular values are given to hundredths of a millimetre.

From the two parts of the table we then have

$$\text{Vapor pressure, } f\ (mm) = \text{Table 43} - \text{Table 44.}$$

The temperature in Table 43, corresponding to the vapor pressure thus obtained, is the *dew-point.*

Example:

Given $t = 10°.4\ C.$; $t_1 = 8°.3\ C.$ and $B = 740$ mm., to find the vapor pressure and dew-point.

Table 43, with the argument $t_1 = 8°.3\ C.$, gives $f_1 = 8.15$ mm.

Table 44, with $t - t_1 = 2°.1$ and $B = 740$ as arguments, gives 1.03 mm. as the value of the last term of the expression for f. Hence we have the vapor pressure, $f = 8.15 - 1.03 = 7.12$ mm. The value of the temperature in Table 40, corresponding to this vapor pressure, is the dew-point $d = 6°.3\ C.$

TABLE 45. *Relative humidity — Temperature Centigrade.*

Table 45 gives the relative humidity of the air in hundredths, having given the air temperature t and the dew-point d in Centigrade degrees.

It is computed by the formula

$$\text{Relative humidity} = \frac{f}{F},$$

f and F being the maximum pressures of aqueous vapor corresponding to the temperatures d and t as given in Tables 36 and 43.

The top argument is the dew-point d, extending by 5° intervals from $-15°$ to $30°\ C.$

The side argument is the depression of the dew-point $t - d$, given for every $0°.2\ C.$ from $0°.0$ to $10°.0$; for every $0°.5$ from $10°.0$ to $20°.0$, and for every $1°$ from $20°.0$ to $30°.0$.

Example:

Given the air temperature $21°\ C.$ and the dew-point $17°\ C.$, to determine the relative humidity.

With $t - d = 4°\ C.$ for the side argument, and $d = 17°\ C.$ for the top argument, the table gives 78 per cent as the relative humidity.

TABLE 46.   REDUCTION OF SNOWFALL MEASUREMENT.

The determination of the water equivalent of snowfall has usually been made by one of two methods: (*a*) by dividing the depth of snow by an arbitrary factor ranging from 8 to 16 for snow of different degrees of compactness; (*b*) by melting the snow and measuring the depth of the resulting water. The first of these methods has always been recognized as incapable of giving reliable results, and the second, although much more accurate, is still open to objection. After extended experience in the trial of both these methods, it has been found that the most accurate and most convenient measurement is that of weighing the collected snow, and then converting the weight into depth in inches. The method is equally applicable whether the snow as it falls is caught in the gage, or a section of the fallen snow is taken by collecting it in an inverted gage.

TABLE 46.

TABLE 46. *Depth of water corresponding to the weight of snow (or rain) collected in an 8-inch gage.*

The table gives the depth to hundredths of an inch, corresponding to the weight of snow or rain collected in a gage having a circular collecting mouth 8 inches in diameter — this being the standard size of gage used throughout the United States.

The argument is given in avoirdupois pounds, ounces and quarter ounces in order that it shall be adapted to the customary graduation of commercial scales.

Example:

The weight of snow collected in an 8-inch gage is 2 lbs. 2¼ oz. To find the corresponding depth of water.

The table gives directly 1.18 inches.

TABLE 47.

TABLE 47. *Rate of decrease of vapor pressure with altitude.*

From hygrometric observations made at various mountain stations on the Himalayas, Mount Ararat, Teneriffe, the Alps, and also in balloon ascensions, Dr. J. Hann (*Zeitschrift für Meteorologie*, vol. ix, 1874, p. 193–200) nas deduced the following empirical formula showing the average relation between the vapor pressure f_0 at a lower station and f the vapor pressure at an altitude h metres above it:

$$\frac{f}{f_0} = 10^{-\frac{h}{6517}}.$$

This is of course an average relation for all times and places from which the actual rate of decrease of vapor pressure in any individual case may widely differ.

Table 47 gives the values of the ratio $\frac{f}{f_0}$ for values of h from 200 to 6000 metres. An additional column gives the equivalent values of h in feet.

WIND TABLES.

CALCULATION OF THE MEAN DIRECTION OF THE WIND BY LAMBERT'S FORMULA.

Lambert's formula for the eight principal points of the compass is

$$\tan \alpha = \frac{E - W + (NE + SE - NW - SW)\cos 45°}{N - S + (NE + NW - SE - SW)\cos 45°}.$$

α is the angle of the resultant wind direction with the meridian.

E, NE, N, etc., represent the wind movement from the corresponding directions East, Northeast, North, etc. In practice instead of taking the total wind movement, it is often considered sufficient to take as proportional thereto the number of times the wind has blown from each direction, which is equivalent to considering the wind to have the same mean velocity for all directions.

If directions are observed to sixteen points, half the number belonging to each extra point, should be added to the two octant points between which it lies; for example, $NNE = 6$ should be separated into $N = 3$ and $NE = 3$; $ESE = 4$ into $E = 2$ and $SE = 2$. The result will be approximately identical with that obtained by using the complete formula for sixteen points.

TABLE 48. *Multiples of cos 45°; form for computing the numerator and denominator.*

TABLE 49. *Values of the mean direction (a) or its complement (90° − a).*

Table 48 gives products of cos 45° by numbers up to 209, together with a form for the computation of the numerator and denominator, illustrated by an example. The quadrant in which a lies is determined by the following rule:

When the numerator and denominator are positive, a lies between N and E.

When the numerator is positive and the denominator negative, a lies between S and E.

When the numerator and denominator are negative, a lies between S and W.

When the numerator is negative and the denominator positive, a lies between N and W.

Table 49 * combines the use of a division table and a table of natural tangents. It enables the computer, with the numerator and denominator of Lambert's formula (computed from Table 48) as arguments, to take out directly the mean wind direction a or its complement.

The top argument consists of every fifth number from 10 to 200.

The side argument is given for every unit from 1 to 50 and for every two units from 50 to 150. Tabular values are given to the nearest whole degree.

Rule for using the table:

Enter the table with the larger number (either numerator or denominator) as the top argument.

If the denominator be larger than the numerator, the table gives a.

If the denominator be smaller than the numerator, the table gives $90° - a$.

a is measured from the meridian in the quadrant determined by the rule given with Table 48.

---

* From *Hand-book of Meteorological Tables.* By H. A. Hazen. Washington, 1888. A corrected copy of the table has been kindly furnished for the present volume by the author.

WIND TABLES.

Example:

$$\tan a = \frac{-43}{-27}.$$

Table 49 gives

$$90° - a = 32°$$
$$a = S\ 58°\ W.$$

NOTE.—If the numerator and denominator both exceed 150 or if either exceeds 200, the fraction must be divided by some number which will bring them within the limits of the table. The larger the values, provided they are within these limits, the easier and more accurate will be the computation. For example, let $\tan a = \frac{-18}{14}$. The top argument is not given for 18, but if we multiply by 5 or 10 and obtain $\frac{-90}{70}$ or $\frac{-180}{140}$, the table gives, without interpolation, $90° - a = 38°$ and $a = N\ 52°\ W.$

CONVERSION OF VELOCITIES.

TABLE 50. *Synoptic conversion of velocities.*

This table*, contained on a single page, converts miles per hour into metres per second, feet per second and kilometres per hour. The argument, miles per hour, is given for every half unit from 0 to 78. Tabular values are given to one decimal. For the rapid interconversion of velocities, when extreme precision is not required, this table has proved of marked convenience and utility.

TABLE 51. *Conversion of miles per hour into feet per second.*

The argument is given for every unit up to 149 and the tabular values are given to one decimal.

TABLE 52. *Conversion of feet per second into miles per hour.*

The argument is given for every unit up to 199 and the tabular values are given to one decimal.

TABLE 53. *Conversion of metres per second into miles per hour.*

The argument is given for every tenth of a metre per second up to 60 metres per second, and the tabular values are given to one decimal.

TABLE 54. *Conversion of miles per hour into metres per second.*

The argument is given for every unit up to 149, and the tabular values are given to two decimals.

---

* From *Hand-book of Meteorological Tables*. By H. A. Hazen. Washington, 1888. With permission of the author.

TABLE 55. *Conversion of metres per second into kilometres per hour.*

The argument is given for every tenth of a metre per second up to 60 metres per second, and the tabular values are given to one decimal.

TABLE 56. *Conversion of kilometres per hour into metres per second.*

The argument is given for every unit up to 200, and the tabular values are given to two decimals.

TABLE 57. *Beaufort wind scale and its conversion into velocity.*

The personal observation of the estimated force of the wind on an arbitrary scale is a method that belongs to the simplest meteorological records and is widely practiced. Although anemometers are used at meteorological observatories, the majority of observers are still dependent upon estimates based largely upon their own judgment, and so reliable can such estimates be made that for many purposes they abundantly answer the needs of meteorology as well as of climatology.

A great variety of such arbitrary scales have been adopted by different observers, but the one that has come into the most general use and received the greatest definiteness of application is the duodecimal scale introduced into the British navy by Admiral Beaufort about 1800.

The definitions of the successive grades of the Beaufort scale were made in terms of the effect of the wind on the sails of a full-rigged ship, so that navigators of all nations have generally acquired a very uniform and definite idea of their meaning and a very considerable expertness in the use of the scale. The Table gives the designations of the 12 grades together with several conversions of the scale into wind velocities as made by different meteorologists. A committee appointed by the Royal Meteorological Society to establish a conversion of the Beaufort scale into wind velocity made a preliminary report (*Quart. Journal Roy. Meteorological Soc.*, vol. 13, 1887), but did not consider their work sufficiently complete to present a definite conversion table.†

GEODETICAL TABLES.

TABLE 58. *Relative acceleration of gravity at sea-level at different latitudes.*

The formula adopted for the variation of gravity with latitude is that of Prof. Harkness*

$$g_\phi = g_{45} (1 - 0.002662 \cos 2\phi)$$

in which g_ϕ is the acceleration of the gravity at latitude ϕ, and g_{45} the acceleration at latitude 45°.

The table gives the values of the ratio $\dfrac{g_\phi}{g_{45}}$ to six decimals for every 10′ of latitude from the equator to the pole.

---

*WM. HARKNESS: *The solar parallax and its related constants.* Washington, 1891.

† Modern steamships move with velocities sufficient to affect all wind observations aboard of them.

LENGTH OF A DEGREE OF THE MERIDIAN AND OF ANY PARALLEL.

The dimensions of the earth used in computing lengths of the meridian and of parallels of latitude are those of Clarke's spheroid of 1866.* This spheroid undoubtedly represents very closely the true size and shape of the earth, and is the one to which nearly all geodetic work in the United States is now referred.

The values of the constants are as follows:

a, semi-major axis $= 20926062$ feet; log $a = 7.3206875$.
b, semi-minor axis $= 20855121$ feet; log $b = 7.3192127$.
$e^2 = \dfrac{a^2 - b^2}{a^2} = 0.00676866$; log $e^2 = 7.8305030 - 10$.

With these values for the figure of the earth, the formula for computing any portion of a quadrant of the meridian is

Meridional distance in feet $= [5.5618284] \triangle \phi$ (in degrees),
$\quad - [5.0269880] \cos 2\phi \sin \triangle \phi$,
$\quad + [2.0528] \cos 4\phi \sin 2 \triangle \phi$,

in which $2\phi = \phi_2 + \phi_1$, $\triangle \phi = \phi_2 - \phi_1$, $\phi_1, \phi_2 =$ end latitudes of arc.

For the length of 1 degree, the formula becomes:

1 degree of the meridian, in feet $= 364609.9 - 1857.1 \cos 2\phi + 3.94 \cos 4\phi$.

The length of the parallel is given by the equation

1 degree of the parallel at latitude ϕ, in feet $=$
$\quad 365538.48 \cos \phi - 310.17 \cos 3\phi + 0.39 \cos 5\phi$.

TABLE 59.

TABLE 59. *Length of one degree of the meridian at different latitudes.*

This gives for every degree of latitude the length of one degree of the meridian in statute miles to three decimals, in metres to one decimal, and in geographic miles to three decimals—the geographic mile being here defined to be one minute of arc on the equator. The values in metres are computed from the relation: 1 metre $= 39.3700$ inches. The tabular values represent the length of an arc of one degree, the middle of which is situated at the corresponding latitude. For example, the length of an arc of one degree of the meridian, whose end latitudes are 29° 30′ and 30° 30′, is 68.879 statute miles.

TABLE 60.

TABLE 60. *Length of one degree of the parallel at different latitudes.*

This table is similar to Table 59.

---

* *Comparisons of standards of length, made at the Ordnance Survey office, Southampton, England, by Capt. A. R. Clarke, R. E., 1866.*

TABLE 61. *Duration of sunshine at different latitudes for different values of the sun's declination.*

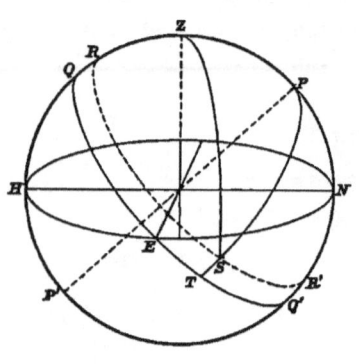

Let Z be the zenith, and NH the horizon of a place in the northern hemisphere.

P the pole;
QEQ' the celestial equator;
RR' the parallel described by the sun on any given day;
S the position of the sun when its upper limit appears on the horizon;
PN the latitude of the place, ϕ.
ST the sun's declination, δ.
PS the sun's polar distance, $90° - \delta$.
ZS the sun's zenith distance, z.
ZPS the hour angle of the sun from meridian, t.
r the mean horizontal refraction $= 34'$ approximately.
s the mean solar semi-diameter $= 16'$

$$z = 90° + r + s = 90° \; 50'$$

In the spherical triangle ZPS, the hour angle ZPS may be computed from the values of the three known side by the formula

$$\sin \tfrac{1}{2} ZPS = \sqrt{\frac{\sin \tfrac{1}{2}(ZS + PZ - PS) \sin \tfrac{1}{2}(ZS + PS - PZ)}{\sin PZ \sin PS}}$$

or

$$\sin \tfrac{1}{2} t = \sqrt{\frac{\sin \tfrac{1}{2}(z + \delta - \phi) \sin \tfrac{1}{2}(z - \delta + \phi)}{\cos \phi \cos \delta}}$$

The hour angle t, converted into mean solar time and multiplied by 2, is the duration of sunshine.

Table 61 has been computed for this volume by Prof. Wm. Libbey, jr. It is a table of double entry with arguments δ and ϕ. For north latitudes northerly declination is considered positive and southerly declination as negative. The table may be used for south latitudes by considering southerly declination as positive and northerly declination as negative.

The top argument is the latitude, given for every 5° from 0° to 40°, for every 2° from 40° to 60°, and for every degree from 60° to 80°.

The side argument is the sun's declination for every 20' from $S\;23°\;27'$ to $N\;23°\;27'$.

The duration of sunshine is given in hours and minutes.

To find the duration of sunshine for a given day at a place whose latitude is known, find the declination of the sun at mean noon for that day in the *Nautical Almanac*, and enter the table with the latitude and declination as arguments.

Example:

To find the duration of sunshine, May 18, 1892, in latitude 49° 30′ North.
From the Nautical Almanac, $\delta = 19° 43′\ N$.
From the table, with $\delta = 19° 43′\ N$ and $\phi = 49° 30′$, the duration of sunshine is found to be $15^h\ 31^m$.

TABLE 62. *Declination of the sun for the year* 1894.

This table is an auxiliary to Table 61, and gives the declination of the sun for every third day of the year 1894. These declinations may be used as approximate values for the corresponding dates of other years when the exact declination can not readily be obtained. Thus, in the preceding example, the declination for May 18 may be taken as approximately the same as that for the same date in 1894, viz. 19° 37′.

RELATIVE INTENSITY OF SOLAR RADIATION AT DIFFERENT LATITUDES FOR DIFFERENT SEASONS OF THE YEAR.

TABLE 63. *Mean vertical intensity for 24 hours of solar radiation J and the solar constant A in terms of the mean solar constant A_0.*

This table is that of Prof. Wm. Ferrel, published in the *Annual Report of the Chief Signal Officer*, 1885, *Part* 2, and in *Professional Papers of the Signal Service*, *No.* 14, *p.* 427, where the formulæ and constants will be found.

It gives the mean vertical intensity for 24 hours of solar radiation J in terms of the mean solar constant A_0 for each tenth parallel of latitude of the northern hemisphere, and for the first and sixteenth day of each month; also the values of the solar constant A in terms of A_0, and the angular motion of the sun in longitude for the given dates.

CONVERSION OF LINEAR MEASURES.

The relation here adopted between the metre and the English measures of length is that used and officially authorized by the U. S. Bureau of Weights and Measures, viz:

1 metre = 39.3700 inches.

TABLE 64. *Inches into millimetres.*

The argument is given for every hundredth of an inch up to 32.00 inches, and the tabular values are given to hundredths of a millimetre. A table of proportional parts for thousandths of an inch is added on each page.

Example:

To convert 24.362 inches to millimetres.
The table gives (p. 184)

$(24.36 + 0.02)$ inches $= (618.75 + 0.05$ mm.$) = 618.80$ mm.

INTRODUCTION.

TABLE 65. *Millimetres into inches.*

From 0 to 400 mm. the argument is given to every millimetre, with subsidiary interpolation tables for tenths and hundredths of a millimetre. The tabular values are given to four decimals. From 400 to 1000 mm., covering the numerical values which are of frequent use in meteorology for the conversion of barometric readings from the metric to the English barometer, the argument is given for every tenth of a millimetre, and the tabular values to three decimals.

Example:

To convert 143.34 mm. to inches.
The table gives

143 + .3 + .04 mm. = 5.6299 + 0.0118 + 0.0016 inches = 5.6433 inches.

TABLE 66. *Feet into metres.*

From the adopted value of the metre, 39.3700 inches—

1 English foot = 0.3048006 metre.

Table 66 gives the value in metres and thousandths (or millimetres) for every foot from 0 to 99 feet; the value to hundredths of a metre (or centimetres) of every 10 feet from 100 to 4000 feet; and the value to tenths of a metre of every 10 feet from 4000 to 9090 feet. In using the latter part, the first line of the table serves to interpolate for single feet.

Example:

To convert 47 feet 7 inches to metres.   47 feet 7 inches = 47.583 feet.
The table gives   47   feet = 14.326 metres.
By moving the decimal point,   0.583 ,, = 0.178
                                          _____
                                          47.583 feet = 14.504 metres.

TABLE 67. *Metres into feet.*

1 metre = 39.3700 inches = 3.280833 + feet.

From 0 to 500 metres the argument is given for every unit, and the tabular values to two decimals; from 500 to 5000 the argument is given to every 10 metres, and the tabular values to one decimal. The conversion for tenths of a metre is added for convenience of interpolation.

Example:

Convert 4327 metres to feet.
The table gives

(4320 + 7) metres = (14173.2 + 23.0) feet = 14196.2 feet.

CONVERSION OF MEASURES OF TIME AND ANGLE. li

TABLE 68.   *Miles into kilometres.*   **TABLE 68.**

$$1 \text{ mile} = 1.609347 \text{ kilometres}.$$

The table extends from 0 to 1000 miles with argument to single miles, and from 1000 to 20000 miles for every 1000 miles. The tabular quantities are given to the nearest kilometre.

TABLE 69.   *Kilometres into miles.*   **TABLE 69.**

$$1 \text{ kilometre} = 0.621370 \text{ mile}.$$

The table extends to 1000 kilometres with argument to single kilometres, and from 1000 to 20000 kilometres for every 1000 kilometres. Tabular values are given to tenths of a mile.

Example:

Convert 3957 kilometres into miles.

The table gives

(3000 + 957) kilometres = (1864.1 + 594.7) miles = 2458.8 miles.

TABLE 70. *Interconversion of nautical and statute miles.* **TABLE 70.**

The definition of the nautical mile here used is that adopted by the U. S. Coast and Geodetic Survey.

A nautical mile is equal to the length of one minute of arc on the great circle of a sphere whose surface is equal to the surface of the earth.

Computed on Clarke's spheroid of 1866, the nautical mile thus defined equals 6080.27 feet. (*Report*, U. S. Coast Survey, 1881, page 354.)

The table gives, for nautical and statute miles from 1 to 9, the equivalent in statute and nautical miles, respectively, to four decimals.

TABLE 71.

TABLE 71. *Continental measures of length with their metric and English equivalents.*

This table gives a miscellaneous list of continental measures of length alphabetically arranged, with the name of the country to which they belong and their metric and English equivalents.

CONVERSION OF MEASURES OF TIME AND ANGLE.

TABLE 72.   *Arc into time.*

$$1° = 4^m; \quad 1' = 4^s; \quad 1'' = \tfrac{1\,s}{15} = 0^s\!.067.$$ **TABLE 72.**

Example:

Change 124° 15′ 24″.7 into time.

From the table,

124°	=	8^h	16^m	0^s
15′	=		1	0
24″	=			1.600
0″.7	=			.047
		8^h	17^m	$1^s\!.647$

lii   INTRODUCTION.

TABLE 73.   *Time into arc.*

$$1^h = 15°; \quad 1^m = 15'; \quad 1^s = 15''.$$

Example:
  Change $8^h\ 17^m\ 1\overset{s}{.}647$ into arc.

From the table,	8^h =	120°	
	17^m =	4	15'
	1^s =		15''
	0.64 =		9.60
By moving the decimal point,	.007 =		0.10
		124°	15' 24''.7

TABLE 74.   *Days into decimals of a year and angle.*

The table gives for the beginning of each day the corresponding decimal of the year to five places. Thus, at the epoch represented by the beginning of the 15th day, the decimal of the year that has elapsed since January 1.0 is computed from the fraction $\dfrac{14}{365.25}$. The corresponding value in angle obtained by multiplying this fraction by 360°, is given to the nearest minute.

Two additional columns serve to enter the table with the day of the month either of the common or the bissextile year as the argument, and may be used also for converting the day of the month to the day of the year, and *vice versa*.

Example:
  To find the number of days and the decimal of a year between February 12 and August 27 in a bissextile year.

  Aug. 27 : Day of year = 240 ; decimal of a year   = 0.65435
  Feb. 12 :   "   "   "   43 ;   "   "   "   = 0.11499
  ─────────────────────
  Interval in days   = 197 ; interval in decimal of a year = 0.53936

The decimal of the year corresponding to the interval 197 days may also be taken from the table by entering with the argument 198.

TABLE 75.   *Hours, minutes and seconds into decimals of a day.*

The tabular values are given to six decimals.

Example:
  Convert $5^h\ 24^m\ 23\overset{s}{.}4$ to the decimal of a day :

5^h =	$0\overset{d}{.}208333$
24^m =	016667
23^s =	266
By interpolation, or by moving the decimal for 4^s 0.4 =	5
	$0\overset{d}{.}225271$

CONVERSION OF MEASURES OF TIME AND ANGLE.   liii

TABLE 76.   *Decimals of a day into hours, minutes and seconds.*

Example:

Convert $0^d.225\,271$ to hours, minutes and seconds:

$$
\begin{aligned}
0.22 \quad &\text{day} = 4^h\,48^m + 28^m\,48^s = 5^h\,16^m\,48^s \\
0.0052 \quad &\text{day} = 7^m\,12^s + 17^s.28 = 7\,29.28 \\
0.000071 \quad &\text{day} = 6^s.05 + 0.09 = 6.14 \\
\hline
& \phantom{\text{day} = 4^h\,48^m + 28^m\,48^s =\;} 5^h\,24^m\,23^s.4
\end{aligned}
$$

TABLE 77.   *Minutes and seconds into decimals of an hour.*

The tabular values are given to six decimals.

Example:

Convert $34^m\,28^s.7$ to decimals of an hour.

$$
\begin{aligned}
34^m &= 0^h.566667 \\
28^s &= 7778 \\
0^s.7 &= 194 \\
\hline
&0.574639
\end{aligned}
$$

TABLE 78.   *Mean time at apparent noon.*

This table gives the time that should be shown by a clock when the sun crosses the meridian, on the 1st, 8th, 16th, and 24th days of each month. The table is useful in correcting a clock by means of a sun-dial or noon-mark.

Example:

To find the correct mean time when the sun crosses the meridian on December 15, 1891.

The table gives for December 16, $11^h\,56^m$. By interpolating, it is seen that the change to December 15 would be less than one-half minute; the correct clock time is therefore 4 minutes before 12 o'clock noon.

TABLE 79.   *Sidereal time into mean solar time.*
TABLE 80.   *Mean solar time into sidereal time.*

According to Bessel, the length of the tropical year is 365.24222 mean solar days,* whence

$$365.24222 \text{ solar days} = 366.24222 \text{ sidereal days.}$$

Any interval of mean time may therefore be changed into sidereal time by increasing it by its $\dfrac{1}{365.24222}$ part, and any interval of sidereal time may be changed into mean time by diminishing it by its $\dfrac{1}{366.24222}$ part.

---

*The length of the tropical year is not absolutely constant. The value here given is for the year 1800. Its decrease in 100 years is about 0.6 s.

Table 79 gives the quantities to be subtracted from the hours, minutes and seconds of a sidereal interval to obtain the corresponding mean time interval, and Table 80 gives the quantities to be added to the hours, minutes and seconds of a mean time interval to obtain the corresponding sidereal interval. The correction for seconds is sensibly the same for either a sidereal or a mean time interval and is therefore given but once, thus forming a part of each table.

Examples:

Change $14^h\ 25^m\ 36^s.2$ sidereal time into mean solar time.

Given sidereal time			$14^h\ 25^m\ 36^s.2$	
Correction for 14^h	$= -2^m\ 17^s.61$			
25^m	$= -\ \ \ \ \ 4.10$			
$36^s.2$	$= -\ \ \ \ \ \ \ .10$			
	$-2\ \ 21.81$		$-2\ \ 21.8$	
Corresponding mean time	$=$		$14\ \ 23\ \ 14.4$	

2. Change $13^h\ 37^m\ 22^s.7$ mean solar time into sidereal time.

Given mean time	$=$		$13^h\ 37^m\ 22^s.7$
Correction for 12^h	$= +2^m\ 8^s.13$		
37^m	$= +\ \ \ \ \ 6.08$		
$22^s.7$	$= +\ \ \ \ \ 0.06$		
	$+2\ \ 14.27$		$+2\ \ 14.3$
Corresponding sidereal time	$=$		$13\ \ 39\ \ 37.0$

MISCELLANEOUS TABLES.

DENSITY OF AIR AT DIFFERENT TEMPERATURES, HUMIDITIES AND PRESSURES.

The following tables (81 to 86) give the factors for computing the density of air at different temperatures, humidities and pressures.

The formula from which they have been computed is, in metric measures,

$$\delta = \frac{0.00129305\ [7.1116153]}{1 + 0.00367\ t} \left(\frac{b - 0.378 e}{760}\right)$$

in which δ is the weight of a cubic centimetre of air expressed in grammes, under the standard value of gravity at latitude 45° and sea level.

b is the barometric pressure in millimetres.

e is the pressure of aqueous vapor in millimetres.

t is the temperature in Centigrade degrees.

For dry atmospheric air (containing 0.0004 of its weight of carbonic acid) at a pressure of 760 mm. and temperature 0° C., the absolute density, or the weight of one cubic centimetre, is 0.00129305 gramme. (International Bureau of Weights and Measures: *Travaux et Mémoires*, t. I, p. A 54.)

In English measures, the formula becomes

$$\delta = \frac{0.00129305}{1 \times 0.0020389\,(t-32°)} \left(\frac{b - 0.378\,e}{29.921}\right)$$

where δ is defined as before, but b and e are expressed in inches and t in Fahrenheit degrees. Thus by the use of tables based on these two formulæ, lines of equal atmospheric density may be drawn for the whole world (neglecting slight variations in gravity), whether the original observations are in English or metric measures. Prof. Cleveland Abbe has kindly furnished for the present volume the logarithms of the density given in the accompanying tables (81 to 86).

TABLE 81.

TABLE 81. *Density of air at different temperatures Fahrenheit.*

This table gives the values and logarithms of the expression

$$\frac{0.00129305}{1 + 0.0020389\,(t - 32°)}$$

for values of t extending from $-45°$ $F.$ to $140°$ $F.$, the intervals between $0°$ $F.$ and $110°$ $F.$ being $1°$.

The tabular values are given to five significant figures.

TABLES 82, 83.

Density of air at different humidities and pressures—English measures.

TABLE 82. *Term for humidity; auxiliary to Table 83.*

TABLE 83. *Values of* $\dfrac{b - 0.378\,e}{29.921}$.

Table 82 gives values of $0.378\,e$ to three decimal places as an aid to the use of Table 83.

The argument is the dew-point given for every degree from $-40°$ $F.$ to $140°$ $F.$ A second column gives the corresponding values of the vapor pressure (e) according to Broch.

Table 83 gives values and logarithms of $\dfrac{h}{29.921} = \dfrac{b - 0.378\,e}{29.921}$ for values of h extending from 10.0 to 31.7 inches. The logarithms are given to five significant figures and the corresponding numbers to four decimals.

Example:

The air temperature is 68° $F.$, the pressure is 29.36 inches and the dew-point 51° $F.$ Find the logarithm of the density.

Table 81, for $t = 68°$ $F.$, gives	7.08085 − 10
Table 82, for dew-point 51°, gives $0.378\,e = 0.141$ inch,	
Table 83, for $h = b - 0.378\,e = 29.36 - 0.14 = 29.22$, gives	9.98941 − 10
	30
Logarithm of density =	7.07056 − 10

TABLE 84. *Density of air at different temperatures Centigrade.*

This gives values and logarithms of the expression

$$\delta_{t,\,760} = \frac{0.00129305}{1 + 0.00367\,t}$$

for values of t extending from $-34°$ C. to $69°$ C. The tabular values are given to five significant figures.

Density of air at different humidities and pressures—Metric measures.

TABLE 85. *Term for humidity: values of $0.378\,e$.*

TABLE 86. *Values of $\dfrac{h}{760} = \dfrac{b - 0.378\,e}{760}$.*

Table 85 gives values of $0.378\,e$ to hundredths of a millimetre for dew-points extending by intervals of $1°$ from $-30°$ C. to $50°$ C. The values of Broch's vapor pressures (e) corresponding to these dew-points are given in a second column to hundredths of a millimetre. The table is thus conveniently used when either the vapor pressure or the dew-point is known.

Table 86 gives values and logarithms of $\dfrac{h}{760} = \dfrac{b - 0.378\,e}{760}$ for values of h extending from 300 to 800 mm. The barometric pressure b is the barometer reading corrected for temperature and $0.378\,e$ is the term for humidity obtained from Table 85. The logarithms are given to five significant figures and the corresponding numbers to four decimal places.

TABLE 87. *Conversion of avoirdupois pounds and ounces into kilogrammes.*

The latest comparisons made by the International Bureau of Weights and Measures between the Imperial standard pound and the "kilogramme proto-type" result in the relation :

$$1 \text{ pound avoirdupois} = 453.592\,427\,7 \text{ grammes.}$$

This value has been adopted by the United States Bureau of Weights and Measures and is here used.

For the conversion of pounds, Table 87 gives the argument for every tenth of a pound up to 9.9, and the tabular conversion values to ten-thousandths of a kilogramme.

For the conversion of ounces, the argument is given for every tenth of an ounce up to 15.9, and the tabular values to ten-thousandths of a kilogramme.

TABLE 88. *Conversion of kilogrammes into avoirdupois pounds and ounces.*

From the above relation between the pound and the kilogramme,

$$1 \text{ kilogramme} = 2.204622 \text{ avoirdupois pounds.}$$
$$= 35.274 \quad \text{avoirdupois ounces.}$$

MISCELLANEOUS TABLES. lvii

The table gives the value to thousandths of a pound of every tenth of a kilogramme up to 9.9; the values of tenths of kilogrammes in ounces to four decimals; and the values of hundredths of a kilogramme in pounds and ounces to three and two decimals respectively.

TABLES 89, 90.

TABLE 89. *Conversion of grains into grammes.*
TABLE 90. *Conversion of grammes into grains.*

From the above relation between the pound and the kilogramme,

    1 gramme = 15.432356 grains.
    1 grain   = 0.06479892 gramme.

Table 89 gives to ten-thousandths of a gramme the value of every grain from 1 to 99, and also the conversion of tenths and hundredths of a grain for convenience in interpolating.

Table 90 gives to hundredths of a grain the value of every tenth of a gramme from 0.1 to 9.9, and the value of every gramme from 1 to 99. The values of hundredths and thousandths of a gramme are added as an aid to interpolation.

The computation of these two tables has been furnished by Professor William Libbey, who has used the relation, 1 gramme = 15.432 531 grains. This value is practically identical with the relation above adopted, differing from it by about 1 part in 3,000,000.

TABLE 91

TABLE 91. *Conversion of units of magnetic intensity.*

This table gives the conversion factors from 1 to 9 for converting English measures of magnetic intensity into C. G. S. measures, and vice versa.

The English unit of magnetic intensity is the force which, acting for 1 second on a unit of magnetism associated with a mass of 1 grain, produces a velocity of 1 foot per second.

The C. G. S. unit of magnetic intensity is the dyne—the force which, acting upon one gramme for 1 second, generates a velocity of 1 centimetre per second. The Gaussian unit of magnetic intensity, which has been extensively used, is a force which, acting upon a mass of 1 milligramme for 1 second, generates a velocity of 1 millimetre per second.

By using the dimensions of magnetic intensity $[M^{\frac{1}{2}}/L^{\frac{1}{2}}T]$, the interconversion of these units is easily made.

$$1 \text{ C. G. S. unit} = \sqrt{\frac{1000 \text{ M}}{10 \text{ L}}} \text{ Gaussian units}$$

$$= 10 \quad \text{Gaussian units}$$

$$1 \text{ C. G. S. unit} = \sqrt{\frac{15.432356 \text{ M}}{.03280833 \text{ L}}} \text{ English units}$$

$$= 21.6882 \quad \text{English units}$$

TABLE 92. *Quantity of water corresponding to given depths of rainfall.*

This table gives for different depths of rainfall over an acre and a square mile the total quantity of water measured in imperial gallons and tons respectively.

TABLE 93. *Dates of Dove's pentades.*

For tabulating and averaging meteorological data, Dove divided the year into seventy-three intervals of five days each, which have been called Dove's pentades, and this system of averaging has been used in the publication of a very considerable amount of meteorological data. Table 93 gives the initial and terminal dates of each pentade throughout the year.

TABLE 94. *Division by 28 of numbers from 28 to 867 972.*
TABLE 95. *Division by 29 of numbers from 29 to 898 971.*
TABLE 96. *Division by 31 of numbers from 31 to 960 969.*

The frequent occasion in meteorological work to divide by the numbers 28, 29 and 31 renders useful the division tables compiled by Mr. H. A. Hazen (*Handbook of Meteorological Tables*, Washington, D. C., 1888), the use of which has been kindly granted.

As here printed, the dividend is given in plain type and the quotient in heavy-face type, and in order that one shall never be mistaken for the other, a column is given containing the letters D and Q successively, which designates that all figures on a line with D are dividends, and all on a line with Q are quotients. The four columns to the right of this D-Q column give the last two figures of the dividend and of the quotient, namely, the units and tens. The ten columns to the left side of the D-Q column give the preceeding figures of the dividend, namely, the hundreds, thousands, and tens of thousands. These two parts of the dividend—to the left and right of the D-Q column—are always to be taken on the same horizontal line.

Each dividend is an exact multiple of the divisor, hence each quotient is exact or without remainder.

For example, the dividend 17360 in Table 94 is found in two parts; 173 is found in the column headed 600 on the left-hand side of the D-Q column, and 60 in the same horizontal row in the third column on the right-hand side.

The hundreds figure of the quotient is given in bold-face type at the top, middle and bottom of the page, and each one obtains for all the dividend figures in its own column. The units and tens figures of the quotient are found, as already stated, on the right side of the D-Q column directly under the last two figures of the dividend. Thus in the above example, for dividend 17360 the hundreds figure of the quotient is 6 and the units and tens will be 20, or the quotient of 17360 divided by 28 is 620. When any given dividend is not an exact multiple of the divisor, the nearest even multiple as given in the table must be used.

For example, $23979 \div 28 = 856$; the 8 is in the 9th column above 239 and the 56 is under 68, the nearest figure to 79 in the right-hand part of the table.

The last column, which is separated from the rest of the table by a triple line, is to be used when the quotient exceeds three figures, or 999.

The bold-face figures in this column give the thousands and tens of thousands figures of the quotient, and the plain figures are the multiples thereof by the divisor. To use the column, find in it the number which, with three ciphers added, comes nearest to (but is less than) the dividend; the heavy-face figures beneath it will be the first figures of the quotient. Subtract this multiple number from the given dividend, and with the remainder enter the main body of the table to obtain the last three figures of the quotient as already described.

For example: Divide 833885 by 28. The nearest figure to 833000 in the last column is 812000 and the quotient 29000. $833885 - 812000 = 21885$. Under 218 we have 7, and under 96, the nearest figure to 85 on the right, we find 82. $833885 \div 28 = 29782$.

TABLE 97. *Natural sines and cosines.*

TABLE 98. *Natural tangents and cotangents.*

TABLE 99. *Logarithms of numbers.*

TABLE 100. *List of meteorological stations.*

This list of meteorological stations has been compiled for this volume from data furnished by the United States Weather Bureau.

A geographical arrangement has been adopted as being most serviceable for the purposes for which the table will most generally be used.

In making the selection of stations from the vast number available, the object has been to choose such of the higher order stations as will fairly represent the varied climatic conditions of each country. With few exceptions, the stations are active; in all cases there are published observations, which may generally be found in the monthly and annual reports of the national meteorological services of the countries in which the stations are situated, or by which they are politically controlled.

So far as known, the list contains all first order stations, *i. e.*, those at which the principal meteorological elements are either recorded continuously and automatically, or are observed at hourly or bi-hourly intervals; such stations are designated by an asterisk (*).

The names of the stations have been given in the native orthography, which is in all cases the form adopted by the national meteorological service in its official publications.

GEORGE E. CURTIS

THERMOMETRICAL TABLES.

Conversion of thermometric scales —

    Reaumur scale to Fahrenheit and Centigrade . . . . TABLE 1

    Fahrenheit scale to Centigrade . . . . TABLE 2

    Centigrade scale to Fahrenheit . . . . TABLE 3

    Centigrade scale to Fahrenheit, near the boiling point of water . . . . TABLE 4

    Differences Fahrenheit to differences Centigrade . . . . TABLE 5

    Differences Centigrade to differences Fahrenheit . . . . TABLE 6

Reduction of temperature to sea level — English measures . TABLE 7

Reduction of temperature to sea level — Metric measures . . TABLE 8

Correction for the temperature of the mercury in the thermometer stem. For Fahrenheit and Centigrade thermometers . . . . TABLE 9

TABLE 1.

REAUMUR SCALE TO FAHRENHEIT AND CENTIGRADE.

Reaumur	Fahrenheit	Centigrade	Reaumur	Fahrenheit	Centigrade	Reaumur	Fahrenheit	Centigrade	Prop. Parts
+80°	+212°.00	+100°.00	+40°	+122°.00	+50°.00	± 0°	+32°.00	± 0°.00	
79	209.75	98.75	39	119.75	48.75	− 1	29.75	− 1.25	
78	207.50	97.50	38	117.50	47.50	2	27.50	2.50	
77	205.25	96.25	37	115.25	46.25	3	25.25	3.75	R. | F.
76	203.00	95.00	36	113.00	45.00	4	23.00	5.00	0.1 | 0.225
									.2 | .450
+75	+200.75	+ 93.75	+35	+110.75	+43.75	− 5	+20.75	− 6.25	.3 | .675
74	198.50	92.50	34	108.50	42.50	6	18.50	7.50	.4 | .900
73	196.25	91.25	33	106.25	41.25	7	16.25	8.75	.5 | 1.125
72	194.00	90.00	32	104.00	40.00	8	14.00	10.00	.6 | 1.350
71	191.75	88.75	31	101.75	38.75	9	11.75	11.25	.7 | 1.575
									.8 | 1.800
									0.9 | 2.025
+70	+189.50	+ 87.50	+30	+ 99.50	+37.50	−10	+ 9.50	−12.50	
69	187.25	86.25	29	97.25	36.25	11	7.25	13.75	
68	185.00	85.00	28	95.00	35.00	12	5.00	15.00	
67	182.75	83.75	27	92.75	33.75	13	2.75	16.25	R. | C.
66	180.50	82.50	26	90.50	32.50	14	+ 0.50	17.50	0.1 | 0.125
									.2 | .250
+65	+178.25	+ 81.25	+25	+ 88.25	+31.25	−15	− 1.75	−18.75	.3 | .375
64	176.00	80.00	24	86.00	30.00	16	4.00	20.00	.4 | .500
63	173.75	78.75	23	83.75	28.75	17	6.25	21.25	.5 | .625
62	171.50	77.50	22	81.50	27.50	18	8.50	22.50	.6 | .750
61	169.25	76.25	21	79.25	26.25	19	10.75	23.75	.7 | .875
									.8 | 1.000
									0.9 | 1.125
+60	+167.00	+ 75.00	+20	+ 77.00	+25.00	−20	−13.00	−25.00	
59	164.75	73.75	19	74.75	23.75	21	15.25	26.25	
58	162.50	72.50	18	72.50	22.50	22	17.50	27.50	F. | °C.
57	160.25	71.25	17	70.25	21.25	23	19.75	28.75	0.25 | 0.14
56	158.00	70.00	16	68.00	20.00	24	22.00	30.00	.50 | .28
									.75 | .42
+55	+155.75	+ 68.75	+15	+ 65.75	+18.75	−25	−24.25	−31.25	1.00 | .56
54	153.50	67.50	14	63.50	17.50	26	26.50	32.50	1.25 | .69
53	151.25	66.25	13	61.25	16.25	27	28.75	33.75	1.50 | .83
52	149.00	65.00	12	59.00	15.00	28	31.00	35.00	1.75 | .97
51	146.75	63.75	11	56.75	13.75	29	33.25	36.25	2.00 | 1.11
+50	+144.50	+ 62.50	+10	+ 54.50	+12.50	−30	−35.50	−37.50	
49	142.25	61.25	9	52.25	11.25	31	37.75	38.75	
48	140.00	60.00	8	50.00	10.00	32	40.00	40.00	C. | F.
47	137.75	58.75	7	47.75	8.75	33	42.25	41.25	0.05 | 0.09
46	135.50	57.50	6	45.50	7.50	34	44.50	42.50	.10 | .18
									.15 | .27
									.20 | .36
+45	+133.25	+ 56.25	+ 5	+ 43.25	+ 6.25	−35	−46.75	−43.75	.25 | .45
44	131.00	55.00	4	41.00	5.00	36	49.00	45.00	.50 | .90
43	128.75	53.75	3	38.75	3.75	37	51.25	46.25	.75 | 1.35
42	126.50	52.50	2	36.50	2.50	38	53.50	47.50	1.00 | 1.80
41	124.25	51.25	+ 1	34.25	+ 1.25	39	55.75	48.75	
+40	+122.00	+ 50.00	± 0	+ 32.00	± 0.00	−40	−58.00	−50.00	

$$F° = \frac{9}{5} C° + 32° \qquad C° = \frac{5}{9}(F° - 32°) \qquad R° = \frac{4}{9}(F° - 32°)$$

$$= \frac{9}{4} R° + 32° \qquad = \frac{5}{4} R° \qquad = \frac{4}{5} C°$$

* Smithsonian Tables.

TABLE 2.

FAHRENHEIT SCALE TO CENTIGRADE.

Fahren-heit.	.0	.1	.2	.3	.4	.5	.6	.7	.8	.9
	C.	C.	C.	C.	C.	C.	C.	C.	C.	C.
+130°	+54.44	+54.50	+54.56	+54.61	+54.67	+54.72	+54.78	+54.83	+54.89	+54.94
129	53.89	53.94	54.00	54.06	54.11	54.17	54.22	54.28	54.33	54.39
128	53.33	53.39	53.44	53.50	53.56	53.61	53.67	53.72	53.78	53.83
127	52.78	52.83	52.89	52.94	53.00	53.06	53.11	53.17	53.22	53.28
126	52.22	52.28	52.33	52.39	52.44	52.50	52.56	52.61	52.67	52.72
+125	+51.67	+51.72	+51.78	+51.83	+51.89	+51.94	+52.00	+52.06	+52.11	+52.17
124	51.11	51.17	51.22	51.28	51.33	51.39	51.44	51.50	51.56	51.61
123	50.56	50.61	50.67	50.72	50.78	50.83	50.89	50.94	51.00	51.06
122	50.00	50.06	50.11	50.17	50.22	50.28	50.33	50.39	50.44	50.50
121	49.44	49.50	49.56	49.61	49.67	49.72	49.78	49.83	49.89	49.94
+120	+48.89	+49.94	+49.00	+49.06	+49.11	+49.17	+49.22	+49.28	+49.33	+49.39
119	48.33	48.39	48.44	48.50	48.56	48.61	48.67	48.72	48.78	48.83
118	47.78	47.83	47.89	47.94	48.00	48.06	48.11	48.17	48.22	48.28
117	47.22	47.28	47.33	47.39	47.44	47.50	47.56	47.61	47.67	47.72
116	46.67	46.72	46.78	46.83	46.89	46.94	47.00	47.06	47.11	47.17
+115	+46.11	+46.17	+46.22	+46.28	+46.33	+46.39	+46.44	+46.50	+46.56	+46.61
114	45.56	45.61	45.67	45.72	45.78	45.83	45.89	45.94	46.00	46.06
113	45.00	45.06	45.11	45.17	45.22	45.28	45.33	45.39	45.44	45.50
112	44.44	44.50	44.56	44.61	44.67	44.72	44.78	44.83	44.89	44.94
111	43.89	43.94	44.00	44.06	44.11	44.17	44.22	44.28	44.33	44.39
+110	+43.33	+43.39	+43.44	+43.50	+43.56	+43.61	+43.67	+43.72	+43.78	+43.83
109	42.78	42.83	42.89	42.94	43.00	43.06	43.11	43.17	43.22	43.28
108	42.22	42.28	42.33	42.39	42.44	42.50	42.56	42.61	42.67	42.72
107	41.67	41.72	41.78	41.83	41.89	41.94	42.00	42.06	42.11	42.17
106	41.11	41.17	41.22	41.28	41.33	41.39	41.44	41.50	41.56	41.61
+105	+40.56	+40.61	+40.67	+40.72	+40.78	+40.83	+40.89	+40.94	+41.00	+41.06
104	40.00	40.06	40.11	40.17	40.22	40.28	40.33	40.39	40.44	40.50
103	39.44	39.50	39.56	39.61	39.67	39.72	39.78	39.83	39.89	39.94
102	38.89	38.94	39.00	39.06	39.11	39.17	39.22	39.28	39.33	39.39
101	38.33	38.39	38.44	38.50	38.56	38.61	38.67	38.72	38.78	38.83
+100	+37.78	+37.83	+37.89	+37.94	+38.00	+38.06	+38.11	+38.17	+38.22	+38.28
99	37.22	37.28	37.33	37.39	37.44	37.50	37.56	37.61	37.67	37.72
98	36.67	36.72	36.78	36.83	36.89	36.94	37.00	37.06	37.11	37.17
97	36.11	36.17	36.22	36.28	36.33	36.39	36.44	36.50	36.56	36.61
96	35.56	35.61	35.67	35.72	35.78	35.83	35.89	35.94	36.00	36.06
+95	+35.00	+35.06	+35.11	+35.17	+35.22	+35.28	+35.33	+35.39	+35.44	+35.50
94	34.44	34.50	34.56	34.61	34.67	34.72	34.78	34.83	34.89	34.94
93	33.89	33.94	34.00	34.06	34.11	34.17	34.22	34.28	34.33	34.39
92	33.33	33.39	33.44	33.50	33.56	33.61	33.67	33.72	33.78	33.83
91	32.78	32.83	32.89	32.94	33.00	33.06	33.11	33.17	33.22	33.28
+90	+32.22	+32.28	+32.33	+32.39	+32.44	+32.50	+32.56	+32.61	+32.67	+32.72
89	31.67	31.72	31.78	31.83	31.89	31.94	32.00	32.06	32.11	33.17
88	31.11	31.17	31.22	31.28	31.33	31.39	31.44	31.50	31.56	31.61
87	30.56	30.61	30.67	30.72	30.78	30.83	30.89	30.94	31.00	31.06
86	30.00	30.06	30.11	30.17	30.22	30.28	30.33	30.39	30.44	30.50
+85	+29.44	+29.50	+29.56	+29.61	+29.67	+29.72	+29.78	+29.83	+29.89	+29.94
84	28.89	28.94	29.00	29.06	29.11	29.17	29.22	29.28	29.33	29.39
83	28.33	28.39	28.44	28.50	28.56	28.61	28.67	28.72	28.78	28.83
82	27.78	27.83	27.89	27.94	28.00	28.06	28.11	28.17	28.22	28.28
81	27.22	27.28	27.33	27.39	27.44	27.50	27.56	27.61	27.67	27.72
+80	+26.67	+26.72	+26.78	+26.83	+26.89	+26.94	+27.00	+27.06	+27.11	+27.17
	.0	.1	.2	.3	.4	.5	.6	.7	.8	.9

SMITHSONIAN TABLES.

TABLE 2.

FAHRENHEIT SCALE TO CENTIGRADE.

Fahren-heit.	.0	.1	.2	.3	.4	.5	.6	.7	.8	.9
	C.	C.	C.	C.	C.	C.	C.	C.	C.	C.
+80°	+26.67	+26.72	+26.78	+26.83	+26.89	+26.94	+27.00	+27.06	+27.11	+27.17
79	26.11	26.17	26.22	26.28	26.33	26.39	26.44	26.50	26.56	26.61
78	25.56	25.61	25.67	25.72	25.78	25.83	25.89	25.94	26.00	26.06
77	25.00	25.06	25.11	25.17	25.22	25.28	25.33	25.39	25.44	25.50
76	24.44	24.50	24.56	24.61	24.67	24.72	24.78	24.83	24.89	24.94
+75	+23.89	+23.94	+24.00	+24.06	+24.11	+24.17	+24.22	+24.28	+24.33	+24.39
74	23.33	23.39	23.44	23.50	23.56	23.61	23.67	23.72	23.78	23.83
73	22.78	22.83	22.89	22.94	23.00	23.06	23.11	23.17	23.22	23.28
72	22.22	22.28	22.33	22.39	22.44	22.50	22.56	22.61	22.67	22.72
71	21.67	21.72	21.78	21.83	21.89	21.94	22.00	22.06	22.11	22.17
+70	+21.11	+21.17	+21.22	+21.28	+21.33	+21.39	+21.44	+21.50	+21.56	+21.61
69	20.56	20.61	20.67	20.72	20.78	20.83	20.89	20.94	21.00	21.06
68	20.00	20.06	20.11	20.17	20.22	20.28	20.33	20.39	20.44	20.50
67	19.44	19.50	19.56	19.61	19.67	19.72	19.78	19.83	19.89	19.94
66	18.89	18.94	19.00	19.06	19.11	19.17	19.22	19.28	19.33	19.39
+65	+18.33	+18.39	+18.44	+18.50	+18.56	+18.61	+18.67	+18.72	+18.78	+18.83
64	17.78	17.83	17.89	17.94	18.00	18.06	18.11	18.17	18.22	18.28
63	17.22	17.28	17.33	17.39	17.44	17.50	17.56	17.61	17.67	17.72
62	16.67	16.72	16.78	16.83	16.89	16.94	17.00	17.06	17.11	17.17
61	16.11	16.17	16.22	16.28	16.33	16.39	16.44	16.50	16.56	16.61
+60	+15.56	+15.61	+15.67	+15.72	+15.78	+15.83	+15.89	+15.94	+16.00	+16.06
59	15.00	15.06	15.11	15.17	15.22	15.28	15.33	15.39	15.44	15.50
58	14.44	14.50	14.56	14.61	14.67	14.72	14.78	14.83	14.89	14.94
57	13.89	13.94	14.00	14.06	14.11	14.17	14.22	14.28	14.33	14.39
56	13.33	13.39	13.44	13.50	13.56	13.61	13.67	13.72	13.78	13.83
+55	+12.78	+12.83	+12.89	+12.94	+13.00	+13.06	+13.11	+13.17	+13.22	+13.28
54	12.22	12.28	12.33	12.39	12.44	12.50	12.56	12.61	12.67	12.72
53	11.67	11.72	11.78	11.83	11.89	11.94	12.00	12.06	12.11	12.17
52	11.11	11.17	11.22	11.28	11.33	11.39	11.44	11.50	11.56	11.61
51	10.56	10.61	10.67	10.72	10.78	10.83	10.89	10.94	11.00	11.06
+50	+10.00	+10.06	+10.11	+10.17	+10.22	+10.28	+10.33	+10.39	+10.44	+10.50
49	9.44	9.50	9.56	9.61	9.67	9.72	9.78	9.83	9.89	9.94
48	8.89	8.94	9.00	9.06	9.11	9.17	9.22	9.28	9.33	9.39
47	8.33	8.39	8.44	8.50	8.56	8.61	8.67	8.72	8.78	8.83
46	7.78	7.83	7.89	7.94	8.00	8.06	8.11	8.17	8.22	8.28
+45	+ 7.22	+ 7.28	+ 7.33	+ 7.39	+ 7.44	+ 7.50	+ 7.56	+ 7.61	+ 7.67	+ 7.72
44	6.67	6.72	6.78	6.83	6.89	6.94	7.00	7.06	7.11	7.17
43	6.11	6.17	6.22	6.28	6.33	6.39	6.44	6.50	6.56	6.61
42	5.56	5.61	5.67	5.72	5.78	5.83	5.89	5.94	6.00	6.06
41	5.00	5.06	5.11	5.17	5.22	5.28	5.33	5.39	5.44	5.50
+40	+ 4.44	+ 4.50	+ 4.56	+ 4.61	+ 4.67	+ 4.72	+ 4.78	+ 4.83	+ 4.89	+ 4.94
39	3.89	3.94	4.00	4.06	4.11	4.17	4.22	4.28	4.33	4.39
38	3.33	3.39	3.44	3.50	3.56	3.61	3.67	3.72	3.78	3.83
37	2.78	2.83	2.89	2.94	3.00	3.06	3.11	3.17	3.22	3.28
36	2.22	2.28	2.33	2.39	2.44	2.50	2.56	2.61	2.67	2.72
+35	+ 1.67	+ 1.72	+ 1.78	+ 1.83	+ 1.89	+ 1.94	+ 2.00	+ 2.06	+ 2.11	+ 2.17
34	+ 1.11	+ 1.17	+ 1.22	+ 1.28	+ 1.33	+ 1.39	+ 1.44	+ 1.50	+ 1.56	+ 1.61
33	+ 0.56	+ 0.61	+ 0.67	+ 0.72	+ 0.78	+ 0.83	+ 0.89	+ 0.94	+ 1.00	+ 1.06
32	0.00	+ 0.06	+ 0.11	+ 0.17	+ 0.22	+ 0.28	+ 0.33	+ 0.39	+ 0.44	+ 0.50
31	− 0.56	− 0.50	− 0.44	− 0.39	− 0.33	− 0.28	− 0.22	− 0.17	− 0.11	− 0.06
+30	− 1.11	− 1.06	− 1.00	− 0.94	− 0.89	− 0.83	− 0.78	− 0.72	− 0.67	− 0.61
	.0	.1	.2	.3	.4	.5	.6	.7	.8	.9

TABLE 2.

FAHRENHEIT SCALE TO CENTIGRADE.

Fahren-heit.	.0	.1	.2	.3	.4	.5	.6	.7	.8	.9
	C.	C.	C.	C.	C.	C.	C.	C.	C.	C.
+30°	− 1°.11	− 1°.06	− 1°.00	− 0°.94	− 0°.89	− 0°.83	− 0°.78	− 0°.72	− 0°.67	− 0°.61
29	1.67	1.61	1.56	1.50	1.44	1.39	1.33	1.28	1.22	1.17
28	2.22	2.17	2.11	2.06	2.00	1.94	1.89	1.83	1.78	1.72
27	2.78	2.72	2.67	2.61	2.56	2.50	2.44	2.39	2.33	2.28
26	3.33	3.28	3.22	3.17	3.11	3.06	3.00	2.94	2.89	2.83
+25	− 3.89	− 3.83	− 3.78	− 3.72	− 3.67	− 3.61	− 3.56	− 3.50	− 3.44	− 3.39
24	4.44	4.39	4.33	4.28	4.22	4.17	4.11	4.06	4.00	3.94
23	5.00	4.94	4.89	4.83	4.78	4.72	4.67	4.61	4.56	4.50
22	5.56	5.50	5.44	5.39	5.33	5.28	5.22	5.17	5.11	5.06
21	6.11	6.06	6.00	5.94	5.89	5.83	5.78	5.72	5.67	5.61
+20	− 6.67	− 6.61	− 6.56	− 6.50	− 6.44	− 6.39	− 6.33	− 6.28	− 6.22	− 6.17
19	7.22	7.17	7.11	7.06	7.00	6.94	6.89	6.83	6.78	6.72
18	7.78	7.72	7.67	7.61	7.56	7.50	7.44	7.39	7.33	7.28
17	8.33	8.28	8.22	8.17	8.11	8.06	8.00	7.94	7.89	7.83
16	8.89	8.83	8.78	8.72	8.67	8.61	8.56	8.50	8.44	8.39
+15	− 9.44	− 9.39	− 9.33	− 9.28	− 9.22	− 9.17	− 9.11	− 9.06	− 9.00	− 8.94
14	10.00	9.94	9.89	9.83	9.78	9.72	9.67	9.61	9.56	9.50
13	10.56	10.50	10.44	10.39	10.33	10.28	10.22	10.17	10.11	10.06
12	11.11	11.06	11.00	10.94	10.89	10.83	10.78	10.72	10.67	10.61
11	11.67	11.61	11.56	11.50	11.44	11.39	11.33	11.28	11.22	11.17
+10	−12.22	−12.17	−12.11	−12.06	−12.00	−11.94	−11.89	−11.83	−11.78	−11.72
9	12.78	12.72	12.67	12.61	12.56	12.50	12.44	12.39	12.33	12.28
8	13.33	13.28	13.22	13.17	13.11	13.06	13.00	12.94	12.89	12.83
7	13.89	13.83	13.78	13.72	13.67	13.61	13.56	13.50	13.44	13.39
6	14.44	14.39	14.33	14.28	14.22	14.17	14.11	14.06	14.00	13.94
+5	−15.00	−14.94	−14.89	−14.83	−14.78	−14.72	−14.67	−14.61	−14.56	−14.50
4	15.56	15.50	15.44	15.39	15.33	15.28	15.22	15.17	15.11	15.06
3	16.11	16.06	16.00	15.94	15.89	15.83	15.78	15.72	15.67	15.61
2	16.67	16.61	16.56	16.50	16.44	16.39	16.33	16.28	16.22	16.17
1	17.22	17.17	17.11	17.06	17.00	16.94	16.89	16.83	16.78	16.72
+0	17.78	17.72	17.67	17.61	17.56	17.50	17.44	17.39	17.33	17.28
−0	−17.78	−17.83	−17.89	−17.94	−18.00	−18.06	−18.11	−18.17	−18.22	−18.28
1	18.33	18.39	18.44	18.50	18.56	18.61	18.67	18.72	18.78	18.83
2	18.89	18.94	19.00	19.06	19.11	19.17	19.22	19.28	19.33	19.39
3	19.44	19.50	19.56	19.61	19.67	19.72	19.78	19.83	19.89	19.94
4	20.00	20.06	20.11	20.17	20.22	20.28	20.33	20.39	20.44	20.50
−5	−20.56	−20.61	−20.67	−20.72	−20.78	−20.83	−20.89	−20.94	−21.00	−21.06
6	21.11	21.17	21.22	21.28	21.33	21.39	21.44	21.50	21.56	21.61
7	21.67	21.72	21.78	21.83	21.89	21.94	22.00	22.06	22.11	22.17
8	22.22	22.28	22.33	22.39	22.44	22.50	22.56	22.61	22.67	22.72
9	22.78	22.83	22.89	22.94	23.00	23.06	23.11	23.17	23.22	23.28
−10	−23.33	−23.39	−23.44	−23.50	−23.56	−23.61	−23.67	−23.72	−23.78	−23.83
11	23.89	23.94	24.00	24.06	24.11	24.17	24.22	24.28	24.33	24.39
12	24.44	24.50	24.56	24.61	24.67	24.72	24.78	24.83	24.89	24.94
13	25.00	25.06	25.11	25.17	25.22	25.28	25.33	25.39	25.44	25.50
14	25.56	25.61	25.67	25.72	25.78	25.83	25.89	25.94	26.00	26.06
−15	−26.11	−26.17	−26.22	−26.28	−26.33	−26.39	−26.44	−26.50	−26.56	−26.61
16	26.67	26.72	26.78	26.83	26.89	26.94	27.00	27.06	27.11	27.17
17	27.22	27.28	27.33	27.39	27.44	27.50	27.56	27.61	27.67	27.72
18	27.78	27.83	27.89	27.94	28.00	28.06	28.11	28.17	28.22	28.28
19	28.33	28.39	28.44	28.50	28.56	28.61	28.67	28.72	28.78	28.83
−20	−28.89	−28.94	−29.00	−29.06	−29.11	−29.17	−29.22	−29.28	−29.33	−29.39
	.0	.1	.2	.3	.4	.5	.6	.7	.8	.9

SMITHSONIAN TABLES.

TABLE 2.

FAHRENHEIT SCALE TO CENTIGRADE.

Fahrenheit.	.0	.1	.2	.3	.4	.5	.6	.7	.8	.9
	C.	C.	C.	C.	C.	C.	C.	C.	C.	C.
−20°	−28.89	−28.94	−29.00	−29.06	−29.11	−29.17	−29.22	−29.28	−29.33	−29.39
21	29.44	29.50	29.56	29.61	29.67	29.72	29.78	29.83	29.89	29.94
22	30.00	30.06	30.11	30.17	30.22	30.28	30.33	30.39	30.44	30.50
23	30.56	30.61	30.67	30.72	30.78	30.83	30.89	30.94	31.00	31.06
24	31.11	31.17	31.22	31.28	31.33	31.39	31.44	31.50	31.56	31.61
−25	−31.67	−31.72	−31.78	−31.83	−31.89	−31.94	−32.00	−32.06	−32.11	−32.17
26	32.22	32.28	32.33	32.39	32.44	32.50	32.56	32.61	32.67	32.72
27	32.78	32.83	32.89	32.94	33.00	33.06	33.11	33.17	33.22	33.28
28	33.33	33.39	33.44	33.50	33.56	33.61	33.67	33.72	33.78	33.83
29	33.89	33.94	34.00	34.06	34.11	34.17	34.22	34.28	34.33	34.39
−30	−34.44	−34.50	−34.56	−34.61	−34.67	−34.72	−34.78	−34.83	−34.89	−34.94
31	35.00	35.06	35.11	35.17	35.22	35.28	35.33	35.39	35.44	35.50
32	35.56	35.61	35.67	35.72	35.78	35.83	35.89	35.94	36.00	36.06
33	36.11	36.17	36.22	36.28	36.33	36.39	36.44	36.50	36.56	36.61
34	36.67	36.72	36.78	36.83	36.89	36.94	37.00	37.06	37.11	37.17
−35	−37.22	−37.28	−37.33	−37.39	−37.44	−37.50	−37.56	−37.61	−37.67	−37.72
36	37.78	37.83	37.89	37.94	38.00	38.06	38.11	38.17	38.22	38.28
37	38.33	38.39	38.44	38.50	38.56	38.61	38.67	38.72	38.78	38.83
38	38.89	38.94	39.00	39.06	39.11	39.17	39.22	39.28	39.33	39.39
39	39.44	39.50	39.56	39.61	39.67	39.72	39.78	39.83	39.89	39.94
−40	−40.00	−40.06	−40.11	−40.17	−40.22	−40.28	−40.33	−40.39	−40.44	−40.50
41	40.56	40.61	40.67	40.72	40.78	40.83	40.89	40.94	41.00	41.06
42	41.11	41.17	41.22	41.28	41.33	41.39	41.44	41.50	41.56	41.61
43	41.67	41.72	41.78	41.83	41.89	41.94	42.00	42.06	42.11	42.17
44	42.22	42.28	42.33	42.39	42.44	42.50	42.56	42.61	42.67	42.72
−45	−42.78	−42.83	−42.89	−42.94	−43.00	−43.06	−43.11	−43.17	−43.22	−43.28
46	43.33	43.39	43.44	43.50	43.56	43.61	43.67	43.72	43.78	43.83
47	43.89	43.94	44.00	44.06	44.11	44.17	44.22	44.28	44.33	44.39
48	44.44	44.50	44.56	44.61	44.67	44.72	44.78	44.83	44.89	44.94
49	45.00	45.06	45.11	45.17	45.22	45.28	45.33	45.39	45.44	45.50
−50	−45.56	−45.61	−45.67	−45.72	−45.78	−45.83	−45.89	−45.94	−46.00	−46.06
51	46.11	46.17	46.22	46.28	46.33	46.39	46.44	46.50	46.56	46.61
52	46.67	46.72	46.78	46.83	46.89	46.94	47.00	47.06	47.11	47.17
53	47.22	47.28	47.33	47.39	47.44	47.50	47.56	47.61	47.67	47.72
54	47.78	47.83	47.89	47.94	48.00	48.06	48.11	48.17	48.22	48.28
−55	−48.33	−48.39	−48.44	−48.50	−48.56	−48.61	−48.67	−48.72	−48.78	−48.83
56	48.89	48.94	49.00	49.06	49.11	49.17	49.22	49.28	49.33	49.39
57	49.44	49.50	49.56	49.61	49.67	49.72	49.78	49.83	49.89	49.94
58	50.00	50.06	50.11	50.17	50.22	50.28	50.33	50.39	50.44	50.50
59	50.56	50.61	50.67	50.72	50.78	50.83	50.89	50.94	51.00	51.06
−60	−51.11	−51.17	−51.22	−51.28	−51.33	−51.39	−51.44	−51.50	−51.56	−51.61
61	51.67	51.72	51.78	51.83	51.89	51.94	52.00	52.06	52.11	52.17
62	52.22	52.28	52.33	52.39	52.44	52.50	52.56	52.61	52.67	52.72
63	52.78	52.83	52.89	52.94	53.00	53.06	53.11	53.17	53.22	53.28
64	53.33	53.39	53.44	53.50	53.56	53.61	53.67	53.72	53.78	53.83
−65	−53.89	−53.94	−54.00	−54.06	−54.11	−54.17	−54.22	−54.28	−54.33	−54.39
66	54.44	54.50	54.56	54.61	54.67	54.72	54.78	54.83	54.89	54.94
67	55.00	55.06	55.11	55.17	55.22	55.28	55.33	55.39	55.44	55.50
68	55.56	55.61	55.67	55.72	55.78	55.83	55.89	55.94	56.00	56.06
69	56.11	56.17	56.22	56.28	56.33	56.39	56.44	56.50	56.56	56.61
−70	−56.67	−56.72	−56.78	−56.83	−56.89	−56.94	−57.00	−57.06	−57.11	−57.17
	.0	.1	.2	.3	.4	.5	.6	.7	.8	.9

SMITHSONIAN TABLES.

TABLE 3.

CENTIGRADE SCALE TO FAHRENHEIT.

Centi-grade.	.0	.1	.2	.3	.4	.5	.6	.7	.8	.9
	F.	F.	F.	F.	F.	F.	F.	F.	F.	F.
+50°	+122.00	+122.18	+122.36	+122.54	+122.72	+122.90	+123.08	+123.26	+123.44	+123.62
49	120.20	120.38	120.56	120.74	120.92	121.10	121.28	121.46	121.64	121.82
48	118.40	118.58	118.76	118.94	119.12	119.30	119.48	119.66	119.84	120.02
47	116.60	116.78	116.96	117.14	117.32	117.50	117.68	117.86	118.04	118.22
46	114.80	114.98	115.16	115.34	115.52	115.70	115.88	116.06	116.24	116.42
+45	+113.00	+113.18	+113.36	+113.54	+113.72	+113.90	+114.08	+114.26	+114.44	+114.62
44	111.20	111.38	111.56	111.74	111.92	112.10	112.28	112.46	112.64	112.82
43	109.40	109.58	109.76	109.94	110.12	110.30	110.48	110.66	110.84	111.02
42	107.60	107.78	107.96	108.14	108.32	108.50	108.68	108.86	109.04	109.22
41	105.80	105.98	106.16	106.34	106.52	106.70	106.88	107.06	107.24	107.42
+40	+104.00	+104.18	+104.36	+104.54	+104.72	+104.90	+105.08	+105.26	+105.44	+105.62
39	102.20	102.38	102.56	102.74	102.92	103.10	103.28	103.46	103.64	103.82
38	100.40	100.58	100.76	100.94	101.12	101.30	101.48	101.66	101.84	102.02
37	98.60	98.78	98.96	99.14	99.32	99.50	99.68	99.86	100.04	100.22
36	96.80	96.98	97.16	97.34	97.52	97.70	97.88	98.06	98.24	98.42
+35	+95.00	+95.18	+95.36	+95.54	+95.72	+95.90	+96.08	+96.26	+96.44	+96.62
34	93.20	93.38	93.56	93.74	93.92	94.10	94.28	94.46	94.64	94.82
33	91.40	91.58	91.76	91.94	92.12	92.30	92.48	92.66	92.84	93.02
32	89.60	89.78	89.96	90.14	90.32	90.50	90.68	90.86	91.04	91.22
31	87.80	87.98	88.16	88.34	88.52	88.70	88.88	89.06	89.24	89.42
+30	+86.00	+86.18	+86.36	+86.54	+86.72	+86.90	+87.08	+87.26	+87.44	+87.62
29	84.20	84.38	84.56	84.74	84.92	85.10	85.28	85.46	85.64	85.82
28	82.40	82.58	82.76	82.94	83.12	83.30	83.48	83.66	83.84	84.02
27	80.60	80.78	80.96	81.14	81.32	81.50	81.68	81.86	82.04	82.22
26	78.80	78.98	79.16	79.34	79.52	79.70	79.88	80.06	80.24	80.42
+25	+77.00	+77.18	+77.36	+77.54	+77.72	+77.90	+78.08	+78.26	+78.44	+78.62
24	75.20	75.38	75.56	75.74	75.92	76.10	76.28	76.46	76.64	76.82
23	73.40	73.58	73.76	73.94	74.12	74.30	74.48	74.66	74.84	75.02
22	71.60	71.78	71.96	72.14	72.32	72.50	72.68	72.86	73.04	73.22
21	69.80	69.98	70.16	70.34	70.52	70.70	70.88	71.06	71.24	71.42
+20	+68.00	+68.18	+68.36	+68.54	+68.72	+68.90	+69.08	+69.26	+69.44	+69.62
19	66.20	66.38	66.56	66.74	66.92	67.10	67.28	67.46	67.64	67.82
18	64.40	64.58	64.76	64.94	65.12	65.30	65.48	65.66	65.84	66.02
17	62.60	62.78	62.96	63.14	63.32	63.50	63.68	63.86	64.04	64.22
16	60.80	60.98	61.16	61.34	61.52	61.70	61.88	62.06	62.24	62.42
+15	+59.00	+59.18	+59.36	+59.54	+59.72	+59.90	+60.08	+60.26	+60.44	+60.62
14	57.20	57.38	57.56	57.74	57.92	58.10	58.28	58.46	58.64	58.82
13	55.40	55.58	55.76	55.94	56.12	56.30	56.48	56.66	56.84	57.02
12	53.60	53.78	53.96	54.14	54.32	54.50	54.68	54.86	55.04	55.22
11	51.80	51.98	52.16	52.34	52.52	52.70	52.88	53.06	53.24	53.42
+10	+50.00	+50.18	+50.36	+50.54	+50.72	+50.90	+51.08	+51.26	+51.44	+51.62
9	48.20	48.38	48.56	48.74	48.92	49.10	49.28	49.46	49.64	49.82
8	46.40	46.58	46.76	46.94	47.12	47.30	47.48	47.66	47.84	48.02
7	44.60	44.78	44.96	45.14	45.32	45.50	45.68	45.86	46.04	46.22
6	42.80	42.98	43.16	43.34	43.52	43.70	43.88	44.06	44.24	44.42
+5	+41.00	+41.18	+41.36	+41.54	+41.72	+41.90	+42.08	+42.26	+42.44	+42.62
4	39.20	39.38	39.56	39.74	39.92	40.10	40.28	40.46	40.64	40.82
3	37.40	37.58	37.76	37.94	38.12	38.30	38.48	38.66	38.84	39.02
2	35.60	35.78	35.96	36.14	36.32	36.50	36.68	36.86	37.04	37.22
1	33.80	33.98	34.16	34.34	34.52	34.70	34.88	35.06	35.24	35.42
+0	+32.00	+32.18	+32.36	+32.54	+32.72	+32.90	+33.08	+33.26	+33.44	+33.62
	.0	.1	.2	.3	.4	.5	.6	.7	.8	.9

TABLE 3.

CENTIGRADE SCALE TO FAHRENHEIT

Centi-grade.	.0	.1	.2	.3	.4	.5	.6	.7	.8	.9
	F.	F.	F.	F.	F.	F.	F.	F.	F.	F.
− 0°	+32.00	+31.82	+31.64	+31.46	+31.28	+31.10	+30.92	+30.74	+30.56	+30.38
1	30.20	30.02	29.84	29.66	29.48	29.30	29.12	28.94	28.76	28.58
2	28.40	28.22	28.04	27.86	27.68	27.50	27.32	27.14	26.96	26.78
3	26.60	26.42	26.24	26.06	25.88	25.70	25.52	25.34	25.16	24.98
4	24.80	24.62	24.44	24.26	24.08	23.90	23.72	23.54	23.36	23.18
− 5	+23.00	+22.82	+22.64	+22.46	+22.28	+22.10	+21.92	+21.74	+21.56	+21.38
6	21.20	21.02	20.84	20.66	20.48	20.30	20.12	19.94	19.76	19.58
7	19.40	19.22	19.04	18.86	18.68	18.50	18.32	18.14	17.96	17.78
8	17.60	17.42	17.24	17.06	16.88	16.70	16.52	16.34	16.16	15.98
9	15.80	15.62	15.44	15.26	15.08	14.90	14.72	14.54	14.36	14.18
−10	+14.00	+13.82	+13.64	+13.46	+13.28	+13.10	+12.92	+12.74	+12.56	+12.38
11	12.20	12.02	11.84	11.66	11.48	11.30	11.12	10.94	10.76	10.58
12	10.40	10.22	10.04	9.86	9.68	9.50	9.32	9.14	8.96	8.78
13	8.60	8.42	8.24	8.06	7.88	7.70	7.52	7.34	7.16	6.98
14	6.80	6.62	6.44	6.26	6.08	5.90	5.72	5.54	5.36	5.18
−15	+ 5.00	+ 4.82	+ 4.64	+ 4.46	+ 4.28	+ 4.10	+ 3.92	+ 3.74	+ 3.56	+ 3.38
16	+ 3.20	+ 3.02	+ 2.84	+ 2.66	+ 2.48	+ 2.30	+ 2.12	+ 1.94	+ 1.76	+ 1.58
17	+ 1.40	+ 1.22	+ 1.04	+ 0.86	+ 0.68	+ 0.50	+ 0.32	+ 0.14	− 0.04	− 0.22
18	− 0.40	− 0.58	− 0.76	− 0.94	− 1.12	− 1.30	− 1.48	− 1.66	− 1.84	− 2.02
19	− 2.20	− 2.38	− 2.56	− 2.74	− 2.92	− 3.10	− 3.28	− 3.46	− 3.64	− 3.82
−20	− 4.00	− 4.18	− 4.36	− 4.54	− 4.72	− 4.90	− 5.08	− 5.26	− 5.44	− 5.62
21	5.80	5.98	6.16	6.34	6.52	6.70	6.88	7.06	7.24	7.42
22	7.60	7.78	7.96	8.14	8.32	8.50	8.68	8.86	9.04	9.22
23	9.40	9.58	9.76	9.94	10.12	10.30	10.48	10.66	10.84	11.02
24	11.20	11.38	11.56	11.74	11.92	12.10	12.28	12.46	12.64	12.82
−25	−13.00	−13.18	−13.36	−13.54	−13.72	−13.90	−14.08	−14.26	−14.44	−14.62
26	14.80	14.98	15.16	15.34	15.52	15.70	15.88	16.06	16.24	16.42
27	16.60	16.78	16.96	17.14	17.32	17.50	17.68	17.86	18.04	18.22
28	18.40	18.58	18.76	18.94	19.12	19.30	19.48	19.66	19.84	20.02
29	20.20	20.38	20.56	20.74	20.92	21.10	21.28	21.46	21.64	21.82
−30	−22.00	−22.18	−22.36	−22.54	−22.72	−22.90	−23.08	−23.26	−23.44	−23.62
31	23.80	23.98	24.16	24.34	24.52	24.70	24.88	25.06	25.24	25.42
32	25.60	25.78	25.96	26.14	26.32	26.50	26.68	26.86	27.04	27.22
33	27.40	27.58	27.76	27.94	28.12	28.30	28.48	28.66	28.84	29.02
34	29.20	29.38	29.56	29.74	29.92	30.10	30.28	30.46	30.64	30.82
−35	−31.00	−31.18	−31.36	−31.54	−31.72	−31.90	−32.08	−32.26	−32.44	−32.62
36	32.80	32.98	33.16	33.34	33.52	33.70	33.88	34.06	34.24	34.42
37	34.60	34.78	34.96	35.14	35.32	35.50	35.68	35.86	36.04	36.22
38	36.40	36.58	36.76	36.94	37.12	37.30	37.48	37.66	37.84	38.02
39	38.20	38.38	38.56	38.74	38.92	39.10	39.28	39.46	39.64	39.82
−40	−40.00	−40.18	−40.36	−40.54	−40.72	−40.90	−41.08	−41.26	−41.44	−41.62
41	41.80	41.98	42.16	42.34	42.52	42.70	42.88	43.06	43.24	43.42
42	43.60	43.78	43.96	44.14	44.32	44.50	44.68	44.86	45.04	45.22
43	45.40	45.58	45.76	45.94	46.12	46.30	46.48	46.66	46.84	47.02
44	47.20	47.38	47.56	47.74	47.92	48.10	48.28	48.46	48.64	48.82
−45	−49.00	−49.18	−49.36	−49.54	−49.72	−49.90	−50.08	−50.26	−50.44	−50.62
46	50.80	50.98	51.16	51.34	51.52	51.70	51.88	52.06	52.24	52.42
47	52.60	52.78	52.96	53.14	53.32	53.50	53.68	53.86	54.04	54.22
48	54.40	54.58	54.76	54.94	55.12	55.30	55.48	55.66	55.84	56.02
49	56.20	56.38	56.56	56.74	56.92	57.10	57.28	57.46	57.64	57.82
−50	−58.00	−58.18	−58.36	−58.54	−58.72	−58.90	−59.08	−59.26	−59.44	−59.62
	.0	.1	.2	.3	.4	.5	.6	.7	.8	.9

TABLE 4.
CENTIGRADE SCALE TO FAHRENHEIT—Near the Boiling Point.

Centi-grade.	.0	.1	.2	.3	.4	.5	.6	.7	.8	.9
	F.	F.	F.	F.	F.	F.	F.	F.	F.	F.
100°	212.00	212.18	212.36	212.54	212.72	212.90	213.08	213.26	213.44	213.62
99	210.20	210.38	210.56	210.74	210.92	211.10	211.28	211.46	211.64	211.82
98	208.40	208.58	208.76	208.94	209.12	209.30	209.48	209.66	209.84	210.02
97	206.60	206.78	206.96	207.14	207.32	207.50	207.68	207.86	208.04	208.22
96	204.80	204.98	205.16	205.34	205.52	205.70	205.88	206.06	206.24	206.42
95	203.00	203.18	203.36	203.54	203.72	203.90	204.08	204.26	204.44	204.62
94	201.20	201.38	201.56	201.74	201.92	202.10	202.28	202.46	202.64	202.82
93	199.40	199.58	199.76	199.94	200.12	200.30	200.48	200.66	200.84	201.02
92	197.60	197.78	197.96	198.14	198.32	198.50	198.68	198.86	199.04	199.22
91	195.80	195.98	196.16	196.34	196.52	196.70	196.88	197.06	197.24	197.42
90	194.00	194.18	194.36	194.54	194.72	194.90	195.08	195.26	195.44	195.62

TABLE 5.
DIFFERENCES FAHRENHEIT TO DIFFERENCES CENTIGRADE.

Fahren-heit.	.0	.1	.2	.3	.4	.5	.6	.7	.8	.9
	C.	C.	C.	C.	C.	C.	C.	C.	C.	C.
0°	0.00	0.06	0.11	0.17	0.22	0.28	0.33	0.39	0.44	0.50
1	0.56	0.61	0.67	0.72	0.78	0.83	0.89	0.94	1.00	1.06
2	1.11	1.17	1.22	1.28	1.33	1.39	1.44	1.50	1.56	1.61
3	1.67	1.72	1.78	1.83	1.89	1.94	2.00	2.06	2.11	2.17
4	2.22	2.28	2.33	2.39	2.44	2.50	2.56	2.61	2.67	2.72
5	2.78	2.83	2.89	2.94	3.00	3.06	3.11	3.17	3.22	3.28
6	3.33	3.39	3.44	3.50	3.56	3.61	3.67	3.72	3.78	3.83
7	3.89	3.94	4.00	4.06	4.11	4.17	4.22	4.28	4.33	4.39
8	4.44	4.50	4.56	4.61	4.67	4.72	4.78	4.83	4.89	4.94
9	5.00	5.06	5.11	5.17	5.22	5.28	5.33	5.39	5.44	5.50
10	5.56	5.61	5.67	5.72	5.78	5.83	5.89	5.94	6.00	6.06
11	6.11	6.17	6.22	6.28	6.33	6.39	6.44	6.50	6.56	6.61
12	6.67	6.72	6.78	6.83	6.89	6.94	7.00	7.06	7.11	7.17
13	7.22	7.28	7.33	7.39	7.44	7.50	7.56	7.61	7.67	7.72
14	7.78	7.83	7.89	7.94	8.00	8.06	8.11	8.17	8.22	8.28
15	8.33	8.39	8.44	8.50	8.56	8.61	8.67	8.72	8.78	8.83
16	8.89	8.94	9.00	9.06	9.11	9.17	9.22	9.28	9.33	9.39
17	9.44	9.50	9.56	9.61	9.67	9.72	9.78	9.83	9.89	9.94
18	10.00	10.06	10.11	10.17	10.22	10.28	10.33	10.39	10.44	10.50
19	10.56	10.61	10.67	10.72	10.78	10.83	10.89	10.94	11.00	11.06
20	11.11	11.17	11.22	11.28	11.33	11.39	11.44	11.50	11.56	11.61

TABLE 6.
DIFFERENCES CENTIGRADE TO DIFFERENCES FAHRENHEIT.

Centi-grade.	.0	.1	.2	.3	.4	.5	.6	.7	.8	.9
	F.	F.	F.	F.	F.	F.	F.	F.	F.	F.
0°	0.00	0.18	0.36	0.54	0.72	0.90	1.08	1.26	1.44	1.62
1	1.80	1.98	2.16	2.34	2.52	2.70	2.88	3.06	3.24	3.42
2	3.60	3.78	3.96	4.14	4.32	4.50	4.68	4.86	5.04	5.22
3	5.40	5.58	5.76	5.94	6.12	6.30	6.48	6.66	6.84	7.02
4	7.20	7.38	7.56	7.74	7.92	8.10	8.28	8.46	8.64	8.82
5	9.00	9.18	9.36	9.54	9.72	9.90	10.08	10.26	10.44	10.62
6	10.80	10.98	11.16	11.34	11.52	11.70	11.88	12.06	12.24	12.42
7	12.60	12.78	12.96	13.14	13.32	13.50	13.68	13.86	14.04	14.22
8	14.40	14.58	14.76	14.94	15.12	15.30	15.48	15.66	15.84	16.02
9	16.20	16.38	16.56	16.74	16.92	17.10	17.28	17.46	17.64	17.82

SMITHSONIAN TABLES.

TABLE 7.
REDUCTION OF TEMPERATURE TO SEA LEVEL.
ENGLISH MEASURES.

Rate of decrease of temperature. 1°F. for every	DIFFERENCES BETWEEN THE TEMPERATURE AT ANY ALTITUDE AND AT SEA LEVEL. ALTITUDE IN FEET.												
	100	200	300	400	500	600	700	800	900	1000	2000	3000	4000	5000
Feet.	F.	F.	F.	F.	F.	F.	F.	F.	F.	F.	F.	F.	F.	F.
200	0.50	1.00	1.50	2.00	2.50	3.00	3.50	4.00	4.50	5.00	10.00	15.00	20.00	25.00
205	0.49	0.98	1.46	1.95	2.44	2.93	3.41	3.90	4.39	4.88	9.76	14.63	19.51	24.39
210	0.48	0.95	1.43	1.90	2.38	2.86	3.33	3.81	4.29	4.76	9.52	14.29	19.05	23.81
215	0.47	0.93	1.40	1.86	2.33	2.79	3.26	3.72	4.19	4.65	9.30	13.95	18.60	23.26
220	0.45	0.91	1.36	1.82	2.27	2.73	3.18	3.64	4.09	4.55	9.09	13.63	18.18	22.72
230	0.43	0.87	1.30	1.74	2.17	2.61	3.04	3.48	3.91	4.35	8.70	13.04	17.39	21.74
240	0.42	0.83	1.25	1.67	2.08	2.50	2.92	3.33	3.75	4.17	8.33	12.50	16.67	20.83
250	0.40	0.80	1.20	1.60	2.00	2.40	2.80	3.20	3.60	4.00	8.00	12.00	16.00	20.00
260	0.38	0.77	1.15	1.54	1.92	2.31	2.69	3.08	3.46	3.85	7.69	11.54	15.38	19.23
270	0.37	0.74	1.11	1.48	1.85	2.22	2.59	2.96	3.33	3.70	7.41	11.11	14.81	18.52
280	0.36	0.71	1.07	1.43	1.79	2.14	2.50	2.86	3.21	3.57	7.14	10.71	14.29	17.86
290	0.34	0.69	1.03	1.38	1.73	2.07	2.41	2.76	3.10	3.45	6.90	10.34	13.79	17.24
300	0.33	0.67	1.00	1.33	1.67	2.00	2.33	2.67	3.00	3.33	6.67	10.00	13.33	16.67
310	0.32	0.65	0.97	1.29	1.61	1.94	2.26	2.58	2.90	3.23	6.45	9.68	12.90	16.13
320	0.31	0.62	0.94	1.25	1.56	1.87	2.19	2.50	2.81	3.12	6.25	9.37	12.50	15.62
340	0.29	0.59	0.88	1.18	1.47	1.76	2.06	2.35	2.65	2.94	5.88	8.82	11.76	14.71
360	0.28	0.56	0.83	1.11	1.39	1.67	1.94	2.22	2.50	2.78	5.56	8.33	11.11	13.89
380	0.26	0.53	0.79	1.05	1.32	1.58	1.84	2.10	2.37	2.63	5.26	7.89	10.53	13.16
400	0.25	0.50	0.75	1.00	1.25	1.50	1.75	2.00	2.25	2.50	5.00	7.50	10.00	12.50
420	0.24	0.48	0.71	0.95	1.19	1.43	1.67	1.90	2.14	2.38	4.76	7.14	9.52	11.90
440	0.23	0.45	0.68	0.91	1.14	1.36	1.59	1.82	2.05	2.27	4.55	6.82	9.09	11.36
460	0.22	0.43	0.65	0.87	1.09	1.30	1.52	1.74	1.96	2.17	4.35	6.52	8.70	10.87
480	0.21	0.42	0.62	0.83	1.04	1.25	1.46	1.67	1.87	2.08	4.17	6.25	8.33	10.42
500	0.20	0.40	0.60	0.80	1.00	1.20	1.40	1.60	1.80	2.00	4.00	6.00	8.00	10.00
520	0.19	0.38	0.58	0.77	0.96	1.15	1.35	1.54	1.73	1.92	3.85	5.77	7.69	9.62
540	0.19	0.37	0.56	0.74	0.93	1.11	1.30	1.48	1.67	1.85	3.70	5.56	7.41	9.26
560	0.18	0.36	0.54	0.71	0.89	1.07	1.25	1.43	1.61	1.79	3.57	5.36	7.14	8.93
580	0.17	0.34	0.52	0.69	0.86	1.03	1.21	1.38	1.55	1.72	3.45	5.17	6.90	8.62
600	0.17	0.33	0.50	0.67	0.83	1.00	1.17	1.33	1.50	1.67	3.33	5.00	6.67	8.33
620	0.16	0.32	0.48	0.65	0.81	0.97	1.13	1.29	1.45	1.61	3.23	4.84	6.45	8.06
650	0.15	0.31	0.46	0.62	0.77	0.92	1.08	1.23	1.38	1.54	3.08	4.62	6.15	7.69
700	0.14	0.29	0.43	0.57	0.71	0.86	1.00	1.14	1.29	1.43	2.86	4.29	5.71	7.14
750	0.13	0.27	0.40	0.53	0.67	0.80	0.93	1.07	1.20	1.33	2.67	4.00	5.33	6.67
800	0.12	0.25	0.37	0.50	0.62	0.75	0.87	1.00	1.12	1.25	2.50	3.75	5.00	6.25
850	0.12	0.24	0.35	0.47	0.59	0.71	0.82	0.94	1.06	1.18	2.35	3.53	4.71	5.88
900	0.11	0.22	0.33	0.44	0.56	0.67	0.78	0.89	1.00	1.11	2.22	3.33	4.44	5.56

Tabular values are to be added to the observed temperature to obtain the temperature at sea level.

TABLE 8.

REDUCTION OF TEMPERATURE TO SEA LEVEL.
METRIC MEASURES.

Rate of decrease of temperature. 1°C. for every	DIFFERENCES BETWEEN THE TEMPERATURE AT ANY ALTITUDE AND AT SEA LEVEL. ALTITUDE IN METRES.										
	100	200	300	400	500	600	700	800	900	1000	2000	3000
m.	c.	c.	c.	c.	c.	c.	c.	c.	c.	c.	c.	c.
100	1.00	2.00	3.00	4.00	5.00	6.00	7.00	8.00	9.00	10.00	20.00	30.00
102	0.98	1.96	2.94	3.92	4.90	5.88	6.86	7.84	8.82	9.80	19.61	29.41
104	0.96	1.92	2.88	3.85	4.81	5.77	6.73	7.69	8.65	9.62	19.23	28.85
106	0.94	1.89	2.83	3.77	4.72	5.66	6.60	7.55	8.49	9.43	18.87	28.30
108	0.93	1.85	2.78	3.70	4.63	5.56	6.48	7.41	8.33	9.26	18.52	27.78
110	0.91	1.82	2.73	3.64	4.55	5.45	6.36	7.27	8.18	9.09	18.18	27.27
115	0.87	1.74	2.61	3.48	4.35	5.22	6.09	6.96	7.83	8.70	17.39	26.09
120	0.83	1.67	2.50	3.33	4.17	5.00	5.83	6.67	7.50	8.33	16.67	25.00
125	0.80	1.60	2.40	3.20	4.00	4.80	5.60	6.40	7.20	8.00	16.00	24.00
130	0.77	1.54	2.31	3.08	3.85	4.62	5.38	6.15	6.92	7.69	15.38	23.08
135	0.74	1.48	2.22	2.96	3.70	4.44	5.19	5.93	6.66	7.41	14.81	22.22
140	0.71	1.43	2.14	2.86	3.57	4.29	5.00	5.71	6.43	7.14	14.29	21.43
145	0.69	1.38	2.07	2.76	3.45	4.14	4.83	5.52	6.21	6.90	13.79	20.69
150	0.67	1.33	2.00	2.67	3.33	4.00	4.67	5.33	6.00	6.67	13.33	20.00
155	0.65	1.29	1.94	2.58	3.23	3.87	4.52	5.16	5.81	6.45	12.90	19.35
160	0.62	1.25	1.87	2.50	3.12	3.75	4.37	5.00	5.62	6.25	12.50	18.75
170	0.59	1.18	1.76	2.35	2.94	3.53	4.12	4.70	5.29	5.88	11.76	17.65
180	0.56	1.11	1.67	2.22	2.78	3.33	3.89	4.44	5.00	5.56	11.11	16.67
190	0.53	1.05	1.58	2.10	2.63	3.16	3.68	4.21	4.74	5.26	10.53	15.79
200	0.50	1.00	1.50	2.00	2.50	3.00	3.50	4.00	4.50	5.00	10.00	15.00
210	0.48	0.95	1.43	1.90	2.38	2.86	3.33	3.81	4.29	4.76	9.52	14.29
220	0.45	0.91	1.36	1.82	2.27	2.73	3.18	3.64	4.09	4.55	9.09	13.64
230	0.43	0.87	1.30	1.74	2.17	2.61	3.04	3.48	3.91	4.35	8.70	13.04
240	0.42	0.83	1.25	1.67	2.08	2.50	2.92	3.33	3.75	4.17	8.33	12.50
250	0.40	0.80	1.20	1.60	2.00	2.40	2.80	3.20	3.60	4.00	8.00	12.00
260	0.38	0.77	1.15	1.54	1.92	2.31	2.69	3.08	3.46	3.85	7.69	11.54
270	0.37	0.74	1.11	1.48	1.85	2.22	2.59	2.96	3.33	3.70	7.41	11.11
280	0.36	0.71	1.07	1.43	1.79	2.14	2.50	2.86	3.21	3.57	7.14	10.71
290	0.34	0.69	1.03	1.38	1.72	2.07	2.41	2.76	3.10	3.45	6.90	10.34
300	0.33	0.67	1.00	1.33	1.67	2.00	2.33	2.67	3.00	3.33	6.67	10.00
320	0.31	0.62	0.94	1.25	1.56	1.87	2.19	2.50	2.81	3.12	6.25	9.37
340	0.29	0.59	0.88	1.18	1.47	1.76	2.06	2.35	2.65	2.94	5.88	8.82
360	0.28	0.56	0.83	1.11	1.39	1.67	1.94	2.22	2.50	2.78	5.56	8.33
380	0.26	0.53	0.79	1.05	1.32	1.58	1.84	2.10	2.37	2.63	5.26	7.89
400	0.25	0.50	0.75	1.00	1.25	1.50	1.75	2.00	2.25	2.50	5.00	7.50
420	0.24	0.48	0.71	0.95	1.19	1.43	1.67	1.90	2.14	2.38	4.76	7.14
440	0.23	0.45	0.68	0.91	1.14	1.36	1.59	1.82	2.05	2.27	4.55	6.82
460	0.22	0.43	0.65	0.87	1.09	1.30	1.52	1.74	1.96	2.17	4.35	6.52
480	0.21	0.42	0.62	0.83	1.04	1.25	1.46	1.67	1.87	2.08	4.17	6.25
500	0.20	0.40	0.60	0.80	1.00	1.20	1.40	1.60	1.80	2.00	4.00	6.00

Tabular values are to be added to the observed temperature to obtain the temperature at sea level.

TABLE 9.

CORRECTION FOR THE TEMPERATURE OF THE MERCURY IN THE THERMOMETER STEM.

$T = t - 0.0000795\ n\ (t' - t)$ — Fahrenheit temperatures.
$T = t - 0.000143\ n\ (t' - t)$ — Centigrade temperatures.
$T =$ Corrected temperature.
$t =$ Observed temperature.
$t' =$ Mean temperature of the glass stem and mercury column.
$n =$ Length of mercury in the stem in scale degrees.

CORRECTION FOR FAHRENHEIT THERMOMETERS.

Values of $0.0000795\ n\ (t' - t)$

n	\multicolumn{10}{c}{$t' - t$}									
	10°	20°	30°	40°	50°	60°	70°	80°	90°	100°
	F.	F.	F.	F.	F.	F.	F.	F.	F.	F.
10°	0.01	0.02	0.02	0.03	0.04	0.05	0.06	0.06	0.07	0.08
20	0.02	0.03	0.05	0.06	0.08	0.10	0.11	0.13	0.14	0.16
30	0.02	0.05	0.07	0.10	0.12	0.14	0.17	0.19	0.21	0.24
40	0.03	0.06	0.10	0.13	0.16	0.19	0.22	0.25	0.29	0.32
50	0.04	0.08	0.12	0.16	0.20	0.24	0.28	0.32	0.36	0.40
60	0.05	0.10	0.14	0.19	0.24	0.29	0.33	0.38	0.43	0.48
70	0.06	0.11	0.17	0.22	0.28	0.33	0.39	0.45	0.50	0.56
80	0.06	0.13	0.19	0.25	0.32	0.38	0.45	0.51	0.57	0.64
90	0.07	0.14	0.21	0.29	0.36	0.43	0.50	0.57	0.64	0.72
100	0.08	0.16	0.24	0.32	0.40	0.48	0.56	0.64	0.72	0.79
110	0.09	0.17	0.26	0.35	0.44	0.52	0.61	0.70	0.79	0.87
120	0.10	0.19	0.29	0.38	0.48	0.57	0.67	0.76	0.86	0.95
130	0.10	0.21	0.31	0.41	0.52	0.62	0.72	0.83	0.93	1.03

CORRECTION FOR CENTIGRADE THERMOMETERS.

Values of $0.000143\ n\ (t' - t)$

n	\multicolumn{8}{c}{$t' - t$}							
	10°	20°	30°	40°	50°	60°	70°	80°
	C.	C.	C.	C.	C.	C.	C.	C.
10°	0.01	0.03	0.04	0.06	0.07	0.09	0.10	0.11
20	0.03	0.06	0.09	0.11	0.14	0.17	0.20	0.23
30	0.04	0.09	0.13	0.17	0.21	0.26	0.30	0.34
40	0.06	0.11	0.17	0.23	0.29	0.34	0.40	0.46
50	0.07	0.14	0.21	0.29	0.36	0.43	0.50	0.57
60	0.09	0.17	0.26	0.34	0.43	0.51	0.60	0.69
70	0.10	0.20	0.30	0.40	0.50	0.60	0.70	0.80
80	0.11	0.23	0.34	0.46	0.57	0.69	0.80	0.92
90	0.13	0.26	0.39	0.51	0.64	0.77	0.90	1.03
100	0.14	0.29	0.43	0.57	0.72	0.86	1.00	1.14

When t' is { greater / less } than t the correction is to be { subtracted / added }

BAROMETRICAL TABLES.

Reduction of the barometer to standard temperature —
  English measures . . . . TABLE 10
  Metric measures 11
Reduction of the barometer to standard gravity at latitude 45°—
  English measures . . . . TABLE 12
  Metric measures 13
Reduction of the barometer to sea level — English measures.
  Values of $2000\,m$ . . . . TABLE 14
  Correction of $2000\,m$ for latitude 15
  $B_o - B = B(10^m - 1)$ 16
Reduction of the barometer to sea level — Metric measures.
  Values of $2000\,m$ . . . . TABLE 17
  Correction of $2000\,m$ for latitude 18
  $B_o - B = B(10^m - 1)$ 19
Determination of heights by the barometer — English measures.
  Values of $60368\,[1 + 0.0010195 \times 36]\,log\,\frac{29.90}{B}$ . . . . TABLE 20
  Term for temperature 21
  Correction for latitude and weight of mercury 22
  Correction for an average degree of humidity 23
  Correction for the variation of gravity with altitude . . 24
Determination of heights by the barometer — Metric measures.
  Values of $18400\,log\,\frac{760}{B}$ . . . . TABLE 25
  Term for temperature 26
  Correction for humidity 27
  Correction for latitude and weight of mercury 28
  Correction for the variation of gravity with altitude . . 29
Difference of height corresponding to a change of 0.1 inch
  in the barometer — English measures . . . . . TABLE 30
Difference of height corresponding to a change of 1 millimetre
  in the barometer — Metric measures . . . . . . TABLE 31
Determination of heights by the barometer.
  Formula of Babinet . . . . TABLE 32
Barometric pressures corresponding to the temperature of the
  boiling point of water —
  English measures . . . . TABLE 33
  Metric measures 34

TABLE 10.

REDUCTION OF THE BAROMETER TO STANDARD TEMPERATURE.
ENGLISH MEASURES.

Attached Thermometer Fahrenheit.	HEIGHT OF THE BAROMETER IN INCHES.									
	19.0	19.5	20.0	20.5	21.0	21.5	22.0	22.5	23.0	23.5
F.	Inch.	Inch.	Inch.	Inch.	Inch.	Inch.	Inch.	Inch.	Inch.	Inch.
0°.0	+0.050	+0.051	+0.052	+0.053	+0.055	+0.056	+0.057	+0.059	+0.060	+0.061
+0.5	+0.049	+0.050	+0.051	+0.053	+0.054	+0.055	+0.056	+0.058	+0.059	+0.060
1.0	.048	.049	.050	.052	.053	.054	.055	.057	.058	.059
1.5	.047	.048	.049	.051	.052	.053	.054	.056	.057	.058
2.0	.046	.047	.049	.050	.051	.052	.053	.055	.056	.057
2.5	.045	.046	.048	.049	.050	.051	.052	.054	.055	.056
3.0	+0.044	+0.046	+0.047	+0.048	+0.049	+0.050	+0.051	+0.053	+0.054	+0.055
3.5	.043	.045	.046	.047	.048	.049	.050	.051	.053	.054
4.0	.043	.044	.045	.046	.047	.048	.049	.050	.052	.053
4.5	.042	.043	.044	.045	.046	.047	.048	.049	.051	.052
5.0	.041	.042	.043	.044	.045	.046	.047	.048	.049	.051
5.5	+0.040	+0.041	+0.042	+0.043	+0.044	+0.045	+0.046	+0.047	+0.048	+0.049
6.0	.039	.040	.041	.042	.043	.044	.045	.046	.047	.048
6.5	.038	.039	.040	.041	.042	.043	.044	.045	.046	.047
7.0	.037	.038	.039	.040	.041	.042	.043	.044	.045	.046
7.5	.037	.038	.038	.039	.040	.041	.042	.043	.044	.045
8.0	+0.036	+0.037	+0.038	+0.038	+0.039	+0.040	+0.041	+0.042	+0.043	+0.044
8.5	.035	.036	.037	.038	.038	.039	.040	.041	.042	.043
9.0	.034	.035	.036	.037	.038	.038	.039	.040	.041	.042
9.5	.033	.034	.035	.036	.037	.037	.038	.039	.040	.041
10.0	.032	.033	.034	.035	.036	.036	.037	.038	.039	.040
10.5	+0.031	+0.032	+0.033	+0.034	+0.035	+0.035	+0.036	+0.037	+0.038	+0.039
11.0	.030	.031	.032	.033	.034	.034	.035	.036	.037	.038
11.5	.030	.030	.031	.032	.033	.034	.034	.035	.036	.037
12.0	.029	.030	.030	.031	.032	.033	.033	.034	.035	.036
12.5	.028	.029	.029	.030	.031	.032	.032	.033	.034	.034
13.0	+0.027	+0.028	+0.028	+0.029	+0.030	+0.031	+0.031	+0.032	+0.033	+0.033
13.5	.026	.027	.028	.028	.029	.030	.030	.031	.032	.032
14.0	.025	.026	.027	.027	.028	.029	.029	.030	.031	.031
14.5	.024	.025	.026	.026	.027	.028	.028	.029	.030	.030
15.0	.024	.024	.025	.025	.026	.027	.027	.028	.029	.029
15.5	+0.023	+0.023	+0.024	+0.024	+0.025	+0.026	+0.026	+0.027	+0.027	+0.028
16.0	.022	.023	.023	.024	.024	.025	.025	.026	.026	.027
16.5	.021	.022	.022	.023	.023	.024	.024	.025	.025	.026
17.0	.020	.021	.021	.022	.022	.023	.023	.024	.024	.025
17.5	.019	.020	.020	.021	.021	.022	.022	.023	.023	.024
18.0	+0.018	+0.019	+0.019	+0.020	+0.020	+0.021	+0.021	+0.022	+0.022	+0.023
18.5	.017	.018	.018	.019	.019	.020	.020	.021	.021	.022
19.0	.017	.017	.018	.018	.018	.019	.019	.020	.020	.021
19.5	.016	.016	.017	.017	.017	.018	.018	.019	.019	.020
20.0	.015	.015	.016	.016	.016	.017	.017	.018	.018	.018
20.5	+0.014	+0.014	+0.015	+0.015	+0.016	+0.016	+0.016	+0.017	+0.017	+0.017
21.0	.013	.014	.014	.014	.015	.015	.015	.016	.016	.016
21.5	.012	.013	.013	.013	.014	.014	.014	.015	.015	.015
22.0	.011	.012	.012	.012	.013	.013	.013	.014	.014	.014
22.5	.011	.011	.011	.011	.012	.012	.012	.013	.013	.013
23.0	+0.010	+0.010	+0.010	+0.010	+0.011	+0.011	+0.011	+0.012	+0.012	+0.012
23.5	.009	.009	.009	.010	.010	.010	.010	.011	.011	.011
24.0	.008	.008	.008	.009	.009	.009	.009	.010	.010	.010
24.5	.007	.007	.008	.008	.008	.008	.008	.009	.009	.009
25.0	.006	.006	.007	.007	.007	.007	.007	.008	.008	.008

TABLE 10.

REDUCTION OF THE BAROMETER TO STANDARD TEMPERATURE.
ENGLISH MEASURES.

Attached Thermometer Fahrenheit.	HEIGHT OF THE BAROMETER IN INCHES.									
	19.0	19.5	20.0	20.5	21.0	21.5	22.0	22.5	23.0	23.5
F.	Inch.	Inch.	Inch.	Inch.	Inch.	Inch.	Inch.	Inch.	Inch.	Inch.
25.5	+0.005	+0.006	+0.006	+0.006	+0.006	+0.006	+0.006	+0.006	+0.007	+0.007
26.0	.005	.005	.005	.005	.005	.005	.005	.005	.005	.006
26.5	.004	.004	.004	.004	.004	.004	.004	.004	.004	.005
27.0	.003	.003	.003	.003	.003	.003	.003	.003	.003	.003
27.5	.002	.002	.002	.002	.002	.002	.002	.002	.002	.002
28.0	+0.001	+0.001	+0.001	+0.001	+0.001	+0.001	+0.001	+0.001	+0.001	+0.001
28.5	0.000	0.000	0.000	0.000	0.000	0.000	0.000	0.000	0.000	0.000
29.0	−0.001	−0.001	−0.001	−0.001	−0.001	−0.001	−0.001	−0.001	−0.001	−0.001
29.5	.002	.002	.002	.002	.002	.002	.002	.002	.002	.002
30.0	.002	.002	.002	.003	.003	.003	.003	.003	.003	.003
30.5	−0.003	−0.003	−0.003	−0.003	−0.004	−0.004	−0.004	−0.004	−0.004	−0.004
31.0	.004	.004	.004	.004	.005	.005	.005	.005	.005	.005
31.5	.005	.005	.005	.005	.005	.006	.006	.006	.006	.006
32.0	.006	.006	.006	.006	.006	.007	.007	.007	.007	.007
32.5	.007	.007	.007	.007	.007	.008	.008	.008	.008	.008
33.0	−0.008	−0.008	−0.008	−0.008	−0.008	−0.009	−0.009	−0.009	−0.009	−0.009
33.5	.008	.009	.009	.009	.009	.010	.010	.010	.010	.010
34.0	.009	.010	.010	.010	.010	.010	.011	.011	.011	.011
34.5	.010	.010	.011	.011	.011	.011	.012	.012	.012	.013
35.0	.011	.011	.012	.012	.012	.012	.013	.013	.013	.014
35.5	−0.012	−0.012	−0.012	−0.013	−0.013	−0.013	−0.014	−0.014	−0.014	−0.015
36.0	.013	.013	.013	.014	.014	.014	.015	.015	.015	.016
36.5	.014	.014	.014	.015	.015	.015	.016	.016	.016	.017
37.0	.014	.015	.015	.016	.016	.016	.017	.017	.017	.018
37.5	.015	.016	.016	.017	.017	.017	.018	.018	.019	.019
38.0	−0.016	−0.017	−0.017	−0.017	−0.018	−0.018	−0.019	−0.019	−0.020	−0.020
38.5	.017	.017	.018	.018	.019	.019	.020	.020	.021	.021
39.0	.018	.018	.019	.019	.020	.020	.021	.021	.022	.022
39.5	.019	.019	.020	.020	.021	.021	.022	.022	.023	.023
40.0	.020	.020	.021	.021	.022	.022	.023	.023	.024	.024
40.5	−0.020	−0.021	−0.022	−0.022	−0.023	−0.023	−0.024	−0.024	−0.025	−0.025
41.0	.021	.022	.022	.023	.024	.024	.025	.025	.026	.026
41.5	.022	.023	.023	.024	.025	.025	.026	.026	.027	.027
42.0	.023	.024	.024	.025	.025	.026	.027	.027	.028	.029
42.5	.024	.025	.025	.026	.026	.027	.028	.028	.029	.030
43.0	−0.025	−0.025	−0.026	−0.027	−0.027	−0.028	−0.029	−0.029	−0.030	−0.031
43.5	.026	.026	.027	.028	.028	.029	.030	.030	.031	.032
44.0	.026	.027	.028	.029	.029	.030	.031	.031	.032	.033
44.5	.027	.028	.029	.030	.030	.031	.032	.032	.033	.034
45.0	.028	.029	.030	.030	.031	.032	.033	.033	.034	.035
45.5	−0.029	−0.030	−0.031	−0.031	−0.032	−0.033	−0.034	−0.034	−0.035	−0.036
46.0	.030	.031	.031	.032	.033	.034	.035	.035	.036	.037
46.5	.031	.032	.032	.033	.034	.035	.036	.036	.037	.038
47.0	.032	.032	.033	.034	.035	.036	.037	.037	.038	.039
47.5	.033	.033	.034	.035	.036	.037	.038	.038	.039	.040
48.0	−0.033	−0.034	−0.035	−0.036	−0.037	−0.038	−0.039	−0.040	−0.040	−0.041
48.5	.034	.035	.036	.037	.038	.039	.040	.041	.041	.042
49.0	.035	.036	.037	.038	.039	.040	.041	.042	.042	.043
49.5	.036	.037	.038	.039	.040	.041	.042	.043	.044	.044
50.0	.037	.038	.039	.040	.041	.042	.043	.044	0.45	.046

SMITHSONIAN TABLES.

TABLE 10.

REDUCTION OF THE BAROMETER TO STANDARD TEMPERATURE.
ENGLISH MEASURES.

Attached Thermometer Fahrenheit.	HEIGHT OF THE BAROMETER IN INCHES.									
	19.0	19.5	20.0	20.5	21.0	21.5	22.0	22.5	23.0	23.5
F.	Inch.	Inch.	Inch.	Inch.	Inch.	Inch.	Inch.	Inch.	Inch.	Inch.
50.5	−0.038	−0.039	−0.040	−0.041	−0.042	−0.043	−0.044	−0.045	−0.046	−0.047
51.0	.039	.040	.041	.042	.043	.044	.045	.046	0.47	.048
51.5	.039	.040	.041	.042	.044	.045	.046	.047	.048	.049
52.0	.040	.041	.042	.043	.044	.046	.047	.048	.049	.050
52.5	.041	.042	.043	.044	.045	.047	.048	.049	.050	.051
53.0	−0.042	−0.043	−0.044	−0.045	−0.046	−0.047	−0.049	−0.050	−0.051	−0.052
53.5	.043	.044	.045	.046	.047	.048	.050	.051	.052	.053
54.0	.044	.045	.046	.047	.048	.049	.051	.052	.053	.054
54.5	.045	.046	.047	.048	.049	.050	.052	.053	.054	.055
55.0	.045	.047	.048	.049	.050	.051	.053	.054	.055	.056
55.5	−0.046	−0.047	−0.049	−0.050	−0.051	−0.052	−0.054	−0.055	−0.056	−0.057
56.0	.047	.048	.050	.051	.052	.053	.055	.056	.057	.058
56.5	.048	.049	.050	.052	.053	.054	.056	.057	.058	.059
57.0	.049	.050	.051	.053	.054	.055	.057	.058	.059	.060
57.5	.050	.051	.052	.054	.055	.056	.058	.059	.060	.061
58.0	−0.051	−0.052	−0.053	−0.055	−0.056	−0.057	−0.059	−0.060	−0.061	−0.063
58.5	.051	.053	.054	.055	.057	.058	.060	.061	.062	.064
59.0	.052	.054	.055	.056	.058	.059	.061	.062	.063	.065
59.5	.053	.055	.056	.057	.059	.060	.061	.063	.064	.066
60.0	.054	.055	.057	.058	.060	.061	.062	.064	.065	.067
60.5	−0.055	−0.056	−0.058	−0.059	−0.061	−0.062	−0.063	−0.065	−0.066	−0.068
61.0	.056	.057	.059	.060	.062	.063	.064	.066	.067	.069
61.5	.057	.058	.060	.061	.062	.064	.065	.067	.068	.070
62.0	.057	.059	.060	.062	.063	.065	.066	.068	.069	.071
62.5	.058	.060	.061	.063	.064	.066	.067	.069	.071	.072
63.0	−0.059	−0.061	−0.062	−0.064	−0.065	−0.067	−0.068	−0.070	−0.072	−0.073
63.5	.060	.062	.063	.065	.066	0.68	.069	.071	.073	.074
64.0	.061	.062	.064	.066	.067	.069	.070	.072	.074	.075
64.5	.062	.063	.065	.067	.068	.070	.071	.073	.075	.076
65.0	.063	.064	.066	.067	.069	.071	.072	.074	.076	.077
65.5	−0.063	−0.065	−0.067	−0.068	−0.070	−0.072	−0.073	−0.075	−0.077	−0.078
66.0	.064	.066	.068	.069	.071	.073	.074	.076	.078	.079
66.5	.065	.067	.069	.070	.072	.074	.075	.077	.079	.081
67.0	.066	.068	.069	.071	.073	.075	.076	.078	.080	.082
67.5	.067	.069	.070	.072	.074	.076	.077	.079	.081	.083
68.0	−0.068	−0.069	−0.071	−0.073	−0.075	−0.077	−0.078	−0.080	−0.082	−0.084
68.5	.069	.070	.072	.074	.076	.078	.079	.081	.083	.085
69.0	.069	.071	.073	.075	.077	.079	.080	.082	.084	.086
69.5	.070	.072	.074	.076	.078	.079	.081	.083	.085	.087
70.0	.071	.073	.075	.077	.079	.080	.082	.084	.086	.088
70.5	−0.072	−0.074	−0.076	−0.078	−0.080	−0.081	−0.083	−0.085	−0.087	−0.089
71.0	.073	.075	.077	.079	.080	.082	.084	.086	.088	.090
71.5	.074	.076	.078	.079	.081	.083	.085	.087	.089	.091
72.0	.075	.076	.078	.080	.082	.084	.086	.088	.090	.092
72.5	.075	.077	.079	.081	.083	.085	.087	.089	.091	.093
73.0	−0.076	−0.078	−0.080	−0.082	−0.084	−0.086	−0.088	−0.090	−0.092	−0.094
73.5	.077	.079	.081	.083	.085	.087	.089	.091	.093	.095
74.0	.078	.080	.082	.084	.086	.088	.090	.092	.094	.096
74.5	.079	.081	.083	.085	.087	.089	.091	.093	.095	.097
75.0	.080	.082	.084	.086	.088	.090	.092	.094	.096	.099

SMITHSONIAN TABLES.

TABLE 10

REDUCTION OF THE BAROMETER TO STANDARD TEMPERATURE.
ENGLISH MEASURES.

Attached Thermometer Fahrenheit.	HEIGHT OF THE BAROMETER IN INCHES.									
	19.0	19.5	20.0	20.5	21.0	21.5	22.0	22.5	23.0	23.5
F.	Inch.	Inch.	Inch.	Inch.	Inch.	Inch.	Inch.	Inch.	Inch.	Inch.
75.5	−0.081	−0.083	−0.085	−0.087	−0.089	−0.091	−0.093	−0.095	−0.097	−0.100
76.0	.081	.084	.086	.088	.090	.092	.094	.096	.098	.101
76.5	.082	.084	.087	.089	.091	.093	.095	.097	.100	.102
77.0	.083	.085	.087	.090	.092	.094	.096	.098	.101	.103
77.5	.084	.086	.088	.091	.093	.095	.097	.099	.102	.104
78.0	−0.085	−0.087	−0.089	−0.091	−0.094	−0.096	−0.098	−0.100	−0.103	−0.105
78.5	.086	.088	.090	.092	.095	.097	.099	.101	.104	.106
79.0	.086	.089	.091	.093	.096	.098	.100	.102	.105	.107
79.5	.087	.090	.092	.094	.097	.099	.101	.103	.106	.108
80.0	.088	.091	.093	.095	.097	.100	.102	.104	.107	.109
80.5	−0.089	−0.091	−0.094	−0.096	−0.098	−0.101	−0.103	−0.105	−0.108	−0.110
81.0	.090	.092	.095	.097	.099	.102	.104	.106	.109	.111
81.5	.091	.093	.096	.098	.100	.103	.105	.107	.110	.112
82.0	.092	.094	.096	.099	.101	.104	.106	.108	.111	.113
82.5	.092	.095	.097	.100	.102	.105	.107	.109	.112	.114
83.0	−0.093	−0.096	−0.098	−0.101	−0.103	−0.106	−0.108	−0.111	−0.113	−0.115
83.5	.094	.097	.099	.102	.104	.107	.109	.112	.114	.117
84.0	.095	.098	.100	.103	.105	.108	.110	.113	.115	.118
84.5	.096	.098	.101	.103	.106	.108	.111	.114	.116	.119
85.0	.097	.099	.102	.104	.107	.109	.112	.115	.117	.120
85.5	−0.098	−0.100	−0.103	−0.105	−0.108	−0.110	−0.113	−0.116	−0.118	−0.121
86.0	.098	.101	.104	.106	.109	.111	.114	.117	.119	.122
86.5	.099	.102	.105	.107	.110	.112	.115	.118	.120	.123
87.0	.100	.103	.105	.108	.111	.113	.116	.119	.121	.124
87.5	.101	.104	.106	.109	.112	.114	.117	.120	.122	.125
88.0	−0.102	−0.105	−0.107	−0.110	−0.113	−0.115	−0.118	−0.121	−0.123	−0.126
88.5	.103	.105	.108	.111	.114	.116	.119	.122	.124	.127
89.0	.104	.106	.109	.112	.114	.117	.120	.123	.125	.128
89.5	.104	.107	.110	.113	.115	.118	.121	.124	.126	.129
90.0	.105	.108	.111	.114	.116	.119	.122	.125	.127	.130
90.5	−0.106	−0.109	−0.112	−0.114	−0.117	−0.120	−0.123	−0.126	−0.128	−0.131
91.0	.107	.110	.113	.115	.118	.121	.124	.127	.129	.132
91.5	.108	.111	.113	.116	.119	.122	.125	.128	.131	.133
92.0	.109	.112	.114	.117	.120	.123	.126	.129	.132	.134
92.5	.110	.112	.115	.118	.121	.124	.127	.130	.133	.135
93.0	−0.110	−0.113	−0.116	−0.119	−0.122	−0.125	−0.128	−0.131	−0.134	−0.137
93.5	.111	.114	.117	.120	.123	.126	.129	.132	.135	.138
94.0	.112	.115	.118	.121	.124	.127	.130	.133	.136	.139
94.5	.113	.116	.119	.122	.125	.128	.131	.134	.137	.140
95.0	.114	.117	.120	.123	.126	.129	.132	.135	.138	.141
95.5	−0.115	−0.118	−0.121	−0.124	−0.127	−0.130	−0.133	−0.136	−0.139	−0.142
96.0	.115	.119	.122	.125	.128	.131	.134	.137	.140	.143
96.5	.116	.119	.122	.126	.129	.132	.135	.138	.141	.144
97.0	.117	.120	.123	.126	.130	.133	.136	.139	.142	.145
97.5	.118	.121	.124	.127	.130	.134	.137	.140	.143	.146
98.0	−0.119	−0.122	−0.125	−0.128	−0.131	−0.135	−0.138	−0.141	−0.144	−0.147
98.5	.120	.123	.126	.129	.132	.135	.139	.142	.145	.148
99.0	.121	.124	.127	.130	.133	.136	.140	.143	.146	.149
99.5	.121	.125	.128	.131	.134	.137	.141	.144	.147	.150
100.0	.122	.126	.129	.132	.135	.138	.142	.145	.148	.151

SMITHSONIAN TABLES.

TABLE 10.
REDUCTION OF THE BAROMETER TO STANDARD TEMPERATURE.
ENGLISH MEASURES.

Attached Thermometer Fahrenheit.	HEIGHT OF THE BAROMETER IN INCHES.								
	24.0	24.2	24.4	24.6	24.8	25.0	25.2	25.4	25.6	25.8
F.	Inch.	Inch.	Inch.	Inch.	Inch.	Inch.	Inch.	Inch.	Inch.	Inch.
0.°0	+0.063	+0.063	+0.064	+0.064	+0.065	+0.065	+0.066	+0.066	+0.067	+0.067
+0.5	+0.061	+0.062	+0.063	+0.063	+0.064	+0.064	+0.065	+0.065	+0.066	+0.066
1.0	.060	.061	.061	.062	.062	.063	.063	.064	.064	.065
1.5	.059	.060	.060	.061	.061	.062	.062	.063	.063	.064
2.0	.058	.059	.059	.060	.060	.061	.061	.062	.062	.063
2.5	.057	.058	.058	.059	.059	.059	.060	.060	.061	.061
3.0	+0.056	+0.056	+0.057	+0.057	+0.058	+0.058	+0.059	+0.059	+0.060	+0.060
3.5	.055	.055	.056	.056	.057	.057	.058	.058	.059	.059
4.0	.054	.054	.055	.055	.056	.056	.057	.057	.057	.058
4.5	.053	.053	.054	.054	.054	.055	.055	.056	.056	.057
5.0	.052	.052	.052	.053	.053	.054	.054	.055	.055	.056
5.5	+0.051	+0.051	+0.051	+0.052	+0.052	+0.053	+0.053	+0.053	+0.054	+0.054
6.0	.049	.050	.050	.051	.051	.052	.052	.052	.053	.053
6.5	.048	.049	.049	.050	.050	.050	.051	.051	.052	.052
7.0	.047	.048	.048	.048	.049	.049	.050	.050	.050	.051
7.5	.046	.047	.047	.047	.048	.048	.048	.049	.049	.050
8.0	+0.045	+0.045	+0.046	+0.046	+0.047	+0.047	+0.047	+0.048	+0.048	+0.048
8.5	.044	.044	.045	.045	.045	.046	.046	.047	.047	.047
9.0	.043	.043	.044	.044	.044	.045	.045	.045	.046	.046
9.5	.042	.042	.042	.043	.043	.044	.044	.044	.045	.045
10.0	.041	.041	.041	.042	.042	.042	.043	.043	.043	.044
10.5	+0.040	+0.040	+0.040	+0.041	+0.041	+0.041	+0.042	+0.042	+0.042	+0.043
11.0	.039	.039	.039	.039	.040	.040	.040	.041	.041	.041
11.5	.037	.038	.038	.038	.039	.039	.039	.040	.040	.040
12.0	.036	.037	.037	.037	.038	.038	.038	.038	.039	.039
12.5	.035	.036	.036	.036	.036	.037	.037	.037	.038	.038
13.0	+0.034	+0.034	+0.035	+0.035	+0.035	+0.036	+0.036	+0.036	+0.036	+0.037
13.5	.033	.033	.034	.034	.034	.034	.035	.035	.035	.036
14.0	.032	.032	.032	.033	.033	.033	.034	.034	.034	.034
14.5	.031	.031	.031	.032	.032	.032	.032	.032	.033	.033
15.0	.030	.030	.030	.030	.031	.031	.031	.031	.032	.032
15.5	+0.029	+0.029	+0.029	+0.029	+0.030	+0.030	+0.030	+0.030	+0.031	+0.031
16.0	.028	.028	.028	.028	.028	.029	.029	.029	.029	.030
16.5	.026	.027	.027	.027	.027	.028	.028	.028	.028	.028
17.0	.025	.026	.026	.026	.026	.026	.027	.027	.027	.027
17.5	.024	.024	.025	.025	.025	.025	.026	.026	.026	.026
18.0	+0.023	+0.023	+0.024	+0.024	+0.024	+0.024	+0.024	+0.025	+0.025	+0.025
18.5	.022	.022	.022	.023	.023	.023	.023	.023	.024	.024
19.0	.021	.021	.021	.022	.022	.022	.022	.022	.022	.023
19.5	.020	.020	.020	.020	.021	.021	.021	.021	.021	.021
20.0	.019	.019	.019	.019	.019	.020	.020	.020	.020	.020
20.5	+0.018	+0.018	+0.018	+0.018	+0.018	+0.018	+0.019	+0.019	+0.019	+0.019
21.0	.017	.017	.017	.017	.017	.017	.017	.018	.018	.018
21.5	.016	.016	.016	.016	.016	.016	.016	.016	.017	.017
22.0	.014	.015	.015	.015	.015	.015	.015	.015	.015	.016
22.5	.013	.013	.014	.014	.014	.014	.014	.014	.014	.014
23.0	+0.012	+0.012	+0.012	+0.013	+0.013	+0.013	+0.013	+0.013	+0.013	+0.013
23.5	.011	.011	.011	.011	.012	.012	.012	.012	.012	.012
24.0	.010	.010	.010	.010	.010	.011	.011	.011	.011	.011
24.5	.009	.009	.009	.009	.009	.009	.009	.010	.010	.010
25.0	.008	.008	.008	.008	.008	.008	.008	.008	.008	.009

SMITHSONIAN TABLES.

TABLE 10.
REDUCTION OF THE BAROMETER TO STANDARD TEMPERATURE.
ENGLISH MEASURES.

Attached Thermometer Fahrenheit.	HEIGHT OF THE BAROMETER IN INCHES.								
	24.0	24.2	24.4	24.6	24.8	25.0	25.2	25.4	25.6	25.8
F.	Inch.	Inch.	Inch.	Inch.	Inch.	Inch.	Inch.	Inch.	Inch.	Inch.
25.5	+0.007	+0.007	+0.007	+0.007	+0.007	+0.007	+0.007	+0.007	+0.007	+0.007
26.0	.006	.006	.006	.006	.006	.006	.006	.006	.006	.006
26.5	.005	.005	.005	.005	.005	.005	.005	.005	.005	.005
27.0	.004	.004	.004	.004	.004	.004	.004	.004	.004	.004
27.5	.002	.002	.003	.003	.003	.003	.003	.003	.003	.003
28.0	+0.001	+0.001	+0.001	+0.001	+0.001	+0.001	+0.001	+0.001	+0.001	+0.001
28.5	0.000	0.000	0.000	0.000	0.000	0.000	0.000	0.000	0.000	0.000
29.0	−0.001	−0.001	−0.001	−0.001	−0.001	−0.001	−0.001	−0.001	−0.001	−0.001
29.5	.002	.002	.002	.002	.002	.002	.002	.002	.002	.002
30.0	.003	.003	.003	.003	.003	.003	.003	.003	.003	.003
30.5	−0.004	−0.004	−0.004	−0.004	−0.004	−0.004	−0.004	−0.004	−0.004	−0.004
31.0	.005	.005	.005	.005	.005	.005	.005	.005	.006	.006
31.5	.006	.006	.006	.006	.006	.007	.007	.007	.007	.007
32.0	.007	.007	.007	.008	.008	.008	.008	.008	.008	.008
32.5	.008	.009	.009	.009	.009	.009	.009	.009	.009	.009
33.0	−0.010	−0.010	−0.010	−0.010	−0.010	−0.010	−0.010	−0.010	−0.010	−0.010
33.5	.011	.011	.011	.011	.011	.011	.011	.011	.011	.011
34.0	.012	.012	.012	.012	.012	.012	.012	.012	.012	.013
34.5	.013	.013	.013	.013	.013	.013	.013	.014	.014	.014
35.0	.014	.014	.014	.014	.014	.014	.015	.015	.015	.015
35.5	−0.015	−0.015	−0.015	−0.015	−0.015	−0.016	−0.016	−0.016	−0.016	−0.016
36.0	.016	.016	.016	.016	.017	.017	.017	.017	.017	.017
36.5	.017	.017	.017	.018	.018	.018	.018	.018	.018	.018
37.0	.018	.018	.019	.019	.019	.019	.019	.019	.019	.019
37.5	.019	.019	.020	.020	.020	.020	.020	.020	.021	.021
38.0	−0.020	−0.021	−0.021	−0.021	−0.021	−0.021	−0.021	−0.022	−0.022	−0.022
38.5	.021	.022	.022	.022	.022	.022	.023	.023	.023	.023
39.0	.023	.023	.023	.023	.023	.024	.024	.024	.024	.024
39.5	.024	.024	.024	.024	.024	.025	.025	.025	.025	.025
40.0	.025	.025	.025	.025	.026	.026	.026	.026	.026	.027
40.5	−0.026	−0.026	−0.026	−0.026	−0.027	−0.027	−0.027	−0.027	−0.028	−0.028
41.0	.027	.027	.027	.028	.028	.028	.028	.029	.029	.029
41.5	.028	.028	.028	.029	.029	.029	.029	.030	.030	.030
42.0	.029	.029	.030	.030	.030	.030	.031	.031	.031	.031
42.5	.030	.030	.031	.031	.031	.031	.032	.032	.032	.032
43.0	−0.031	−0.032	−0.032	−0.032	−0.032	−0.033	−0.033	−0.033	−0.033	−0.034
43.5	.032	.033	.033	.033	.033	.034	.034	.034	.035	.035
44.0	.033	.034	.034	.034	.035	.035	.035	.035	.036	.036
44.5	.035	.035	.035	.035	.036	.036	.036	.037	.037	.037
45.0	.036	.036	.036	.037	.037	.037	.037	.038	.038	.038
45.5	−0.037	−0.037	−0.037	−0.038	−0.038	−0.038	−0.039	−0.039	−0.039	−0.039
46.0	.038	.038	.038	.039	.039	.039	.040	.040	.040	.041
46.5	.039	.039	.040	.040	.040	.041	.041	.041	.041	.042
47.0	.040	.040	.041	.041	.041	.042	.042	.042	.043	.043
47.5	.041	.041	.042	.042	.042	.043	.043	.043	.044	.044
48.0	−0.042	−0.042	−0.043	−0.043	−0.044	−0.044	−0.044	−0.045	−0.045	−0.045
48.5	.043	.044	.044	.044	.045	.045	.045	.046	.046	.046
49.0	.044	.045	.045	.045	.046	.046	.047	.047	.047	.048
49.5	.045	.046	.046	.047	.047	.047	.048	.048	.048	.049
50.0	.046	.047	.047	.048	.048	.048	.049	.049	.050	.050

SMITHSONIAN TABLES.

TABLE 10.

REDUCTION OF THE BAROMETER TO STANDARD TEMPERATURE.
ENGLISH MEASURES.

Attached Thermometer Fahrenheit.	HEIGHT OF THE BAROMETER IN INCHES.								
	24.0	24.2	24.4	24.6	24.8	25.0	25.2	25.4	25.6	25.8
F.	Inch.	Inch.	Inch.	Inch.	Inch.	Inch.	Inch.	Inch.	Inch.	Inch.
50.°5	−0.048	−0.048	−0.048	−0.049	−0.049	−0.050	−0.050	−0.050	−0.051	−0.051
51.0	.049	.049	.049	.050	.050	.051	.051	.051	.052	.052
51.5	.050	.050	.051	.051	.051	.052	.052	.053	.053	.053
52.0	.051	.051	.052	.052	.053	.053	.053	.054	.054	.055
52.5	.052	.052	.053	.053	.054	.054	.055	.055	.055	.056
53.0	−0.053	−0.053	−0.054	−0.054	−0.055	−0.055	−0.056	−0.056	−0.057	−0.057
53.5	.054	.055	.055	.055	.056	.056	.057	.057	.058	.058
54.0	.055	.056	.056	.057	.057	.057	.058	.058	.059	.059
54.5	.056	.057	.057	.058	.058	.059	.059	.060	.060	.060
55.0	.057	.058	.058	.059	.059	.060	.060	.061	.061	.062
55.5	−0.058	−0.059	−0.059	−0.060	−0.060	−0.061	−0.061	−0.062	−0.062	−0.063
56.0	.060	.060	.060	.061	.061	.062	.062	.063	.063	.064
56.5	.061	.061	.062	.062	.063	.063	.064	.064	.065	.065
57.0	.062	.062	.063	.063	.064	.064	.065	.065	.066	.066
57.5	.063	.063	.064	.064	.065	.065	.066	.066	.067	.067
58.0	−0.064	−0.064	−0.065	−0.065	−0.066	−0.066	−0.067	−0.068	−0.068	−0.069
58.5	.065	.065	.066	.067	.067	.068	.068	.069	.069	.070
59.0	.066	.067	.067	.068	.068	.069	.069	.070	.070	.071
59.5	.067	.068	.068	.069	.069	.070	.070	.071	.072	.072
60.0	.068	.069	.069	.070	.070	.071	.072	.072	.073	.073
60.5	−0.069	−0.070	−0.070	−0.071	−0.072	−0.072	−0.073	−0.073	−0.074	−0.074
61.0	.070	.071	.072	.072	.073	.073	.074	.074	.075	.076
61.5	.071	.072	.073	.073	.074	.074	.075	.076	.076	.077
62.0	.073	.073	.074	.074	.075	.076	.076	.077	.077	.078
62.5	.074	.074	.075	.075	.076	.077	.077	.078	.078	.079
63.0	−0.075	−0.075	−0.076	−0.077	−0.077	−0.078	−0.078	−0.079	−0.080	−0.080
63.5	.076	.076	.077	.078	.078	.079	.080	.080	.081	.081
64.0	.077	.077	.078	.079	.079	.080	.081	.081	.082	.082
64.5	.078	.079	.079	.080	.081	.081	.082	.082	.083	.084
65.0	.079	.080	.080	.081	.082	.082	.083	.084	.084	.085
65.5	−0.080	−0.081	−0.081	−0.082	−0.083	−0.083	−0.084	−0.085	−0.085	−0.086
66.0	.081	.082	.083	.083	.084	.085	.085	.086	.087	.087
66.5	.082	.083	.084	.084	.085	.086	.086	.087	.088	.088
67.0	.083	.084	.085	.085	.086	.087	.087	.088	.089	.090
67.5	.084	.085	.086	.087	.087	.088	.089	.089	.090	.091
68.0	−0.085	−0.086	−0.087	−0.088	−0.088	−0.089	−0.090	−0.090	−0.091	−0.092
68.5	.087	.087	.088	.089	.089	.090	.091	.092	.092	.093
69.0	.088	.088	.089	.090	.091	.091	.092	.093	.093	.094
69.5	.089	.089	.090	.091	.092	.092	.093	.094	.095	.095
70.0	.090	.091	.091	.092	.093	.094	.094	.095	.096	.097
70.5	−0.091	−0.092	−0.092	−0.093	−0.094	−0.095	−0.095	−0.096	−0.097	−0.098
71.0	.092	.093	.094	.094	.095	.096	.097	.097	.098	.099
71.5	.093	.094	.095	.095	.096	.097	.098	.098	.099	.100
72.0	.094	.095	.096	.096	.097	.098	.099	.100	.100	.101
72.5	.095	.096	.097	.098	.098	.099	.100	.101	.102	.102
73.0	−0.096	−0.097	−0.098	−0.099	−0.100	−0.100	−0.101	−0.102	−0.103	−0.104
73.5	.097	.098	.099	.100	.101	.101	.102	.103	.104	.105
74.0	.098	.099	.100	.101	.102	.103	.103	.104	.105	.106
74.5	.100	.100	.101	.102	.103	.104	.105	.105	.106	.107
75.0	.101	.101	.102	.103	.104	.105	.106	.106	.107	.108

SMITHSONIAN TABLES.

TABLE 10.

REDUCTION OF THE BAROMETER TO STANDARD TEMPERATURE.
ENGLISH MEASURES.

Attached Thermometer Fahrenheit.	HEIGHT OF THE BAROMETER IN INCHES.								
	24.0	24.2	24.4	24.6	24.8	25.0	25.2	25.4	25.6	25.8
F.	Inch.	Inch.	Inch.	Inch.	Inch.	Inch.	Inch.	Inch.	Inch.	Inch.
75.5	−0.102	−0.103	−0.103	−0.104	−0.105	−0.106	−0.107	−0.108	−0.108	−0.109
76.0	.103	.104	.104	.105	.106	.107	.108	.109	.110	.110
76.5	.104	.105	.106	.106	.107	.108	.109	.110	.111	.112
77.0	.105	.106	.107	.108	.108	.109	.110	.111	.112	.113
77.5	.106	.107	.108	.109	.110	.110	.111	.112	.113	.114
78.0	−0.107	−0.108	−0.109	−0.110	−0.111	−0.112	−0.112	−0.113	−0.114	−0.115
78.5	.108	.109	.110	.111	.112	.113	.114	.114	.115	.116
79.0	.109	.110	.111	.112	.113	.114	.115	.116	.117	.117
79.5	.110	.111	.112	.113	.114	.115	.116	.117	.118	.119
80.0	.111	.112	.113	.114	.115	.116	.117	.118	.119	.120
80.5	−0.112	−0.113	−0.114	−0.115	−0.116	−0.117	−0.118	−0.119	−0.120	−0.121
81.0	.114	.115	.115	.116	.117	.118	.119	.120	.121	.122
81.5	.115	.116	.117	.118	.118	.119	.120	.121	.122	.123
82.0	.116	.117	.118	.119	.120	.121	.122	.122	.123	.124
82.5	.117	.118	.119	.120	.121	.122	.123	.124	.125	.126
83.0	−0.118	−0.119	−0.120	−0.121	−0.122	−0.123	−0.124	−0.125	−0.126	−0.127
83.5	.119	.120	.121	.122	.123	.124	.125	.126	.127	.128
84.0	.120	.121	.122	.123	.124	.125	.126	.127	.128	.129
84.5	.121	.122	.123	.124	.125	.126	.127	.128	.129	.130
85.0	.122	.123	.124	.125	.126	.127	.128	.129	.130	.131
85.5	−0.123	−0.124	−0.125	−0.126	−0.127	−0.128	−0.129	−0.130	−0.131	−0.133
86.0	.124	.125	.126	.127	.128	.130	.131	.132	.133	.134
86.5	.125	.126	.128	.129	.130	.131	.132	.133	.134	.135
87.0	.126	.128	.129	.130	.131	.132	.133	.134	.135	.136
87.5	.128	.129	.130	.131	.132	.133	.134	.135	.136	.137
88.0	−0.129	−0.130	−0.131	−0.132	−0.133	−0.134	−0.135	−0.136	−0.137	−0.138
88.5	.130	.131	.132	.133	.134	.135	.136	.137	.138	.139
89.0	.131	.132	.133	.134	.135	.136	.137	.138	.140	.141
89.5	.132	.133	.134	.135	.136	.137	.138	.140	.141	.142
90.0	.133	.134	.135	.136	.137	.138	.140	.141	.142	.143
90.5	−0.134	−0.135	−0.136	−0.137	−01.39	−0.140	−0.141	−0.142	−0.143	−0.144
91.0	.135	.136	.137	.138	.140	.141	.142	.143	.144	.145
91.5	.136	.137	.138	.140	.141	.142	.143	.144	.145	.146
92.0	.137	.138	.140	.141	.142	.143	.144	.145	.146	.148
92.5	.138	.139	.141	.142	.143	.144	.145	.146	.148	.149
93.0	−0.139	−0.141	−0.142	−0.143	−0.144	−0.145	−0.146	−0.148	−0.149	−0.150
93.5	.140	.142	.143	.144	.145	.146	.148	.149	.150	.151
94.0	.142	.143	.144	.145	.146	.147	.149	.150	.151	.152
94.5	.143	.144	.145	.146	.147	.149	.150	.151	.152	.153
95.0	.144	.145	.146	.147	.149	.150	.151	.152	.153	.154
95.5	−0.145	−0.146	−0.147	−0.148	−0.150	−0.151	−0.152	−0.153	−0.154	−0.156
96.0	.146	.147	.148	.150	.151	.152	.153	.154	.156	.157
96.5	.147	.148	.149	.151	.152	.153	.154	.156	.157	.158
97.0	.148	.149	.150	.152	.153	.154	.155	.157	.158	.159
97.5	.149	.150	.152	.153	.154	.155	.157	.158	.159	.160
98.0	−0.150	−0.151	−0.153	−0.154	−0.155	−0.156	−0.158	−0.159	−0.160	−0.161
98.5	.151	.153	.154	.155	.156	.158	.159	.160	.161	.163
99.0	.152	.154	.155	.156	.157	.159	.160	.161	.162	.164
99.5	.153	.155	.156	.157	.159	.160	.161	.162	.164	.165
100.0	.154	.156	.157	.158	.160	.161	.162	.163	.165	.166

SMITHSONIAN TABLES.

TABLE 10.
REDUCTION OF THE BAROMETER TO STANDARD TEMPERATURE.
ENGLISH MEASURES.

Attached Thermometer Fahrenheit.	HEIGHT OF THE BAROMETER IN INCHES.									
	26.0	26.2	26.4	26.6	26.8	27.0	27.2	27.4	27.6	27.8
F.	Inch.	Inch.	Inch.	Inch.	Inch.	Inch.	Inch.	Inch.	Inch.	Inch.
0.°0	+0.068	+0.068	+0.069	+0.069	+0.070	+0.070	+0.071	+0.071	+0.072	+0.072
+0.5	+0.067	+0.067	+0.068	+0.068	+0.069	+0.069	+0.070	+0.070	+0.071	+0.071
1.0	.065	.066	.066	.067	.067	.068	.068	.069	.069	.070
1.5	.064	.065	.065	.066	.066	.067	.067	.068	.068	.069
2.0	.063	.064	.064	.065	.065	.065	.066	.066	.067	.067
2.5	.062	.062	.063	.063	.064	.064	.065	.065	.066	.066
3.0	+0.061	+0.061	+0.062	+0.062	+0.063	+0.063	+0.063	+0.064	+0.064	+0.065
3.5	.059	.060	.060	.061	.061	.062	.062	.063	.063	.064
4.0	.058	.059	.059	.060	.060	.061	.061	.061	.062	.062
4.5	.057	.058	.058	.058	.059	.059	.060	.060	.061	.061
5.0	.056	.056	.057	.057	.058	.058	.059	.059	.059	.060
5.5	+0.055	+0.055	+0.056	+0.056	+0.056	+0.057	+0.057	+0.058	+0.058	+0.059
6.0	.054	.054	.054	.055	.055	.056	.056	.056	.057	.057
6.5	.052	.053	.053	.054	.054	.054	.055	.055	.056	.056
7.0	.051	.052	.052	.052	.053	.053	.054	.054	.054	.055
7.5	.050	.050	.051	.051	.052	.052	.052	.053	.053	.053
8.0	+0.049	+0.049	+0.050	+0.050	+0.050	+0.051	+0.051	+0.051	+0.052	+0.052
8.5	.048	.048	.048	.049	.049	.049	.050	.050	.051	.051
9.0	.046	.047	.047	.048	.048	.048	.049	.049	.049	.050
9.5	.045	.046	.046	.046	.047	.047	.047	.048	.048	.048
10.0	.044	.044	.045	.045	.045	.046	.046	.046	.047	.047
10.5	+0.043	+0.043	+0.044	+0.044	+0.044	+0.045	+0.045	+0.045	+0.046	+0.046
11.0	.042	.042	.042	.043	.043	.043	.044	.044	.044	.045
11.5	.041	.041	.041	.041	.042	.042	.042	.043	.043	.043
12.0	.039	.040	.040	.040	.041	.041	.041	.041	.042	.042
12.5	.038	.038	.039	.039	.039	.040	.040	.040	.040	.041
13.0	+0.037	+0.037	+0.038	+0.038	+0.038	+0.038	+0.039	+0.039	+0.039	+0.040
13.5	.036	.036	.036	.037	.037	.037	.037	.038	.038	.038
14.0	.035	.035	.035	.035	.036	.036	.036	.036	.037	.037
14.5	.033	.034	.034	.034	.034	.035	.035	.035	.035	.036
15.0	.032	.032	.033	.033	.033	.033	.034	.034	.034	.034
15.5	+0.031	+0.031	+0.032	+0.032	+0.032	+0.032	+0.032	+0.033	+0.033	+0.033
16.0	.030	.030	.030	.031	.031	.031	.031	.031	.032	.032
16.5	.029	.029	.029	.029	.030	.030	.030	.030	.030	.031
17.0	.027	.028	.028	.028	.028	.029	.029	.029	.029	.029
17.5	.026	.027	.027	.027	.027	.027	.028	.028	.028	.028
18.0	+0.025	+0.025	+0.026	+0.026	+0.026	+0.026	+0.026	+0.026	+0.027	+0.027
18.5	.024	.024	.024	.024	.025	.025	.025	.025	.025	.026
19.0	.023	.023	.023	.023	.023	.024	.024	.024	.024	.024
19.5	.022	.022	.022	.022	.022	.022	.023	.023	.023	.023
20.0	.020	.021	.021	.021	.021	.021	.021	.021	.022	.022
20.5	+0.019	+0.019	+0.020	+0.020	+0.020	+0.020	+0.020	+0.020	+0.020	+0.021
21.0	.018	.018	.018	.018	.019	.019	.019	.019	.019	.019
21.5	.017	.017	.017	.017	.017	.017	.018	.018	.018	.018
22.0	.016	.016	.016	.016	.016	.016	.016	.017	.017	.017
22.5	.014	.015	.015	.015	.015	.015	.015	.015	.015	.015
23.0	+0.013	+0.013	+0.014	+0.014	+0.014	+0.014	+0.014	+0.014	+0.014	+0.014
23.5	.012	.012	.012	.012	.012	.013	.013	.013	.013	.013
24.0	.011	.011	.011	.011	.011	.011	.011	.012	.012	.012
24.5	.010	.010	.010	.010	.010	.010	.010	.010	.010	.110
25.0	.009	.009	.009	.009	.009	.009	.009	.009	.009	.009

SMITHSONIAN TABLES.

TABLE 10.

REDUCTION OF THE BAROMETER TO STANDARD TEMPERATURE.
ENGLISH MEASURES.

Attached Thermometer Fahrenheit.	HEIGHT OF THE BAROMETER IN INCHES.									
	26.0	26.2	26.4	26.6	26.8	27.0	27.2	27.4	27.6	27.8
F.	Inch.	Inch.	Inch.	Inch.	Inch.	Inch.	Inch.	Inch.	Inch.	Inch.
25°.5	+0.007	+0.007	+0.008	+0.008	+0.008	+0.008	+0.008	+0.008	+0.008	+0.008
26.0	.006	.006	.006	.006	.006	.006	.006	.007	.007	.007
26.5	.005	.005	.005	.005	.005	.005	.005	.005	.005	.005
27.0	.004	.004	.004	.004	.004	.004	.004	.004	.004	.004
27.5	.003	.003	.003	.003	.003	.003	.003	.003	.003	.003
28.0	+0.001	+0.001	+0.002	+0.002	+0.002	+0.002	+0.002	+0.002	+0.002	+0.002
28.5	0.000	0.000	0.000	0.000	0.000	0.000	0.000	0.000	0.000	0.000
29.0	−0.001	−0.001	−0.001	−0.001	−0.001	−0.001	−0.001	−0.001	−0.001	−0.001
29.5	.002	.002	.002	.002	.002	.002	.002	.002	.002	.002
30.0	.003	.003	.003	.003	.003	.003	.003	.003	.003	.003
30.5	−0.004	−0.004	−0.004	−0.005	−0.005	−0.005	−0.005	−0.005	−0.005	−0.005
31.0	.006	.006	.006	.006	.006	.006	.006	.006	.006	.006
31.5	.007	.007	.007	.007	.007	.007	.007	.007	.007	.007
32.0	.008	.008	.008	.008	.008	.008	.008	.008	.008	.009
32.5	.009	.009	.009	.009	.009	.009	.010	.010	.010	.010
33.0	−0.010	−0.010	−0.010	−0.011	−0.011	−0.011	−0.011	−0.011	−0.011	−0.011
33.5	.011	.012	.012	.012	.012	.012	.012	.012	.012	.012
34.0	.013	.013	.013	.013	.013	.013	.013	.013	.013	.014
34.5	.014	.014	.014	.014	.014	.014	.014	.015	.015	.015
35.0	.015	.015	.015	.015	.015	.016	.016	.016	.016	.016
35.5	−0.016	−0.016	−0.016	−0.017	−0.017	−0.017	−0.017	−0.017	−0.017	−0.017
36.0	.017	.018	.018	.018	.018	.018	.018	.018	.018	.019
36.5	.019	.019	.019	.019	.019	.019	.019	.020	.020	.020
37.0	.020	.020	.020	.020	.020	.021	.021	.021	.021	.021
37.5	.021	.021	.021	.021	.022	.022	.022	.022	.022	.022
38.0	−0.022	−0.022	−0.022	−0.023	−0.023	−0.023	−0.023	−0.023	−0.023	−0.024
38.5	.023	.023	.024	.024	.024	.024	.024	.025	.025	.025
39.0	.024	.025	.025	.025	.025	.025	.026	.026	.026	.026
39.5	.026	.026	.026	.026	.026	.027	.027	.027	.027	.027
40.0	.027	.027	.027	.027	.028	.028	.028	.028	.028	.029
40.5	−0.028	−0.028	−0.028	−0.029	−0.029	−0.029	−0.029	−0.030	−0.030	−0.030
41.0	.029	.029	.030	.030	.030	.030	.031	.031	.031	.031
41.5	.030	.031	.031	.031	.031	.032	.032	.032	.032	.032
42.0	.032	.032	.032	.032	.033	.033	.033	.033	.033	.034
42.5	.033	.033	.033	.033	.034	.034	.034	.034	.035	.035
43.0	−0.034	−0.034	−0.034	−0.035	−0.035	−0.035	−0.035	−0.036	−0.036	−0.036
43.5	.035	.035	.036	.036	.036	.036	.037	.037	.037	.037
44.0	.036	.037	.037	.037	.037	.038	.038	.038	.038	.039
44.5	.037	.038	.038	.038	.039	.039	.039	.039	.040	.040
45.0	.039	.039	.039	.039	.040	.040	.040	.041	.041	.041
45.5	−0.040	−0.040	−0.040	−0.041	−0.041	−0.041	−0.042	−0.042	−0.042	−0.043
46.0	.041	.041	.042	.042	.042	.043	.043	.043	.043	.044
46.5	.042	.042	.043	.043	.043	.044	.044	.044	.045	.045
47.0	.043	.044	.044	.044	.045	.045	.045	.046	.046	.046
47.5	.045	.045	.045	.046	.046	.046	.047	.047	.047	.048
48.0	−0.046	−0.046	−0.046	−0.047	−0.047	−0.047	−0.048	−0.048	−0.048	−0.049
48.5	.047	.047	.048	.048	.048	.049	.049	.049	.050	.050
49.0	.048	.048	.049	.049	.049	.050	.050	.051	.051	.051
49.5	.049	.050	.050	.050	.051	.051	.051	.052	.052	.053
50.0	.050	.051	.051	.052	.052	.052	.053	.053	.053	.054

SMITHSONIAN TABLES.

TABLE 10.
REDUCTION OF THE BAROMETER TO STANDARD TEMPERATURE.
ENGLISH MEASURES.

Attached Thermometer Fahrenheit.	HEIGHT OF THE BAROMETER IN INCHES.								
	26.0	26.2	26.4	26.6	26.8	27.0	27.2	27.4	27.6	27.8
F.	Inch.	Inch.	Inch.	Inch.	Inch.	Inch.	Inch.	Inch.	Inch.	Inch.
50.5	−0.052	−0.052	−0.052	−0.053	−0.053	−0.054	−0.054	−0.054	−0.055	−0.055
51.0	.053	.053	.054	.054	.054	.055	.055	.056	.056	.056
51.5	.054	.054	.055	.055	.056	.056	.056	.057	.057	.058
52.0	.055	.055	.056	.056	.057	.057	.058	.058	.058	.059
52.5	.056	.057	.057	.058	.058	.058	.059	.059	.060	.060
53.0	−0.057	−0.058	−0.058	−0.059	−0.059	−0.060	−0.060	−0.061	−0.061	−0.061
53.5	.059	.059	.059	.060	.060	.061	.061	.062	.062	.063
54.0	.060	.060	.061	.061	.062	.062	.063	.063	.063	.064
54.5	.061	.061	.062	.062	.063	.063	.064	.064	.065	.065
55.0	.062	.063	.063	.064	.064	.064	.065	.065	.066	.066
55.5	−0.063	−0.064	−0.064	−0.065	−0.065	−0.066	−0.066	−0.067	−0.067	−0.068
56.0	.064	.065	.065	.066	.066	.067	.067	.068	.068	.069
56.5	.066	.066	.067	.067	.068	.068	.069	.069	.070	.070
57.0	.067	.067	.068	.068	.069	.069	.070	.070	.071	.071
57.5	.068	.069	.069	.070	.070	.071	.071	.072	.072	.073
58.0	−0.069	−0.070	−0.070	−0.071	−0.071	−0.072	−0.072	−0.073	−0.073	−0.074
58.5	.070	.071	.071	.072	.072	.073	.074	.074	.075	.075
59.0	.072	.072	.073	.073	.074	.074	.075	.075	.076	.076
59.5	.073	.073	.074	.074	.075	.075	.076	.077	.077	.078
60.0	.074	.074	.075	.076	.076	.077	.077	.078	.078	.079
60.5	−0.075	−0.076	−0.076	−0.077	−0.077	−0.078	−0.078	−0.079	−0.080	−0.080
61.0	.076	.077	.077	.078	.079	.079	.080	.080	.081	.081
61.5	.077	.078	.079	.079	.080	.080	.081	.082	.082	.083
62.0	.079	.079	.080	.080	.081	.082	.082	.083	.083	.084
62.5	.080	.080	.081	.082	.082	.083	.083	.084	.085	.085
63.0	−0.081	−0.082	−0.082	−0.083	−0.083	−0.084	−0.085	−0.085	−0.086	−0.086
63.5	.082	.083	.083	.084	.085	.085	.086	.086	.087	.088
64.0	.083	.084	.085	.085	.086	.086	.087	.087	.088	.089
64.5	.084	.085	.086	.086	.087	.088	.088	.089	.090	.090
65.0	.086	.086	.087	.088	.088	.089	.090	.090	.091	.092
65.5	−0.087	−0.087	−0.088	−0.089	−0.089	−0.090	−0.091	−0.091	−0.092	−0.093
66.0	.088	.089	.089	.090	.091	.091	.092	.093	.093	.094
66.5	.089	.090	.090	.091	.092	.093	.093	.094	.095	.095
67.0	.090	.091	.092	.092	.093	.094	.094	.095	.096	.097
67.5	.092	.092	.093	.094	.094	.095	.096	.096	.097	.098
68.0	−0.093	−0.093	−0.094	−0.095	−0.095	−0.096	−0.097	−0.098	−0.098	−0.099
68.5	.094	.095	.095	.096	.097	.097	.098	.099	.100	.100
69.0	.095	.096	.096	.097	.098	.099	.099	.100	.101	.102
69.5	.096	.097	.098	.098	.099	.100	.101	.101	.102	.103
70.0	.097	.098	.099	.100	.100	.101	.102	.103	.103	.104
70.5	−0.098	−0.099	−0.100	−0.101	−0.101	−0.102	−0.103	−0.104	−0.105	−0.105
71.0	.100	.100	.101	.102	.103	.103	.104	.105	.106	.107
71.5	.101	.102	.102	.103	.104	.105	.105	.106	.107	.108
72.0	.102	.103	.104	.104	.105	.106	.107	.107	.108	.109
72.5	.103	.104	.105	.106	.106	.107	.108	.109	.109	.110
73.0	−0.104	−0.105	−0.106	−0.107	−0.108	−0.108	−0.109	−0.110	−0.111	−0.112
73.5	.105	.106	.107	.108	.109	.110	.110	.111	.112	.113
74.0	.107	.107	.108	.109	.110	.111	.112	.112	.113	.114
74.5	.108	.109	.109	.110	.111	.112	.113	.114	.114	.115
75.0	.109	.110	.111	.112	.112	.113	.114	.115	.116	.117

SMITHSONIAN TABLES.

TABLE 10.

REDUCTION OF THE BAROMETER TO STANDARD TEMPERATURE.
ENGLISH MEASURES.

Attached Thermometer Fahrenheit.	HEIGHT OF THE BAROMETER IN INCHES.									
	26.0	26.2	26.4	26.6	26.8	27.0	27.2	27.4	27.6	27.8
F.	Inch.	Inch.	Inch.	Inch.	Inch.	Inch.	Inch.	Inch.	Inch.	Inch.
75.5	−0.110	−0.111	−0.112	−0.113	−0.114	−0.114	−0.115	−0.116	−0.117	−0.118
76.0	.111	.112	.113	.114	.115	.116	.116	.117	.118	.119
76.5	.113	.113	.114	.115	.116	.117	.118	.119	.119	.120
77.0	.114	.115	.115	.116	.117	.118	.119	.120	.121	.122
77.5	.115	.116	.117	.117	.118	.119	.120	.121	.122	.123
78.0	−0.116	−0.117	−0.118	−0.119	−0.120	−0.120	−0.121	−0.122	−0.123	−0.124
78.5	.117	.118	.119	.120	.121	.122	.123	.123	.124	.125
79.0	.118	.119	.120	.121	.122	.123	.124	.125	.126	.127
79.5	.120	.120	.121	.122	.123	.124	.125	.126	.127	.128
80.0	.121	.122	.123	.123	.124	.125	.126	.127	.128	.129
80.5	−0.122	−0.123	−0.124	−0.125	−0.126	−0.127	−0.127	−0.128	−0.129	−0.130
81.0	.123	.124	.125	.126	.127	.128	.129	.130	.131	.132
81.5	.124	.125	.126	.127	.128	.129	.130	.131	.132	.133
82.0	.125	.126	.127	.128	.129	.130	.131	.132	.133	.134
82.5	.127	.128	.128	.129	.130	.131	.132	.133	.134	.135
83.0	−0.128	−0.129	−0.130	−0.131	−0.132	−0.133	−0.134	−0.135	−0.136	−0.137
83.5	.129	.130	.131	.132	.133	.134	.135	.136	.137	.138
84.0	.130	.131	.132	.133	.134	.135	.136	.137	.138	.139
84.5	.131	.132	.133	.134	.135	.136	.137	.138	.139	.140
85.0	.132	.133	.134	.135	.136	.137	.138	.139	.141	.142
85.5	−0.134	−0.135	−0.136	−0.137	−0.138	−0.139	−0.140	−0.141	−0.142	−0.143
86.0	.135	.136	.137	.138	.139	.140	.141	.142	.143	.144
86.5	.136	.137	.138	.139	.140	.141	.142	.143	.144	.145
87.0	.137	.138	.139	.140	.141	.142	.143	.144	.145	.147
87.5	.138	.139	.140	.141	.142	.144	.145	.146	.147	.148
88.0	−0.139	−0.140	−0.142	−0.143	−0.144	−0.145	−0.146	−0.147	−0.148	−0.149
88.5	.141	.142	.143	.144	.145	.146	.147	.148	.149	.150
89.0	.142	.143	.144	.145	.146	.147	.148	.149	.150	.152
89.5	.143	.144	.145	.146	.147	.148	.149	.151	.152	.153
90.0	.144	.145	.146	.147	.148	.150	.151	.152	.153	.154
90.5	−0.145	−0.146	−0.147	−0.149	−0.150	−0.151	−0.152	−0.153	−0.154	−0.155
91.0	.146	.147	.149	.150	.151	.152	.153	.154	.155	.157
91.5	.148	.149	.150	.151	.152	.153	.154	.155	.157	.158
92.0	.149	.150	.151	.152	.153	.154	.156	.157	.158	.159
92.5	.150	.151	.152	.153	.154	.156	.157	.158	.159	.160
93.0	−0.151	−0.152	−0.153	−0.155	−0.156	−0.157	−0.158	−0.159	−0.160	−0.161
93.5	.152	.153	.155	.156	.157	.158	.159	.160	.162	.163
94.0	.153	.155	.156	.157	.158	.159	.160	.162	.163	.164
94.5	.155	.156	.157	.158	.159	.160	.162	.163	.164	.165
95.0	.156	.157	.158	.159	.160	.162	.163	.164	.165	.166
95.5	−0.157	−0.158	−0.159	−0.160	−0.162	−0.163	−0.164	−0.165	−0.167	−0.168
96.0	.158	.159	.160	.162	.163	.164	.165	.167	.168	.169
96.5	.159	.160	.162	.163	.164	.165	.167	.168	.169	.170
97.0	.160	.162	.163	.164	.165	.167	.168	.169	.170	.171
97.5	.162	.163	.164	.165	.166	.168	.169	.170	.171	.173
98.0	−0.163	−0.164	−0.165	−0.166	−0.168	−0.169	−0.170	−0.171	−0.173	−0.174
98.5	.164	.165	.166	.168	.169	.170	.171	.173	.174	.175
99.0	.165	.166	.168	.169	.170	.171	.173	.174	.175	.176
99.5	.166	.167	.169	.170	.171	.173	.174	.175	.176	.178
100.0	.167	.169	.170	.171	.172	.174	.175	.176	.178	.179

SMITHSONIAN TABLES.

TABLE 10.
REDUCTION OF THE BAROMETER TO STANDARD TEMPERATURE.
ENGLISH MEASURES.

Attached Thermometer Fahrenheit.	HEIGHT OF THE BAROMETER IN INCHES.									
	28.0	28.2	28.4	28.6	28.8	29.0	29.2	29.4	29.6	29.8
F.	Inch.	Inch.	Inch.	Inch.	Inch.	Inch.	Inch.	Inch.	Inch.	Inch.
0°.0	+0.073	+0.074	+0.074	+0.075	+0.075	+0.076	+0.076	+0.077	+0.077	+0.078
+0.5	+0.072	+0.072	+0.073	+0.073	+0.074	+0.074	+0.075	+0.075	+0.076	+0.076
1.0	.070	.071	.071	.072	.072	.073	.073	.074	.074	.075
1.5	.069	.070	.070	.071	.071	.072	.072	.073	.073	.074
2.0	.068	.068	.069	.069	.070	.070	.071	.071	.072	.072
2.5	.067	.067	.068	.068	.069	.069	.069	.070	.070	.071
3.0	+0.065	+0.066	+0.066	+0.067	+0.067	+0.068	+0.068	+0.069	+0.069	+0.070
3.5	.064	.065	.065	.065	.066	.066	.067	.067	.068	.068
4.0	.063	.063	.064	.064	.065	.065	.065	.066	.066	.067
4.5	.062	.062	.062	.063	.063	.064	.064	.065	.065	.065
5.0	.060	.061	.061	.062	.062	.062	.063	.063	.064	.064
5.5	+0.059	+0.059	+0.060	+0.060	+0.061	+0.061	+0.062	+0.062	+0.062	+0.063
6.0	.058	.058	.059	.059	.059	.060	.060	.061	.061	.061
6.5	.056	.057	.057	.058	.058	.058	.059	.059	.060	.060
7.0	.055	.056	.056	.056	.057	.057	.057	.058	.058	.059
7.5	.054	.054	.055	.055	.055	.056	.056	.057	.057	.057
8.0	+0.053	+0.053	+0.053	+0.054	+0.054	+0.054	+0.055	+0.055	+0.056	+0.056
8.5	.051	.052	.052	.052	.053	.053	.053	.054	.054	.055
9.0	.050	.050	.051	.051	.051	.052	.052	.053	.053	.053
9.5	.049	.049	.049	.050	.050	.050	.051	.051	.052	.052
10.0	.047	.048	.048	.048	.049	.049	.050	.050	.050	.051
10.5	+0.046	+0.047	+0.047	+0.047	+0.048	+0.048	+0.048	+0.049	+0.049	+0.049
11.0	.045	.045	.046	.046	.046	.047	.047	.047	.047	.048
11.5	.044	.044	.044	.045	.045	.045	.046	.046	.046	.046
12.0	.042	.043	.043	.043	.044	.044	.044	.044	.045	.045
12.5	.041	.041	.042	.042	.042	.043	.043	.043	.043	.044
13.0	+0.040	+0.040	+0.040	+0.041	+0.041	+0.041	+0.042	+0.042	+0.042	+0.042
13.5	.039	.039	.039	.039	.040	.040	.040	.040	.041	.041
14.0	.037	.038	.038	.038	.038	.039	.039	.039	.039	.040
14.5	.036	.036	.037	.037	.037	.037	.038	.038	.038	.038
15.0	.035	.035	.035	.035	.036	.036	.036	.036	.037	.037
15.5	+0.033	+0.034	+0.034	+0.034	+0.034	+0.035	+0.035	+0.035	+0.035	+0.036
16.0	.032	.032	.033	.033	.033	.033	.034	.034	.034	.034
16.5	.031	.031	.031	.032	.032	.032	.032	.032	.033	.033
17.0	.030	.030	.030	.030	.030	.031	.031	.031	.031	.032
17.5	.028	.029	.029	.029	.029	.029	.030	.030	.030	.030
18.0	+0.027	+0.027	+0.027	+0.028	+0.028	+0.028	+0.028	+0.028	+0.029	+0.029
18.5	.026	.026	.026	.026	.027	.027	.027	.027	.027	.027
19.0	.025	.025	.025	.025	.025	.025	.026	.026	.026	.026
19.5	.023	.023	.024	.024	.024	.024	.024	.024	.025	.025
20.0	.022	.022	.022	.022	.023	.023	.023	.023	.023	.023
20.5	+0.021	+0.021	+0.021	+0.021	+0.021	+0.021	+0.022	+0.022	+0.022	+0.022
21.0	.019	.020	.020	.020	.020	.020	.020	.020	.021	.021
21.5	.018	.018	.018	.019	.019	.019	.019	.019	.019	.019
22.0	.017	.017	.017	.017	.017	.017	.018	.018	.018	.018
22.5	.016	.016	.016	.016	.016	.016	.016	.016	.016	.017
23.0	+0.014	+0.014	+0.015	+0.015	+0.015	+0.015	+0.015	+0.015	+0.015	+0.015
23.5	.013	.013	.013	.013	.013	.014	.014	.014	.014	.014
24.0	.012	.012	.012	.012	.012	.012	.012	.012	.012	.013
24.5	.011	.011	.011	.011	.011	.011	.011	.011	.011	.011
25.0	.009	.009	.009	.009	.009	.010	.010	.010	.010	.010

SMITHSONIAN TABLES.

TABLE 10.

REDUCTION OF THE BAROMETER TO STANDARD TEMPERATURE.
ENGLISH MEASURES.

Attached Thermometer Fahrenheit.	HEIGHT OF THE BAROMETER IN INCHES.								
	28.0	28.2	28.4	28.6	28.8	29.0	29.2	29.4	29.6	29.8
F.	Inch.	Inch.	Inch.	Inch.	Inch.	Inch.	Inch.	Inch.	Inch.	Inch.
25.5	+0.008	+0.008	+0.008	+0.008	+0.008	+0.008	+0.008	+0.008	+0.008	+0.008
26.0	.007	.007	.007	.007	.007	.007	.007	.007	.007	.007
26.5	.005	.005	.005	.006	.006	.006	.006	.006	.006	.006
27.0	.004	.004	.004	.004	.004	.004	.004	.004	.004	.004
27.5	.003	.003	.003	.003	.003	.003	.003	.003	.003	.003
28.0	+0.002	+0.002	+0.002	+0.002	+0.002	+0.002	+0.002	+0.002	+0.002	+0.002
28.5	0.000	0.000	0.000	0.000	0.000	0.000	0.000	0.000	0.000	0.000
29.0	−0.001	−0.001	−0.001	−0.001	−0.001	−0.001	−0.001	−0.001	−0.001	−0.001
29.5	.002	.002	.002	.002	.002	.002	.002	.002	.002	.002
30.0	.003	.004	.004	.004	.004	.004	.004	.004	.004	.004
30.5	−0.005	−0.005	−0.005	−0.005	−0.005	−0.005	−0.005	−0.005	−0.005	−0.005
31.0	.006	.006	.006	.006	.006	.006	.006	.006	.006	.006
31.5	.007	.007	.007	.007	.008	.008	.008	.008	.008	.008
32.0	.009	.009	.009	.009	.009	.009	.009	.009	.009	.009
32.5	.010	.010	.010	.010	.010	.010	.010	.010	.010	.010
33.0	−0.011	−0.011	−0.011	−0.011	−0.011	−0.012	−0.012	−0.012	−0.012	−0.012
33.5	.012	.012	.013	.013	.013	.013	.013	.013	.013	.013
34.0	.014	.014	.014	.014	.014	.014	.014	.014	.014	.015
34.5	.015	.015	.015	.015	.015	.015	.016	.016	.016	.016
35.0	.016	.016	.016	.017	.017	.017	.017	.017	.017	.017
35.5	−0.017	−0.018	−0.018	−0.018	−0.018	−0.018	−0.018	−0.018	−0.018	−0.019
36.0	.019	.019	.019	.019	.019	.019	.020	.020	.020	.020
36.5	.020	.020	.020	.020	.021	.021	.021	.021	.021	.021
37.0	.021	.021	.022	.022	.022	.022	.022	.022	.022	.023
37.5	.023	.023	.023	.023	.023	.023	.024	.024	.024	.024
38.0	−0.024	−0.024	−0.024	−0.024	−0.024	−0.025	−0.025	−0.025	−0.025	−0.025
38.5	.025	.025	.025	.026	.026	.026	.026	.026	.027	.027
39.0	.026	.027	.027	.027	.027	.027	.027	.028	.028	.028
39.5	.028	.028	.028	.028	.028	.029	.029	.029	.029	.029
40.0	.029	.029	.029	.030	.030	.030	.030	.030	.031	.031
40.5	−0.030	−0.030	−0.031	−0.031	−0.031	−0.031	−0.031	−0.032	−0.032	−0.032
41.0	.031	.032	.032	.032	.032	.033	.033	.033	.033	.033
41.5	.033	.033	.033	.033	.034	.034	.034	.034	.035	.035
42.0	.034	.034	.034	.035	.035	.035	.035	.036	.036	.036
42.5	.035	.035	.036	.036	.036	.036	.037	.037	.037	.037
43.0	−0.036	−0.037	−0.037	−0.037	−0.038	−0.038	−0.038	−0.038	−0.039	−0.039
43.5	.038	.038	.038	.039	.039	.039	.039	.040	.040	.040
44.0	.039	.039	.040	.040	.040	.040	.041	.041	.041	.042
44.5	.040	.041	.041	.041	.041	.042	.042	.042	.043	.043
45.0	.042	.042	.042	.042	.043	.043	.043	.044	.044	.044
45.5	−0.043	−0.043	−0.043	−0.044	−0.044	−0.044	−0.045	−0.045	−0.045	−0.046
46.0	.044	.044	.045	.045	.045	.046	.046	.046	.047	.047
46.5	.045	.046	.046	.046	.047	.047	.047	.048	.048	.048
47.0	.047	.047	.047	.048	.048	.048	.049	.049	.049	.050
47.5	.048	.048	.049	.049	.049	.050	.050	.050	.051	.051
48.0	−0.049	−0.050	−0.050	−0.050	−0.051	−0.051	−0.051	−0.052	−0.052	−0.052
48.5	.050	.051	.051	.052	.052	.052	.053	.053	.053	.054
49.0	.052	.052	.052	.053	.053	.054	.054	.054	.055	.055
49.5	.053	.053	.054	.054	.054	.055	.055	.056	.056	.056
50.0	.054	.055	.055	.055	.056	.056	.057	.057	.057	.058

SMITHSONIAN TABLES.

TABLE 10.
REDUCTION OF THE BAROMETER TO STANDARD TEMPERATURE.
ENGLISH MEASURES.

Attached Thermometer Fahrenheit.	HEIGHT OF THE BAROMETER IN INCHES.								
	28.0	28.2	28.4	28.6	28.8	29.0	29.2	29.4	29.6	29.8
F.	Inch.	Inch.	Inch.	Inch.	Inch.	Inch.	Inch.	Inch.	Inch.	Inch.
50.5	−0.055	−0.056	−0.056	−0.057	−0.057	−0.057	−0.058	−0.058	−0.059	−0.059
51.0	.057	.057	.058	.058	.058	.059	.059	.060	.060	.060
51.5	.058	.058	.059	.059	.060	.060	.061	.061	.061	.062
52.0	.059	.060	.060	.061	.061	.061	.062	.062	.063	.063
52.5	.061	.061	.061	.062	.062	.063	.063	.064	.064	.064
53.0	−0.062	−0.062	−0.063	−0.063	−0.064	−0.064	−0.064	−0.065	−0.065	−0.066
53.5	.063	.064	.064	.064	.065	.065	.066	.066	.067	.067
54.0	.064	.065	.065	.066	.066	.067	.067	.068	.068	.068
54.5	.066	.066	.067	.067	.067	.068	.068	.069	.069	.070
55.0	.067	.067	.068	.068	.069	.069	.070	.070	.071	.071
55.5	−0.068	−0.069	−0.069	−0.070	−0.070	−0.071	−0.071	−0.072	−0.072	−0.073
56.0	.069	.070	.070	.071	.071	.072	.072	.073	.073	.074
56.5	.071	.071	.072	.072	.073	.073	.074	.074	.075	.075
57.0	.072	.072	.073	.073	.074	.075	.075	.076	.076	.077
57.5	.073	.074	.074	.075	.075	.076	.076	.077	.077	.078
58.0	−0.074	−0.075	−0.076	−0.076	−0.077	−0.077	−0.078	−0.078	−0.079	−0.079
58.5	.076	.076	.077	.077	.078	.078	.079	.080	.080	.081
59.0	.077	.078	.078	.079	.079	.080	.080	.081	.081	.082
59.5	.078	.079	.079	.080	.081	.081	.082	.082	.083	.083
60.0	.080	.080	.081	.081	.082	.082	.083	.084	.084	.085
60.5	−0.081	−0.081	−0.082	−0.083	−0.083	−0.084	−0.084	−0.085	−0.085	−0.086
61.0	.082	.083	.083	.084	.084	.085	.086	.086	.087	.087
61.5	.083	.084	.085	.085	.086	.086	.087	.087	.088	.089
62.0	.085	.085	.086	.086	.087	.088	.088	.089	.089	.090
62.5	.086	.086	.087	.088	.088	.089	.090	.090	.091	.091
63.0	−0.087	−0.088	−0.088	−0.089	−0.090	−0.090	−0.091	−0.091	−0.092	−0.093
63.5	.088	.089	.090	.090	.091	.092	.092	.093	.093	.094
64.0	.090	.090	.091	.092	.092	.093	.093	.094	.095	.095
64.5	.091	.092	.092	.093	.093	.094	.095	.095	.096	.097
65.0	.092	.093	.093	.094	.095	.095	.096	.097	.097	.098
65.5	−0.093	−0.094	−0.095	−0.095	−0.096	−0.097	−0.097	−0.098	−0.099	−0.099
66.0	.095	.095	.096	.097	.097	.098	.099	.099	.100	.101
66.5	.096	.097	.097	.098	.099	.099	.100	.101	.101	.102
67.0	.097	.098	.099	.099	.100	.101	.101	.102	.103	.103
67.5	.098	.099	.100	.101	.101	.102	.103	.103	.104	.105
68.0	−0.100	−0.100	−0.101	−0.102	−0.103	−0.103	−0.104	−0.105	−0.105	−0.106
68.5	.101	.102	.102	.103	.104	.105	.105	.106	.107	.107
69.0	.102	.103	.104	.104	.105	.106	.107	.107	.108	.109
69.5	.104	.104	.105	.106	.106	.107	.108	.109	.109	.110
70.0	.105	.106	.106	.107	.108	.109	.109	.110	.111	.112
70.5	−0.106	−0.107	−0.108	−0.108	−0.109	−0.110	−0.111	−0.111	−0.112	−0.113
71.0	.107	.108	.109	.110	.110	.111	.112	.113	.113	.114
71.5	.109	.109	.110	.111	.112	.112	.113	.114	.115	.116
72.0	.110	.111	.111	.112	.113	.114	.115	.115	.116	.117
72.5	.111	.112	.113	.113	.114	.115	.116	.117	.117	.118
73.0	−0.112	−0.113	−0.114	−0.115	−0.116	−0.116	−0.117	−0.118	−0.119	−0.120
73.5	.114	.114	.115	.116	.117	.118	.118	.119	.120	.121
74.0	.115	.116	.117	.117	.118	.119	.120	.121	.121	.122
74.5	.116	.117	.118	.119	.119	.120	.121	.122	.123	.124
75.0	.117	.118	.119	.120	.121	.122	.122	.123	.124	.125

TABLE 10.

REDUCTION OF THE BAROMETER TO STANDARD TEMPERATURE.
ENGLISH MEASURES.

Attached Thermometer Fahrenheit.	HEIGHT OF THE BAROMETER IN INCHES.									
	28.0	28.2	28.4	28.6	28.8	29.0	29.2	29.4	29.6	29.8
F.	Inch.	Inch.	Inch.	Inch.	Inch.	Inch.	Inch.	Inch.	Inch.	Inch.
75.5	−0.119	−0.119	−0.120	−0.121	−0.122	−0.123	−0.124	−0.125	−0.125	−0.126
76.0	.120	.121	.122	.122	.123	.124	.125	.126	.127	.128
76.5	.121	.122	.123	.124	.125	.125	.126	.127	.128	.129
77.0	.122	.123	.124	.125	.126	.127	.128	.129	.129	.130
77.5	.124	.125	.125	.126	.127	.128	.129	.130	.131	.132
78.0	−0.125	−0.126	−0.127	−0.128	−0.129	−0.129	−0.130	−0.131	−0.132	−0.133
78.5	.126	.127	.128	.129	.130	.131	.132	.133	.133	.134
79.0	.127	.128	.129	.130	.131	.132	.133	.134	.135	.136
79.5	.129	.130	.131	.131	.132	.133	.134	.135	.136	.137
80.0	.130	.131	.132	.133	.134	.135	.136	.136	.137	.138
80.5	−0.131	−0.132	−0.133	−0.134	−0.135	−0.136	−0.137	−0.138	−0.139	−0.140
81.0	.132	.133	.134	.135	.136	.137	.138	.139	.140	.141
81.5	.134	.135	.136	.137	.138	.139	.139	.140	.141	.142
82.0	.135	.136	.137	.138	.139	.140	.141	.142	.143	.144
82.5	.136	.137	.138	.139	.140	.141	.142	.143	.144	.145
83.0	−0.138	−0.139	−0.139	−0.140	−0.141	−0.142	−0.143	−0.144	−0.145	−0.146
83.5	.139	.140	.141	.142	.143	.144	.145	.146	.147	.148
84.0	.140	.141	.142	.143	.144	.145	.146	.147	.148	.149
84.5	.141	.142	.143	.144	.145	.146	.147	.148	.149	.150
85.0	.143	.144	.145	.146	.147	.148	.149	.150	.151	.152
85.5	−0.144	−0.145	−0.146	−0.147	−0.148	−0.149	−0.150	−0.151	−0.152	−0.153
86.0	.145	.146	.147	.148	.149	.150	.151	.152	.153	.154
86.5	.146	.147	.148	.149	.151	.152	.153	.154	.155	.156
87.0	.148	.149	.150	.151	.152	.153	.154	.155	.156	.157
87.5	.149	.150	.151	.152	.153	.154	.155	.156	.157	.158
88.0	−0.150	−0.151	−0.152	−0.153	−0.154	−0.155	−0.157	−0.158	−0.159	−0.160
88.5	.151	.152	.154	.155	.156	.157	.158	.159	.160	.161
89.0	.153	.154	.155	.156	.157	.158	.159	.160	.161	.162
89.5	.154	.155	.156	.157	.158	.159	.160	.162	.163	.164
90.0	.155	.156	.157	.158	.160	.161	.162	.163	.164	.165
90.5	−0.156	−0.157	−0.159	−0.160	−0.161	−0.162	−0.163	−0.164	−0.165	−0.166
91.0	.158	.159	.160	.161	.162	.163	.164	.166	.167	.168
91.5	.159	.160	.161	.162	.163	.165	.166	.167	.168	.169
92.0	.160	.161	.162	.164	.165	.166	.167	.168	.169	.170
92.5	.161	.163	.164	.165	.166	.167	.168	.169	.171	.172
93.0	−0.163	−0.164	−0.165	−0.166	−0.167	−0.168	−0.170	−0.171	−0.172	−0.173
93.5	.164	.165	.166	.167	.169	.170	.171	.172	.173	.174
94.0	.165	.166	.168	.169	.170	.171	.172	.173	.175	.176
94.5	.166	.168	.169	.170	.171	.172	.174	.175	.176	.177
95.0	.168	.169	.170	.171	.172	.174	.175	.176	.177	.178
95.5	−0.169	−0.170	−0.171	−0.173	−0.174	−0.175	−0.176	−0.177	−0.179	−0.180
96.0	.170	.171	.173	.174	.175	.176	.177	.179	.180	.181
96.5	.171	.173	.174	.175	.176	.178	.179	.180	.181	.182
97.0	.173	.174	.175	.176	.178	.179	.180	.181	.183	.184
97.5	.174	.175	.176	.178	.179	.180	.181	.183	.184	.185
98.0	−0.175	−0.176	−0.178	−0.179	−0.180	−0.181	−0.183	−0.184	−0.185	−0.186
98.5	.176	.178	.179	.180	.181	.183	.184	.185	.187	.188
99.0	.178	.179	.180	.182	.183	.184	.185	.187	.188	.189
99.5	.179	.180	.182	.183	.184	.185	.187	.188	.189	.190
100.0	.180	.182	.183	.184	.185	.187	.188	.189	.191	.192

SMITHSONIAN TABLES.

TABLE 10.

REDUCTION OF THE BAROMETER TO STANDARD TEMPERATURE.
ENGLISH MEASURES.

Attached Thermometer Fahrenheit.	HEIGHT OF THE BAROMETER IN INCHES.									
	29.8	30.0	30.2	30.4	30.6	30.8	31.0	31.2	31.4	31.6
F.	Inch.	Inch.	Inch.	Inch.	Inch.	Inch.	Inch.	Inch.	Inch.	Inch.
0.0	+0.078	+0.078	+0.079	+0.079	+0.080	+0.080	+0.081	+0.081	+0.082	+0.082
0.5	+0.076	+0.077	+0.077	+0.078	+0.078	+0.079	+0.079	+0.080	+0.080	+0.081
1.0	.075	.076	.076	.077	.077	.078	.078	.079	.079	.080
1.5	.074	.074	.075	.075	.076	.076	.077	.077	.078	.078
2.0	.072	.073	.073	.074	.074	.075	.075	.076	.076	.077
2.5	.071	.071	.072	.072	.073	.073	.074	.074	.075	.075
3.0	+0.070	+0.070	+0.070	+0.071	+0.071	+0.072	+0.072	+0.073	+0.073	+0.074
3.5	.068	.069	.069	.070	.070	.070	.071	.071	.072	.072
4.0	.067	.067	.068	.068	.069	.069	.070	.070	.070	.071
4.5	.065	.066	.066	.067	.067	.068	.068	.069	.069	.069
5.0	.064	.065	.065	.065	.066	.066	.067	.067	.068	.068
5.5	+0.063	+0.063	+0.064	+0.064	+0.064	+0.065	+0.065	+0.066	+0.066	+0.067
6.0	.061	.062	.062	.063	.063	.063	.064	.064	.065	.065
6.5	.060	.060	.061	.061	.062	.062	.062	.063	.063	.064
7.0	.059	.059	.059	.060	.060	.061	.061	.061	.062	.062
7.5	.057	.058	.058	.058	.059	.059	.060	.060	.060	.061
8.0	+0.056	+0.056	+0.057	+0.057	+0.057	+0.058	+0.058	+0.059	+0.059	+0.059
8.5	.055	.055	.055	.056	.056	.056	.057	.057	.058	.058
9.0	.053	.054	.054	.054	.055	.055	.055	.056	.056	.056
9.5	.052	.052	.053	.053	.053	.054	.054	.054	.055	.055
10.0	.051	.051	.051	.052	.052	.052	.053	.053	.053	.054
10.5	+0.049	+0.049	+0.050	+0.050	+0.050	+0.051	+0.051	+0.051	+0.052	+0.052
11.0	.048	.048	.048	.049	.049	.049	.050	.050	.050	.051
11.5	.046	.047	.047	.047	.048	.048	.048	.049	.049	.049
12.0	.045	.045	.046	.046	.046	.047	.047	.047	.048	.048
12.5	.044	.044	.044	.045	.045	.045	.045	.046	.046	.046
13.0	+0.042	+0.043	+0.043	+0.043	+0.044	+0.044	+0.044	+0.044	+0.045	+0.045
13.5	.041	.041	.042	.042	.042	.042	.043	.043	.043	.043
14.0	.040	.040	.040	.040	.041	.041	.041	.042	.042	.042
14.5	.038	.039	.039	.039	.039	.040	.040	.040	.040	.041
15.0	.037	.037	.037	.038	.038	.038	.038	.039	.039	.039
15.5	+0.036	+0.036	+0.036	+0.036	+0.037	+0.037	+0.037	+0.037	+0.037	+0.038
16.0	.034	.034	.035	.035	.035	.035	.036	.036	.036	.036
16.5	.033	.033	.033	.034	.034	.034	.034	.034	.035	.035
17.0	.032	.032	.032	.032	.032	.033	.033	.033	.033	.033
17.5	.030	.030	.031	.031	.031	.031	.031	.032	.032	.032
18.0	+0.029	+0.029	+0.029	+0.029	+0.030	+0.030	+0.030	+0.030	+0.030	+0.031
18.5	.027	.028	.028	.028	.028	.028	.029	.029	.029	.029
19.0	.026	.026	.026	.027	.027	.027	.027	.027	.027	.028
19.5	.025	.025	.025	.025	.025	.026	.026	.026	.026	.026
20.0	.023	.024	.024	.024	.024	.024	.024	.024	.025	.025
20.5	+0.022	+0.022	+0.022	+0.022	+0.023	+0.023	+0.023	+0.023	+0.023	+0.023
21.0	.021	.021	.021	.021	.021	.021	.022	.022	.022	.022
21.5	.019	.019	.020	.020	.020	.020	.020	.020	.020	.020
22.0	.018	.018	.018	.018	.018	.019	.019	.019	.019	.019
22.5	.017	.017	.017	.017	.017	.017	.017	.017	.017	.018
23.0	+0.015	+0.015	+0.015	+0.016	+0.016	+0.016	+0.016	+0.016	+0.016	+0.016
23.5	.014	.014	.014	.014	.014	.014	.014	.015	.015	.015
24.0	.013	.013	.013	.013	.013	.013	.013	.013	.013	.013
24.5	.011	.011	.011	.011	.011	.012	.012	.012	.012	.012
25.0	.010	.010	.010	.010	.010	.010	.010	.010	0.10	.010

SMITHSONIAN TABLES.

TABLE 10.
REDUCTION OF THE BAROMETER TO STANDARD TEMPERATURE.
ENGLISH MEASURES.

Attached Thermometer Fahrenheit.	HEIGHT OF THE BAROMETER IN INCHES.									
	29.8	30.0	30.2	30.4	30.6	30.8	31.0	31.2	31.4	31.6
F.	Inch.	Inch.	Inch.	Inch.	Inch.	Inch.	Inch.	Inch.	Inch.	Inch.
25.5	+0.008	+0.009	+0.009	+0.009	+0.009	+0.009	+0.009	+0.009	+0.009	+0.009
26.0	.007	.007	.007	.007	.007	.007	.007	.007	.008	.008
26.5	.006	.006	.006	.006	.006	.006	.006	.006	.006	.006
27.0	.004	.004	.004	.005	.005	.005	.005	.005	.005	.005
27.5	.003	.003	.003	.003	.003	.003	.003	.003	.003	.003
28.0	+0.002	+0.002	+0.002	+0.002	+0.002	+0.002	+0.002	+0.002	+0.002	+0.002
28.5	0.000	0.000	0.000	0.000	0.000	0.000	0.000	0.000	0.000	0.000
29.0	−0.001	−0.001	−0.001	−0.001	−0.001	−0.001	−0.001	−0.001	−0.001	−0.001
29.5	.002	.002	.002	.002	.002	.002	.002	.002	.002	.002
30.0	.004	.004	.004	.004	.004	.004	.004	.004	.004	.004
30.5	−0.005	−0.005	−0.005	−0.005	−0.005	−0.005	−0.005	−0.005	−0.005	−0.005
31.0	.006	.006	.006	.007	.007	.007	.007	.007	.007	.007
31.5	.008	.008	.008	.008	.008	.008	.008	.008	.008	.008
32.0	.009	.009	.009	.009	.009	.009	.009	.010	.010	.010
32.5	.010	.011	.011	.011	.011	.011	.011	.011	.011	.011
33.0	−0.012	−0.012	−0.012	−0.012	−0.012	−0.012	−0.012	−0.012	−0.012	−0.013
33.5	.013	.013	.013	.013	.014	.014	.014	.014	.014	.014
34.0	.015	.015	.015	.015	.015	.015	.015	.015	.015	.015
34.5	.016	.016	.016	.016	.016	.016	.017	.017	.017	.017
35.0	.017	.017	.017	.018	.018	.018	.018	.018	.018	.018
35.5	−0.019	−0.019	−0.019	−0.019	−0.019	−0.019	−0.019	−0.019	−0.020	−0.020
36.0	.020	.020	.020	.020	.020	.021	.021	.021	.021	.021
36.5	.021	.021	.022	.022	.022	.022	.022	.022	.022	.023
37.0	.023	.023	.023	.023	.023	.023	.024	.024	.024	.024
37.5	.024	.024	.024	.024	.025	.025	.025	.025	.025	.025
38.0	−0.025	−0.026	−0.026	−0.026	−0.026	−0.026	−0.026	−0.027	−0.027	−0.027
38.5	.027	.027	.027	.027	.027	.028	.028	.028	.028	.028
39.0	.028	.028	.028	.029	.029	.029	.029	.029	.030	.030
39.5	.029	.030	.030	.030	.030	.030	.031	.031	.031	.031
40.0	.031	.031	.031	.031	.032	.032	.032	.032	.032	.033
40.5	−0.032	−0.032	−0.033	−0.033	−0.033	−0.033	−0.033	−0.034	−0.034	−0.034
41.0	.033	.034	.034	.034	.034	.035	.035	.035	.035	.035
41.5	.035	.035	.035	.035	.036	.036	.036	.036	.037	.037
42.0	.036	.036	.037	.037	.037	.037	.038	.038	.038	.038
42.5	.037	.038	.038	.038	.038	.039	.039	.039	.040	.040
43.0	−0.039	−0.039	−0.039	−0.040	−0.040	−0.040	−0.040	−0.041	−0.041	−0.041
43.5	.040	.040	.041	.041	.041	.042	.042	.042	.042	.043
44.0	.042	.042	.042	.042	.043	.043	.043	.043	.044	.044
44.5	.043	.043	.043	.044	.044	.044	.045	.045	.045	.045
45.0	.044	.045	.045	.045	.045	.046	.046	.046	.047	.047
45.5	−0.046	−0.046	−0.046	−0.047	−0.047	−0.047	−0.047	−0.048	−0.048	−0.048
46.0	.047	.047	.048	.048	.048	.049	.049	.049	.049	.050
46.5	.048	.049	.049	.049	.050	.050	.050	.051	.051	.051
47.0	.050	.050	.050	.051	.051	.051	.052	.052	.052	.053
47.5	.051	.051	.052	.052	.052	.053	.053	.053	.054	.054
48.0	−0.052	−0.053	−0.053	−0.053	−0.054	−0.054	−0.054	−0.055	−0.055	−0.055
48.5	.054	.054	.054	.055	.055	.055	.056	.056	.057	.057
49.0	.055	.055	.056	.056	.057	.057	.057	.058	.058	.058
49.5	.056	.057	.057	.058	.058	.058	.059	.059	.059	.060
50.0	.058	.058	.058	.059	.059	.060	.060	.060	.061	.061

SMITHSONIAN TABLES.

TABLE 10.
REDUCTION OF THE BAROMETER TO STANDARD TEMPERATURE.
ENGLISH MEASURES.

Attached Thermometer Fahrenheit.	HEIGHT OF THE BAROMETER IN INCHES.									
	29.8	30.0	30.2	30.4	30.6	30.8	31.0	31.2	31.4	31.6
F.	Inch.	Inch.	Inch.	Inch.	Inch.	Inch.	Inch.	Inch.	Inch.	Inch.
50.5	−0.059	−0.059	−0.060	−0.060	−0.061	−0.061	−0.061	−0.062	−0.062	−0.063
51.0	.060	.061	.061	.062	.062	.062	.063	.063	.064	.064
51.5	.062	.062	.063	.063	.063	.064	.064	.065	.065	.065
52.0	.063	.064	.064	.064	.065	.065	.066	.066	.066	.067
52.5	.064	.065	.065	.066	.066	.067	.067	.067	.068	.068
53.0	−0.066	−0.066	−0.067	−0.067	−0.068	−0.068	−0.068	−0.069	−0.069	−0.070
53.5	.067	.068	.068	.069	.069	.069	.070	.070	.071	.071
54.0	.068	.069	.069	.070	.070	.071	.071	.072	.072	.073
54.5	.070	.070	.071	.071	.072	.072	.073	.073	.074	.074
55.0	.071	.072	.072	.073	.073	.074	.074	.075	.075	.075
55.5	−0.073	−0.073	−0.074	−0.074	−0.074	−0.075	−0.075	−0.076	−0.076	−0.077
56.0	.074	.074	.075	.075	.076	.076	.077	.077	.078	.078
56.5	.075	.076	.076	.077	.077	.078	.078	.079	.079	.080
57.0	.077	.077	.078	.078	.079	.079	.080	.080	.081	.081
57.5	.078	.078	.079	.079	.080	.081	.081	.082	.082	.083
58.0	−0.079	−0.080	−0.080	−0.081	−0.081	−0.082	−0.082	−0.083	−0.084	−0.084
58.5	.081	.081	.082	.082	.083	.083	.084	.084	.085	.085
59.0	.082	.083	.083	.084	.084	.085	.085	.086	.086	.087
59.5	.083	.084	.084	.085	.086	.086	.087	.087	.088	.088
60.0	.085	.085	.086	.086	.087	.087	.088	.089	.089	.090
60.5	−0.086	−0.087	−0.087	−0.088	−0.088	−0.089	−0.089	−0.090	−0.091	−0.091
61.0	.087	.088	.089	.089	.090	.090	.091	.091	.092	.093
61.5	.089	.089	.090	.090	.091	.092	.092	.093	.093	.094
62.0	.090	.091	.091	.092	.092	.093	.094	.094	.095	.095
62.5	.091	.092	.093	.093	.094	.094	.095	.096	.096	.097
63.0	−0.093	−0.093	−0.094	−0.095	−0.095	−0.096	−0.096	−0.097	−0.098	−0.098
63.5	.094	.095	.095	.096	.097	.097	.098	.098	.099	.100
64.0	.095	.096	.097	.097	.098	.099	.099	.100	.101	.101
64.5	.097	.097	.098	.099	.099	.100	.101	.101	.102	.103
65.0	.098	.099	.099	.100	.101	.101	.102	.103	.103	.104
65.5	−0.099	−0.100	−0.101	−0.101	−0.102	−0.103	−0.103	−0.104	−0.105	−0.105
66.0	.101	.101	.102	.103	.103	.104	.105	.106	.106	.107
66.5	.102	.103	.103	.104	.105	.106	.106	.107	.108	.108
67.0	.103	.104	.105	.106	.106	.107	.108	.108	.109	.110
67.5	.105	.106	.106	.107	.108	.108	.109	.110	.110	.111
68.0	−0.106	−0.107	−0.108	−0.108	−0.109	−0.110	−0.110	−0.111	−0.112	−0.113
68.5	.107	.108	.109	.110	.110	.111	.112	.113	.113	.114
69.0	.109	.110	.110	.111	.112	.112	.113	.114	.115	.115
69.5	.110	.111	.112	.112	.113	.114	.115	.115	.116	.117
70.0	.112	.112	.113	.114	.115	.115	.116	.117	.117	.118
70.5	−0.113	−0.114	−0.114	−0.115	−0.116	−0.117	−0.117	−0.118	−0.119	−0.120
71.0	.114	.115	.116	.116	.117	.118	.119	.120	.120	.121
71.5	.116	.116	.117	.118	.119	.119	.120	.121	.122	.123
72.0	.117	.118	.118	.119	.120	.121	.122	.122	.123	.124
72.5	.118	.119	.120	.121	.121	.122	.123	.124	.125	.125
73.0	−0.120	−0.120	−0.121	−0.122	−0.123	−0.124	−0.124	−0.125	−0.126	−0.127
73.5	.121	.122	.123	.123	.124	.125	.126	.127	.127	.128
74.0	.122	.123	.124	.125	.126	.126	.127	.128	.129	.130
74.5	.124	.124	.125	.126	.127	.128	.129	.129	.130	.131
75.0	.125	.126	.127	.127	.128	.129	.130	.131	.132	.132

SMITHSONIAN TABLES.

TABLE 10.

REDUCTION OF THE BAROMETER TO STANDARD TEMPERATURE.
ENGLISH MEASURES.

Attached Thermometer Fahrenheit.	HEIGHT OF THE BAROMETER IN INCHES.								
	29.8	30.0	30.2	30.4	30.6	30.8	31.0	31.2	31.4	31.6
F.	Inch.	Inch.	Inch.	Inch.	Inch.	Inch.	Inch.	Inch.	Inch.	Inch.
75.5	−0.126	−0.127	−0.128	−0.129	−0.130	−0.131	−0.131	−0.132	−0.133	−0.134
76.0	.128	.128	.129	.130	.131	.132	.133	.134	.134	.135
76.5	.129	.130	.131	.132	.132	.133	.134	.135	.136	.137
77.0	.130	.131	.132	.133	.134	.135	.136	.136	.137	.138
77.5	.132	.133	.133	.134	.135	.136	.137	.138	.139	.140
78.0	−0.133	−0.134	−0.135	−0.136	−0.137	−0.137	−0.138	−0.139	−0.140	−0.141
78.5	.134	.135	.136	.137	.138	.139	.140	.141	.142	.142
79.0	.136	.137	.137	.138	.139	.140	.141	.142	.143	.144
79.5	.137	.138	.139	.140	.141	.142	.143	.143	.144	.145
80.0	.138	.139	.140	.141	.142	.143	.144	.145	.146	.147
80.5	−0.140	−0.141	−0.142	−0.142	−0.143	−0.144	−0.145	−0.146	−0.147	−0.148
81.0	.141	.142	.143	.144	.145	.146	.147	.148	.149	.150
81.5	.142	.143	.144	.145	.146	.147	.148	.149	.150	.151
82.0	.144	.145	.146	.147	.148	.149	.149	.150	.151	.152
82.5	.145	.146	.147	.148	.149	.150	.151	.152	.153	.154
83.0	−0.146	−0.147	−0.148	−0.149	−0.150	−0.151	−0.152	−0.153	−0.154	−0.155
83.5	.148	.149	.150	.151	.152	.153	.154	.155	.156	.157
84.0	.149	.150	.151	.152	.153	.154	.155	.156	.157	.158
84.5	.150	.151	.152	.153	.154	.155	.156	.157	.158	.159
85.0	.152	.153	.154	.155	.156	.157	.158	.159	.160	.161
85.5	−0.153	−0.154	−0.155	−0.156	−0.157	−0.158	−0.159	−0.160	−0.161	−0.162
86.0	.154	.155	.156	.158	.159	.160	.161	.162	.163	.164
86.5	.156	.157	.158	.159	.160	.161	.162	.163	.164	.165
87.0	.157	.158	.159	.160	.161	.162	.163	.164	.166	.167
87.5	.158	.159	.161	.162	.163	.164	.165	.166	.167	.168
88.0	−0.160	−0.161	−0.162	−0.163	−0.164	−0.165	−0.166	−0.167	−0.168	−0.169
88.5	.161	.162	.163	.164	.165	.166	.168	.169	.170	.171
89.0	.162	.164	.165	.166	.167	.168	.169	.170	.171	.172
89.5	.164	.165	.166	.167	.168	.169	.170	.171	.173	.174
90.0	.165	.166	.167	.168	.170	.171	.172	.173	.174	.175
90.5	−0.166	−0.168	−0.169	−0.170	−0.171	−0.172	−0.173	−0.174	−0.175	−0.176
91.0	.168	.169	.170	.171	.172	.173	.175	.176	.177	.178
91.5	.169	.170	.171	.173	.174	.175	.176	.177	.178	.179
92.0	.170	.172	.173	.174	.175	.176	.177	.178	.180	.181
92.5	.172	.173	.174	.175	.176	.178	.179	.180	.181	.182
93.0	−0.173	−0.174	−0.175	−0.177	−0.178	−0.179	−0.180	−0.181	−0.182	−0.184
93.5	.174	.176	.177	.178	.179	.180	.181	.183	.184	.185
94.0	.176	.177	.178	.179	.180	.182	.183	.184	.185	.186
94.5	.177	.178	.179	.181	.182	.183	.184	.185	.187	.188
95.0	.178	.180	.181	.182	.183	.184	.186	.187	.188	.189
95.5	−0.180	−0.181	−0.182	−0.183	−0.185	−0.186	−0.187	−0.188	−0.189	−0.191
96.0	.181	.182	.184	.185	.186	.187	.188	.190	.191	.192
96.5	.182	.184	.185	.186	.187	.189	.190	.191	.192	.193
97.0	.184	.185	.186	.187	.189	.190	.191	.192	.194	.195
97.5	.185	.186	.188	.189	.190	.191	.193	.194	.195	.196
98.0	−0.186	−0.188	−0.189	−0.190	−0.191	−0.193	−0.194	−0.195	−0.196	−0.198
98.5	.188	.189	.190	.192	.193	.194	.195	.197	.198	.199
99.0	.189	.190	.192	.193	.194	.195	.197	.198	.199	.201
99.5	.190	.192	.193	.194	.196	.197	.198	.199	.201	.202
100.0	.192	.193	.194	.196	.197	.198	.200	.201	.202	.203

SMITHSONIAN TABLES.

TABLE 11.
REDUCTION OF THE BAROMETER TO STANDARD TEMPERATURE.
METRIC MEASURES.

FOR TEMPERATURES ABOVE 0° CENTIGRADE, THE CORRECTION TO BE SUBTRACTED.

Attached Thermometer Centigrade.	HEIGHT OF THE BAROMETER IN MILLIMETRES.												
	440	450	460	470	480	490	500	510	520	530	540	550	560
C.	mm.	mm.	mm.	mm.	mm.	mm.	mm.	mm.	mm.	mm.	mm.	mm.	mm.
0.0	0.00	0.00	0.00	0.00	0.00	0.00	0.00	0.00	0.00	0.00	0.00	0.00	0.00
0.5	.04	.04	.04	.04	.04	.04	.04	.04	.04	.04	.04	.04	.05
1.0	.07	.07	.08	.08	.08	.08	.08	.08	.08	.09	.09	.09	.09
1.5	.11	.11	.11	.12	.12	.12	.12	.12	.13	.13	.13	.13	.14
2.0	.14	.15	.15	.15	.16	.16	.16	.17	.17	.17	.18	.18	.18
2.5	0.18	0.18	0.19	0.19	0.20	0.20	0.20	0.21	0.21	0.22	0.22	0.22	0.23
3.0	.22	.22	.23	.23	.24	.24	.24	.25	.25	.26	.26	.27	.27
3.5	.25	.26	.26	.27	.27	.28	.29	.29	.30	.30	.31	.31	.32
4.0	.29	.29	.30	.31	.31	.32	.33	.33	.34	.35	.35	.36	.37
4.5	.32	.33	.34	.35	.35	.36	.37	.37	.38	.39	.40	.40	.41
5.0	0.36	0.37	0.38	0.38	0.39	0.40	0.41	0.42	0.42	0.43	0.44	0.45	0.46
5.5	.40	.40	.41	.42	.43	.44	.45	.46	.47	.48	.48	.49	.50
6.0	.43	.44	.45	.46	.47	.48	.49	.50	.51	.52	.53	.54	.55
6.5	.47	.48	.49	.50	.51	.52	.53	.54	.55	.56	.57	.58	.59
7.0	.50	.51	.53	.54	.55	.56	.57	.58	.59	.61	.62	.63	.64
7.5	0.54	0.55	0.56	0.58	0.59	0.60	0.61	0.62	0.64	0.65	0.66	0.67	0.69
8.0	.57	.59	.60	.61	.63	.64	.65	.67	.68	.69	.70	.72	.73
8.5	.61	.62	.64	.65	.67	.68	.69	.71	.72	.73	.75	.76	.78
9.0	.65	.66	.68	.69	.70	.72	.73	.75	.76	.78	.79	.81	.82
9.5	.68	.70	.71	.73	.74	.76	.77	.79	.81	.82	.84	.85	.87
10.0	0.72	0.73	0.75	0.77	0.78	0.80	0.82	0.83	0.85	0.86	0.88	0.90	0.91
10.5	.75	.77	.79	.80	.82	.84	.86	.87	.89	.91	.92	.94	.96
11.0	.79	.81	.83	.84	.86	.88	.90	.91	.93	.95	.97	.99	1.00
11.5	.83	.84	.86	.88	.90	.92	.94	.96	.98	.99	1.01	1.03	1.05
12.0	.86	.88	.90	.92	.94	.96	.98	1.00	1.02	1.04	1.06	1.08	1.10
13.0	0.93	0.95	0.97	1.00	1.02	1.04	1.06	1.08	1.10	1.12	1.14	1.17	1.19
14.0	1.00	1.03	1.05	1.07	1.10	1.12	1.14	1.16	1.19	1.21	1.23	1.25	1.28
15.0	1.08	1.10	1.12	1.15	1.17	1.20	1.22	1.25	1.27	1.30	1.32	1.34	1.37
16.0	1.15	1.17	1.20	1.23	1.25	1.28	1.30	1.33	1.36	1.38	1.41	1.43	1.46
17.0	1.22	1.25	1.27	1.30	1.33	1.36	1.38	1.41	1.44	1.47	1.50	1.52	1.55
18.0	1.29	1.32	1.35	1.38	1.41	1.44	1.47	1.50	1.52	1.55	1.58	1.61	1.64
19.0	1.36	1.39	1.42	1.45	1.49	1.52	1.55	1.58	1.61	1.64	1.67	1.70	1.73
20.0	1.43	1.47	1.50	1.53	1.56	1.60	1.63	1.66	1.69	1.73	1.76	1.79	1.82
21.0	1.50	1.54	1.57	1.61	1.64	1.67	1.71	1.74	1.78	1.81	1.85	1.88	1.91
22.0	1.58	1.61	1.65	1.68	1.72	1.75	1.79	1.83	1.86	1.90	1.93	1.97	2.01
23.0	1.65	1.68	1.72	1.76	1.80	1.83	1.87	1.91	1.95	1.98	2.02	2.06	2.10
24.0	1.72	1.76	1.80	1.84	1.87	1.91	1.95	1.99	2.03	2.07	2.11	2.15	2.19
25.0	1.79	1.83	1.87	1.91	1.95	1.99	2.03	2.07	2.11	2.16	2.20	2.24	2.28
26.0	1.86	1.90	1.95	1.99	2.03	2.07	2.11	2.16	2.20	2.24	2.28	2.33	2.37
27.0	1.93	1.98	2.02	2.06	2.11	2.15	2.20	2.24	2.28	2.33	2.37	2.41	2.46
28.0	2.00	2.05	2.09	2.14	2.18	2.23	2.28	2.32	2.37	2.41	2.46	2.50	2.55
29.0	2.07	2.12	2.17	2.22	2.26	2.31	2.36	2.40	2.45	2.50	2.55	2.59	2.64
30.0	2.15	2.19	2.24	2.29	2.34	2.39	2.44	2.49	2.54	2.58	2.63	2.68	2.73
31.0	2.22	2.27	2.32	2.37	2.42	2.47	2.52	2.57	2.62	2.67	2.72	2.77	2.82
32.0	2.29	2.34	2.39	2.44	2.50	2.55	2.60	2.65	2.70	2.76	2.81	2.86	2.91
33.0	2.36	2.41	2.47	2.52	2.57	2.63	2.68	2.73	2.79	2.84	2.89	2.95	3.00
34.0	2.43	2.48	2.54	2.60	2.65	2.71	2.76	2.82	2.87	2.93	2.98	3.04	3.09
35.0	2.50	2.55	2.61	2.67	2.73	2.78	2.84	2.90	2.96	3.01	3.07	3.13	3.18

TABLE 11.
REDUCTION OF THE BAROMETER TO STANDARD TEMPERATURE.
METRIC MEASURES.

FOR TEMPERATURES ABOVE 0° CENTIGRADE, THE CORRECTION IS TO BE SUBTRACTED.

Attached Thermometer.	HEIGHT OF THE BAROMETER 560 mm.					HEIGHT OF THE BAROMETER 570 mm.				
	0°.0	0°.2	0°.4	0°.6	0°.8	0°.0	0°.2	0°.4	0°.6	0°.8
C.	mm.	mm.	mm.	mm.	mm.	mm.	mm.	mm.	mm.	mm.
0°	0.00	0.02	0.04	0.05	0.07	0.00	0.02	0.04	0.06	0.07
1	.09	.11	.13	.15	.16	.09	.11	.13	.15	.17
2	.18	.20	.22	.24	.26	.19	.20	.22	.24	.26
3	.27	.29	.31	.33	.35	.28	.30	.32	.34	.35
4	.37	.38	.40	.42	.44	.37	.39	.41	.43	.45
5	0.46	0.48	0.49	0.51	0.53	0.47	0.48	0.50	0.52	0.54
6	.55	.57	.58	.60	.62	.56	.58	.60	.61	.63
7	.64	.66	.68	.69	.71	.65	.67	.69	.71	.73
8	.73	.75	.77	.79	.80	.74	.76	.78	.80	.82
9	.82	.84	.86	.88	.90	.84	.86	.87	.89	.91
10	0.91	0.93	0.95	0.97	0.99	0.93	0.95	0.97	0.99	1.00
11	1.00	1.02	1.04	1.06	1.08	1.02	1.04	1.06	1.08	1.10
12	1.10	1.11	1.13	1.15	1.17	1.12	1.13	1.15	1.17	1.19
13	1.19	1.20	1.22	1.24	1.26	1.21	1.23	1.25	1.26	1.28
14	1.28	1.30	1.31	1.33	1.35	1.30	1.32	1.34	1.36	1.37
15	1.37	1.39	1.41	1.42	1.44	1.39	1.41	1.43	1.45	1.47
16	1.46	1.48	1.50	1.51	1.53	1.49	1.50	1.52	1.54	1.56
17	1.55	1.57	1.59	1.61	1.62	1.58	1.60	1.62	1.63	1.65
18	1.64	1.66	1.68	1.70	1.71	1.67	1.69	1.71	1.73	1.75
19	1.73	1.75	1.77	1.79	1.81	1.76	1.78	1.80	1.82	1.84
20	1.82	1.84	1.86	1.88	1.90	1.86	1.87	1.89	1.91	1.93
21	1.91	1.93	1.95	1.97	1.99	1.95	1.97	1.99	2.00	2.02
22	2.01	2.02	2.04	2.06	2.08	2.04	2.06	2.08	2.10	2.11
23	2.10	2.11	2.13	2.15	2.17	2.13	2.15	2.17	2.19	2.21
24	2.19	2.20	2.22	2.24	2.26	2.23	2.24	2.26	2.28	2.30
25	2.28	2.30	2.31	2.33	2.35	2.32	2.34	2.35	2.37	2.39
26	2.37	2.39	2.40	2.42	2.44	2.41	2.43	2.45	2 47	2.48
27	2.46	2.48	2.49	2.51	2.53	2.50	2.52	2.54	2.56	2.58
28	2.55	2.57	2.59	2.60	2.62	2.59	2.61	2.63	2.65	2.67
29	2.64	2.66	2.68	2.69	2.71	2.69	2.71	2.72	2.74	2.76
30	2.73	2.75	2.77	2.78	2.80	2.78	2.80	2.82	2.83	2.85
31	2.82	2.84	2.86	2.87	2.89	2.87	2.89	2.91	2.93	2.94
32	2.91	2.93	2.95	2.97	2.98	2.96	2.98	3.00	3.02	3.04
33	3.00	3.02	3.04	3.06	3.07	3.06	3.07	3.09	3.11	3.13
34	3.09	3.11	3.13	3.15	3.16	3.15	3.17	3.18	3.20	3.22
35	3.18	3.20	3.22	3.24	3.25	3.24	3.26	3.28	3.29	3.31

SMITHSONIAN TABLES.

Table 11.
REDUCTION OF THE BAROMETER TO STANDARD TEMPERATURE.
METRIC MEASURES.

FOR TEMPERATURES ABOVE 0° CENTIGRADE, THE CORRECTION IS TO BE SUBTRACTED.

Attached Thermometer.	HEIGHT OF THE BAROMETER 580 mm.					HEIGHT OF THE BAROMETER 590 mm.				
	0°.0	0°.2	0°.4	0°.6	0°.8	0°.0	0°.2	0°.4	0°.6	0°.8
C.	mm.	mm.	mm.	mm.	mm.	mm.	mm.	mm.	mm.	mm.
0°	0.00	0.02	0.04	0.06	0.08	0.00	0.02	0.04	0.06	0.08
1	.09	.11	.13	.15	.17	.10	.12	.13	.15	.17
2	.19	.21	.23	.25	.27	.19	.21	.23	.25	.27
3	.28	.30	.32	.34	.36	.29	.31	.33	.35	.37
4	.38	.40	.42	.44	.45	.39	.40	.42	.44	.46
5	0.47	0.49	0.51	0.53	0.55	0.48	0.50	0.52	0.54	0.56
6	.57	.59	.61	.62	.64	.58	.60	.62	.64	.65
7	.66	.68	.70	.72	.74	.67	.69	.71	.73	.75
8	.76	.78	.79	.81	.83	.77	.79	.81	.83	.85
9	.85	.87	.89	.91	.93	.87	.89	.90	.92	.94
10	0.95	0.96	0.98	1.00	1.02	0.96	0.98	1.00	1.02	1.04
11	1.04	1.06	1.08	1.10	1.12	1.06	1.08	1.10	1.12	1.14
12	1.13	1.15	1.17	1.19	1.21	1.15	1.17	1.19	1.21	1.23
13	1.23	1.25	1.27	1.29	1.30	1.25	1.27	1.29	1.31	1.33
14	1.32	1.34	1.36	1.38	1.40	1.35	1.37	1.38	1.40	1.42
15	1.42	1.44	1.46	1.47	1.49	1.44	1.46	1.48	1.50	1.52
16	1.51	1.53	1.55	1.57	1.59	1.54	1.56	1.58	1.60	1.61
17	1.61	1.62	1.64	1.66	1.68	1.63	1.65	1.67	1.69	1.71
18	1.70	1.72	1.74	1.76	1.78	1.73	1.75	1.77	1.79	1.81
19	1.79	1.81	1.83	1.85	1.87	1.83	1.84	1.86	1.88	1.90
20	1.89	1.91	1.93	1.95	1.96	1.92	1.94	1.96	1.98	2.00
21	1.98	2.00	2.02	2.04	2.06	2.02	2.04	2.06	2.07	2.09
22	2.08	2.10	2.11	2.13	2.15	2.11	2.13	2.15	2.17	2.19
23	2.17	2.19	2.21	2.23	2.25	2.21	2.23	2.25	2.27	2.28
24	2.26	2.28	2.30	2.32	2.34	2.30	2.32	2.34	2.36	2.38
25	2.36	2.38	2.40	2.41	2.43	2.40	2.42	2.44	2.46	2.48
26	2.45	2.47	2.49	2.51	2.53	2.49	2.51	2.53	2.55	2.57
27	2.55	2.57	2.58	2.60	2.62	2.59	2.61	2.63	2.65	2.67
28	2.64	2.66	2.68	2.70	2.72	2.69	2.70	2.72	2.74	2.76
29	2.73	2.75	2.77	2.79	2.81	2.78	2.80	2.82	2.84	2.86
30	2.83	2.85	2.87	2.88	2.90	2.88	2.90	2.91	2.93	2.95
31	2.92	2.94	2.96	2.98	3.00	2.97	2.99	3.01	3.03	3.05
32	3.02	3.03	3.05	3.07	3.09	3.07	3.09	3.11	3.12	3.14
33	3.11	3.13	3.15	3.16	3.18	3.16	3.18	3.20	3.22	3.24
34	3.20	3.22	3.24	3.26	3.28	3.26	3.28	3.30	3.31	3.33
35	3.30	3.31	3.33	3.35	3.37	3.35	3.37	3.39	3.41	3.43

SMITHSONIAN TABLES.

TABLE 11.
REDUCTION OF THE BAROMETER TO STANDARD TEMPERATURE.
METRIC MEASURES.

FOR TEMPERATURES ABOVE 0° CENTIGRADE, THE CORRECTION IS TO BE SUBTRACTED.

Attached Thermometer.	HEIGHT OF THE BAROMETER 600 mm.					HEIGHT OF THE BAROMETER 605 mm.				
	0°.0	0°.2	0°.4	0°.6	0°.8	0°.0	0°.2	0°.4	0°.6	0°.8
C.	mm.	mm.	mm.	mm.	mm.	mm.	mm.	mm.	mm.	mm.
0°	0.00	0.02	0.04	0.06	0.08	0.00	0.02	0.04	0.06	0.08
1	.10	.12	.14	.16	.18	.10	.12	.14	.16	.18
2	.20	.22	.24	.25	.27	.20	.22	.24	.26	.28
3	.29	.31	.33	.35	.37	.30	.32	.34	.36	.38
4	.39	.41	.43	.45	.47	.40	.41	.43	.45	.47
5	0.49	0.51	0.53	0.55	0.57	0.49	0.51	0.53	0.55	0.57
6	.59	.61	.63	.65	.67	.59	.61	.63	.65	.67
7	.69	.70	.72	.74	.76	.69	.71	.73	.75	.77
8	.78	.80	.82	.84	.86	.79	.81	.83	.85	.87
9	.88	.90	.92	.94	.96	.89	.91	.93	.95	.97
10	0.98	1.00	1.02	1.04	1.06	0.99	1.01	1.03	1.05	1.07
11	1.08	1.10	1.12	1.13	1.15	1.09	1.10	1.12	1.14	1.16
12	1.17	1.19	1.21	1.23	1.25	1.18	1.20	1.22	1.24	1.26
13	1.27	1.29	1.31	1.33	1.35	1.28	1.30	1.32	1.34	1.36
14	1.37	1.39	1.41	1.43	1.45	1.38	1.40	1.42	1.44	1.46
15	1.47	1.49	1.51	1.53	1.54	1.48	1.50	1.52	1.54	1.56
16	1.56	1.58	1.60	1.62	1.64	1.58	1.60	1.62	1.64	1.66
17	1.66	1.68	1.70	1.72	1.74	1.68	1.70	1.71	1.73	1.75
18	1.76	1.78	1.80	1.82	1.84	1.77	1.79	1.81	1.83	1.85
19	1.86	1.88	1.90	1.91	1.93	1.87	1.89	1.91	1.93	1.95
20	1.95	1.97	1.99	2.01	2.03	1.97	1.99	2.01	2.03	2.05
21	2.05	2.07	2.09	2.11	2.13	2.07	2.09	2.11	2.13	2.15
22	2.15	2.17	2.19	2.21	2.23	2.17	2.19	2.21	2.23	2.24
23	2.25	2.26	2.28	2.30	2.32	2.26	2.28	2.30	2.32	2.34
24	2.34	2.36	2.38	2.40	2.42	2.36	2.38	2.40	2.42	2.44
25	2.44	2.46	2.48	2.50	2.52	2.46	2.48	2.50	2.52	2.54
26	2.54	2.56	2.58	2.60	2.61	2.56	2.58	2.60	2.62	2.64
27	2.63	2.65	2.67	2.69	2.71	2.66	2.68	2.70	2.71	2.73
28	2.73	2.75	2.77	2.79	2.81	2.75	2.77	2.79	2.81	2.83
29	2.83	2.85	2.87	2.89	2.91	2.85	2.87	2.89	2.91	2.93
30	2.93	2.94	2.96	2.98	3.00	2.95	2.97	2.99	3.01	3.03
31	3.02	3.04	3.06	3.08	3.10	3.05	3.07	3.09	3.11	3.13
32	3.12	3.14	3.16	3.18	3.20	3.15	3.16	3.18	3.20	3.22
33	3.22	3.24	3.25	3.27	3.29	3.24	3.26	3.28	3.30	3.32
34	3.31	3.33	3.35	3.37	3.39	3.34	3.36	3.38	3.40	3.42
35	3.41	3.43	3.45	3.47	3.49	3.44	3.46	3.48	3.50	3.52

SMITHSONIAN TABLES.

TABLE 11.
REDUCTION OF THE BAROMETER TO STANDARD TEMPERATURE.
METRIC MEASURES.

FOR TEMPERATURES ABOVE 0° CENTIGRADE, THE CORRECTION IS TO BE SUBTRACTED.

Attached Thermometer.	HEIGHT OF THE BAROMETER 610 mm.					HEIGHT OF THE BAROMETER 615 mm.				
	0°.0	0°.2	0°.4	0°.6	0°.8	0°.0	0°.2	0°.4	0°.6	0°.8
C.	mm.	mm.	mm.	mm.	mm.	mm.	mm.	mm.	mm.	mm.
0°	0.00	0.02	0.04	0.06	0.08	0.00	0.02	0.04	0.06	0.08
1	.10	.12	.14	.16	.18	.10	.12	.14	.16	.18
2	.20	.22	.24	.26	.28	.20	.22	.24	.26	.28
3	.30	.32	.34	.36	.38	.30	.32	.34	.36	.38
4	.40	.42	.44	.46	.48	.40	.42	.44	.46	.48
5	0.50	0.52	0.54	0.56	0.58	0.50	0.52	0.54	0.56	0.58
6	.60	.62	.64	.66	.68	.60	.62	.64	.66	.68
7	.70	.72	.74	.76	.78	.70	.72	.74	.76	.78
8	.80	.82	.84	.86	.88	.80	.82	.84	.86	.88
9	.90	.92	.94	.96	.98	.90	.92	.94	.96	.98
10	0.99	1.01	1.03	1.05	1.07	1.00	1.02	1.04	1.06	1.08
11	1.09	1.11	1.13	1.15	1.17	1.10	1.12	1.14	1.16	1.18
12	1.19	1.21	1.23	1.25	1.27	1.20	1.22	1.24	1.26	1.28
13	1.29	1.31	1.33	1.35	1.37	1.30	1.32	1.34	1.36	1.38
14	1.39	1.41	1.43	1.45	1.47	1.40	1.42	1.44	1.46	1.48
15	1.49	1.51	1.53	1.55	1.57	1.50	1.52	1.54	1.56	1.58
16	1.59	1.61	1.63	1.65	1.67	1.60	1.62	1.64	1.66	1.68
17	1.69	1.71	1.73	1.75	1.77	1.70	1.72	1.74	1.76	1.78
18	1.79	1.81	1.83	1.85	1.87	1.80	1.82	1.84	1.86	1.88
19	1.89	1.91	1.93	1.95	1.97	1.90	1.92	1.94	1.96	1.98
20	1.99	2.01	2.03	2.05	2.07	2.00	2.02	2.04	2.06	2.08
21	2.09	2.10	2.12	2.14	2.16	2.10	2.12	2.14	2.16	2.18
22	2.18	2.20	2.22	2.24	2.26	2.20	2.22	2.24	2.26	2.28
23	2.28	2.30	2.32	2.34	2.36	2.30	2.32	2.34	2.36	2.38
24	2.38	2.40	2.42	2.44	2.46	2.40	2.42	2.44	2.46	2.48
25	2.48	2.50	2.52	2.54	2.56	2.50	2.52	2.54	2.56	2.58
26	2.58	2.60	2.62	2.64	2.66	2.60	2.62	2.64	2.66	2.68
27	2.68	2.70	2.72	2.74	2.76	2.70	2.72	2.74	2.76	2.78
28	2.78	2.80	2.82	2.84	2.86	2.80	2.82	2.84	2.86	2.88
29	2.88	2.90	2.91	2.93	2.95	2.90	2.92	2.94	2.96	2.98
30	2.97	2.99	3.01	3.03	3.05	3.00	3.02	3.04	3.06	3.08
31	3.07	3.09	3.11	3.13	3.15	3.10	3.12	3.14	3.16	3.18
32	3.17	3.19	3.21	3.23	3.25	3.20	3.22	3.24	3.26	3.28
33	3.27	3.29	3.31	3.33	3.35	3.30	3.32	3.34	3.36	3.38
34	3.37	3.39	3.41	3.43	3.45	3.40	3.42	3.44	3.46	3.48
35	3.47	3.49	3.51	3.53	3.55	3.49	3.51	3.53	3.55	3.57

SMITHSONIAN TABLES.

TABLE 11.

REDUCTION OF THE BAROMETER TO STANDARD TEMPERATURE.
METRIC MEASURES.

FOR TEMPERATURES ABOVE 0° CENTIGRADE, THE CORRECTION IS TO BE SUBTRACTED.

Attached Thermometer.	HEIGHT OF THE BAROMETER 620 mm.					HEIGHT OF THE BAROMETER 625 mm.				
	0°.0	0°.2	0°.4	0°.6	0°.8	0°.0	0°.2	0°.4	0°.6	0°.8
C.	mm.	mm.	mm.	mm.	mm.	mm.	mm.	mm.	mm.	mm.
0°	0.00	0.02	0.04	0.06	0.08	0.00	0.02	0.04	0.06	0.08
1	.10	.12	.14	.16	.18	.10	.12	.14	.16	.18
2	.20	.22	.24	.26	.28	.20	.22	.24	.27	.29
3	.30	.32	.34	.36	.38	.31	.33	.35	.37	.39
4	.40	.43	.45	.47	.49	.41	.43	.45	.47	.49
5	0.51	0.53	0.55	0.57	0.59	0.51	0.53	0.55	0.57	0.59
6	.61	.63	.65	.67	.69	.61	.63	.65	.67	.69
7	.71	.73	.75	.77	.79	.71	.73	.75	.78	.80
8	.81	.83	.85	.87	.89	.82	.84	.86	.88	.90
9	.91	.93	.95	.97	.99	.92	.94	.96	.98	1.00
10	1.01	1.03	1.05	1.07	1.09	1.02	1.04	1.06	1.08	1.10
11	1.11	1.13	1.15	1.17	1.19	1.12	1.14	1.16	1.18	1.20
12	1.21	1.23	1.25	1.27	1.29	1.22	1.24	1.26	1.28	1.30
13	1.31	1.33	1.35	1.37	1.39	1.32	1.34	1.37	1.39	1.41
14	1.41	1.43	1.46	1.48	1.50	1.43	1.45	1.47	1.49	1.51
15	1.52	1.54	1.56	1.58	1.60	1.53	1.55	1.57	1.59	1.61
16	1.62	1.64	1.66	1.68	1.70	1.63	1.65	1.67	1.69	1.71
17	1.72	1.74	1.76	1.78	1.80	1.73	1.75	1.77	1.79	1.81
18	1.82	1.84	1.86	1.88	1.90	1.83	1.85	1.87	1.89	1.91
19	1.92	1.94	1.96	1.98	2.00	1.93	1.95	1.97	1.99	2.01
20	2.02	2.04	2.06	2.08	2.10	2.04	2.06	2.08	2.10	2.12
21	2.12	2.14	2.16	2.18	2.20	2.14	2.16	2.18	2.20	2.22
22	2.22	2.24	2.26	2.28	2.30	2.24	2.26	2.28	2.30	2.32
23	2.32	2.34	2.36	2.38	2.40	2.34	2.36	2.38	2.40	2.42
24	2.42	2.44	2.46	2.48	2.50	2.44	2.46	2.48	2.50	2.52
25	2.52	2.54	2.56	2.58	2.60	2.54	2.56	2.58	2.60	2.62
26	2.62	2.64	2.66	2.68	2.70	2.64	2.66	2.68	2.70	2.72
27	2.72	2.74	2.76	2.78	2.80	2.74	2.76	2.78	2.80	2.82
28	2.82	2.84	2.86	2.88	2.90	2.85	2.87	2.89	2.91	2.93
29	2.92	2.94	2.96	2.98	3.00	2.95	2.97	2.99	3.01	3.03
30	3.02	3.04	3.06	3.08	3.10	3.05	3.07	3.09	3.11	3.13
31	3.12	3.14	3.16	3.18	3.20	3.15	3.17	3.19	3.21	3.23
32	3.22	3.24	3.26	3.28	3.30	3.25	3.27	3.29	3.31	3.33
33	3.32	3.34	3.36	3.38	3.40	3.35	3.37	3.39	3.41	3.43
34	3.42	3.44	3.46	3.48	3.50	3.45	3.47	3.49	3.51	3.53
35	3.52	3.54	3.56	3.58	3.60	3.55	3.57	3.59	3.61	3.63

SMITHSONIAN TABLES.

TABLE 11.
REDUCTION OF THE BAROMETER TO STANDARD TEMPERATURE.
METRIC MEASURES.

FOR TEMPERATURES ABOVE 0° CENTIGRADE, THE CORRECTION IS TO BE SUBTRACTED.

Attached Thermometer.	HEIGHT OF THE BAROMETER 630 mm.					HEIGHT OF THE BAROMETER 635 mm.				
	0°.0	0°.2	0°.4	0°.6	0°.8	0°.0	0°.2	0°.4	0°.6	0°.8
C.	mm.	mm.	mm.	mm.	mm.	mm.	mm.	mm.	mm.	mm.
0°	0.00	0.02	0.04	0.06	0.08	0.00	0.02	0.04	0.06	0.08
1	.10	.12	.14	.16	.19	.10	.12	.15	.17	.19
2	.21	.23	.25	.27	.29	.21	.23	.25	.27	.29
3	.31	.33	.35	.37	.39	.31	.33	.35	.37	.39
4	.41	.43	.45	.47	.49	.41	.44	.46	.48	.50
5	0.51	0.53	0.56	0.58	0.60	0.52	0.54	0.56	0.58	0.60
6	.62	.64	.66	.68	.70	.62	.64	.66	.68	.70
7	.72	.74	.76	.78	.80	.73	.75	.77	.79	.81
8	.82	.84	.86	.88	.90	.83	.85	.87	.89	.91
9	.92	.95	.97	.99	1.01	.93	.95	.97	.99	1.02
10	1.03	1.05	1.07	1.09	1.11	1.04	1.06	1.08	1.10	1.12
11	1.13	1.15	1.17	1.19	1.21	1.14	1.16	1.18	1.20	1.22
12	1.23	1.25	1.27	1.29	1.31	1.24	1.26	1.28	1.30	1.33
13	1.34	1.36	1.38	1.40	1.42	1.35	1.37	1.39	1.41	1.43
14	1.44	1.46	1.48	1.50	1.52	1.45	1.47	1.49	1.51	1.53
15	1.54	1.56	1.58	1.60	1.62	1.55	1.57	1.59	1.61	1.63
16	1.64	1.66	1.68	1.70	1.72	1.66	1.68	1.70	1.72	1.74
17	1.74	1.77	1.79	1.81	1.83	1.76	1.78	1.80	1.82	1.84
18	1.85	1.87	1.89	1.91	1.93	1.86	1.88	1.90	1.92	1.94
19	1.95	1.97	1.99	2.01	2.03	1.96	1.99	2.01	2.03	2.05
20	2.05	2.07	2.09	2.11	2.13	2.07	2.09	2.11	2.13	2.15
21	2.15	2.17	2.19	2.21	2.24	2.17	2.19	2.21	2.23	2.25
22	2.26	2.28	2.30	2.32	2.34	2.27	2.29	2.31	2.34	2.36
23	2.36	2.38	2.40	2.42	2.44	2.38	2.40	2.42	2.44	2.46
24	2.46	2.48	2.50	2.52	2.54	2.48	2.50	2.52	2.54	2.56
25	2.56	2.58	2.60	2.62	2.64	2.58	2.60	2.62	2.64	2.66
26	2.66	2.68	2.70	2.73	2.75	2.69	2.71	2.73	2.75	2.77
27	2.77	2.79	2.81	2.83	2.85	2.79	2.81	2.83	2.85	2.87
28	2.87	2.89	2.91	2.93	2.95	2.89	2.91	2.93	2.95	2.97
29	2.97	2.99	3.01	3.03	3.05	2.99	3.01	3.03	3.05	3.08
30	3.07	3.09	3.11	3.13	3.15	3.10	3.12	3.14	3.16	3.18
31	3.17	3.19	3.21	3.23	3.25	3.20	3.22	3.24	3.26	3.28
32	3.28	3.30	3.32	3.34	3.36	3.30	3.32	3.34	3.36	3.38
33	3.38	3.40	3.42	3.44	3.46	3.40	3.42	3.44	3.47	3.49
34	3.48	3.50	3.52	3.54	3.56	3.51	3.53	3.55	3.57	3.59
35	3.58	3.60	3.62	3.64	3.66	3.61	3.63	3.65	3.67	3.69

Smithsonian Tables.

TABLE 11.
REDUCTION OF THE BAROMETER TO STANDARD TEMPERATURE.
METRIC MEASURES.

FOR TEMPERATURES ABOVE 0° CENTIGRADE, THE CORRECTION IS TO BE SUBTRACTED.

Attached Thermometer.	HEIGHT OF THE BAROMETER 640 mm.					HEIGHT OF THE BAROMETER 645 mm.				
	0°.0	0°.2	0°.4	0°.6	0°.8	0°.0	0°.2	0°.4	0°.6	0°.8
C.	mm.	mm.	mm.	mm.	mm.	mm.	mm.	mm.	mm.	mm.
0°	0.00	0.02	0.04	0.06	0.08	0.00	0.02	0.04	0.06	0.08
1	.10	.13	.15	.17	.19	.11	.13	.15	.17	.19
2	.21	.23	.25	.27	.29	.21	.23	.25	.27	.29
3	.31	.33	.36	.38	.40	.32	.34	.36	.38	.40
4	.42	.44	.46	.48	.50	.42	.44	.46	.48	.51
5	0.52	0.54	0.56	0.59	0.61	0.53	0.55	0.57	0.59	0.61
6	.63	.65	.67	.69	.71	.63	.65	.67	.69	.72
7	.73	.75	.77	.79	.81	.74	.76	.78	.80	.82
8	.84	.86	.88	.90	.92	.84	.86	.88	.90	.93
9	.94	.96	.98	1.00	1.02	.95	.97	.99	1.01	1.03
10	1.04	1.06	1.09	1.11	1.13	1.05	1.07	1.09	1.12	1.14
11	1.15	1.17	1.19	1.21	1.23	1.16	1.18	1.20	1.22	1.24
12	1.25	1.27	1.29	1.31	1.34	1.26	1.28	1.30	1.32	1.35
13	1.36	1.38	1.40	1.42	1.44	1.37	1.39	1.41	1.43	1.45
14	1.46	1.48	1.50	1.52	1.54	1.47	1.49	1.51	1.53	1.56
15	1.56	1.59	1.61	1.63	1.65	1.58	1.60	1.62	1.64	1.66
16	1.67	1.69	1.71	1.73	1.75	1.68	1.70	1.72	1.74	1.77
17	1.77	1.79	1.81	1.83	1.86	1.79	1.81	1.83	1.85	1.87
18	1.88	1.90	1.92	1.94	1.96	1.89	1.91	1.93	1.95	1.97
19	1.98	2.00	2.02	2.04	2.06	2.00	2.02	2.04	2.06	2.08
20	2.08	2.10	2.13	2.15	2.17	2.10	2.12	2.14	2.16	2.18
21	2.19	2.21	2.23	2.25	2.27	2.20	2.23	2.25	2.27	2.29
22	2.29	2.31	2.33	2.35	2.37	2.31	2.33	2.35	2.37	2.39
23	2.40	2.42	2.44	2.46	2.48	2.41	2.43	2.46	2.48	2.50
24	2.50	2.52	2.54	2.56	2.58	2.52	2.54	2.56	2.58	2.60
25	2.60	2.62	2.64	2.66	2.69	2.62	2.64	2.66	2.69	2.71
26	2.71	2.73	2.75	2.77	2.79	2.73	2.75	2.77	2.79	2.81
27	2.81	2.83	2.85	2.87	2.89	2.83	2.85	2.87	2.89	2.92
28	2.91	2.93	2.95	2.98	3.00	2.94	2.96	2.98	3.00	3.02
29	3.02	3.04	3.06	3.08	3.10	3.04	3.06	3.08	3.10	3.12
30	3.12	3.14	3.16	3.18	3.20	3.14	3.17	3.19	3.21	3.23
31	3.22	3.24	3.27	3.29	3.31	3.25	3.27	3.29	3.31	3.33
32	3.33	3.35	3.37	3.39	3.41	3.35	3.37	3.39	3.42	3.44
33	3.43	3.45	3.47	3.49	3.51	3.46	3.48	3.50	3.52	3.54
34	3.53	3.55	3.58	3.60	3.62	3.56	3.58	3.60	3.62	3.64
35	3.64	3.66	3.68	3.70	3.72	3.67	3.69	3.71	3.73	3.75

TABLE 11.

REDUCTION OF THE BAROMETER TO STANDARD TEMPERATURE.
METRIC MEASURES.

FOR TEMPERATURES ABOVE 0° CENTIGRADE, THE CORRECTION IS TO BE SUBTRACTED.

Attached Thermometer.	HEIGHT OF THE BAROMETER 650 mm.					HEIGHT OF THE BAROMETER 655 mm.				
	0°.0	0°.2	0°.4	0°.6	0°.8	0°.0	0°.2	0°.4	0°.6	0°.8
C.	mm.	mm.	mm.	mm.	mm.	mm.	mm.	mm.	mm.	mm.
0°	0.00	0.02	0.04	0.06	0.08	0.00	0.02	0.04	0.06	0.09
1	.11	.13	.15	.17	.19	.11	.13	.15	.17	.19
2	.21	.23	.25	.28	.30	.21	.24	.26	.28	.30
3	.32	.34	.36	.38	.40	.32	.34	.36	.39	.41
4	.42	.45	.47	.49	.51	.43	.45	.47	.49	.51
5	0.53	0.55	0.57	0.59	0.62	0.53	0.56	0.58	0.60	0.62
6	.64	.66	.68	.70	.72	.64	.66	.68	.71	.73
7	.74	.76	.78	.81	.83	.75	.77	.79	.81	.83
8	.85	.87	.89	.91	.93	.85	.88	.90	.92	.94
9	.95	.98	1.00	1.02	1.04	.96	.98	1.00	1.03	1.05
10	1.06	1.08	1.10	1.12	1.14	1.07	1.09	1.11	1.13	1.15
11	1.17	1.19	1.21	1.23	1.25	1.17	1.20	1.22	1.24	1.26
12	1.27	1.29	1.31	1.34	1.36	1.28	1.30	1.32	1.35	1.37
13	1.38	1.40	1.42	1.44	1.46	1.39	1.41	1.43	1.45	1.47
14	1.48	1.50	1.53	1.55	1.57	1.49	1.52	1.54	1.56	1.58
15	1.59	1.61	1.63	1.65	1.67	1.60	1.62	1.64	1.66	1.69
16	1.69	1.72	1.74	1.76	1.78	1.71	1.73	1.75	1.77	1.79
17	1.80	1.82	1.84	1.86	1.88	1.81	1.84	1.86	1.88	1.90
18	1.91	1.93	1.95	1.97	1.99	1.92	1.94	1.96	1.98	2.01
19	2.01	2.03	2.05	2.07	2.10	2.03	2.05	2.07	2.09	2.11
20	2.12	2.14	2.16	2.18	2.20	2.13	2.15	2.18	2.20	2.22
21	2.22	2.24	2.26	2.29	2.31	2.24	2.26	2.28	2.30	2.32
22	2.33	2.35	2.37	2.39	2.41	2.35	2.37	2.39	2.41	2.43
23	2.43	2.45	2.47	2.50	2.52	2.45	2.47	2.49	2.52	2.54
24	2.54	2.56	2.58	2.60	2.62	2.56	2.58	2.60	2.62	2.64
25	2.64	2.66	2.69	2.71	2.73	2.66	2.68	2.71	2.73	2.75
26	2.75	2.77	2.79	2.81	2.83	2.77	2.79	2.81	2.83	2.85
27	2.85	2.87	2.90	2.92	2.94	2.88	2.90	2.92	2.94	2.96
28	2.96	2.98	3.00	3.02	3.04	2.98	3.00	3.02	3.05	3.07
29	3.06	3.08	3.11	3.13	3.15	3.09	3.11	3.13	3.15	3.17
30	3.17	3.19	3.21	3.23	3.25	3.19	3.21	3.24	3.26	3.28
31	3.27	3.30	3.32	3.34	3.36	3.30	3.32	3.34	3.36	3.38
32	3.38	3.40	3.42	3.44	3.46	3.41	3.43	3.45	3.47	3.49
33	3.48	3.51	3.53	3.55	3.57	3.51	3.53	3.55	3.57	3.60
34	3.59	3.61	3.63	3.65	3.67	3.62	3.64	3.66	3.68	3.70
35	3.69	3.71	3.74	3.76	3.78	3.72	3.74	3.76	3.79	3.81

SMITHSONIAN TABLES.

TABLE 11.

REDUCTION OF THE BAROMETER TO STANDARD TEMPERATURE.
METRIC MEASURES.

FOR TEMPERATURES ABOVE 0° CENTIGRADE, THE CORRECTION IS TO BE SUBTRACTED.

Attached Thermometer.	HEIGHT OF THE BAROMETER 660 mm.					HEIGHT OF THE BAROMETER 665 mm.				
	0°.0	0°.2	0°.4	0°.6	0°.8	0°.0	0°.2	0°.4	0°.6	0°.8
C.	mm.	mm.	mm.	mm.	mm.	mm.	mm.	mm.	mm.	mm.
0°	0.00	0.02	0.04	0.06	0.09	0.00	0.02	0.04	0.07	0.09
1	.11	.13	.15	.17	.19	.11	.13	.15	.17	.20
2	.22	.24	.26	.28	.30	.22	.24	.26	.28	.30
3	.32	.34	.37	.39	.41	.33	.35	.37	.39	.41
4	.43	.45	.47	.50	.52	.43	.46	.48	.50	.52
5	0.54	0.56	0.58	0.60	0.62	0.54	0.56	0.59	0.61	0.63
6	.65	.67	.69	.71	.73	.65	.67	.69	.72	.74
7	.75	.78	.80	.82	.84	.76	.78	.80	.82	.85
8	.86	.88	.90	.93	.95	.87	.89	.91	.93	.95
9	.97	.99	1.01	1.03	1.05	.98	1.00	1.02	1.04	1.06
10	1.08	1.10	1.12	1.14	1.16	1.08	1.11	1.13	1.15	1.17
11	1.18	1.21	1.23	1.25	1.27	1.19	1.21	1.24	1.26	1.28
12	1.29	1.31	1.33	1.36	1.38	1.30	1.32	1.34	1.37	1.39
13	1.40	1.42	1.44	1.46	1.48	1.41	1.43	1.45	1.47	1.50
14	1.51	1.53	1.55	1.57	1.59	1.52	1.54	1.56	1.58	1.60
15	1.61	1.63	1.66	1.68	1.70	1.63	1.65	1.67	1.69	1.71
16	1.72	1.74	1.76	1.78	1.81	1.73	1.76	1.78	1.80	1.82
17	1.83	1.85	1.87	1.89	1.91	1.84	1.86	1.88	1.91	1.93
18	1.93	1.96	1.98	2.00	2.02	1.95	1.97	1.99	2.01	2.04
19	2.04	2.06	2.08	2.11	2.13	2.06	2.08	2.10	2.12	2.14
20	2.15	2.17	2.19	2.21	2.23	2.17	2.19	2.21	2.23	2.25
21	2.26	2.28	2.30	2.32	2.34	2.27	2.29	2.32	2.34	2.36
22	2.36	2.38	2.41	2.43	2.45	2.38	2.40	2.42	2.45	2.47
23	2.47	2.49	2.51	2.53	2.56	2.49	2.51	2.53	2.55	2.57
24	2.58	2.60	2.62	2.64	2.66	2.60	2.62	2.64	2.66	2.68
25	2.68	2.71	2.73	2.75	2.77	2.70	2.73	2.75	2.77	2.79
26	2.79	2.81	2.83	2.85	2.88	2.81	2.83	2.85	2.88	2.90
27	2.90	2.92	2.94	2.96	2.98	2.92	2.94	2.96	2.98	3.01
28	3.00	3.03	3.05	3.07	3.09	3.03	3.05	3.07	3.09	3.11
29	3.11	3.13	3.15	3.18	3.20	3.13	3.16	3.18	3.20	3.22
30	3.22	3.24	3.26	3.28	3.30	3.24	3.26	3.29	3.31	3.33
31	3.32	3.35	3.37	3.39	3.41	3.35	3.37	3.39	3.41	3.44
32	3.43	3.45	3.47	3.49	3.52	3.46	3.48	3.50	3.52	3.54
33	3.54	3.56	3.58	3.60	3.62	3.56	3.59	3.61	3.63	3.65
34	3.64	3.67	3.69	3.71	3.73	3.67	3.69	3.71	3.74	3.76
35	3.75	3.77	3.79	3.81	3.84	3.78	3.80	3.82	3.84	3.86

SMITHSONIAN TABLES.

TABLE 11.
REDUCTION OF THE BAROMETER TO STANDARD TEMPERATURE.
METRIC MEASURES.

FOR TEMPERATURES ABOVE 0° CENTIGRADE, THE CORRECTION IS TO BE SUBTRACTED.

Attached Thermometer.	HEIGHT OF THE BAROMETER 670 mm.					HEIGHT OF THE BAROMETER 675 mm.				
	0°.0	0°.2	0°.4	0°.6	0°.8	0°.0	0°.2	0°.4	0°.6	0°.8
C.	mm.	mm.	mm.	mm.	mm.	mm.	mm.	mm.	mm.	mm.
0°	0.00	0.02	0.04	0.07	0.09	0.00	0.02	0.04	0.07	0.09
1	.11	.13	.15	.18	.20	.11	.13	.15	.18	.20
2	.22	.24	.26	.28	.31	.22	.24	.26	.29	.31
3	.33	.35	.37	.39	.42	.33	.35	.37	.40	.42
4	.44	.46	.48	.50	.53	.44	.46	.48	.51	.53
5	0.55	0.57	0.59	0.61	0.63	0.55	0.57	0.60	0.62	0.64
6	.66	.68	.70	.72	.74	.66	.68	.71	.73	.75
7	.77	.79	.81	.83	.85	.77	.79	.82	.84	.86
8	.87	.90	.92	.94	.96	.88	.90	.93	.95	.97
9	.98	1.01	1.03	1.05	1.07	.99	1.01	1.04	1.06	1.08
10	1.09	1.11	1.14	1.16	1.18	1.10	1.12	1.14	1.17	1.19
11	1.20	1.22	1.25	1.27	1.29	1.21	1.23	1.25	1.28	1.30
12	1.31	1.33	1.35	1.38	1.40	1.32	1.34	1.36	1.39	1.41
13	1.42	1.44	1.46	1.49	1.51	1.43	1.45	1.47	1.50	1.52
14	1.53	1.55	1.57	1.59	1.62	1.54	1.56	1.58	1.61	1.63
15	1.64	1.66	1.68	1.70	1.72	1.65	1.67	1.69	1.72	1.74
16	1.75	1.77	1.79	1.81	1.83	1.76	1.78	1.80	1.83	1.85
17	1.86	1.88	1.90	1.92	1.94	1.87	1.89	1.91	1.94	1.96
18	1.96	1.99	2.01	2.03	2.05	1.98	2.00	2.02	2.04	2.07
19	2.07	2.09	2.12	2.14	2.16	2.09	2.11	2.13	2.15	2.18
20	2.18	2.20	2.23	2.25	2.27	2.20	2.22	2.24	2.26	2.29
21	2.29	2.31	2.33	2.36	2.38	2.31	2.33	2.35	2.37	2.39
22	2.40	2.42	2.44	2.46	2.49	2.42	2.44	2.46	2.48	2.50
23	2.51	2.53	2.55	2.57	2.59	2.53	2.55	2.57	2.59	2.61
24	2.62	2.64	2.66	2.68	2.70	2.64	2.66	2.68	2.70	2.72
25	2.72	2.75	2.77	2.79	2.81	2.74	2.77	2.79	2.81	2.83
26	2.83	2.85	2.88	2.90	2.92	2.85	2.88	2.90	2.92	2.94
27	2.94	2.96	2.98	3.01	3.03	2.96	2.99	3.01	3.03	3.05
28	3.05	3.07	3.09	3.11	3.14	3.07	3.09	3.12	3.14	3.16
29	3.16	3.18	3.20	3.22	3.24	3.18	3.20	3.23	3.25	3.27
30	3.27	3.29	3.31	3.33	3.35	3.29	3.31	3.33	3.36	3.38
31	3.37	3.40	3.42	3.44	3.46	3.40	3.42	3.44	3.47	3.49
33	3.48	3.50	3.53	3.55	3.57	3.51	3.53	3.55	3.57	3.60
33	3.59	3.61	3.63	3.66	3.68	3.62	3.64	3.66	3.68	3.71
34	3.70	3.72	3.74	3.76	3.79	3.73	3.75	3.77	3.79	3.81
35	3.81	3.83	3.85	3.87	3.89	3.84	3.86	3.88	3.90	3.92

SMITHSONIAN TABLES.

TABLE 11.

REDUCTION OF THE BAROMETER TO STANDARD TEMPERATURE.
METRIC MEASURES.

FOR TEMPERATURES ABOVE 0° CENTIGRADE, THE CORRECTION IS TO BE SUBTRACTED.

Attached Thermometer.	HEIGHT OF THE BAROMETER 680 mm.					HEIGHT OF THE BAROMETER 685 mm.				
	0°.0	0°.2	0°.4	0°.6	0°.8	0°.0	0°.2	0°.4	0°.6	0°.8
C.	mm.	mm.	mm.	mm.	mm.	mm.	mm.	mm.	mm.	mm.
0°	0.00	0.02	0.04	0.07	0.09	0.00	0.02	0.04	0.07	0.09
1	.11	.13	.16	.18	.20	.11	.13	.16	.18	.20
2	.22	.24	.27	.29	.31	.22	.25	.27	.29	.31
3	.33	.36	.38	.40	.42	.34	.36	.38	.40	.43
4	.44	.47	.49	.51	.53	.45	.47	.49	.51	.54
5	0.56	0.58	0.60	0.62	0.64	0.56	0.58	0.60	0.63	0.65
6	.67	.69	.71	.73	.75	.67	.69	.72	.74	.76
7	.78	.80	.82	.84	.87	.78	.80	.83	.85	.87
8	.89	.91	.93	.95	.98	.89	.92	.94	.96	.98
9	1.00	1.02	1.04	1.06	1.09	1.01	1.03	1.05	1.07	1.09
10	1.11	1.13	1.15	1.18	1.20	1.12	1.14	1.16	1.18	1.21
11	1.22	1.24	1.26	1.29	1.31	1.23	1.25	1.27	1.30	1.32
12	1.33	1.35	1.37	1.40	1.42	1.34	1.36	1.38	1.41	1.43
13	1.44	1.46	1.49	1.51	1.53	1.45	1.47	1.50	1.52	1.54
14	1.55	1.57	1.60	1.62	1.64	1.56	1.59	1.61	1.63	1.65
15	1.66	1.68	1.71	1.73	1.75	1.67	1.70	1.72	1.74	1.76
16	1.77	1.79	1.82	1.84	1.86	1.79	1.81	1.83	1.85	1.87
17	1.88	1.91	1.93	1.95	1.97	1.90	1.92	1.94	1.96	1.99
18	1.99	2.02	2.04	2.06	2.08	2.01	2.03	2.05	2.07	2.10
19	2.10	2.13	2.15	2.17	2.19	2.12	2.14	2.16	2.19	2.21
20	2.21	2.24	2.26	2.28	2.30	2.23	2.25	2.27	2.30	2.32
21	2.32	2.35	2.37	2.39	2.41	2.34	2.36	2.39	2.41	2.43
22	2.43	2.46	2.48	2.50	2.52	2.45	2.47	2.50	2.52	2.54
23	2.54	2.57	2.59	2.61	2.63	2.56	2.59	2.61	2.63	2.65
24	2.66	2.68	2.70	2.72	2.74	2.67	2.70	2.72	2.74	2.76
25	2.77	2.79	2.81	2.83	2.85	2.79	2.81	2.83	2.85	2.87
26	2.88	2.90	2.92	2.94	2.96	2.90	2.92	2.94	2.96	2.99
27	2.99	3.01	3.03	3.05	3.07	3.01	3.03	3.05	3.07	3.10
28	3.10	3.12	3.14	3.16	3.18	3.12	3.14	3.16	3.18	3.21
29	3.21	3.23	3.25	3.27	3.29	3.23	3.25	3.27	3.30	3.32
30	3.32	3.34	3.36	3.38	3.40	3.34	3.36	3.38	3.41	3.43
31	3.43	3.45	3.47	3.49	3.51	3.45	3.47	3.49	3.52	3.54
32	3.54	3.56	3.58	3.60	3.62	3.56	3.58	3.61	3.63	3.65
33	3.64	3.67	3.69	3.71	3.73	3.67	3.69	3.72	3.74	3.76
34	3.75	3.78	3.80	3.82	3.84	3.78	3.80	3.83	3.85	3.87
35	3.86	3.89	3.91	3.93	3.95	3.89	3.91	3.94	3.96	3.98

SMITHSONIAN TABLES.

TABLE 11.

REDUCTION OF THE BAROMETER TO STANDARD TEMPERATURE.
METRIC MEASURES.

FOR TEMPERATURES ABOVE 0° CENTIGRADE, THE CORRECTION IS TO BE SUBTRACTED.

Attached Thermometer.	HEIGHT OF THE BAROMETER 690 mm.					HEIGHT OF THE BAROMETER 695 mm.				
	0°.0	0°.2	0°.4	0°.6	0°.8	0°.0	0°.2	0°.4	0°.6	0°.8
C.	mm.	mm.	mm.	mm.	mm.	mm.	mm.	mm.	mm.	mm.
0°	0.00	0.02	0.05	0.07	0.09	0.00	0.02	0.05	0.07	0.09
1	.11	.14	.16	.18	.20	.11	.14	.16	.18	.20
2	.23	.25	.27	.29	.32	.23	.25	.27	.30	.32
3	.34	.36	.38	.41	.43	.34	.36	.39	.41	.43
4	.45	.47	.50	.52	.54	.45	.48	.50	.52	.54
5	0.56	0.59	0.61	0.63	0.65	0.57	0.59	0.61	0.64	0.66
6	.68	.70	.72	.74	.77	.68	.70	.73	.75	.77
7	.79	.81	.83	.86	.88	.79	.82	.84	.86	.88
8	.90	.92	.95	.97	.99	.91	.93	.95	.98	1.00
9	1.01	1.04	1.06	1.08	1.10	1.02	1.04	1.07	1.09	1.11
10	1.13	1.15	1.17	1.19	1.22	1.13	1.16	1.18	1.20	1.22
11	1.24	1.26	1.28	1.31	1.33	1.25	1.27	1.29	1.31	1.34
12	1.35	1.37	1.39	1.42	1.44	1.36	1.38	1.41	1.43	1.45
13	1.46	1.48	1.51	1.53	1.55	1.47	1.50	1.52	1.54	1.56
14	1.57	1.60	1.62	1.64	1.66	1.59	1.61	1.63	1.65	1.68
15	1.69	1.71	1.73	1.75	1.78	1.70	1.72	1.74	1.77	1.79
16	1.80	1.82	1.84	1.87	1.89	1.81	1.83	1.86	1.88	1.90
17	1.91	1.93	1.96	1.98	2.00	1.92	1.95	1.97	1.99	2.01
18	2.02	2.05	2.07	2.09	2.11	2.04	2.06	2.08	2.11	2.13
19	2.13	2.16	2.18	2.20	2.22	2.15	2.17	2.20	2.22	2.24
20	2.25	2.27	2.29	2.31	2.34	2.26	2.29	2.31	2.33	2.35
21	2.36	2.38	2.40	2.43	2.45	2.38	2.40	2.42	2.44	2.47
22	2.47	2.49	2.52	2.54	2.56	2.49	2.51	2.53	2.56	2.58
23	2.58	2.60	2.63	2.65	2.67	2.60	2.62	2.65	2.67	2.69
24	2.69	2.72	2.74	2.76	2.78	2.71	2.74	2.76	2.78	2.80
25	2.81	2.83	2.85	2.87	2.90	2.83	2.85	2.87	2.89	2.92
26	2.92	2.94	2.96	2.99	3.01	2.94	2.96	2.98	3.01	3.03
27	3.03	3.05	3.07	3.10	3.12	3.05	3.07	3.10	3.12	3.14
28	3.14	3.16	3.19	3.21	3.23	3.16	3.19	3.21	3.23	3.25
29	3.25	3.27	3.30	3.32	3.34	3.28	3.30	3.32	3.34	3.37
30	3.36	3.39	3.41	3.43	3.45	3.39	3.41	3.43	3.46	3.48
31	3.48	3.50	3.52	3.54	3.56	3.50	3.52	3.55	3.57	3.59
32	3.59	3.61	3.63	3.65	3.68	3.61	3.64	3.66	3.68	3.70
33	3.70	3.72	3.74	3.77	3.79	3.73	3.75	3.77	3.79	3.81
34	3.81	3.83	3.85	3.88	3.90	3.84	3.86	3.88	3.90	3.93
35	3.92	3.94	3.97	3.99	4.01	3.95	3.97	3.99	4.02	4.04

SMITHSONIAN TABLES.

TABLE 11.

REDUCTION OF THE BAROMETER TO STANDARD TEMPERATURE.
METRIC MEASURES.

FOR TEMPERATURES ABOVE 0° CENTIGRADE, THE CORRECTION IS TO BE SUBTRACTED.

Attached Thermometer.	HEIGHT OF THE BAROMETER 700 mm.					HEIGHT OF THE BAROMETER 705 mm.				
	0°.0	0°.2	0°.4	0°.6	0°.8	0°.0	0°.2	0°.4	0°.6	0°.8
C.	mm.	mm.	mm.	mm.	mm.	mm.	mm.	mm.	mm.	mm.
0°	0.00	0.02	0.05	0.07	0.09	0.00	0.02	0.05	0.07	0.09
1	.11	.14	.16	.18	.21	.12	.14	.16	.18	.21
2	.23	.25	.27	.30	.32	.23	.25	.28	.30	.32
3	.34	.37	.39	.41	.43	.35	.37	.39	.41	.44
4	.46	.48	.50	.53	.55	.46	.48	.51	.53	.55
5	0.57	0.59	0.62	0.64	0.66	0.58	0.60	0.62	0.64	0.67
6	.69	.71	.73	.75	.78	.69	.71	.74	.76	.78
7	.80	.82	.85	.87	.89	.81	.83	.85	.87	.90
8	.91	.94	.96	.98	1.00	.92	.94	.97	.99	1.01
9	1.03	1.05	1.07	1.10	1.12	1.04	1.06	1.08	1.10	1.13
10	1.14	1.16	1.19	1.21	1.23	1.15	1.17	1.20	1.22	1.24
11	1.26	1.28	1.30	1.32	1.35	1.26	1.29	1.31	1.33	1.36
12	1.37	1.39	1.42	1.44	1.46	1.38	1.40	1.43	1.45	1.47
13	1.48	1.51	1.53	1.55	1.57	1.49	1.52	1.54	1.56	1.59
14	1.60	1.62	1.64	1.67	1.69	1.61	1.63	1.65	1.68	1.70
15	1.71	1.73	1.76	1.78	1.80	1.72	1.75	1.77	1.79	1.81
16	1.82	1.85	1.87	1.89	1.92	1.84	1.86	1.88	1.91	1.93
17	1.94	1.96	1.98	2.01	2.03	1.95	1.98	2.00	2.02	2.04
18	2.05	2.07	2.10	2.12	2.14	2.07	2.09	2.11	2.14	2.16
19	2.17	2.19	2.21	2.23	2.26	2.18	2.20	2.23	2.25	2.27
20	2.28	2.30	2.32	2.35	2.37	2.30	2.32	2.34	2.36	2.39
21	2.39	2.42	2.44	2.46	2.48	2.41	2.43	2.46	2.48	2.50
22	2.51	2.53	2.55	2.57	2.60	2.52	2.55	2.57	2.59	2.62
23	2.62	2.64	2.67	2.69	2.71	2.64	2.66	2.68	2.71	2.73
24	2.73	2.76	2.78	2.80	2.82	2.75	2.78	2.80	2.82	2.84
25	2.85	2.87	2.89	2.91	2.94	2.87	2.89	2.91	2.94	2.96
26	2.96	2.98	3.01	3.03	3.05	2.98	3.00	3.03	3.05	3.07
27	3.07	3.10	3.12	3.14	3.16	3.10	3.12	3.14	3.16	3.19
28	3.19	3.21	3.23	3.25	3.28	3.21	3.23	3.25	3.28	3.30
29	3.30	3.32	3.34	3.37	3.39	3.32	3.35	3.37	3.39	3.41
30	3.41	3.44	3.46	3.48	3.50	3.44	3.46	3.48	3.51	3.53
31	3.53	3.55	3.57	3.59	3.62	3.55	3.57	3.60	3.62	3.64
32	3.64	3.66	3.68	3.71	3.73	3.66	3.69	3.71	3.73	3.76
33	3.75	3.77	3.80	3.82	3.84	3.78	3.80	3.82	3.85	3.87
34	3.87	3.89	3.91	3.93	3.96	3.89	3.92	3.94	3.96	3.98
35	3.98	4.00	4.02	4.05	4.07	4.01	4.03	4.05	4.07	4.10

SMITHSONIAN TABLES.

TABLE 11.

REDUCTION OF THE BAROMETER TO STANDARD TEMPERATURE.
METRIC MEASURES.

FOR TEMPERATURES ABOVE 0° CENTIGRADE, THE CORRECTION IS TO BE SUBTRACTED.

Attached Thermometer.	HEIGHT OF THE BAROMETER 710 mm.					HEIGHT OF THE BAROMETER 715 mm.				
	0°.0	0°.2	0°.4	0°.6	0°.8	0°.0	0°.2	0°.4	0°.6	0°.8
C.	mm.	mm.	mm.	mm.	mm.	mm.	mm.	mm.	mm.	mm.
0°	0.00	0.02	0.05	0.07	0.09	0.00	0.02	0.05	0.07	0.09
1	.12	.14	.16	.19	.21	.12	.14	.16	.19	.21
2	.23	.26	.28	.30	.32	.23	.26	.28	.30	.33
3	.35	.37	.39	.42	.44	.35	.37	.40	.42	.44
4	.46	.49	.51	.53	.56	.47	.49	.51	.54	.56
5	0.58	0.60	0.63	0.65	0.67	0.58	0.61	0.63	0.65	0.68
6	.70	.72	.74	.76	.79	.70	.72	.75	.77	.79
7	.81	.83	.86	.88	.90	.82	.84	.86	.89	.91
8	.93	.95	.97	1.00	1.02	.93	.96	.98	1.00	1.03
9	1.04	1.07	1.09	1.11	1.13	1.05	1.07	1.10	1.12	1.14
10	1.16	1.18	1.20	1.23	1.25	1.17	1.19	1.21	1.24	1.26
11	1.27	1.30	1.32	1.34	1.37	1.28	1.31	1.33	1.35	1.38
12	1.39	1.41	1.44	1.46	1.48	1.40	1.42	1.45	1.47	1.49
13	1.50	1.53	1.55	1.57	1.60	1.52	1.54	1.56	1.58	1.61
14	1.62	1.64	1.67	1.69	1.71	1.63	1.65	1.68	1.70	1.72
15	1.74	1.76	1.78	1.80	1.83	1.75	1.77	1.79	1.82	1.84
16	1.85	1.87	1.90	1.92	1.94	1.86	1.89	1.91	1.93	1.96
17	1.97	1.99	2.01	2.04	2.06	1.98	2.00	2.03	2.05	2.07
18	2.08	2.10	2.13	2.15	2.17	2.10	2.12	2.14	2.17	2.19
19	2.20	2.22	2.24	2.27	2.29	2.21	2.24	2.26	2.28	2.30
20	2.31	2.33	2.36	2.38	2.40	2.33	2.35	2.37	2.40	2.42
21	2.43	2.45	2.47	2.50	2.52	2.44	2.47	2.49	2.51	2.54
22	2.54	2.57	2.59	2.61	2.63	2.56	2.58	2.61	2.63	2.65
23	2.66	2.68	2.70	2.73	2.75	2.68	2.70	2.72	2.75	2.77
24	2.77	2.80	2.82	2.84	2.86	2.79	2.81	2.84	2.86	2.88
25	2.89	2.91	2.93	2.96	2.98	2.91	2.93	2.95	2.98	3.00
26	3.00	3.03	3.05	3.07	3.09	3.02	3.05	3.07	3.09	3.12
27	3.12	3.14	3.16	3.19	3.21	3.14	3.16	3.19	3.21	3.23
28	3.23	3.25	3.28	3.30	3.32	3.25	3.28	3.30	3.32	3.35
29	3.35	3.37	3.39	3.42	3.44	3.37	3.39	3.42	3.44	3.46
30	3.46	3.48	3.51	3.53	3.55	3.49	3.51	3.53	3.56	3.58
31	3.58	3.60	3.62	3.65	3.67	3.60	3.62	3.65	3.67	3.69
32	3.69	3.71	3.74	3.76	3.78	3.72	3.74	3.76	3.79	3.81
33	3.81	3.83	3.85	3.87	3.90	3.83	3.86	3.88	3.90	3.92
34	3.92	3.94	3.97	3.99	4.01	3.95	3.97	3.99	4.02	4.04
35	4.03	4.06	4.08	4.10	4.13	4.06	4.09	4.11	4.13	4.16

SMITHSONIAN TABLES.

TABLE 11.

REDUCTION OF THE BAROMETER TO STANDARD TEMPERATURE.
METRIC MEASURES.

FOR TEMPERATURES ABOVE 0° CENTIGRADE, THE CORRECTION IS TO BE SUBTRACTED.

Attached Thermometer.	HEIGHT OF THE BAROMETER 720 mm.					HEIGHT OF THE BAROMETER 725 mm.				
	0°.0	0°.2	0°.4	0°.6	0°.8	0°.0	0°.2	0°.4	0°.6	0°.8
C.	mm.	mm.	mm.	mm.	mm.	mm.	mm.	mm.	mm.	mm.
0°	0.00	0.02	0.05	0.07	0.09	0.00	0.02	0.05	0.07	0.09
1	.12	.14	.16	.19	.21	.12	.14	.17	.19	.21
2	.24	.26	.28	.31	.33	.24	.26	.28	.31	.33
3	.35	.38	.40	.42	.45	.36	.38	.40	.43	.45
4	.47	.49	.52	.54	.56	.47	.50	.52	.54	.57
5	0.59	0.61	0.63	0.66	0.68	0.59	0.62	0.64	0.66	0.69
6	.71	.73	.75	.78	.80	.71	.73	.76	.78	.80
7	.82	.85	.87	.89	.92	.83	.85	.88	.90	.92
8	.94	.96	.99	1.01	1.03	.95	.97	.99	1.02	1.04
9	1.06	1.08	1.10	1.13	1.15	1.06	1.09	1.11	1.14	1.16
10	1.17	1.20	1.22	1.24	1.27	1.18	1.21	1.23	1.25	1.28
11	1.29	1.31	1.34	1.36	1.39	1.30	1.32	1.35	1.37	1.39
12	1.41	1.43	1.46	1.48	1.50	1.42	1.44	1.47	1.49	1.51
13	1.53	1.55	1.57	1.60	1.62	1.54	1.56	1.58	1.61	1.63
14	1.64	1.67	1.69	1.71	1.74	1.65	1.68	1.70	1.73	1.75
15	1.76	1.78	1.81	1.83	1.85	1.77	1.80	1.82	1.84	1.87
16	1.88	1.90	1.92	1.95	1.97	1.89	1.91	1.94	1.96	1.98
17	1.99	2.02	2.04	2.06	2.09	2.01	2.03	2.05	2.08	2.10
18	2.11	2.13	2.16	2.18	2.20	2.13	2.15	2.17	2.20	2.22
19	2.23	2.25	2.27	2.30	2.32	2.24	2.27	2.29	2.31	2.34
20	2.34	2.37	2.39	2.41	2.44	2.36	2.38	2.41	2.43	2.45
21	2.46	2.48	2.51	2.53	2.55	2.48	2.50	2.53	2.55	2.57
22	2.58	2.60	2.62	2.65	2.67	2.60	2.62	2.64	2.67	2.69
23	2.69	2.72	2.74	2.76	2.79	2.71	2.74	2.76	2.78	2.81
24	2.81	2.83	2.86	2.88	2.90	2.83	2.85	2.88	2.90	2.92
25	2.93	2.95	2.97	3.00	3.02	2.95	2.97	3.00	3.02	3.04
26	3.04	3.07	3.09	3.11	3.14	3.07	3.09	3.11	3.14	3.16
27	3.16	3.18	3.21	3.23	3.25	3.18	3.21	3.23	3.25	3.28
28	3.28	3.30	3.32	3.35	3.37	3.30	3.32	3.35	3.37	3.39
29	3.39	3.42	3.44	3.46	3.49	3.42	3.44	3.46	3.49	3.51
30	3.51	3.53	3.56	3.58	3.60	3.53	3.56	3.58	3.60	3.63
31	3.63	3.65	3.67	3.70	3.72	3.65	3.68	3.70	3.72	3.75
32	3.74	3.77	3.79	3.81	3.84	3.77	3.79	3.82	3.84	3.86
33	3.86	3.88	3.91	3.93	3.95	3.89	3.91	3.93	3.96	3.98
24	3.98	4.00	4.02	4.05	4.07	4.00	4.03	4.05	4.07	4.10
35	4.09	4.11	4.14	4.16	4.18	4.12	4.14	4.17	4.19	4.21

SMITHSONIAN TABLES.

TABLE 11.
REDUCTION OF THE BAROMETER TO STANDARD TEMPERATURE.
METRIC MEASURES.

FOR TEMPERATURES ABOVE 0° CENTIGRADE, THE CORRECTION IS TO BE SUBTRACTED.

Attached Thermometer.	HEIGHT OF THE BAROMETER 730 mm.					HEIGHT OF THE BAROMETER 735 mm.				
	0°.0	0°.2	0°.4	0°.6	0°.8	0°.0	0°.2	0°.4	0°.6	0°.8
C.	mm.	mm.	mm.	mm.	mm.	mm.	mm.	mm.	mm.	mm.
0°	0.00	0.02	0.05	0.07	0.10	0.00	0.02	0.05	0.07	0.10
1	.12	.14	.17	.19	.21	.12	.14	.17	.19	.22
2	.24	.26	.29	.31	.33	.24	.26	.29	.31	.34
3	.36	.38	.41	.43	.45	.36	.38	.41	.43	.46
4	.48	.50	.52	.55	.57	.48	.50	.53	.55	.58
5	0.60	0.62	0.64	0.67	0.69	0.60	0.62	0.65	0.67	0.70
6	.71	.74	.76	.79	.81	.72	.74	.77	.79	.82
7	.83	.86	.88	.91	.93	.84	.86	.89	.91	.94
8	.95	.98	1.00	1.02	1.05	.96	.98	1.01	1.03	1.06
9	1.07	1.10	1.12	1.14	1.17	1.08	1.10	1.13	1.15	1.17
10	1.19	1.21	1.24	1.26	1.29	1.20	1.22	1.25	1.27	1.29
11	1.31	1.33	1.36	1.38	1.40	1.32	1.34	1.37	1.39	1.41
12	1.43	1.45	1.48	1.50	1.52	1.44	1.46	1.49	1.51	1.53
13	1.55	1.57	1.59	1.62	1.64	1.56	1.58	1.61	1.63	1.65
14	1.67	1.69	1.71	1.74	1.76	1.68	1.70	1.72	1.75	1.77
15	1.78	1.81	1.83	1.86	1.88	1.80	1.82	1.84	1.87	1.89
16	1.90	1.93	1.95	1.97	2.00	1.92	1.94	1.96	1.99	2.01
17	2.02	2.05	2.07	2.09	2.12	2.04	2.06	2.08	2.11	2.13
18	2.14	2.16	2.19	2.21	2.23	2.15	2.18	2.20	2.23	2.25
19	2.26	2.28	2.31	2.33	2.35	2.27	2.30	2.32	2.35	2.37
20	2.38	2.40	2.42	2.45	2.47	2.39	2.42	2.44	2.46	2.49
21	2.50	2.52	2.54	2.57	2.59	2.51	2.54	2.56	2.58	2.61
22	2.61	2.64	2.66	2.68	2.71	2.63	2.66	2.68	2.70	2.73
23	2.73	2.76	2.78	2.80	2.83	2.75	2.77	2.80	2.82	2.85
24	2.85	2.87	2.90	2.92	2.94	2.87	2.89	2.92	2.94	2.97
25	2.97	2.99	3.02	3.04	3.06	2.99	3.01	3.04	3.06	3.08
26	3.09	3.11	3.13	3.16	3.18	3.11	3.13	3.16	3.18	3.20
27	3.20	3.23	3.25	3.28	3.30	3.23	3.25	3.27	3.30	3.32
28	3.32	3.35	3.37	3.39	3.42	3.35	3.37	3.39	3.42	3.44
29	3.44	3.46	3.49	3.51	3.54	3.46	3.49	3.51	3.54	3.56
30	3.56	3.58	3.61	3.63	3.65	3.58	3.61	3.63	3.65	3.68
31	3.68	3.70	3.72	3.75	3.77	3.70	3.73	3.75	3.77	3.80
32	3.79	3.82	3.84	3.87	3.89	3.82	3.84	3.87	3.89	3.92
33	3.91	3.94	3.96	3.98	4.01	3.94	3.96	3.99	4.01	4.03
34	4.03	4.05	4.08	4.10	4.12	4.06	4.08	4.11	4.13	4.15
35	4.15	4.17	4.20	4.22	4.24	4.18	4.20	4.22	4.25	4.27

SMITHSONIAN TABLES.

TABLE 11.

REDUCTION OF THE BAROMETER TO STANDARD TEMPERATURE.
METRIC MEASURES.

FOR TEMPERATURES ABOVE 0° CENTIGRADE, THE CORRECTION IS TO BE SUBTRACTED.

Attached Thermometer.	HEIGHT OF THE BAROMETER 740 mm.					HEIGHT OF THE BAROMETER 745 mm.				
	0°.0	0°.2	0°.4	0°.6	0°.8	0°.0	0°.2	0°.4	0°.6	0°.8
C.	mm.	mm.	mm.	mm.	mm.	mm.	mm.	mm.	mm.	mm.
0°	0.00	0.02	0.05	0.07	0.10	0.00	0.02	0.05	0.07	0.10
1	.12	.15	.17	.19	.22	.12	.15	.17	.19	.22
2	.24	.27	.29	.31	.34	.24	.27	.29	.32	.34
3	.36	.39	.41	.44	.46	.37	.39	.41	.44	.46
4	.48	.51	.53	.56	.58	.49	.51	.54	.56	.58
5	0.60	0.63	0.65	0.68	0.70	0.61	0.63	0.66	0.68	0.71
6	.72	.75	.77	.80	.82	.73	.75	.78	.80	.83
7	.85	.87	.89	.92	.94	.85	.88	.90	.92	.95
8	.97	.99	1.01	1.04	1.06	.97	1.00	1.02	1.05	1.07
9	1.09	1.11	1.13	1.16	1.18	1.09	1.12	1.14	1.17	1.19
10	1.21	1.23	1.26	1.28	1.30	1.22	1.24	1.26	1.29	1.31
11	1.33	1.35	1.38	1.40	1.42	1.34	1.36	1.38	1.41	1.43
12	1.45	1.47	1.50	1.52	1.54	1.46	1.48	1.51	1.53	1.55
13	1.57	1.59	1.62	1.64	1.66	1.58	1.60	1.63	1.65	1.68
14	1.69	1.71	1.74	1.76	1.78	1.70	1.72	1.75	1.77	1.80
15	1.81	1.83	1.86	1.88	1.90	1.82	1.85	1.87	1.89	1.92
16	1.93	1.95	1.98	2.00	2.03	1.94	1.97	1.99	2.01	2.04
17	2.05	2.07	2.10	2.12	2.15	2.06	2.09	2.11	2.14	2.16
18	2.17	2.19	2.22	2.24	2.27	2.18	2.21	2.23	2.26	2.28
19	2.29	2.31	2.34	2.36	2.39	2.31	2.33	2.35	2.38	2.40
20	2.41	2.43	2.46	2.48	2.51	2.43	2.45	2.47	2.50	2.52
21	2.53	2.55	2.58	2.60	2.63	2.55	2.57	2.59	2.62	2.64
22	2.65	2.67	2.70	2.72	2.75	2.67	2.69	2.72	2.74	2.76
23	2.77	2.79	2.82	2.84	2.87	2.79	2.81	2.84	2.86	2.88
24	2.89	2.91	2.94	2.96	2.99	2.91	2.93	2.96	2.98	3.01
25	3.01	3.03	3.06	3.08	3.11	3.03	3.05	3.08	3.10	3.13
26	3.13	3.15	3.18	3.20	3.22	3.15	3.17	3.20	3.22	3.25
27	3.25	3.27	3.30	3.32	3.34	3.27	3.29	3.32	3.34	3.37
28	3.37	3.39	3.42	3.44	3.46	3.39	3.42	3.44	3.46	3.49
29	3.49	3.51	3.54	3.56	3.58	3.51	3.54	3.56	3.58	3.61
30	3.61	3.63	3.66	3.68	3.70	3.63	3.66	3.68	3.70	3.73
31	3.73	3.75	3.78	3.80	3.82	3.75	3.78	3.80	3.82	3.85
32	3.85	3.87	3.89	3.92	3.94	3.87	3.90	3.92	3.95	3.97
33	3.97	3.99	4.01	4.04	4.06	3.99	4.02	4.04	4.07	4.09
34	4.09	4.11	4.13	4.16	4.18	4.11	4.14	4.16	4.19	4.21
35	4.21	4.23	4.25	4.28	4.30	4.23	4.26	4.28	4.31	4.33

SMITHSONIAN TABLES.

TABLE 11.
REDUCTION OF THE BAROMETER TO STANDARD TEMPERATURE.
METRIC MEASURES.

FOR TEMPERATURES ABOVE 0° CENTIGRADE, THE CORRECTION IS TO BE SUBTRACTED.

Attached Thermometer.	HEIGHT OF THE BAROMETER 750 mm.					HEIGHT OF THE BAROMETER 755 mm.				
	0°.0	0°.2	0°.4	0°.6	0°.8	0°.0	0°.2	0°.4	0°.6	0°.8
C.	mm.	mm.	mm.	mm.	mm.	mm.	mm.	mm.	mm.	mm.
0°	0.00	0.02	0.05	0.07	0.10	0.00	0.02	0.05	0.07	0.10
1	.12	.15	.17	.20	.22	.12	.15	.17	.20	.22
2	.25	.27	.29	.32	.34	.25	.27	.30	.32	.35
3	.37	.39	.42	.44	.47	.37	.39	.42	.44	.47
4	.49	.51	.54	.56	.59	.49	.52	.54	.57	.59
5	0.61	0.64	0.66	0.69	0.71	0.62	0.64	0.67	0.69	0.71
6	.73	.76	.78	.81	.83	.74	.76	.79	.81	.84
7	.86	.88	.91	.93	.95	.86	.89	.91	.94	.96
8	.98	1.00	1.03	1.05	1.08	.99	1.01	1.03	1.06	1.08
9	1.10	1.13	1.15	1.17	1.20	1.11	1.13	1.16	1.18	1.21
10	1.22	1.25	1.27	1.30	1.32	1.23	1.26	1.28	1.31	1.33
11	1.35	1.37	1.39	1.42	1.44	1.35	1.38	1.40	1.43	1.45
12	1.47	1.49	1.52	1.54	1.56	1.48	1.50	1.53	1.55	1.58
13	1.59	1.61	1.64	1.66	1.69	1.60	1.62	1.65	1.67	1.70
14	1.71	1.74	1.76	1.78	1.81	1.72	1.75	1.77	1.80	1.82
15	1.83	1.86	1.88	1.91	1.93	1.85	1.87	1.89	1.92	1.94
16	1.96	1.98	2.00	2.03	2.05	1.97	1.99	2.02	2.04	2.07
17	2.08	2.10	2.13	2.15	2.17	2.09	2.12	2.14	2.16	2.19
18	2.20	2.22	2.25	2.27	2.30	2.21	2.24	2.26	2.29	2.31
19	2.32	2.34	2.37	2.39	2.42	2.34	2.36	2.38	2.41	2.43
20	2.44	2.47	2.49	2.52	2.54	2.46	2.48	2.51	2.53	2.56
21	2.56	2.59	2.61	2.64	2.66	2.58	2.61	2.63	2.65	2.68
22	2.69	2.71	2.73	2.76	2.78	2.70	2.73	2.75	2.78	2.80
23	2.81	2.83	2.86	2.88	2.90	2.83	2.85	2.87	2.90	2.92
24	2.93	2.95	2.98	3.00	3.03	2.95	2.97	3.00	3.02	3.05
25	3.05	3.07	3.10	3.12	3.15	3.07	3.09	3.12	3.14	3.17
26	3.17	3.20	3.22	3.24	3.27	3.19	3.22	3.24	3.27	3.29
27	3.29	3.32	3.34	3.37	3.39	3.31	3.34	3.36	3.39	3.41
28	3.41	3.44	3.46	3.49	3.51	3.44	3.46	3.49	3.51	3.53
29	3.54	3.56	3.58	3.61	3.63	3.56	3.58	3.61	3.63	3.66
30	3.66	3.68	3.71	3.73	3.75	3.68	3.71	3.73	3.75	3.78
31	3.78	3.80	3.83	3.85	3.87	3.80	3.83	3.85	3.88	3.90
32	3.90	3.92	3.95	3.97	4.00	3.92	3.95	3.97	4.00	4.02
33	4.02	4.04	4.07	4.09	4.12	4.05	4.07	4.10	4.12	4.14
34	4.14	4.17	4.19	4.21	4.24	4.17	4.19	4.22	4.24	4.27
35	4.26	4.29	4.31	4.33	4.36	4.29	4.31	4.34	4.36	4.39

SMITHSONIAN TABLES.

TABLE 11.

REDUCTION OF THE BAROMETER TO STANDARD TEMPERATURE.
METRIC MEASURES.

FOR TEMPERATURES ABOVE 0° CENTIGRADE, THE CORRECTION IS TO BE SUBTRACTED.

Attached Thermometer.	HEIGHT OF THE BAROMETER 760 mm.					HEIGHT OF THE BAROMETER 765 mm.				
	0°.0	0°.2	0°.4	0°.6	0°.8	0°.0	0°.2	0°.4	0°.6	0°.8
C.	mm.	mm.	mm.	mm.	mm.	mm.	mm.	mm.	mm.	mm.
0°	0.00	0.02	0.05	0.07	0.10	0.00	0.03	0.05	0.07	0.10
1	.12	.15	.17	.20	.22	.13	.15	.17	.20	.22
2	.25	.27	.30	.32	.35	.25	.27	.30	.32	.35
3	.37	.40	.42	.45	.47	.37	.40	.42	.45	.47
4	.50	.52	.55	.57	.60	.50	.52	.55	.57	.60
5	0.62	0.65	0.67	0.69	0.72	0.62	0.65	0.67	0.70	0.72
6	.74	.77	.79	.82	.84	.75	.77	.80	.82	.85
7	.87	.89	.92	.94	.97	.87	.90	.92	.95	.97
8	.99	1.02	1.04	1.07	1.09	1.00	1.02	1.05	1.07	1.10
9	1.12	1.14	1.17	1.19	1.21	1.12	1.15	1.17	1.20	1.22
10	1.24	1.26	1.29	1.31	1.34	1.25	1.27	1.30	1.32	1.35
11	1.36	1.39	1.41	1.44	1.46	1.37	1.40	1.42	1.45	1.47
12	1.49	1.51	1.54	1.56	1.59	1.50	1.52	1.55	1.57	1.60
13	1.61	1.64	1.66	1.68	1.71	1.62	1.65	1.67	1.70	1.72
14	1.73	1.76	1.78	1.81	1.83	1.75	1.77	1.80	1.82	1.85
15	1.86	1.88	1.91	1.93	1.96	1.87	1.89	1.92	1.94	1.97
16	1.98	2.01	2.03	2.06	2.08	1.99	2.02	2.04	2.07	2.09
17	2.10	2.13	2.15	2.18	2.20	2.12	2.14	2.17	2.19	2.22
18	2.23	2.25	2.28	2.30	2.33	2.24	2.27	2.29	2.32	2.34
19	2.35	2.38	2.40	2.43	2.45	2.37	2.39	2.42	2.44	2.47
20	2.47	2.50	2.52	2.55	2.57	2.49	2.52	2.54	2.57	2.59
21	2.60	2.62	2.65	2.67	2.70	2.62	2.64	2.66	2.69	2.71
22	2.72	2.75	2.77	2.80	2.82	2.74	2.76	2.79	2.81	2.84
23	2.84	2.87	2.89	2.92	2.94	2.86	2.89	2.91	2.94	2.96
24	2.97	2.99	3.02	3.04	3.07	2.99	3.01	3.04	3.06	3.09
25	3.09	3.12	3.14	3.16	3.19	3.11	3.14	3.16	3.19	3.21
26	3.21	3.24	3.26	3.29	3.31	3.23	3.26	3.28	3.31	3.33
27	3.34	3.36	3.39	3.41	3.43	3.36	3.38	3.41	3.43	3.46
28	3.46	3.48	3.51	3.53	3.56	3.48	3.51	3.53	3.56	3.58
29	3.58	3.61	3.63	3.66	3.68	3.61	3.63	3.66	3.68	3.70
30	3.71	3.73	3.75	3.78	3.80	3.73	3.75	3.78	3.80	3.83
31	3.83	3.85	3.88	3.90	3.93	3.85	3.88	3.90	3.93	3.95
32	3.95	3.98	4.00	4.02	4.05	3.98	4.00	4.03	4.05	4.08
33	4.07	4.10	4.12	4.15	4.17	4.10	4.13	4.15	4.17	4.20
34	4.20	4.22	4.25	4.27	4.29	4.22	4.25	4.27	4.30	4.32
35	4.32	4.34	4.37	4.39	4.42	4.35	4.37	4.40	4.42	4.45

SMITHSONIAN TABLES.

TABLE 11.

REDUCTION OF THE BAROMETER TO STANDARD TEMPERATURE.
METRIC MEASURES.

FOR TEMPERATURES ABOVE 0° CENTIGRADE, THE CORRECTION IS TO BE SUBTRACTED.

Attached Thermometer.	HEIGHT OF THE BAROMETER 770 mm.					HEIGHT OF THE BAROMETER 775 mm.				
	0°.0	0°.2	0°.4	0°.6	0°.8	0°.0	0°.2	0°.4	0°.6	0°.8
C.	mm.	mm.	mm.	mm.	mm.	mm.	mm.	mm.	mm.	mm.
0°	0.00	0.03	0.05	0.08	0.10	0.00	0.03	0.05	0.08	0.10
1	.13	.15	.18	.20	.23	.13	.15	.18	.20	.23
2	.25	.28	.30	.33	.35	.25	.28	.30	.33	.35
3	.38	.40	.43	.45	.48	.38	.40	.43	.46	.48
4	.50	.53	.55	.58	.60	.51	.53	.56	.58	.61
5	0.63	0.65	0.68	0.70	0.73	0.63	0.66	0.68	0.71	0.73
6	.75	.78	.80	.83	.85	.76	.78	.81	.83	.86
7	.88	.90	.93	.95	.98	.89	.91	.94	.96	.99
8	1.01	1.03	1.06	1.08	1.11	1.01	1.04	1.06	1.09	1.11
9	1.13	1.16	1.18	1.21	1.23	1.14	1.16	1.19	1.21	1.24
10	1.26	1.28	1.31	1.33	1.36	1.26	1.29	1.31	1.34	1.36
11	1.38	1.41	1.43	1.46	1.48	1.39	1.42	1.44	1.47	1.49
12	1.51	1.53	1.56	1.58	1.61	1.52	1.54	1.57	1.59	1.62
13	1.63	1.66	1.68	1.71	1.73	1.64	1.67	1.69	1.72	1.74
14	1.76	1.78	1.81	1.83	1.86	1.77	1.79	1.82	1.84	1.87
15	1.88	1.91	1.93	1.96	1.98	1.89	1.92	1.94	1.97	2.00
16	2.01	2.03	2.06	2.08	2.11	2.02	2.05	2.07	2.10	2.12
17	2.13	2.16	2.18	2.21	2.23	2.15	2.17	2.20	2.22	2.25
18	2.26	2.28	2.31	2.33	2.36	2.27	2.30	2.32	2.35	2.37
19	2.38	2.41	2.43	2.46	2.48	2.40	2.42	2.45	2.47	2.50
20	2.51	2.53	2.56	2.58	2.61	2.52	2.55	2.57	2.60	2.62
21	2.63	2.66	2.68	2.71	2.73	2.65	2.67	2.70	2.72	2.75
22	2.76	2.78	2.81	2.83	2.86	2.77	2.80	2.83	2.85	2.88
23	2.88	2.91	2.93	2.96	2.98	2.90	2.93	2.95	2.98	3.00
24	3.01	3.03	3.06	3.08	3.11	3.03	3.05	3.08	3.10	3.13
25	3.13	3.16	3.18	3.21	3.23	3.15	3.18	3.20	3.23	3.25
26	3.26	3.28	3.31	3.33	3.36	3.28	3.30	3.33	3.35	3.38
27	3.38	3.41	3.43	3.46	3.48	3.40	3.43	3.45	3.48	3.50
28	3.51	3.53	3.56	3.58	3.60	3.53	3.55	3.58	3.60	3.63
29	3.63	3.65	3.68	3.70	3.73	3.65	3.68	3.70	3.73	3.75
30	3.75	3.78	3.80	3.83	3.85	3.78	3.80	3.83	3.85	3.88
31	3.88	3.90	3.93	3.95	3.98	3.90	3.93	3.95	3.98	4.00
32	4.00	4.03	4.05	4.08	4.10	4.03	4.05	4.08	4.10	4.13
33	4.13	4.15	4.18	4.20	4.23	4.15	4.18	4.20	4.23	4.25
34	4.25	4.28	4.30	4.33	4.35	4.28	4.30	4.33	4.35	4.38
35	4.38	4.40	4.43	4.45	4.48	4.40	4.43	4.45	4.48	4.50

SMITHSONIAN TABLES.

TABLE 11.

REDUCTION OF THE BAROMETER TO STANDARD TEMPERATURE.
METRIC MEASURES.

FOR TEMPERATURES ABOVE 0° CENTIGRADE, THE CORRECTION IS TO BE SUBTRACTED.

Attached Thermometer.	HEIGHT OF THE BAROMETER 780 mm.					HEIGHT OF THE BAROMETER 785 mm.				
	0°.0	0°.2	0°.4	0°.6	0°.8	0°.0	0°.2	0°.4	0°.6	0°.8
C.	mm.	mm.	mm.	mm.	mm.	mm.	mm.	mm.	mm.	mm.
0°	0.00	0.03	0.05	0.08	0.10	0.00	0.03	0.05	0.08	0.10
1	.13	.15	.18	.20	.23	.13	.15	.18	.21	.23
2	.25	.28	.31	.33	.36	.26	.28	.31	.33	.36
3	.38	.41	.43	.46	.48	.38	.41	.44	.46	.49
4	.51	.53	.56	.59	.61	.51	.54	.56	.59	.62
5	0.64	0.66	0.69	0.71	0.74	0.64	0.67	0.69	0.72	0.74
6	.76	.79	.81	.84	.87	.77	.79	.82	.85	.87
7	.89	.92	.94	.97	.99	.90	.92	.95	.97	1.00
8	1.02	1.04	1.07	1.09	1.12	1.02	1.05	1.08	1.10	1.13
9	1.15	1.17	1.20	1.22	1.25	1.15	1.18	1.20	1.23	1.25
10	1.27	1.30	1.32	1.35	1.37	1.28	1.31	1.33	1.36	1.38
11	1.40	1.42	1.45	1.48	1.50	1.41	1.43	1.46	1.48	1.51
12	1.53	1.55	1.58	1.60	1.63	1.54	1.56	1.59	1.61	1.64
13	1.65	1.68	1.70	1.73	1.75	1.66	1.69	1.71	1.74	1.77
14	1.78	1.81	1.83	1.86	1.88	1.79	1.82	1.84	1.87	1.89
15	1.91	1.93	1.96	1.98	2.01	1.92	1.94	1.97	2.00	2.02
16	2.03	2.06	2.08	2.11	2.13	2.05	2.07	2.10	2.12	2.15
17	2.16	2.19	2.21	2.24	2.26	2.17	2.20	2.22	2.25	2.28
18	2.29	2.31	2.34	2.36	2.39	2.30	2.33	2.35	2.38	2.40
19	2.41	2.44	2.46	2.49	2.51	2.43	2.45	2.48	2.51	2.53
20	2.54	2.57	2.59	2.62	2.64	2.56	2.58	2.61	2.63	2.66
21	2.67	2.69	2.72	2.74	2.77	2.68	2.71	2.73	2.76	2.79
22	2.79	2.82	2.84	2.87	2.89	2.81	2.84	2.86	2.89	2.91
23	2.92	2.94	2.97	3.00	3.02	2.94	2.96	2.99	3.01	3.04
24	3.05	3.07	3.10	3.12	3.15	3.07	3.09	3.12	3.14	3.17
25	3.17	3.20	3.22	3.25	3.27	3.19	3.22	3.24	3.27	3.29
26	3.30	3.32	3.35	3.37	3.40	3.32	3.34	3.37	3.40	3.42
27	3.42	3.45	3.47	3.50	3.53	3.45	3.47	3.50	3.52	3.55
28	3.55	3.58	3.60	3.63	3.65	3.57	3.60	3.62	3.65	3.67
29	3.68	3.70	3.73	3.75	3.78	3.70	3.73	3.75	3.78	3.80
30	3.80	3.83	3.85	3.88	3.90	3.83	3.85	3.88	3.90	3.93
31	3.93	3.95	3.98	4.00	4.03	3.95	3.98	4.00	4.03	4.06
32	4.05	4.08	4.11	4.13	4.16	4.08	4.11	4.13	4.16	4.18
33	4.18	4.21	4.23	4.26	4.28	4.21	4.23	4.26	4.28	4.31
34	4.31	4.33	4.36	4.38	4.41	4.33	4.36	4.39	4.41	4.44
35	4.43	4.46	4.48	4.51	4.53	4.46	4.49	4.51	4.54	4.56

SMITHSONIAN TABLES.

TABLE 11.
REDUCTION OF THE BAROMETER TO STANDARD TEMPERATURE.
METRIC MEASURES.

FOR TEMPERATURES ABOVE 0° CENTIGRADE, THE CORRECTION IS TO BE SUBTRACTED.

Attached Thermometer.	HEIGHT OF THE BAROMETER 790 mm.					HEIGHT OF THE BAROMETER 795 mm.				
	0°.0	0°.2	0°.4	0°.6	0°.8	0°.0	0°.2	0°.4	0°.6	0°.8
C.	mm.	mm.	mm.	mm.	mm.	mm.	mm.	mm.	mm.	mm.
0°	0.00	0.03	0.05	0.08	0.10	0.00	0.03	0.05	0.08	0.10
1	.13	.15	.18	.21	.23	.13	.16	.18	.21	.23
2	.26	.28	.31	.34	.36	.26	.29	.31	.34	.36
3	.39	.41	.44	.46	.49	.39	.42	.44	.47	.49
4	.52	.54	.57	.59	.62	.52	.55	.57	.60	.62
5	0.64	0.67	0.70	0.72	0.75	0.65	0.67	0.70	0.73	0.75
6	.77	.80	.83	.85	.88	.78	.80	.83	.86	.88
7	.90	.93	.95	.98	1.01	.91	.93	.96	.99	1.01
8	1.03	1.06	1.08	1.11	1.13	1.04	1.06	1.09	1.12	1.14
9	1.16	1.19	1.21	1.24	1.26	1.17	1.19	1.22	1.24	1.27
10	1.29	1.31	1.34	1.37	1.39	1.30	1.32	1.35	1.37	1.40
11	1.42	1.44	1.47	1.49	1.52	1.43	1.45	1.48	1.50	1.53
12	1.55	1.57	1.60	1.62	1.65	1.56	1.58	1.61	1.63	1.66
13	1.67	1.70	1.73	1.75	1.78	1.68	1.71	1.74	1.76	1.79
14	1.80	1.83	1.85	1.88	1.91	1.81	1.84	1.87	1.89	1.92
15	1.93	1.96	1.98	2.01	2.03	1.94	1.97	1.99	2.02	2.05
16	2.06	2.09	2.11	2.14	2.16	2.07	2.10	2.12	2.15	2.18
17	2.19	2.21	2.24	2.26	2.29	2.20	2.23	2.25	2.28	2.30
18	2.32	2.34	2.37	2.39	2.42	2.33	2.36	2.38	2.41	2.43
19	2.44	2.47	2.50	2.52	2.55	2.46	2.49	2.51	2.54	2.56
20	2.57	2.60	2.62	2.65	2.67	2.59	2.61	2.64	2.67	2.69
21	2.70	2.73	2.75	2.78	2.80	2.72	2.74	2.77	2.79	2.82
22	2.83	2.85	2.88	2.91	2.93	2.85	2.87	2.90	2.92	2.95
23	2.96	2.98	3.01	3.03	3.06	2.98	3.00	3.03	3.05	3.08
24	3.08	3.11	3.14	3.16	3.19	3.10	3.13	3.16	3.18	3.21
25	3.21	3.24	3.26	3.29	3.31	3.23	3.26	3.28	3.31	3.34
26	3.34	3.37	3.39	3.42	3.44	3.36	3.39	3.41	3.44	3.46
27	3.47	3.49	3.52	3.54	3.57	3.49	3.52	3.54	3.57	3.59
28	3.60	3.62	3.65	3.67	3.70	3.62	3.64	3.67	3.70	3.72
29	3.72	3.75	3.77	3.80	3.83	3.75	3.77	3.80	3.82	3.85
30	3.85	3.88	3.90	3.93	3.95	3.88	3.90	3.93	3.95	3.98
31	3.98	4.00	4.03	4.06	4.08	4.00	4.03	4.06	4.08	4.11
32	4.11	4.13	4.16	4.18	4.21	4.13	4.16	4.18	4.21	4.24
33	4.23	4.26	4.29	4.31	4.34	4.26	4.29	4.31	4.34	4.36
34	4.36	4.39	4.41	4.44	4.46	4.39	4.42	4.44	4.47	4.49
35	4.49	4.51	4.54	4.57	4.59	4.52	4.54	4.57	4.59	4.62

SMITHSONIAN TABLES.

TABLE 12.
REDUCTION OF THE BAROMETER TO STANDARD GRAVITY.
ENGLISH MEASURES.
Reduction to Latitude 45°.

From latitude 0° to 45°, the correction is to be subtracted.
From latitude 90° to 45°, the correction is to be added.

Latitude.		HEIGHT OF THE BAROMETER IN INCHES.										
		19	20	21	22	23	24	25	26	27	28	29	30
0°	90°	Inch. 0.051	Inch. 0.053	Inch. 0.056	Inch. 0.059	Inch. 0.061	Inch. 0.064	Inch. 0.067	Inch. 0.069	Inch. 0.072	Inch. 0.074	Inch. 0.077	Inch. 0.080
5	85	0.050	0.052	0.055	0.058	0.060	0.063	0.066	0.068	0.071	0.073	0.076	0.079
6	84	.049	.052	.055	.057	.060	.062	.065	.068	.070	.073	.076	.078
7	83	.049	.052	.054	.057	.059	.062	.065	.067	.070	.072	.075	.077
8	82	.049	.051	.054	.056	.059	.061	.064	.067	.069	.072	.074	.077
9	81	.048	.051	.053	.056	.058	.061	.063	.066	.068	.071	.073	.076
10	80	0.048	0.050	0.053	0.055	0.058	0.060	0.063	0.065	0.068	0.070	0.073	0.075
11	79	.047	.049	.052	.054	.057	.059	.062	.064	.067	.069	.072	.074
12	78	.046	.049	.051	.054	.056	.058	.061	.063	.066	.068	.071	.073
13	77	.045	.048	.050	.053	.055	.057	.060	.062	.065	.067	.069	.072
14	76	.045	.047	.049	.052	.054	.056	.059	.061	.063	.066	.068	.071
15	75	0.044	0.046	0.048	0.051	0.053	0.055	0.058	0.060	0.062	0.065	0.067	0.069
16	74	.043	.045	.047	.050	.052	.054	.056	.059	.061	.063	.065	.068
17	73	.042	.044	.046	.049	.051	.053	.055	.057	.060	.062	.064	.066
18	72	.041	.043	.045	.047	.050	.052	.054	.056	.058	.060	.062	.065
19	71	.040	.042	.044	.046	.048	.050	.052	.055	.057	.059	.061	.063
20	70	0.039	0.041	0.043	0.045	0.047	0.049	0.051	0.053	0.055	0.057	0.059	0.061
21	69	.038	.040	.042	.044	.045	.047	.049	.051	.053	.055	.057	.059
22	68	.036	.038	.040	.042	.044	.046	.048	.050	.052	.054	.056	.057
23	67	.035	.037	.039	.041	.043	.044	.046	.048	.050	.052	.054	.055
24	66	.034	.036	.037	.039	.041	.043	.045	.046	.048	.050	.052	.053
25	65	0.033	0.034	0.036	0.038	0.039	0.041	0.043	0.044	0.046	0.048	0.050	0.051
26	64	.031	.033	.034	.036	.038	.039	.041	.043	.044	.046	.048	.049
27	63	.030	.031	.033	.034	.036	.038	.039	.041	.042	.044	.045	.047
28	62	.028	.030	.031	.033	.034	.036	.037	.039	.040	.042	.043	.045
29	61	.027	.028	.030	.031	.032	.034	.035	.037	.038	.039	.041	.042
30	60	0.025	0.027	0.028	0.029	0.031	0.032	0.033	0.035	0.036	0.037	0.039	0.040
31	59	.024	.025	.026	.027	.029	.030	.031	.032	.034	.035	.036	.037
32	58	.022	.023	.025	.026	.027	.028	.029	.030	.032	.033	.034	.035
33	57	.021	.022	.023	.024	.025	.026	.027	.028	.029	.030	.031	.032
34	56	.019	.020	.021	.022	.023	.024	.025	.026	.027	.028	.029	.030
35	55	0.017	0.018	0.019	0.020	0.021	0.022	0.023	0.024	0.025	0.025	0.026	0.027
36	54	.016	.016	.017	.018	.019	.020	.021	.021	.022	.023	.024	.025
37	53	.014	.015	.015	.016	.017	.018	.018	.019	.020	.021	.021	.022
38	52	.012	.013	.014	.014	.015	.015	.016	.017	.017	.018	.019	.019
39	51	.011	.011	.012	.012	.013	.013	.014	.014	.015	.015	.016	.017
40	50	0.009	0.009	0.010	0.010	0.011	0.011	0.012	0.012	0.012	0.013	0.013	0.014
41	49	.007	.007	.008	.008	.009	.009	.009	.010	.010	.010	.011	.011
42	48	.005	.006	.006	.006	.006	.007	.007	.007	.008	.008	.008	.008
43	47	.004	.004	.004	.004	.004	.004	.005	.005	.005	.005	.005	.006
44	46	.002	.002	.002	.002	.002	.002	.002	.002	.003	.003	.003	.003
45	45	0.000	0.000	0.000	0.000	0.000	0.000	0.000	0.000	0.000	0.000	0.000	0.000

SMITHSONIAN TABLES.

TABLE 13.

REDUCTION OF THE BAROMETER TO STANDARD GRAVITY.
METRIC MEASURES.
Reduction to Latitude 45°.

From latitude 0° to 45°, the correction is to be subtracted.
From latitude 90° to 45°, the correction is to be added.

Latitude.		HEIGHT OF THE BAROMETER IN MILLIMETRES.												
		520	540	560	580	600	620	640	660	680	700	720	740	760	780
0°	90°	mm. 1.38	mm. 1.44	mm. 1.49	mm. 1.54	mm. 1.60	mm. 1.65	mm. 1.70	mm. 1.76	mm. 1.81	mm. 1.86	mm. 1.92	mm. 1.97	mm. 2.02	mm. 2.08
5	85	1.36	1.42	1.47	1.52	1.57	1.63	1.68	1.73	1.78	1.84	1.89	1.94	1.99	2.04
6	84	1.35	1.41	1.46	1.51	1.56	1.61	1.67	1.72	1.77	1.82	1.87	1.93	1.98	2.03
7	83	1.34	1.39	1.45	1.50	1.55	1.60	1.65	1.70	1.76	1.81	1.86	1.91	1.96	2.01
8	82	1.33	1.38	1.43	1.48	1.54	1.59	1.64	1.69	1.74	1.79	1.84	1.89	1.94	2.00
9	81	1.32	1.37	1.42	1.47	1.52	1.57	1.62	1.67	1.72	1.77	1.82	1.87	1.92	1.97
10	80	1.30	1.35	1.40	1.45	1.50	1.55	1.60	1.65	1.70	1.75	1.80	1.85	1.90	1.95
11	79	1.28	1.33	1.38	1.43	1.48	1.53	1.58	1.63	1.68	1.73	1.78	1.83	1.88	1.93
12	78	1.26	1.31	1.36	1.41	1.46	1.51	1.56	1.60	1.65	1.70	1.75	1.80	1.85	1.90
13	77	1.24	1.29	1.34	1.39	1.44	1.48	1.53	1.58	1.63	1.67	1.72	1.77	1.82	1.87
14	76	1.22	1.27	1.32	1.36	1.41	1.46	1.50	1.55	1.60	1.65	1.69	1.74	1.79	1.83
15	75	1.20	1.24	1.29	1.34	1.38	1.43	1.48	1.52	1.57	1.61	1.66	1.71	1.75	1.80
16	74	1.17	1.22	1.26	1.31	1.35	1.40	1.44	1.49	1.54	1.58	1.63	1.67	1.72	1.76
17	73	1.15	1.19	1.24	1.28	1.32	1.37	1.41	1.45	1.50	1.54	1.59	1.63	1.68	1.72
18	72	1.12	1.16	1.21	1.25	1.29	1.34	1.38	1.42	1.46	1.51	1.55	1.59	1.64	1.68
19	71	1.09	1.13	1.17	1.22	1.26	1.30	1.34	1.38	1.43	1.47	1.51	1.55	1.59	1.64
20	70	1.06	1.10	1.14	1.18	1.22	1.26	1.31	1.35	1.39	1.43	1.47	1.51	1.55	1.59
21	69	1.03	1.07	1.11	1.15	1.19	1.23	1.27	1.31	1.35	1.38	1.42	1.46	1.50	1.54
22	68	1.00	1.03	1.07	1.11	1.15	1.19	1.23	1.26	1.30	1.34	1.38	1.42	1.46	1.49
23	67	0.96	1.00	1.04	1.07	1.11	1.15	1.18	1.22	1.26	1.29	1.33	1.37	1.41	1.44
24	66	.93	.96	1.00	1.03	1.07	1.10	1.14	1.18	1.21	1.25	1.28	1.32	1.35	1.39
25	65	0.89	0.92	0.96	0.99	1.03	1.06	1.10	1.13	1.16	1.20	1.23	1.27	1.30	1.33
26	64	.85	.88	.92	.95	0.98	1.02	1.05	1.08	1.11	1.15	1.18	1.21	1.25	1.28
27	63	.81	.84	.88	.91	.94	0.97	1.00	1.03	1.06	1.10	1.13	1.16	1.19	1.22
28	62	.77	.80	.83	.86	.89	.92	0.95	0.98	1.01	1.04	1.07	1.10	1.13	1.16
29	61	.73	.76	.79	.82	.85	.87	.90	.93	0.96	0.99	1.02	1.04	1.07	1.10
30	60	0.69	0.72	0.75	0.77	0.80	0.83	0.85	0.88	0.91	0.94	0.96	0.98	1.01	1.04
31	59	.65	.67	.70	.72	.75	.77	.80	.82	.85	.87	.90	.92	0.95	0.97
32	58	.61	.63	.65	.68	.70	.72	.75	.77	.79	.82	.84	.86	.89	.91
33	57	.56	.58	.61	.63	.65	.67	.69	.71	.74	.76	.78	.80	.82	.84
34	56	.52	.54	.56	.58	.60	.62	.64	.66	.68	.70	.72	.74	.76	.78
35	55	0.47	0.49	0.51	0.53	0.55	0.56	0.58	0.60	0.62	0.64	0.66	0.67	0.69	0.71
36	54	.43	.44	.46	.48	.49	.51	.53	.54	.56	.58	.59	.61	.63	.64
37	53	.38	.40	.41	.43	.44	.45	.47	.48	.50	.51	.53	.54	.56	.57
38	52	.33	.35	.36	.37	.39	.40	.41	.43	.44	.45	.46	.48	.49	.50
39	51	.29	.30	.31	.32	.33	.34	.35	.37	.38	.39	.40	.41	.42	.43
40	50	0.24	0.25	0.26	0.27	0.28	0.29	0.30	0.31	0.31	0.32	0.33	0.34	0.35	0.36
41	49	.19	.20	.21	.21	.22	.23	.24	.24	.25	.26	.27	.27	.28	.29
42	48	.14	.15	.16	.16	.17	.17	.18	.18	.19	.19	.20	.21	.21	.22
43	47	.10	.10	.10	.11	.11	.12	.12	.12	.13	.13	.13	.14	.14	.14
44	46	.05	.05	.05	.05	.06	.06	.06	.06	.06	.07	.07	.07	.07	.07
45	45	0.00	0.00	0.00	0.00	0.00	0.00	0.00	0.00	0.00	0.00	0.00	0.00	0.00	0.00

SMITHSONIAN TABLES.

TABLE 14.
REDUCTION OF THE BAROMETER TO SEA LEVEL.
ENGLISH MEASURES.

Values of $2000 \times m$.

$$m = \frac{Z(1-\beta)}{56573 + 123.1\theta + .003Z}$$

Mean Temperature of air column. θ Fahr.	ALTITUDE OF STATION IN FEET (Z).										Difference for 100 Feet.
	100	200	300	400	500	600	700	800	900	1000	
−20°	3.7	7.4	11.1	14.8	18.5	22.2	25.9	29.6	33.3	37.0	3.7
−16	3.7	7.3	11.0	14.6	18.3	22.0	25.6	29.3	33.0	36.6	3.7
−12	3.6	7.3	10.9	14.5	18.1	21.8	25.4	29.0	32.7	36.3	3.6
−8	3.6	7.2	10.8	14.4	18.0	21.6	25.2	28.8	32.4	36.0	3.6
−6	3.6	7.2	10.7	14.3	17.9	21.5	25.1	28.6	32.2	35.8	3.6
−4	3.6	7.1	10.7	14.3	17.8	21.4	25.0	28.5	32.1	35.6	3.6
−2	3.5	7.1	10.7	14.2	17.7	21.3	24.8	28.4	31.9	35.4	3.5
0	3.5	7.1	10.6	14.1	17.7	21.2	24.7	28.3	31.8	35.3	3.5
+2	3.5	7.0	10.6	14.1	17.6	21.1	24.6	28.1	31.7	35.2	3.5
4	3.5	7.0	10.5	14.0	17.5	21.0	24.5	28.0	31.5	35.0	3.5
6	3.5	7.0	10.5	13.9	17.4	20.9	24.4	27.9	31.4	34.9	3.5
8	3.5	6.9	10.4	13.9	17.4	20.8	24.3	27.8	31.2	34.7	3.5
10	3.5	6.9	10.4	13.8	17.3	20.7	24.2	27.7	31.1	34.6	3.5
12	3.4	6.9	10.3	13.8	17.2	20.6	24.1	27.5	31.0	34.4	3.4
14	3.4	6.9	10.3	13.7	17.1	20.6	24.0	27.4	30.8	34.3	3.4
16	3.4	6.8	10.2	13.6	17.1	20.5	23.9	27.3	30.7	34.1	3.4
18	3.4	6.8	10.2	13.6	17.0	20.4	23.8	27.2	30.6	34.0	3.4
20	3.4	6.8	10.1	13.5	16.9	20.3	23.7	27.1	30.4	33.8	3.4
22	3.4	6.7	10.1	13.5	16.8	20.2	23.6	26.9	30.3	33.7	3.4
24	3.4	6.7	10.1	13.4	16.8	20.1	23.5	26.8	30.2	33.5	3.4
26	3.3	6.7	10.0	13.4	16.7	20.0	23.4	26.7	30.1	33.4	3.3
28	3.3	6.7	10.0	13.3	16.6	20.0	23.3	26.6	29.9	33.3	3.3
30	3.3	6.6	9.9	13.2	16.6	19.9	23.2	26.5	29.8	33.1	3.3
32	3.3	6.6	9.9	13.2	16.5	19.8	23.1	26.4	29.7	33.0	3.3
34	3.3	6.6	9.9	13.1	16.4	19.7	23.0	26.3	29.6	32.8	3.3
36	3.3	6.5	9.8	13.1	16.4	19.6	22.9	26.2	29.4	32.7	3.3
38	3.3	6.5	9.8	13.0	16.3	19.5	22.8	26.0	29.3	32.6	3.3
40	3.2	6.5	9.7	13.0	16.2	19.5	22.7	25.9	29.2	32.4	3.2
42	3.2	6.5	9.7	12.9	16.1	19.4	22.6	25.8	29.1	32.3	3.2
44	3.2	6.4	9.6	12.9	16.1	19.3	22.5	25.7	28.9	32.1	3.2
46	3.2	6.4	9.6	12.8	16.0	19.2	22.4	25.6	28.8	32.0	3.2
48	3.2	6.4	9.6	12.7	15.9	19.1	22.3	25.5	28.7	31.9	3.2
50	3.2	6.3	9.5	12.7	15.9	19.0	22.2	25.4	28.6	31.7	3.2
52	3.2	6.3	9.5	12.6	15.8	19.0	22.1	25.3	28.4	31.6	3.2
54	3.1	6.3	9.4	12.6	15.7	18.9	22.0	25.2	28.3	31.5	3.1
56	3.1	6.3	9.4	12.5	15.7	18.8	21.9	25.1	28.2	31.3	3.1
58	3.1	6.2	9.4	12.5	15.6	18.7	21.8	25.0	28.1	31.2	3.1
60	3.1	6.2	9.3	12.4	15.5	18.6	21.7	24.8	28.0	31.1	3.1
62	3.1	6.2	9.3	12.4	15.5	18.6	21.6	24.7	27.8	30.9	3.1
64	3.1	6.2	9.2	12.3	15.4	18.5	21.6	24.6	27.7	30.8	3.1
66	3.1	6.1	9.2	12.3	15.3	18.4	21.5	24.5	27.6	30.7	3.1
68	3.1	6.1	9.2	12.2	15.3	18.3	21.4	24.4	27.5	30.5	3.1
70	3.0	6.1	9.1	12.2	15.2	18.2	21.3	24.3	27.4	30.4	3.0
72	3.0	6.1	9.1	12.1	15.1	18.2	21.2	24.2	27.3	30.3	3.0
76	3.0	6.0	9.0	12.0	15.0	18.0	21.0	24.0	27.0	30.0	3.0
80	3.0	6.0	8.9	11.9	14.9	17.9	20.9	23.8	26.8	29.8	3.0
84	3.0	5.9	8.9	11.8	14.8	17.7	20.7	23.6	26.6	29.6	3.0
88	2.9	5.9	8.8	11.7	14.7	17.6	20.5	23.5	26.4	29.3	2.9
92	2.9	5.8	8.7	11.6	14.5	17.4	20.4	23.3	26.2	29.1	2.9
96	2.9	5.8	8.7	11.5	14.4	17.3	20.2	23.1	26.0	28.9	2.9

SMITHSONIAN TABLES.

TABLE 14.

REDUCTION OF THE BAROMETER TO SEA LEVEL.
ENGLISH MEASURES.

Values of 2000 × m.

$$m = \frac{Z(1-\beta)}{56573 + 123.1\theta + .003Z}$$

Mean Temperature of air column. θ Fahr.	ALTITUDE OF STATION IN FEET (Z).										Difference for 100 Feet.
	1100	1200	1300	1400	1500	1600	1700	1800	1900	2000	
−20°	40.7	44.3	48.0	51.7	55.4	59.1	62.8	66.5	70.2	73.9	3.7
−16	40.3	43.9	47.6	51.3	54.9	58.6	62.2	65.9	69.6	73.2	3.7
−12	39.9	43.5	47.2	50.8	54.4	58.1	61.7	65.3	68.9	72.6	3.6
−8	39.6	43.2	46.7	50.3	53.9	57.5	61.1	64.7	68.3	71.9	3.6
−6	39.4	43.0	46.5	50.1	53.7	57.3	60.9	64.4	68.0	71.6	3.6
−4	39.2	42.8	46.3	49.9	53.5	57.0	60.6	64.2	67.7	71.3	3.6
−2	39.0	42.6	46.1	49.7	53.2	56.8	60.3	63.9	67.4	71.0	3.6
0	38.9	42.4	45.9	49.5	53.0	56.5	60.1	63.6	67.1	70.6	3.5
+2	38.7	42.2	45.7	49.2	52.8	56.3	59.8	63.3	66.8	70.3	3.5
4	38.5	42.0	45.5	49.0	52.5	56.0	59.5	63.0	66.5	70.0	3.5
6	38.4	41.8	45.3	48.8	52.3	55.8	59.3	62.8	66.2	69.7	3.5
8	38.2	41.7	45.1	48.6	52.1	55.5	59.0	62.5	66.0	69.4	3.5
10	38.0	41.5	44.9	48.4	51.8	55.3	58.8	62.2	65.7	69.1	3.5
12	37.9	41.3	44.7	48.2	51.6	55.1	58.5	61.9	65.4	68.8	3.4
14	37.7	41.1	44.5	48.0	51.4	54.8	58.2	61.7	65.1	68.5	3.4
16	37.5	40.9	44.4	47.8	51.2	54.6	58.0	61.4	64.8	68.2	3.4
18	37.4	40.8	44.2	47.6	51.0	54.4	57.8	61.1	64.5	67.9	3.4
20	37.2	40.6	44.0	47.4	50.7	54.1	57.5	60.9	64.3	67.7	3.4
22	37.1	40.4	43.8	47.2	50.5	53.9	57.3	60.6	64.0	67.4	3.4
24	36.9	40.3	43.6	47.0	50.3	53.7	57.0	60.4	63.7	67.1	3.4
26	36.7	40.1	43.4	46.8	50.1	53.4	56.8	60.1	63.5	66.8	3.3
28	36.6	39.9	43.2	46.6	49.9	53.2	56.5	59.9	63.2	66.5	3.3
30	36.4	39.7	43.1	46.4	49.7	53.0	56.3	59.6	62.9	66.2	3.3
32	36.3	39.6	42.9	46.2	49.5	52.8	56.1	59.4	62.7	66.0	3.3
34	36.1	39.4	42.7	46.0	49.3	52.5	55.8	59.1	62.4	65.7	3.3
36	36.0	39.2	42.5	45.8	49.0	52.3	55.6	58.9	62.1	65.4	3.3
38	35.8	39.1	42.3	45.6	48.8	52.1	55.3	58.6	61.9	65.1	3.3
40	35.7	38.9	42.1	45.4	48.6	51.9	55.1	58.4	61.6	64.8	3.2
42	35.5	38.7	42.0	45.2	48.4	51.6	54.9	58.1	61.3	64.6	3.2
44	35.4	38.6	41.8	45.0	48.2	51.4	54.6	57.9	61.1	64.3	3.2
46	35.2	38.4	41.6	44.8	48.0	51.2	54.4	57.6	60.8	64.0	3.2
48	35.1	38.2	41.4	44.6	47.8	51.0	54.2	57.4	60.5	63.7	3.2
50	34.9	38.1	41.2	44.4	47.6	50.8	53.9	57.1	60.3	63.4	3.2
52	34.8	37.9	41.1	44.2	47.4	50.5	53.7	56.9	60.0	63.2	3.2
54	34.6	37.7	40.9	44.0	47.2	50.3	53.5	56.6	59.8	62.9	3.1
56	34.5	37.6	40.7	43.9	47.0	50.1	53.2	56.4	59.5	62.6	3.1
58	34.3	37.4	40.5	43.7	46.8	49.9	53.0	56.1	59.3	62.4	3.1
60	34.2	37.3	40.4	43.5	46.6	49.7	52.8	55.9	59.0	62.1	3.1
62	34.0	37.1	40.2	43.3	46.4	49.5	52.6	55.7	58.8	61.9	3.1
64	33.9	37.0	40.0	43.1	46.2	49.3	52.4	55.4	58.5	61.6	3.1
66	33.7	36.8	39.9	42.9	46.0	49.1	52.1	55.2	58.3	61.3	3.1
68	33.6	36.6	39.7	42.8	45.8	48.9	51.9	55.0	58.0	61.1	3.1
70	33.5	36.5	39.5	42.6	45.6	48.7	51.7	54.7	57.8	60.8	3.0
72	33.3	36.3	39.4	42.4	45.4	48.5	51.5	54.5	57.5	60.6	3.0
76	33.0	36.0	39.1	42.1	45.1	48.1	51.1	54.1	57.1	60.1	3.0
80	32.8	35.8	38.7	41.7	44.7	47.7	50.6	53.6	56.6	59.6	3.0
84	32.5	35.5	38.4	41.4	44.3	47.3	50.2	53.2	56.2	59.1	3.0
88	32.2	35.2	38.1	41.0	44.0	46.9	49.8	52.8	55.7	58.6	2.9
92	32.0	34.9	37.8	40.7	43.6	46.5	49.4	52.3	55.3	58.2	2.9
96	31.7	34.6	37.5	40.4	43.3	46.2	49.0	51.9	54.8	57.7	2.9

TABLE 14.
REDUCTION OF THE BAROMETER TO SEA LEVEL.
ENGLISH MEASURES.

Values of $2000 \times m$.

$$m = \frac{Z(1-\beta)}{56573 + 123.1\theta + .003Z}$$

Mean Temperature of air column. θ Fahr.	ALTITUDE OF STATION IN FEET (z).										Difference for 100 Feet.
	2100	2200	2300	2400	2500	2600	2700	2800	2900	3000	
−20°	77.6	81.3	85.0	88.7	92.4	96.1	99.8	103.5	107.2	110.9	3.7
−16	76.9	80.6	84.2	87.9	91.5	95.2	98.9	102.5	106.2	109.8	3.7
−12	76.2	79.8	83.5	87.1	90.7	94.3	98.0	101.6	105.2	108.8	3.6
−8	75.5	79.1	82.7	86.3	89.9	93.5	97.1	100.7	104.3	107.9	3.6
−6	75.2	78.8	82.3	85.9	89.5	93.1	96.6	100.2	103.8	107.4	3.6
−4	74.8	78.4	82.0	85.5	89.1	92.7	96.2	99.8	103.3	106.9	3.6
−2	74.5	78.1	81.6	85.2	88.7	92.2	95.8	99.3	102.9	106.4	3.5
0	74.2	77.7	81.2	84.8	88.3	91.8	95.4	98.9	102.5	106.0	3.5
+2	73.9	77.4	80.9	84.4	87.9	91.4	95.0	98.5	102.0	105.5	3.5
4	73.5	77.0	80.5	84.0	87.5	91.0	94.5	98.0	101.5	105.0	3.5
6	73.2	76.7	80.2	83.7	87.2	90.6	94.1	97.6	101.1	104.6	3.5
8	72.9	76.4	79.8	83.3	86.8	90.2	93.7	97.2	100.7	104.1	3.5
10	72.6	76.0	79.5	82.9	86.4	89.9	93.3	96.8	100.2	103.7	3.5
12	72.3	75.7	79.1	82.6	86.0	89.5	92.9	96.3	99.8	103.2	3.4
14	72.0	75.4	78.8	82.2	85.7	89.1	92.5	95.9	99.4	102.8	3.4
16	71.6	75.1	78.5	81.9	85.3	88.7	92.1	95.5	98.9	102.3	3.4
18	71.3	74.7	78.1	81.5	84.9	88.3	91.7	95.1	98.5	101.9	3.4
20	71.0	74.4	77.8	81.2	84.6	87.9	91.3	94.7	98.1	101.5	3.4
22	70.7	74.1	77.5	80.8	84.2	87.6	90.9	94.3	97.7	101.0	3.4
24	70.4	73.8	77.1	80.5	83.8	87.2	90.6	93.9	97.3	100.6	3.3
26	70.1	73.5	76.8	80.2	83.5	86.8	90.2	93.5	96.9	100.2	3.3
28	69.8	73.2	76.5	79.8	83.1	86.5	89.8	93.1	96.4	99.8	3.3
30	69.5	72.9	76.2	79.5	82.8	86.1	89.4	92.7	96.0	99.3	3.3
32	69.2	72.5	75.8	79.1	82.4	85.7	89.0	92.3	95.6	98.9	3.3
34	69.0	72.2	75.5	78.8	82.1	85.4	88.7	91.9	95.2	98.5	3.3
36	68.7	71.9	75.2	78.5	81.7	85.0	88.3	91.5	94.8	98.1	3.3
38	68.4	71.6	74.9	78.1	81.4	84.6	87.9	91.2	94.4	97.7	3.3
40	68.1	71.3	74.6	77.8	81.0	84.3	87.5	90.8	94.0	97.2	3.2
42	67.8	71.0	74.2	77.5	80.7	83.9	87.1	90.4	93.6	96.8	3.2
44	67.5	70.7	73.9	77.1	80.3	83.6	86.8	90.0	93.2	96.4	3.2
46	67.2	70.4	73.6	76.8	80.0	83.2	86.4	89.6	92.8	96.0	3.2
48	66.9	70.1	73.3	76.5	79.7	82.8	86.0	89.2	92.4	95.6	3.2
50	66.6	69.8	73.0	76.1	79.3	82.5	85.7	88.8	92.0	95.2	3.2
52	66.3	69.5	72.7	75.8	79.0	82.1	85.3	88.4	91.6	94.8	3.2
54	66.1	69.2	72.3	75.5	78.6	81.8	84.9	88.1	91.2	94.4	3.1
56	65.8	68.9	72.0	75.2	78.3	81.4	84.6	87.7	90.8	94.0	3.1
58	65.5	68.6	71.7	74.9	78.0	81.1	84.2	87.3	90.4	93.6	3.1
60	65.2	68.3	71.4	74.5	77.6	80.7	83.8	87.0	90.1	93.2	3.1
62	64.9	68.0	71.1	74.2	77.3	80.4	83.5	86.6	89.7	92.8	3.1
64	64.7	67.8	70.8	73.9	77.0	80.1	83.1	86.2	89.3	92.4	3.1
66	64.4	67.5	70.5	73.6	76.7	79.7	82.8	85.9	88.9	92.0	3.1
68	64.1	67.2	70.2	73.3	76.3	79.4	82.5	85.5	88.6	91.6	3.1
70	63.9	66.9	69.9	73.0	76.0	79.1	82.1	85.1	88.2	91.2	3.0
72	63.6	66.6	69.7	72.7	75.7	78.7	81.8	84.8	87.8	90.9	3.0
76	63.1	66.1	69.1	72.1	75.1	78.1	81.1	84.1	87.1	90.1	3.0
80	62.6	65.5	68.5	71.5	74.5	77.5	80.5	83.4	86.4	89.4	3.0
84	62.1	65.0	68.0	70.9	73.9	76.8	79.8	82.7	85.7	88.6	3.0
88	61.6	64.5	67.4	70.4	73.3	76.2	79.1	82.1	85.0	87.9	2.9
92	61.1	64.0	66.9	69.8	72.7	75.6	78.5	81.4	84.3	87.2	2.9
96	60.6	63.5	66.4	69.2	72.1	75.0	77.9	80.8	83.7	86.5	2.9

SMITHSONIAN TABLES.

TABLE 14.
REDUCTION OF THE BAROMETER TO SEA LEVEL.
ENGLISH MEASURES.

Values of 2000 × m.

$$m = \frac{Z(1-\beta)}{56573 + 123.1\theta + .003Z}$$

Mean Temperature of air column. θ Fahr.	ALTITUDE OF STATION IN FEET (Z).									Difference for 100 Feet.
	3100	3200	3300	3400	3500	3600	3700	3800	3900	4000	
−20°	114.5	118.2	121.9	125.6	129.3	133.0	136.7	140.4	144.1	147.8	3.7
−16	113.5	117.2	120.8	124.5	128.1	131.8	135.5	139.1	142.8	146.4	3.7
−12	112.5	116.1	119.7	123.3	127.0	130.6	134.2	137.9	141.5	145.1	3.6
−8	111.5	115.1	118.7	122.3	125.9	129.4	133.0	136.6	140.2	143.8	3.6
−6	111.0	114.5	118.1	121.7	125.3	128.9	132.4	136.0	139.6	143.2	3.6
−4	110.5	114.0	117.6	121.2	124.7	128.3	131.9	135.4	139.0	142.5	3.6
−2	110.0	113.5	117.1	120.6	124.2	127.7	131.3	134.8	138.4	141.9	3.5
0	109.5	113.0	116.6	120.1	123.6	127.2	130.7	134.2	137.8	141.3	3.5
+2	109.0	112.5	116.1	119.6	123.1	126.6	130.1	133.6	137.1	140.7	3.5
4	108.5	112.0	115.5	119.0	122.5	126.0	129.5	133.0	136.5	140.0	3.5
6	108.1	111.6	115.0	118.5	122.0	125.5	129.0	132.5	135.9	139.4	3.5
8	107.6	111.1	114.5	118.0	121.5	124.9	128.4	131.9	135.4	138.8	3.5
10	107.1	110.6	114.0	117.5	121.0	124.4	127.9	131.3	134.8	138.2	3.5
12	106.7	110.1	113.6	117.0	120.4	123.9	127.3	130.7	134.2	137.6	3.4
14	106.2	109.6	113.1	116.5	119.9	123.3	126.8	130.2	133.6	137.0	3.4
16	105.8	109.2	112.6	116.0	119.4	122.8	126.2	129.6	133.0	136.5	3.4
18	105.3	108.7	112.1	115.5	118.9	122.3	125.7	129.1	132.5	135.9	3.4
20	104.9	108.2	111.6	115.0	118.4	121.8	125.1	128.5	131.9	135.3	3.4
22	104.4	107.8	111.1	114.5	117.9	121.2	124.6	128.0	131.3	134.7	3.4
24	104.0	107.3	110.7	114.0	117.4	120.7	124.1	127.4	130.8	134.1	3.4
26	103.5	106.9	110.2	113.5	116.9	120.2	123.6	126.9	130.2	133.6	3.3
28	103.1	106.4	109.7	113.1	116.4	119.7	123.0	126.4	129.7	133.0	3.3
30	102.7	106.0	109.3	112.6	115.9	119.2	122.5	125.8	129.1	132.4	3.3
32	102.2	105.5	108.8	112.1	115.4	118.7	122.0	125.3	128.6	131.9	3.3
34	101.8	105.1	108.3	111.6	114.9	118.2	121.5	124.8	128.0	131.3	3.3
36	101.3	104.6	107.9	111.2	114.4	117.7	121.0	124.2	127.5	130.8	3.3
38	100.9	104.1	107.4	110.7	113.9	117.2	120.4	123.7	126.9	130.2	3.3
40	100.5	103.7	107.0	110.2	113.4	116.7	119.9	123.2	126.4	129.6	3.2
42	100.0	103.3	106.5	109.7	113.0	116.2	119.4	122.6	125.9	129.1	3.2
44	99.6	102.8	106.0	109.3	112.5	115.7	118.9	122.1	125.3	128.5	3.2
46	99.2	102.4	105.6	108.8	112.0	115.2	118.4	121.6	124.8	128.0	3.2
48	98.8	101.9	105.1	108.3	111.5	114.7	117.9	121.1	124.2	127.4	3.2
50	98.3	101.5	104.7	107.9	111.0	114.2	117.4	120.5	123.7	126.9	3.2
52	97.9	101.1	104.2	107.4	110.5	113.7	116.9	120.0	123.2	126.3	3.2
54	97.5	100.6	103.8	106.9	110.1	113.2	116.4	119.5	122.7	125.8	3.1
56	97.1	100.2	103.3	106.5	109.6	112.7	115.9	119.0	122.1	125.2	3.1
58	96.7	99.8	102.9	106.0	109.1	112.3	115.4	118.5	121.6	124.7	3.1
60	96.3	99.4	102.5	105.6	108.7	111.8	114.9	118.0	121.1	124.2	3.1
62	95.9	98.9	102.0	105.1	108.2	111.3	114.4	117.5	120.6	123.7	3.1
64	95.5	98.5	101.6	104.7	107.8	110.8	113.9	117.0	120.1	123.1	3.1
66	95.1	98.1	101.2	104.3	107.3	110.4	113.5	116.5	119.6	122.6	3.1
68	94.7	97.7	100.8	103.8	106.9	109.9	113.0	116.0	119.1	122.1	3.1
70	94.3	97.3	100.3	103.4	106.4	109.5	112.5	115.5	118.6	121.6	3.0
72	93.9	96.9	99.9	103.0	106.0	109.0	112.0	115.1	118.1	121.1	3.0
76	93.1	96.1	99.1	102.1	105.1	108.1	111.1	114.1	117.1	120.1	3.0
80	92.3	95.3	98.3	101.3	104.3	107.2	110.2	113.2	116.2	119.2	3.0
84	91.6	94.5	97.5	100.5	103.4	106.4	109.3	112.3	115.2	118.2	3.0
88	90.9	93.8	96.7	99.7	102.6	105.5	108.4	111.4	114.3	117.2	2.9
92	90.1	93.0	96.0	98.9	101.8	104.7	107.6	110.5	113.4	116.3	2.9
96	89.4	92.3	95.2	98.1	101.0	103.8	106.7	109.6	112.5	115.4	2.9

SMITHSONIAN TABLES.

TABLE 14.
REDUCTION OF THE BAROMETER TO SEA LEVEL
ENGLISH MEASURES.

Values of $2000 \times m$.

$$m = \frac{Z(1-\beta)}{56573 + 123.1\theta + .003Z}$$

Mean Temperature of air column. θ Fahr.	ALTITUDE OF STATION IN FEET (z).										Difference for 100 Feet.
	4100	4200	4300	4400	4500	4600	4700	4800	4900	5000	
−20°	151.5	155.2	158.9	162.6	166.3	170.0	173.7	177.3	181.0	184.7	3.7
−16	150.1	153.8	157.4	161.1	164.8	168.4	172.1	175.7	179.4	183.1	3.7
−12	148.8	152.4	156.0	159.6	163.3	166.9	170.5	174.1	177.8	181.4	3.6
−8	147.4	151.0	154.6	158.2	161.8	165.4	169.0	172.6	176.2	179.8	3.6
−6	146.8	150.3	153.9	157.5	161.1	164.7	168.2	171.8	175.4	179.0	3.6
−4	146.1	149.7	153.2	156.8	160.4	163.9	167.5	171.0	174.6	178.2	3.6
−2	145.5	149.0	152.5	156.1	159.6	163.2	166.7	170.3	173.8	177.4	3.5
0	144.8	148.3	151.9	155.4	158.9	162.5	166.0	169.5	173.1	176.6	3.5
+2	144.2	147.7	151.2	154.7	158.2	161.8	165.3	168.8	172.3	175.8	3.5
4	143.5	147.0	150.5	154.0	157.5	161.0	164.5	168.0	171.5	175.0	3.5
6	142.9	146.4	149.9	153.4	156.9	160.3	163.8	167.3	170.8	174.3	3.5
8	142.3	145.8	149.2	152.7	156.2	159.6	163.1	166.6	170.1	173.5	3.5
10	141.7	145.1	148.6	152.0	155.5	159.0	162.4	165.9	169.3	172.8	3.5
12	141.1	144.5	147.9	151.4	154.8	158.3	161.7	165.1	168.6	172.0	3.4
14	140.5	143.9	147.3	150.7	154.2	157.6	161.0	164.4	167.9	171.3	3.4
16	139.9	143.3	146.7	150.1	153.5	156.9	160.3	163.7	167.1	170.6	3.4
18	139.3	142.7	146.1	149.5	152.9	156.2	159.6	162.0	166.4	169.8	3.4
20	138.7	142.1	145.4	148.8	152.2	155.6	159.0	162.3	165.7	169.1	3.4
22	138.1	141.5	144.8	148.2	151.6	154.9	158.3	161.7	165.0	168.4	3.4
24	137.5	140.9	144.2	147.6	150.9	154.3	157.6	161.0	164.3	167.7	3.4
26	136.9	140.3	143.6	146.9	150.3	153.6	157.0	160.3	163.6	167.0	3.3
28	136.3	139.7	143.0	146.3	149.6	153.0	156.3	159.6	162.9	166.3	3.3
30	135.8	139.1	142.4	145.7	149.0	152.3	155.6	158.9	162.2	165.6	3.3
32	135.2	138.5	141.8	145.1	148.4	151.7	155.0	158.3	161.6	164.8	3.3
34	134.6	137.9	141.2	144.5	147.7	151.0	154.3	157.6	160.9	164.1	3.3
36	134.0	137.3	140.6	143.8	147.1	150.4	153.6	156.9	160.2	163.4	3.3
38	133.5	136.7	140.0	143.2	146.5	149.7	153.0	156.2	159.5	162.7	3.3
40	132.9	136.1	139.4	142.6	145.8	149.1	152.3	155.6	158.8	162.0	3.2
42	132.3	135.5	138.8	142.0	145.2	148.4	151.7	154.9	158.1	161.4	3.2
44	131.7	135.0	138.2	141.4	144.6	147.8	151.0	154.2	157.4	160.7	3.2
46	131.2	134.4	137.6	140.8	144.0	147.2	150.4	153.6	156.8	160.0	3.2
48	130.6	133.8	137.0	140.2	143.4	146.5	149.7	152.9	156.1	159.3	3.2
50	130.1	133.2	136.4	139.6	142.7	145.9	149.1	152.2	155.4	158.6	3.2
52	129.5	132.6	135.8	139.0	142.1	145.3	148.4	151.6	154.8	157.9	3.2
54	129.0	132.1	135.2	138.4	141.5	144.7	147.8	151.0	154.1	157.2	3.1
56	128.4	131.5	134.7	137.8	140.9	144.0	147.2	150.3	153.4	156.6	3.1
58	127.9	131.0	134.1	137.2	140.3	143.4	146.6	149.7	152.8	155.9	3.1
60	127.3	130.4	133.5	136.6	139.7	142.8	145.9	149.0	152.1	155.2	3.1
62	126.8	129.9	133.0	136.0	139.1	142.2	145.3	148.4	151.5	154.6	3.1
64	126.2	129.3	132.4	135.5	138.6	141.6	144.7	147.8	150.9	153.9	3.1
66	125.7	128.8	131.8	134.9	138.0	141.0	144.1	147.2	150.2	153.3	3.1
68	125.2	128.2	131.3	134.3	137.4	140.5	143.5	146.6	149.6	152.7	3.1
70	124.7	127.7	130.7	133.8	136.8	139.9	142.9	145.9	149.0	152.0	3.0
72	124.2	127.2	130.2	133.2	136.3	139.3	142.3	145.3	148.4	151.4	3.0
76	123.1	126.1	129.1	132.1	135.1	138.1	141.1	144.1	147.2	150.2	3.0
80	122.1	125.1	128.1	131.1	134.0	137.0	140.0	143.0	146.0	148.9	3.0
84	121.1	124.0	127.0	130.0	133.0	135.9	138.9	141.8	144.8	147.7	3.0
88	120.2	123.1	126.0	129.0	131.9	134.8	137.7	140.7	143.6	146.5	2.9
92	119.2	122.1	125.0	127.9	130.8	133.7	136.6	139.6	142.5	145.4	2.9
96	118.3	121.1	124.0	126.9	129.8	132.7	135.6	138.4	141.3	144.2	2.9

SMITHSONIAN TABLES.

TABLE 14.
REDUCTION OF THE BAROMETER TO SEA LEVEL.
ENGLISH MEASURES.

Values of 2000 × m.

$$m = \frac{Z(1-\beta)}{56573 + 123.1\theta + .003Z}$$

Mean Temperature of air column. θ Fahr.	ALTITUDE OF STATION IN FEET (Z).									Difference for 100 Feet.
	5100	5200	5300	5400	5500	5600	5700	5800	5900	6000	
−20°	188.4	192.1	195.8	199.5	203.2	206.9	210.6	214.3	218.0	221.7	3.7
−16	186.7	190.4	194.0	197.7	201.4	205.0	208.7	212.3	216.0	219.7	3.7
−12	185.0	188.7	192.3	195.9	199.5	203.2	206.8	210.4	214.0	217.7	3.6
−8	183.4	187.0	190.5	194.1	197.7	201.3	204.9	208.5	212.1	215.7	3.6
−6	182.5	186.1	189.7	193.3	196.9	200.4	204.0	207.6	211.2	214.7	3.6
−4	181.7	185.3	188.9	192.4	196.0	199.5	203.1	206.7	210.2	213.8	3.6
−2	180.9	184.5	188.0	191.6	195.1	198.7	202.2	205.7	209.3	212.8	3.5
0	180.1	183.7	187.2	190.7	194.3	197.8	201.3	204.8	208.4	211.9	3.5
+2	179.3	182.8	186.4	189.9	193.4	196.9	200.4	203.9	207.5	211.0	3.5
4	178.5	182.0	185.5	189.0	192.5	196.0	199.5	203.0	206.5	210.0	3.5
6	177.8	181.3	184.7	188.2	191.7	195.2	198.7	202.2	205.6	209.1	3.5
8	177.0	180.5	183.9	187.4	190.9	194.3	197.8	201.3	204.8	208.2	3.5
10	176.2	179.7	183.1	186.6	190.0	193.5	197.0	200.4	203.9	207.3	3.5
12	175.5	178.9	182.3	185.8	189.2	192.7	196.1	199.5	203.0	206.4	3.4
14	174.7	178.1	181.6	185.0	188.4	191.8	195.3	198.7	202.1	205.5	3.4
16	174.0	177.4	180.8	184.2	187.6	191.0	194.4	197.8	201.2	204.7	3.4
18	173.2	176.6	180.0	183.4	186.8	190.2	193.6	197.0	200.4	203.8	3.4
20	172.5	175.9	179.2	182.6	186.0	189.4	192.8	196.2	199.5	202.9	3.4
22	171.8	175.1	178.5	181.9	185.2	188.6	192.0	195.3	198.7	202.1	3.4
24	171.0	174.4	177.7	181.1	184.4	187.8	191.1	194.5	197.8	201.2	3.4
26	170.3	173.6	177.0	180.3	183.7	187.0	190.3	193.7	197.0	200.3	3.3
28	169.6	172.9	176.2	179.6	182.9	186.2	189.5	192.9	196.2	199.5	3.3
30	168.9	172.2	175.5	178.8	182.1	185.4	188.7	192.0	195.3	198.7	3.3
32	168.1	171.4	174.7	178.0	181.3	184.6	187.9	191.2	194.5	197.8	3.3
34	167.4	170.7	174.0	177.3	180.6	183.8	187.1	190.4	193.7	197.0	3.3
36	166.7	170.0	173.2	176.5	179.8	183.1	186.3	189.6	192.9	196.1	3.3
38	166.0	169.3	172.5	175.8	179.0	182.3	185.5	188.8	192.0	195.3	3.3
40	165.3	168.5	171.8	175.0	178.2	181.5	184.7	188.0	191.2	194.4	3.2
42	164.6	167.8	171.0	174.3	177.5	180.7	183.9	187.2	190.4	193.6	3.2
44	163.9	167.1	170.3	173.5	176.7	179.9	183.1	186.4	189.6	192.8	3.2
46	163.2	166.4	169.6	172.8	176.0	179.2	182.4	185.6	188.7	192.0	3.2
48	162.5	165.6	168.8	172.0	175.2	178.4	181.6	184.8	187.9	191.1	3.2
50	161.8	164.9	168.1	171.3	174.4	177.6	180.8	184.0	187.1	190.3	3.2
52	161.1	164.2	167.4	170.5	173.7	176.9	180.0	183.2	186.3	189.5	3.2
54	160.4	163.5	166.7	169.8	173.0	176.1	179.2	182.4	185.5	188.7	3.1
56	159.7	162.8	166.0	169.1	172.2	175.4	178.5	181.6	184.7	187.9	3.1
58	159.0	162.1	165.3	168.4	171.5	174.6	177.7	180.8	184.0	187.1	3.1
60	158.4	161.5	164.6	167.7	170.8	173.9	177.0	180.1	183.2	186.3	3.1
62	157.7	160.8	163.9	167.0	170.1	173.1	176.2	179.3	182.4	185.5	3.1
64	157.0	160.1	163.2	166.3	169.3	172.4	175.5	178.6	181.6	184.7	3.1
66	156.4	159.4	162.5	165.6	168.6	171.7	174.8	177.8	180.9	184.0	3.1
68	155.7	158.8	161.8	164.9	167.9	171.0	174.0	177.1	180.1	183.2	3.1
70	155.1	158.1	161.1	164.2	167.2	170.3	173.3	176.3	179.4	182.4	3.0
72	154.4	157.5	160.5	163.5	166.5	169.6	172.6	175.6	178.6	181.7	3.0
76	153.2	156.2	159.2	162.2	165.2	168.2	171.2	174.2	177.2	180.2	3.0
80	151.9	154.9	157.9	160.8	163.8	166.8	169.8	172.8	175.7	178.7	3.0
84	150.7	153.6	156.6	159.5	162.5	165.5	168.4	171.4	174.3	177.3	3.0
88	149.5	152.4	155.3	158.3	161.2	164.1	167.0	170.0	172.9	175.8	2.9
92	148.3	151.2	154.1	157.0	159.9	162.8	165.7	168.6	171.5	174.4	2.9
96	147.1	150.0	152.9	155.7	158.6	161.5	164.4	167.3	170.2	173.0	2.9

TABLE 14.
REDUCTION OF THE BAROMETER TO SEA LEVEL
ENGLISH MEASURES.

Values of $2000 \times m$.

$$m = \frac{Z(1-\beta)}{56573 + 123.1\theta + .003Z}$$

Mean Temperature of air column. θ Fahr.	ALTITUDE OF STATION IN FEET (z).										Difference for 100 feet.
	6100	6200	6300	6400	6500	6600	6700	6800	6900	7000	
−20°	225.4	229.1	232.8	236.4	240.1	243.8	247.5	251.2	254.9	258.6	3.7
−16	223.3	227.0	230.6	234.3	237.9	241.6	245.3	248.9	252.6	256.2	3.7
−12	221.3	224.9	228.5	232.2	235.8	239.4	243.0	246.7	250.3	253.9	3.6
−8	219.3	222.9	226.5	230.1	233.7	237.3	240.9	244.5	248.1	251.6	3.6
−6	218.3	221.9	225.5	229.1	232.6	236.2	239.8	243.4	246.9	250.5	3.6
−4	217.4	220.9	224.5	228.0	231.6	235.2	238.7	242.3	245.8	249.4	3.6
−2	216.4	219.9	223.5	227.0	230.6	234.1	237.7	241.2	244.8	248.3	3.5
0	215.4	219.0	222.5	226.0	229.6	233.1	236.6	240.1	243.7	247.2	3.5
+2	214.5	218.0	221.5	225.0	228.5	232.1	235.6	239.1	242.6	246.1	3.5
4	213.5	217.0	220.5	224.0	227.5	231.0	234.5	238.0	241.5	245.0	3.5
6	212.6	216.1	219.6	223.1	226.6	230.0	233.5	237.0	240.5	244.0	3.5
8	211.7	215.2	218.6	222.1	225.6	229.0	232.5	236.0	239.4	242.9	3.5
10	210.8	214.2	217.7	221.1	224.6	228.0	231.5	235.0	238.4	241.9	3.5
12	209.9	213.3	216.7	220.2	223.6	227.1	230.5	233.9	237.4	240.8	3.4
14	209.0	212.4	215.8	219.2	222.7	226.1	229.5	232.9	236.4	239.8	3.4
16	208.1	211.5	214.9	218.3	221.7	225.1	228.5	231.9	235.3	238.8	3.4
18	207.2	210.6	214.0	217.4	220.8	224.2	227.6	230.9	234.3	237.7	3.4
20	206.3	209.7	213.1	216.4	219.8	223.2	226.6	230.0	233.3	236.7	3.4
22	205.4	208.8	212.2	215.5	218.9	222.3	225.6	229.0	232.4	235.7	3.4
24	204.6	207.9	211.3	214.6	218.0	221.3	224.7	228.0	231.4	234.7	3.4
26	203.7	207.0	210.4	213.7	217.0	220.4	223.7	227.0	230.4	233.7	3.3
28	202.8	206.2	209.5	212.8	216.1	219.4	222.8	226.1	229.4	232.7	3.3
30	202.0	205.3	208.6	211.9	215.2	218.5	221.8	225.1	228.4	231.8	3.3
32	201.1	204.4	207.7	211.0	214.3	217.6	220.9	224.2	227.5	230.8	3.3
34	200.2	203.5	206.8	210.1	213.4	216.7	219.9	223.2	226.5	229.8	3.3
36	199.4	202.7	205.9	209.2	212.5	215.7	219.0	222.3	225.5	228.8	3.3
38	198.5	201.8	205.0	208.3	211.6	214.8	218.1	221.3	224.6	227.8	3.3
40	197.7	200.9	204.2	207.4	210.6	213.9	217.1	220.4	223.6	226.8	3.2
42	196.8	200.1	203.3	206.5	209.7	213.0	216.2	219.4	222.6	225.9	3.2
44	196.0	199.2	202.4	205.6	208.8	212.1	215.3	218.4	221.7	224.9	3.2
46	195.2	198.4	201.5	204.7	207.9	211.1	214.3	217.5	220.7	223.9	3.2
48	194.3	197.5	200.7	203.9	207.0	210.2	213.4	216.6	219.8	223.0	3.2
50	193.5	196.6	199.8	203.0	206.2	209.3	212.5	215.7	218.8	222.0	3.2
52	192.6	195.8	199.0	202.1	205.3	208.4	211.6	214.7	217.9	221.1	3.2
54	191.8	195.0	198.1	201.3	204.4	207.5	210.7	213.8	217.0	220.1	3.1
56	191.0	194.1	197.3	200.4	203.5	206.7	209.8	212.9	216.0	219.2	3.1
58	190.2	193.3	196.4	199.5	202.7	205.8	208.9	212.0	215.1	218.3	3.1
60	189.4	192.5	195.6	198.7	201.8	204.9	208.0	211.1	214.2	217.3	3.1
62	188.6	191.7	194.8	197.9	201.0	204.1	207.2	210.2	213.3	216.4	3.1
64	187.8	190.9	194.0	197.0	200.1	203.2	206.3	209.3	212.4	215.5	3.1
66	187.0	190.1	193.1	196.2	199.3	202.3	205.4	208.5	211.5	214.6	3.1
68	186.2	189.3	192.3	195.4	198.4	201.5	204.6	207.6	210.7	213.7	3.0
70	185.5	188.5	191.5	194.6	197.6	200.7	203.7	206.7	209.8	212.8	3.0
72	184.7	187.7	190.8	193.8	196.8	199.8	202.9	205.9	208.9	211.9	3.0
76	183.2	186.2	189.2	192.2	195.2	198.2	201.2	204.2	207.2	210.2	3.0
80	181.7	184.7	187.6	190.6	193.6	196.6	199.6	202.5	205.5	208.5	3.0
84	180.2	183.2	186.1	189.1	192.0	195.0	197.9	200.9	203.8	206.8	3.0
88	178.8	181.7	184.6	187.6	190.5	193.4	196.3	199.3	202.2	205.1	2.9
92	177.3	180.2	183.2	186.1	189.0	191.9	194.8	197.7	200.6	203.5	2.9
96	175.9	178.8	181.7	184.6	187.5	190.3	193.2	196.1	199.0	201.9	2.9

TABLE 14.
REDUCTION OF THE BAROMETER TO SEA LEVEL
ENGLISH MEASURES.

Values of 2000 × m.

$$m = \frac{Z(1-\beta)}{56573 + 123.1\theta + .003Z}$$

Mean Temperature of air column. θ Fahr.	\multicolumn{10}{c	}{ALTITUDE OF STATION IN FEET (Z).}	Difference for 100 Feet.
	7100	7200	7300	7400	7500	7600	7700	7800	7900	8000	
−20°	262.3	266.0	269.7	273.4	277.1	280.8	284.5	288.1	291.8	295.5	3.7
−16	259.9	263.6	267.2	270.9	274.5	278.2	281.9	285.5	289.2	292.8	3.7
−12	257.6	261.2	264.8	268.4	272.1	275.7	279.3	282.9	286.6	290.2	3.6
−8	255.2	258.8	262.4	266.0	269.6	273.2	276.8	280.4	284.0	287.6	3.6
−6	254.1	257.7	261.3	264.8	268.4	272.0	275.6	279.1	282.7	286.3	3.6
−4	253.0	256.5	260.1	263.7	267.2	270.8	274.3	277.9	281.5	285.0	3.6
−2	251.8	255.4	258.9	262.5	266.0	269.6	273.1	276.7	280.2	283.8	3.5
0	250.7	254.3	257.8	261.3	264.9	268.4	271.9	275.4	279.0	282.5	3.5
+2	249.6	253.1	256.7	260.2	263.7	267.2	270.7	274.2	277.8	281.3	3.5
4	248.5	252.0	255.5	259.0	262.5	266.0	269.5	273.0	276.5	280.0	3.5
6	247.5	250.9	254.4	257.9	261.4	264.9	268.4	271.8	275.3	278.8	3.5
8	246.4	249.8	253.3	256.8	260.3	263.7	267.2	270.7	274.1	277.6	3.5
10	245.3	248.8	252.2	255.7	259.1	262.6	266.0	269.5	272.9	276.4	3.5
12	244.3	247.7	251.1	254.6	258.0	261.4	264.9	268.3	271.8	275.2	3.4
14	243.2	246.6	250.1	253.5	256.9	260.3	263.8	267.2	270.6	274.0	3.4
16	242.2	245.6	249.0	252.4	255.8	259.2	262.6	266.0	269.4	272.8	3.4
18	241.1	244.5	247.9	251.3	254.7	258.1	261.5	264.9	268.3	271.7	3.4
20	240.1	243.5	246.9	250.2	253.6	257.0	260.4	263.8	267.1	270.5	3.4
22	239.1	242.4	245.8	249.2	252.5	255.9	259.3	262.6	266.0	269.4	3.4
24	238.1	241.4	244.8	248.1	251.5	254.8	258.2	261.5	264.9	268.2	3.4
26	237.1	240.4	243.7	247.1	250.4	253.8	257.1	260.4	263.8	267.1	3.3
28	236.1	239.4	242.7	246.0	249.4	252.7	256.0	259.3	262.7	266.0	3.3
30	235.1	238.4	241.7	245.0	248.3	251.6	254.9	258.2	261.5	264.8	3.3
32	234.1	237.4	240.7	243.9	247.2	250.5	253.8	257.1	260.4	263.7	3.3
34	233.1	236.3	239.6	242.9	246.2	249.5	252.8	256.0	259.3	262.6	3.3
36	232.1	235.3	238.6	241.9	245.1	248.4	251.7	254.9	258.2	261.5	3.3
38	231.1	234.3	237.6	240.8	244.1	247.3	250.6	253.9	257.1	260.4	3.3
40	230.1	233.3	236.6	239.8	243.0	246.3	249.5	252.8	256.0	259.2	3.2
42	229.1	232.3	235.5	238.8	242.0	245.2	248.4	251.7	254.9	258.1	3.2
44	228.1	231.3	234.5	237.7	241.0	244.2	247.4	250.6	253.8	257.0	3.2
46	227.1	230.3	233.5	236.7	239.9	243.1	246.3	249.5	252.7	255.9	3.2
48	226.2	229.3	232.5	235.7	238.9	242.1	245.3	248.4	251.6	254.8	3.2
50	225.2	228.4	231.5	234.7	237.9	241.0	244.2	247.4	250.5	253.7	3.2
52	224.2	227.4	230.5	233.7	236.8	240.0	243.2	246.3	249.5	252.6	3.2
54	223.3	226.4	229.5	232.7	235.8	239.0	242.1	245.3	248.4	251.5	3.1
56	222.3	225.4	228.6	231.7	234.8	238.0	241.1	244.2	247.3	250.5	3.1
58	221.4	224.5	227.6	230.7	233.8	236.9	240.1	243.2	246.3	249.4	3.1
60	220.4	223.5	226.6	229.7	232.8	235.9	239.1	242.2	245.3	248.4	3.1
62	219.5	222.6	225.7	228.8	231.9	235.0	238.0	241.1	244.2	247.3	3.1
64	218.6	221.7	224.7	227.8	230.9	234.0	237.0	240.1	243.2	246.3	3.1
66	217.7	220.7	223.8	226.9	229.9	233.0	236.1	239.1	242.2	245.2	3.1
68	216.8	219.8	222.9	225.9	229.0	232.0	235.1	238.1	241.2	244.2	3.0
70	215.9	218.9	221.9	225.0	228.0	231.1	234.1	237.1	240.2	243.2	3.0
72	215.0	218.0	221.0	224.1	227.1	230.1	233.1	236.2	239.2	242.2	3.0
76	213.2	216.2	219.2	222.2	225.2	228.2	231.2	234.2	237.2	240.2	3.0
80	211.5	214.4	217.4	220.4	223.4	226.4	229.3	232.3	235.3	238.3	3.0
84	209.8	212.7	215.7	218.6	221.6	224.5	227.5	230.4	233.4	236.3	2.9
88	208.1	211.0	213.9	216.9	219.8	222.7	225.6	228.6	231.5	234.4	2.9
92	206.4	209.3	212.2	215.1	218.0	220.9	223.8	226.7	229.7	232.6	2.9
96	204.8	207.6	210.5	213.4	216.3	219.2	222.1	224.9	227.8	230.7	2.9

SMITHSONIAN TABLES.

TABLE 14.
REDUCTION OF THE BAROMETER TO SEA LEVEL
ENGLISH MEASURES.

Values of 2000 × m.

$$m = \frac{Z(1-\beta)}{56573 + 123.1\theta + .003Z}$$

Mean Temperature of air column. θ Fahr.	ALTITUDE OF STATION IN FEET (z).									Difference for 100 Feet.
	8100	8200	8300	8400	8500	8600	8700	8800	8900	9000	
−20°	299.2	302.9	306.6	310.3	314.0	317.7	321.4	325.1	328.8	332.5	3.7
−16	296.5	300.2	303.8	307.5	311.1	314.8	318.4	322.1	325.8	329.4	3.7
−12	293.8	297.4	301.1	304.7	308.3	311.9	315.6	319.2	322.8	326.4	3.6
−8	291.2	294.8	298.4	302.0	305.5	309.1	312.7	316.3	319.9	323.5	3.6
−6	289.9	293.5	297.0	300.6	304.2	307.8	311.3	314.9	318.5	322.1	3.6
−4	288.6	292.1	295.7	299.3	302.8	306.4	309.9	313.5	317.1	320.6	3.6
−2	287.3	290.9	294.4	297.9	301.5	305.0	308.6	312.1	315.7	319.2	3.5
0	286.0	289.6	293.1	296.6	300.2	303.7	307.2	310.7	314.3	317.8	3.5
+2	284.8	288.3	291.8	295.3	298.8	302.4	305.9	309.4	312.9	316.4	3.5
4	283.5	287.0	290.5	294.0	297.5	301.0	304.5	308.0	311.5	315.0	3.5
6	282.3	285.8	289.3	292.7	296.2	299.7	303.2	306.7	310.2	313.6	3.5
8	281.1	284.5	288.0	291.5	294.9	298.4	301.9	305.3	308.8	312.3	3.5
10	279.8	283.3	286.8	290.2	293.7	297.1	300.6	304.0	307.5	310.9	3.5
12	278.6	282.1	285.5	289.0	292.4	295.8	299.3	302.7	306.2	309.6	3.4
14	277.5	280.9	284.3	287.7	291.1	294.6	298.0	301.4	304.8	308.3	3.4
16	276.3	279.7	283.1	286.5	289.9	293.3	296.7	300.1	303.5	306.9	3.4
18	275.1	278.5	281.9	285.3	288.7	292.1	295.4	298.8	302.2	305.6	3.4
20	273.9	277.3	280.7	284.0	287.4	290.8	294.2	297.6	300.9	304.3	3.4
22	272.7	276.1	279.5	282.8	286.2	289.6	292.9	296.3	299.7	303.0	3.4
24	271.6	274.9	278.3	281.6	285.0	288.3	291.7	295.0	298.4	301.8	3.4
26	270.4	273.8	277.1	280.5	283.8	287.1	290.5	293.8	297.1	300.5	3.3
28	269.3	272.6	275.9	279.3	282.6	285.9	289.2	292.6	295.9	299.2	3.3
30	268.2	271.5	274.8	278.1	281.4	284.7	288.0	291.3	294.6	297.9	3.3
32	267.0	270.3	273.6	276.9	280.2	283.5	286.8	290.1	293.4	296.7	3.3
34	265.9	269.2	272.4	275.7	279.0	282.3	285.6	288.8	292.1	295.4	3.3
36	264.7	268.0	271.3	274.5	277.8	281.1	284.3	287.6	290.9	294.1	3.3
38	263.6	266.9	270.1	273.4	276.6	279.9	283.1	286.4	289.6	292.9	3.3
40	262.5	265.7	269.0	272.2	275.4	278.7	281.9	285.2	288.4	291.6	3.2
42	261.4	264.6	267.8	271.0	274.3	277.5	280.7	283.9	287.2	290.4	3.2
44	260.2	263.4	266.7	269.9	273.1	276.3	279.5	282.7	285.9	289.1	3.2
46	259.1	262.3	265.5	268.7	271.9	275.1	278.3	281.5	284.7	287.9	3.2
48	258.0	261.2	264.4	267.5	270.7	273.9	277.1	280.3	283.5	286.6	3.2
50	256.9	260.1	263.2	266.4	269.6	272.7	275.9	279.1	282.2	285.4	3.2
52	255.8	258.9	262.1	265.3	268.4	271.6	274.7	277.9	281.0	284.2	3.1
54	254.7	257.8	261.0	264.1	267.3	270.4	273.5	276.7	279.8	283.0	3.1
56	253.6	256.7	259.9	263.0	266.1	269.3	272.4	275.5	278.6	281.8	3.1
58	252.5	255.6	258.8	261.9	265.0	268.1	271.2	274.3	277.5	280.6	3.1
60	251.5	254.6	257.7	260.8	263.9	267.0	270.1	273.2	276.3	279.4	3.1
62	250.4	253.5	256.6	259.7	262.8	265.9	268.9	272.0	275.1	278.2	3.1
64	249.4	252.4	255.5	258.6	261.7	264.7	267.8	270.9	274.0	277.1	3.1
66	248.3	251.4	254.4	257.5	260.6	263.6	266.7	269.8	272.8	275.9	3.1
68	247.3	250.3	253.4	256.4	259.5	262.5	265.6	268.6	271.7	274.7	3.0
70	246.3	249.3	252.3	255.4	258.4	261.4	264.5	267.5	270.6	273.6	3.0
72	245.2	248.3	251.3	254.3	257.3	260.4	263.4	266.4	269.4	272.5	3.0
76	243.2	246.2	249.2	252.2	255.2	258.2	261.2	264.2	267.2	270.2	3.0
80	241.2	244.2	247.2	250.2	253.1	256.1	259.1	262.1	265.1	268.0	3.0
84	239.3	242.2	245.2	248.1	251.1	254.1	257.0	260.0	262.9	265.9	2.9
88	237.4	240.3	243.2	246.1	249.1	252.0	254.9	257.9	260.8	263.7	2.9
92	235.5	238.4	241.3	244.2	247.1	250.0	252.9	255.8	258.7	261.6	2.9
96	233.6	236.5	239.4	242.2	245.1	248.0	250.9	253.8	256.7	259.5	2.9

SMITHSONIAN TABLES.

TABLE 18.

REDUCTION OF THE BAROMETER TO SEA LEVEL.
ENGLISH MEASURES.

Correction of 2000 m for Latitude: $2000 \, m \times 0.002662 \cos 2\phi$.

For latitudes 0° to 45°, the correction is to be subtracted.
For latitudes 45° to 90°, the correction is to be added.

2000 m.	LATITUDE.									
	0°	5°	10°	15°	20°	25°	30°	35°	40°	45°
10	0.0	0.0	0.0	0.0	0.0	0.0	0.0	0.0	0.0	0.0
20	0.1	0.1	0.1	0.0	0.0	0.0	0.0	0.0	0.0	0.0
30	0.1	0.1	0.1	0.1	0.1	0.1	0.0	0.0	0.0	0.0
40	0.1	0.1	0.1	0.1	0.1	0.1	0.1	0.0	0.0	0.0
50	0.1	0.1	0.1	0.1	0.1	0.1	0.1	0.0	0.0	0.0
60	0.2	0.2	0.2	0.1	0.1	0.1	0.1	0.1	0.0	0.0
70	0.2	0.2	0.2	0.2	0.1	0.1	0.1	0.1	0.0	0.0
80	0.2	0.2	0.2	0.2	0.2	0.1	0.1	0.1	0.0	0.0
90	0.2	0.2	0.2	0.2	0.2	0.2	0.1	0.1	0.0	0.0
100	0.3	0.3	0.3	0.2	0.2	0.2	0.1	0.1	0.0	0.0
110	0.3	0.3	0.3	0.3	0.2	0.2	0.1	0.1	0.1	0.0
120	0.3	0.3	0.3	0.3	0.2	0.2	0.2	0.1	0.1	0.0
130	0.3	0.3	0.3	0.3	0.3	0.2	0.2	0.1	0.1	0.0
140	0.4	0.4	0.4	0.3	0.3	0.2	0.2	0.1	0.1	0.0
150	0.4	0.4	0.4	0.3	0.3	0.3	0.2	0.1	0.1	0.0
160	0.4	0.4	0.4	0.4	0.3	0.3	0.2	0.1	0.1	0.0
170	0.5	0.4	0.4	0.4	0.3	0.3	0.2	0.2	0.1	0.0
180	0.5	0.5	0.5	0.4	0.4	0.3	0.2	0.2	0.1	0.0
190	0.5	0.5	0.5	0.4	0.4	0.3	0.3	0.2	0.1	0.0
200	0.5	0.5	0.5	0.5	0.4	0.3	0.3	0.2	0.1	0.0
210	0.6	0.6	0.5	0.5	0.4	0.4	0.3	0.2	0.1	0.0
220	0.6	0.6	0.6	0.5	0.4	0.4	0.3	0.2	0.1	0.0
230	0.6	0.6	0.6	0.5	0.5	0.4	0.3	0.2	0.1	0.0
240	0.6	0.6	0.6	0.6	0.5	0.4	0.3	0.2	0.1	0.0
250	0.7	0.7	0.6	0.6	0.5	0.4	0.3	0.2	0.1	0.0
260	0.7	0.7	0.7	0.6	0.5	0.4	0.3	0.2	0.1	0.0
270	0.7	0.7	0.7	0.6	0.6	0.5	0.4	0.2	0.1	0.0
280	0.7	0.7	0.7	0.6	0.6	0.5	0.4	0.3	0.1	0.0
290	0.8	0.8	0.7	0.7	0.6	0.5	0.4	0.3	0.1	0.0
300	0.8	0.8	0.8	0.7	0.6	0.5	0.4	0.3	0.1	0.0
310	0.8	0.8	0.8	0.7	0.6	0.5	0.4	0.3	0.1	0.0
320	0.9	0.8	0.8	0.7	0.7	0.5	0.4	0.3	0.1	0.0
330	0.9	0.9	0.8	0.8	0.7	0.6	0.4	0.3	0.2	0.0
340	0.9	0.9	0.9	0.8	0.7	0.6	0.5	0.3	0.2	0.0
350	0.9	0.9	0.9	0.8	0.7	0.6	0.5	0.3	0.2	0.0
	90°	85°	80°	75°	70°	65°	60°	55°	50°	45°

TABLE 16.
REDUCTION OF THE BAROMETER TO SEA LEVEL
ENGLISH MEASURES.

$$B_0 - B = B(10^m - 1).$$

Top argument: Height of the barometer (B).
Side argument: Values of 2000 m obtained from Table 14.

2000 m.	HEIGHT OF THE BAROMETER IN INCHES.								
	31.0	30.5	30.0	29.5	29.0	28.5	28.0	27.5	27.0
	Inches.	Inches.	Inches.	Inches.	Inches.	Inches.	Inches.	Inches.	Inches.
1	0.04	0.04	0.03	0.03	0.03	0.03			
2	0.07	0.07	0.07	0.07	0.07	0.07			
3	0.11	0.11	0.10	0.10	0.10	0.10			
4	0.14	0.14	0.14	0.14	0.13	0.13			
5	0.18	0.18	0.17	0.17	0.17	0.16			
6	0.21	0.21	0.21	0.20	0.20	0.20	0.19		
7	0.25	0.25	0.24	0.24	0.23	0.23	0.23		
8	0.29	0.28	0.28	0.27	0.27	0.26	0.26		
9	0.32	0.32	0.31	0.31	0.30	0.30	0.29		
10	0.36	0.35	0.35	0.34	0.34	0.33	0.32		
11	0.40	0.39	0.38	0.38	0.37	0.36	0.36		
12	0.43	0.42	0.42	0.41	0.40	0.40	0.39		
13	0.47	0.46	0.45	0.44	0.44	0.43	0.42		
14	0.50	0.50	0.49	0.48	0.47	0.46	0.46		
15	0.54	0.53	0.52	0.51	0.51	0.50	0.49		
16	0.58	0.57	0.56	0.55	0.54	0.53	0.52		
17	0.61	0.60	0.59	0.58	0.57	0.56	0.55		
18	0.65	0.64	0.63	0.62	0.61	0.60	0.59		
19	0.69	0.67	0.66	0.65	0.64	0.63	0.62		
20	0.72	0.71	0.70	0.69	0.68	0.66	0.65		
21		0.75	0.73	0.72	0.71	0.70	0.69		
22		0.78	0.77	0.76	0.74	0.73	0.72	0.71	
23		0.82	0.80	0.79	0.77	0.76	0.75	0.74	
24		0.85	0.84	0.83	0.81	0.80	0.78	0.77	
25		0.89	0.88	0.86	0.85	0.83	0.82	0.80	
26		0.93	0.91	0.90	0.88	0.87	0.85	0.84	
27		0.96	0.95	0.93	0.92	0.90	0.88	0.87	
28		1.00	0.98	0.97	0.95	0.93	0.92	0.90	
29		1.04	1.02	1.00	0.98	0.97	0.95	0.93	
30		1.07	1.05	1.04	1.02	1.00	0.98	0.97	
31		1.11	1.09	1.07	1.05	1.04	1.02	1.00	
32		1.14	1.13	1.11	1.09	1.07	1.05	1.03	
33		1.18	1.16	1.14	1.12	1.10	1.08	1.06	
34		1.22	1.20	1.18	1.16	1.14	1.12	1.10	
35		1.25	1.23	1.21	1.19	1.17	1.15	1.13	
36			1.27	1.25	1.23	1.21	1.18	1.16	
37			1.31	1.28	1.26	1.24	1.22	1.20	
38			1.34	1.32	1.30	1.27	1.25	1.23	1.21
39			1.38	1.35	1.33	1.31	1.29	1.26	1.24
40			1.41	1.39	1.37	1.34	1.32	1.30	1.27
41			1.45	1.43	1.40	1.38	1.35	1.33	1.30
42			1.49	1.46	1.44	1.41	1.39	1.36	1.34
43			1.52	1.50	1.47	1.45	1.42	1.40	1.37
44			1.56	1.53	1.51	1.48	1.45	1.43	1.40
45			1.60	1.57	1.54	1.52	1.49	.146	1.44

TABLE 16

REDUCTION OF THE BAROMETER TO SEA LEVEL.
ENGLISH MEASURES.
$B_o - B = B(10^m - 1)$.

Top argument: Height of the barometer (B).
Side argument: Values of 2000 m obtained from Table 14.

2000 m.	HEIGHT OF THE BAROMETER IN INCHES.								
	29.5	29.0	28.5	28.0	27.5	27.0	26.5	26.0	25.5
	Inches.	Inches.	Inches.	Inches.	Inches.	Inches.	Inches.	Inches.	Inches.
45	1.57	1.54	1.52	1.49	1.46	1.44			
46	1.60	1.58	1.55	1.52	1.50	1.47			
47	1.64	1.61	1.58	1.56	1.53	1.50			
48	1.68	1.65	1.62	1.59	1.56	1.53			
49	1.71	1.68	1.65	1.62	1.60	1.57			
50	1.75	1.72	1.69	1.66	1.63	1.60			
51	1.78	1.75	1.72	1.69	1.66	1.63			
52	1.82	1.79	1.76	1.73	1.70	1.67			
53	1.86	1.82	1.79	1.76	1.73	1.70			
54	1.89	1.86	1.83	1.80	1.76	1.73	1.70		
55	1.93	1.90	1.86	1.83	1.80	1.76	1.73		
56	1.96	1.93	1.90	1.86	1.83	1.80	1.76		
57	2.00	1.97	1.93	1.90	1.87	1.83	1.80		
58	2.04	2.00	1.97	1.93	1.90	1.86	1.83		
59	2.07	2.04	2.00	1.97	1.93	1.90	1.86		
60	2.11	2.07	2.04	2.00	1.97	1.93	1.90		
61		2.11	2.07	2.04	2.00	1.96	1.93		
62		2.15	2.11	2.07	2.03	2.00	1.96		
63		2.18	2.14	2.11	2.07	2.03	1.99		
64		2.22	2.18	2.14	2.10	2.06	2.03		
65		2.25	2.21	2.18	2.14	2.10	2.06		
66		2.29	2.25	2.21	2.17	2.13	2.09		
67		2.33	2.29	2.25	2.21	2.17	2.13		
68		2.36	2.32	2.28	2.24	2.20	2.16		
69		2.40	2.36	2.32	2.27	2.23	2.19		
70		2.43	2.39	2.35	2.31	2.27	2.22		
71		2.47	2.43	2.38	2.34	2.30	2.26	2.21	
72		2.51	2.46	2.42	2.38	2.33	2.29	2.25	
73		2.54	2.50	2.45	2.41	2.37	2.32	2.28	
74			2.53	2.49	2.45	2.40	2.36	2.31	
75			2.57	2.53	2.48	2.43	2.39	2.34	
76			2.61	2.56	2.51	2.47	2.42	2.38	
77			2.64	2.60	2.55	2.50	2.46	2.41	
78			2.68	2.63	2.58	2.54	2.49	2.44	
79			2.71	2.67	2.62	2.57	2.52	2.48	
80			2.75	2.70	2.65	2.60	2.56	2.51	
81			2.79	2.74	2.69	2.64	2.59	2.54	
82			2.82	2.77	2.72	2.67	2.62	2.57	
83			2.86	2.81	2.76	2.71	2.66	2.61	
84			2.89	2.84	2.79	2.74	2.69	2.64	
85			2.93	2.88	2.83	2.78	2.72	2.67	
86			2.97	2.91	2.86	2.81	2.76	2.71	
87			3.00	2.95	2.90	2.84	2.79	2.74	
88			3.04	2.99	2.93	2.88	2.83	2.77	2.72
89			3.08	3.02	2.97	2.91	2.86	2.81	2.75
90			3.11	3.06	3.00	2.95	2.89	2.84	2.78

SMITHSONIAN TABLES.

TABLE 16.

REDUCTION OF THE BAROMETER TO SEA LEVEL.
ENGLISH MEASURES.
$B_0 - B = B (10^m - 1)$.

Top argument: Height of the barometer (B).
Side argument: Values of $2000\,m$ obtained from Table 14.

2000 m.	HEIGHT OF THE BAROMETER IN INCHES.							
	28.0	27.5	27.0	26.5	26.0	25.5	25.0	24.5
	Inches.	Inches.	Inches.	Inches.	Inches.	Inches.	Inches.	Inches.
90	3.06	3.00	2.95	2.89	2.84	2.78		
91	3.09	3.04	2.98	2.93	2.87	2.82		
92	3.13	3.07	3.02	2.96	2.91	2.85		
93	3.16	3.11	3.05	2.99	2.94	2.88		
94	3.20	3.14	3.09	3.03	2.97	2.91		
95	3.24	3.18	3.12	3.06	3.01	2.95		
96	3.27	3.21	3.16	3.10	3.04	2.98		
97	3.31	3.25	3.19	3.13	3.07	3.01		
98	3.34	3.28	3.22	3.17	3.11	3.05		
99	3.38	3.32	3.26	3.20	3.14	3.08		
100	3.42	3.36	3.29	3.23	3.17	3.11		
101	3.45	3.39	3.33	3.27	3.21	3.14		
102	3.49	3.43	3.36	3.30	3.24	3.18		
103	3.53	3.46	3.40	3.34	3.27	3.21		
104	3.56	3.50	3.43	3.37	3.31	3.24		
105	3.60	3.53	3.47	3.41	3.34	3.28	3.21	
106		3.57	3.50	3.44	3.37	3.31	3.24	
107		3.61	3.54	3.47	3.41	3.34	3.28	
108		3.64	3.57	3.51	3.44	3.38	3.31	
109		3.68	3.61	3.54	3.48	3.41	3.34	
110		3.71	3.65	3.58	3.51	3.44	3.38	
111		3.75	3.68	3.61	3.54	3.48	3.41	
112		3.78	3.72	3.65	3.58	3.51	3.44	
113		3.82	3.75	3.68	3.61	3.54	3.47	
114		3.86	3.79	3.72	3.65	3.58	3.51	
115		3.89	3.82	3.75	3.68	3.61	3.54	
116		3.93	3.86	3.79	3.71	3.64	3.57	
117		3.97	3.89	3.82	3.75	3.68	3.60	
118		4.00	3.93	3.86	3.78	3.71	3.64	
119		4.04	3.96	3.89	3.82	3.74	3.67	
120		4.07	4.00	3.93	3.85	3.78	3.70	
121		4.11	4.04	3.96	3.89	3.81	3.74	
122			4.07	4.00	3.92	3.85	3.77	3.69
123			4.11	4.03	3.96	3.88	3.80	3.73
124			4.14	4.07	3.99	3.91	3.84	3.76
125			4.18	4.10	4.02	3.95	3.87	3.79
126			4.22	4.14	4.06	3.98	3.90	3.82
127			4.25	4.17	4.09	4.01	3.94	3.86
128			4.29	4.21	4.13	4.05	3.97	3.89
129			4.32	4.24	4.16	4.08	4.00	3.92
130			4.36	4.28	4.20	4.12	4.04	3.96
131			4.40	4.31	4.23	4.15	4.07	3.99
132			4.43	4.35	4.27	4.19	4.10	4.02
133			4.47	4.38	4.30	4.22	4.14	4.05
134			4.50	4.42	4.34	4.25	4.17	4.09
135			4.54	4.46	4.37	4.29	4.20	4.12

SMITHSONIAN TABLES.

TABLE 16.

REDUCTION OF THE BAROMETER TO SEA LEVEL
ENGLISH MEASURES.

$B_o - B = B(10^m - 1)$.

Top argument: Height of the barometer (B).
Side argument: Values of 2000 m obtained from Table 14.

2000 m.	HEIGHT OF THE BAROMETER IN INCHES.							
	26.5	26.0	25.5	25.0	24.5	24.0	23.5	23.0
	Inches.	Inches.	Inches.	Inches.	Inches.	Inches.	Inches.	Inches.
135	4.46	4.37	4.29	4.20	4.12			
136	4.49	4.41	4.32	4.24	4.15			
137	4.53	4.44	4.36	4.27	4.19			
138	4.56	4.48	4.39	4.30	4.22			
139	4.60	4.51	4.43	4.34	4.25			
140	4.63	4.55	4.46	4.37	4.28	4.20		
141	4.67	4.58	4.49	4.41	4.32	4.23		
142	4.71	4.62	4.53	4.44	4.35	4.26		
143	4.74	4.65	4.56	4.47	4.38	4.30		
144	4.78	4.69	4.60	4.51	4.42	4.33		
145	4.81	4.72	4.63	4.54	4.45	4.36		
146	4.85	4.76	4.67	4.58	4.48	4.39		
147	4.89	4.79	4.70	4.61	4.52	4.43		
148	4.92	4.83	4.74	4.64	4.55	4.46		
149	4.96	4.87	4.77	4.68	4.58	4.49		
150	5.00	4.90	4.81	4.71	4.62	4.52		
151	5.03	4.94	4.84	4.75	4.65	4.56		
152	5.07	4.97	4.88	4.78	4.69	4.59		
153	5.10	5.01	4.91	4.82	4.72	4.62		
154		5.04	4.95	4.85	4.75	4.66		
155		5.08	4.98	4.88	4.79	4.69		
156		5.12	5.02	4.92	4.82	4.72		
157		5.15	5.05	4.95	4.85	4.75		
158		5.19	5.09	4.99	4.89	4.79		
159		5.22	5.12	5.02	4.92	4.82	4.72	
160		5.26	5.16	5.06	4.96	4.85	4.75	
161		5.29	5.19	5.09	4.99	4.89	4.79	
162		5.33	5.23	5.13	5.02	4.92	4.82	
163		5.37	5.26	5.16	5.06	4.95	4.85	
164		5.40	5.30	5.20	5.09	4.99	4.88	
165		5.44	5.33	5.23	5.13	5.02	4.92	
166		5.48	5.37	5.26	5.16	5.05	4.95	
167		5.51	5.41	5.30	5.19	5.09	4.98	
168		5.55	5.44	5.33	5.23	5.12	5.01	
169		5.58	5.48	5.37	5.26	5.15	5.05	
170		5.62	5.51	5.40	5.30	5.19	5.08	
171			5.55	5.44	5.33	5.22	5.11	
172			5.58	5.47	5.37	5.26	5.15	
173			5.62	5.51	5.40	5.29	5.18	
174			5.66	5.54	5.43	5.32	5.21	
175			5.69	5.58	5.47	5.36	5.25	
176			5.73	5.62	5.50	5.39	5.28	
177			5.76	5.65	5.54	5.42	5.31	5.20
178			5.80	5.69	5.57	5.46	5.34	5.23
179			5.84	5.72	5.61	5.49	5.38	5.26
180			5.87	5.76	5.64	5.53	5.41	5.30

SMITHSONIAN TABLES.

TABLE 16.

REDUCTION OF THE BAROMETER TO SEA LEVEL
ENGLISH MEASURES.
$B_0 - B = B(10^m - 1)$.

Top argument: Height of the barometer (B).
Side argument: Values of 2000 m obtained from Table **14**.

2000 m.	HEIGHT OF THE BAROMETER IN INCHES.							
	25.5	25.0	24.5	24.0	23.5	23.0	22.5	22.0
	Inches.	Inches.	Inches.	Inches.	Inches.	Inches.	Inches.	Inches.
180	5.87	5.76	5.64	5.53	5.41	5.30		
181	5.91	5.79	5.68	5.56	5.44	5.33		
182	5.94	5.83	5.71	5.59	5.48	5.36		
183	5.98	5.86	5.75	5.63	5.51	5.39		
184	6.02	5.90	5.78	5.66	5.54	5.43		
185		5.93	5.82	5.70	5.58	5.46		
186		5.97	5.85	5.73	5.61	5.49		
187		6.01	5.89	5.77	5.65	5.53		
188		6.04	5.92	5.80	5.68	5.56		
189		6.08	5.96	5.83	5.71	5.59		
190		6.11	5.99	5.87	5.75	5.62		
191		6.15	6.03	5.90	5.78	5.66		
192		6.18	6.06	5.94	5.81	5.69		
193		6.22	6.10	5.97	5.85	5.72		
194		6.26	6.13	6.01	5.88	5.76		
195		6.29	6.17	6.04	5.91	5.79		
196		6.33	6.20	6.08	5.95	5.82	5.70	
197		6.36	6.24	6.11	5.98	5.86	5.73	
198		6.40	6.27	6.14	6.02	5.89	5.76	
199		6.44	6.31	6.18	6.05	5.92	5.79	
200		6.47	6.34	6.21	6.08	5.96	5.83	
201		6.51	6.38	6.25	6.12	5.99	5.86	
202		6.55	6.41	6.28	6.15	6.02	5.89	
203		6.58	6.45	6.32	6.19	6.06	5.92	
204		6.62	6.49	6.35	6.22	6.09	5.96	
205			6.52	6.39	6.26	6.12	5.99	
206			6.56	6.42	6.29	6.16	6.02	
207			6.59	6.46	6.32	6.19	6.06	
208			6.63	6.49	6.36	6.22	6.09	
209			6.66	6.53	6.39	6.26	6.12	
210			6.70	6.56	6.43	6.29	6.15	
211			6.74	6.60	6.46	6.32	6.19	
212			6.77	6.63	6.50	6.36	6.22	
213			6.81	6.67	6.53	6.39	6.25	
214			6.84	6.71	6.57	6.43	6.29	
215			6.88	6.74	6.60	6.46	6.32	
216			6.92	6.78	6.63	6.49	6.35	6.21
217			6.95	6.81	6.67	6.53	6.39	6.24
218			6.99	6.85	6.70	6.56	6.42	6.28
219			7.03	6.88	6.74	6.60	6.45	6.31
220				6.92	6.77	6.63	6.49	6.34
221				6.95	6.81	6.66	6.52	6.37
222				6.99	6.84	6.70	6.55	6.41
223				7.02	6.88	6.73	6.59	6.44
224				7.06	6.91	6.77	6.62	6.47
225				7.10	6.95	6.80	6.65	6.51

TABLE 16.

REDUCTION OF THE BAROMETER TO SEA LEVEL.
ENGLISH MEASURES.
$B_0 - B = B(10^m - 1)$.

Top argument: Height of the barometer (B).
Side argument: Values of 2000 m obtained from Table 14.

2000 m.	HEIGHT OF THE BAROMETER IN INCHES.						
	24.0	23.5	23.0	22.5	22.0	21.5	21.0
	Inches.	Inches.	Inches.	Inches.	Inches.	Inches.	Inches.
225	7.10	6.95	6.80	6.65	6.51		
226	7.13	6.98	6.84	6.69	6.54		
227	7.17	7.02	6.87	6.72	6.57		
228	7.20	7.05	6.90	6.75	6.60		
229	7.24	7.09	6.94	6.79	6.64		
230	7.28	7.12	6.97	6.82	6.67		
231	7.31	7.16	7.01	6.86	6.70		
232	7.35	7.20	7.04	6.89	6.74		
233	7.38	7.23	7.08	6.92	6.77		
234	7.42	7.27	7.11	6.96	6.80		
235	7.46	7.30	7.15	6.99	6.84	6.68	
236	7.49	7.34	7.18	7.02	6.87	6.71	
237	7.53	7.37	7.22	7.06	6.90	6.74	
238		7.41	7.25	7.09	6.93	6.78	
239		7.44	7.29	7.13	6.97	6.81	
240		7.48	7.32	7.16	7.00	6.84	
241		7.51	7.35	7.19	7.04	6.88	
242		7.55	7.39	7.23	7.07	6.91	
243		7.59	7.42	7.26	7.10	6.94	
244		7.62	7.46	7.30	7.14	6.97	
245		7.66	7.49	7.33	7.17	7.01	
246		7.69	7.53	7.37	7.20	7.04	
247		7.73	7.57	7.40	7.24	7.07	
248		7.77	7.60	7.44	7.27	7.10	
249		7.80	7.64	7.47	7.30	7.14	
250		7.84	7.67	7.50	7.34	7.17	
251		7.87	7.71	7.54	7.37	7.20	
252		7.91	7.74	7.57	7.41	7.24	
253		7.95	7.78	7.61	7.44	7.27	
254		7.98	7.81	7.64	7.47	7.30	
255		8.02	7.85	7.68	7.51	7.34	
256		8.05	7.88	7.71	7.54	7.37	7.20
257		8.09	7.92	7.75	7.57	7.40	7.23
258		8.13	7.95	7.78	7.61	7.44	7.26
259		8.16	7.99	7.82	7.64	7.47	7.30
260			8.03	7.85	7.68	7.50	7.33
261			8.06	7.89	7.71	7.54	7.36
262			8.10	7.92	7.75	7.57	7.39
263			8.13	7.96	7.78	7.60	7.43
264			8.17	7.99	7.81	7.64	7.46
265			8.21	8.03	7.85	7.67	7.49
266			8.24	8.06	7.88	7.70	7.52
267			8.28	8.10	7.92	7.74	7.56
268			8.31	8.13	7.95	7.77	7.59
269			8.35	8.17	7.99	7.80	7.62
270			8.39	8.20	8.02	7.84	7.66

SMITHSONIAN TABLES.

TABLE 16.
REDUCTION OF THE BAROMETER TO SEA LEVEL
ENGLISH MEASURES.
$B_0 - B = B (10^m - 1)$.

Top argument: Height of the barometer (B).
Side argument: Values of $2000\,m$ obtained from Table 14.

2000 m.	HEIGHT OF THE BAROMETER IN INCHES.						
	23.0	22.5	22.0	21.5	21.0	20.5	20.0
	Inches.	Inches.	Inches.	Inches.	Inches.	Inches.	Inches.
270	8.39	8.20	8.02	7.84	7.66		
271	8.42	8.24	8.06	7.87	7.69		
272	8.46	8.27	8.09	7.91	7.72		
273	8.49	8.31	8.12	7.94	7.76		
274	8.53	8.34	8.16	7.97	7.79		
275	8.57	8.38	8.19	8.01	7.82		
276	8.60	8.42	8.23	8.04	7.85		
277		8.45	8.26	8.08	7.89	7.70	
278		8.49	8.30	8.11	7.92	7.73	
279		8.52	8.33	8.14	7.95	7.77	
280		8.56	8.37	8.18	7.99	7.80	
281		8.59	8.40	8.21	8.02	7.83	
282		8.63	8.44	8.25	8.05	7.86	
283		8.67	8.47	8.28	8.09	7.90	
284		8.70	8.51	8.32	8.12	7.93	
285		8.74	8.54	8.35	8.16	7.96	
286		8.77	8.58	8.38	8.19	7.99	
287		8.81	8.61	8.42	8.22	8.03	
288		8.85	8.65	8.45	8.26	8.06	
289		8.88	8.68	8.49	8.29	8.09	
290		8.92	8.72	8.52	8.32	8.13	
291		8.95	8.76	8.56	8.36	8.16	
292		8.99	8.79	8.59	8.39	8.19	
293		9.03	8.83	8.63	8.43	8.22	
294		9.06	8.86	8.66	8.46	8.26	
295		9.10	8.90	8.70	8.49	8.29	8.09
296		9.14	8.93	8.73	8.53	8.32	8.12
297			8.97	8.76	8.56	8.36	8.15
298			9.00	8.80	8.60	8.39	8.19
299			9.04	8.83	8.63	8.42	8.22
300			9.08	8.87	8.66	8.46	8.25
301			9.11	8.90	8.70	8.49	8.28
302			9.15	8.94	8.73	8.52	8.32
303			9.18	8.97	8.77	8.56	8.35
304			9.22	9.01	8.80	8.59	8.38
305			9.26	9.04	8.83	8.62	8.41
306			9.29	9.08	8.87	8.66	8.45
307			9.33	9.12	8.90	8.69	8.48
308			9.36	9.15	8.94	8.72	8.51
309			9.40	9.19	8.97	8.76	8.54
310			9.44	9.22	9.01	8.79	8.58
311			9.47	9.26	9.04	8.83	8.61
312			9.51	9.29	9.08	8.86	8.64
313			9.54	9.33	9.11	8.89	8.68
314			9.58	9.36	9.15	8.93	8.71
315			9.62	9.40	9.18	8.96	8.74

SMITHSONIAN TABLES.

TABLE 16.
REDUCTION OF THE BAROMETER TO SEA LEVEL.
ENGLISH MEASURES.
$B_0 - B = B(10^m - 1)$.

Top argument: Height of the barometer (B).
Side argument: Values of 2000 m obtained from Table 14.

2000 m.	HEIGHT OF THE BAROMETER IN INCHES.					
	22.0	21.5	21.0	20.5	20.0	19.5
	Inches.	Inches.	Inches.	Inches.	Inches.	Inches.
315	9.62	9.40	9.18	8.96	8.74	8.52
316	9.65	9.43	9.21	9.00	8.78	8.56
317	9.69	9.47	9.25	9.03	8.81	8.59
318	9.73	9.51	9.28	9.06	8.84	8.62
319	9.76	9.54	9.32	9.10	8.88	8.65
320	9.80	9.58	9.35	9.13	8.91	8.69
321		9.61	9.39	9.17	8.94	8.72
322		9.65	9.42	9.20	8.98	8.75
323		9.68	9.46	9.23	9.01	8.78
324		9.72	9.49	9.27	9.04	8.82
325		9.76	9.53	9.30	9.08	8.85
326		9.79	9.56	9.34	9.11	8.88
327		9.83	9.60	9.37	9.14	8.91
328		9.86	9.64	9.41	9.18	8.95
329		9.90	9.67	9.44	9.21	8.98
330		9.94	9.71	9.47	9.24	9.01
331		9.97	9.74	9.51	9.28	9.05
332		10.01	9.78	9.54	9.31	9.08
333		10.05	9.81	9.58	9.34	9.11
334			9.85	9.61	9.38	9.14

TABLE 17.

REDUCTION OF THE BAROMETER TO SEA LEVEL
METRIC MEASURES.

Values of 2000 × m.

$$m = \frac{Z(1-\beta)}{18444 + 67.53\theta + .003Z}$$

Altitude in metres. Z.	MEAN TEMPERATURE OF AIR COLUMN IN CENTIGRADE DEGREES (θ).										
	−16°	−12°	−8°	−4°	−2°	0°	+2°	+4°	+6°	+8°	+10°
10	1.2	1.1	1.1	1.1	1.1	1.1	1.1	1.1	1.1	1.0	1.0
20	2.3	2.3	2.2	2.2	2.2	2.2	2.1	2.1	2.1	2.1	2.1
30	3.5	3.4	3.3	3.3	3.3	3.2	3.2	3.2	3.1	3.1	3.1
40	4.6	4.5	4.5	4.4	4.4	4.3	4.3	4.3	4.2	4.2	4.2
50	5.8	5.7	5.6	5.5	5.5	5.4	5.4	5.3	5.3	5.2	5.2
60	6.9	6.8	6.7	6.6	6.5	6.5	6.4	6.4	6.3	6.3	6.2
70	8.1	7.9	7.8	7.7	7.6	7.6	7.5	7.5	7.4	7.3	7.3
80	9.2	9.1	8.9	8.8	8.7	8.7	8.6	8.5	8.5	8.4	8.3
90	10.4	10.2	10.0	9.9	9.8	9.7	9.7	9.6	9.5	9.4	9.4
100	11.5	11.3	11.2	11.0	10.9	10.8	10.7	10.7	10.6	10.5	10.4
110	12.7	12.5	12.3	12.1	12.0	11.9	11.8	11.7	11.6	11.5	11.4
120	13.8	13.6	13.4	13.2	13.1	13.0	12.9	12.8	12.7	12.6	12.5
130	15.0	14.7	14.5	14.3	14.2	14.1	14.0	13.9	13.7	13.6	13.5
140	16.1	15.9	15.6	15.4	15.3	15.1	15.0	14.9	14.8	14.7	14.6
150	17.3	17.0	16.7	16.5	16.4	16.2	16.1	16.0	15.9	15.7	15.6
160	18.4	18.1	17.8	17.6	17.4	17.3	17.2	17.0	16.9	16.8	16.7
170	19.6	19.3	19.0	18.7	18.5	18.4	18.3	18.1	18.0	17.8	17.7
180	20.7	20.4	20.1	19.8	19.6	19.5	19.3	19.2	19.0	18.9	18.7
190	21.9	21.5	21.2	20.9	20.7	20.6	20.4	20.2	20.1	19.9	19.8
200	23.0	22.7	22.3	22.0	21.8	21.6	21.5	21.3	21.1	21.0	20.8
210	24.2	23.8	23.4	23.1	22.9	22.7	22.6	22.4	22.2	22.0	21.9
220	25.3	24.9	24.5	24.2	24.0	23.8	23.6	23.4	23.3	23.1	22.9
230	26.5	26.1	25.7	25.3	25.1	24.9	24.7	24.5	24.3	24.1	23.9
240	27.6	27.2	26.8	26.4	26.2	26.0	25.8	25.6	25.4	25.2	25.0
250	28.8	28.3	27.9	27.5	27.3	27.0	26.8	26.6	26.4	26.2	26.0
260	29.9	29.5	29.0	28.6	28.3	28.1	27.9	27.7	27.5	27.3	27.1
270	31.1	30.6	30.1	29.7	29.4	29.2	29.0	28.8	28.5	28.3	28.1
280	32.2	31.7	31.2	30.8	30.5	30.3	30.1	29.8	29.6	29.4	29.1
290	33.4	32.9	32.4	31.9	31.6	31.4	31.1	30.9	30.7	30.4	30.2
300	34.5	34.0	33.5	33.0	32.7	32.5	32.2	32.0	31.7	31.5	31.2
310	35.7	35.1	34.6	34.1	33.8	33.5	33.3	33.0	32.8	32.5	32.3
320	36.8	36.3	35.7	35.2	34.9	34.6	34.4	34.1	33.8	33.6	33.3
330	38.0	37.4	36.8	36.3	36.0	35.7	35.4	35.2	34.9	34.6	34.3
340	39.1	38.5	37.9	37.4	37.1	36.8	36.5	36.2	35.9	35.7	35.4
350	40.3	39.7	39.0	38.5	38.2	37.9	37.6	37.3	37.0	36.7	36.4
360	41.4	40.8	40.2	39.5	39.2	38.9	38.6	38.4	38.1	37.8	37.5
370	42.6	41.9	41.3	40.6	40.3	40.0	39.7	39.4	39.1	38.8	38.5
380	43.7	43.1	42.4	41.7	41.4	41.1	40.8	40.5	40.2	39.9	39.6
390	44.9	44.2	43.5	42.8	42.5	42.2	41.9	41.5	41.2	40.9	40.6
400	46.0	45.3	44.6	43.9	43.6	43.3	42.9	42.6	42.3	42.0	41.6
410	47.2	46.4	45.7	45.0	44.7	44.4	44.0	43.7	43.3	43.0	42.7
420	48.3	47.6	46.9	46.1	45.8	45.4	45.1	44.7	44.4	44.1	43.7
430	49.5	48.7	48.0	47.2	46.9	46.5	46.2	45.8	45.5	45.1	44.8
440	50.6	49.8	49.1	48.3	48.0	47.6	47.2	46.9	46.5	46.2	45.8
450	51.8	51.0	50.2	49.4	49.1	48.7	48.3	47.9	47.6	47.2	46.8
460	52.9	52.1	51.3	50.5	50.1	49.8	49.4	49.0	48.6	48.2	47.9
470	54.1	53.2	52.4	51.6	51.2	50.8	50.5	50.1	49.7	49.3	48.9
480	55.2	54.4	53.5	52.7	52.3	51.9	51.5	51.1	50.7	50.3	50.0
490	56.4	55.5	54.7	53.8	53.4	53.0	52.6	52.2	51.8	51.4	51.0
500	57.5	56.6	55.8	54.9	54.5	54.1	53.7	53.3	52.9	52.4	52.0

SMITHSONIAN TABLES.

TABLE 17.

REDUCTION OF THE BAROMETER TO SEA LEVEL.
METRIC MEASURES.

Values of 2000 × m.

$$m = \frac{Z(1-\beta)}{18444 + 67.53\theta + .003Z}$$

Altitude in metres. Z.	MEAN TEMPERATURE OF AIR COLUMN IN CENTIGRADE DEGREES (θ).										
	+12°	+14°	+16°	+18°	+20°	+22°	+24°	+26°	+28°	+32°	+36°
10	1.0	1.0	1.0	1.0	1.0	1.0	1.0	1.0	1.0	1.0	0.9
20	2.1	2.1	2.0	2.0	2.0	2.0	2.0	2.0	1.9	1.9	1.9
30	3.1	3.1	3.1	3.0	3.0	3.0	3.0	2.9	2.9	2.9	2.8
40	4.1	4.1	4.1	4.0	4.0	4.0	3.9	3.9	3.9	3.8	3.8
50	5.2	5.1	5.1	5.0	5.0	5.0	4.9	4.9	4.9	4.8	4.7
60	6.2	6.2	6.1	6.1	6.0	6.0	5.9	5.9	5.8	5.8	5.7
70	7.2	7.2	7.1	7.1	7.0	7.0	6.9	6.9	6.8	6.7	6.6
80	8.3	8.2	8.1	8.1	8.0	8.0	7.9	7.8	7.8	7.7	7.6
90	9.3	9.2	9.2	9.1	9.0	9.0	8.9	8.8	8.8	8.6	8.5
100	10.3	10.3	10.2	10.1	10.0	9.9	9.9	9.8	9.7	9.6	9.5
110	11.4	11.3	11.2	11.1	11.0	10.9	10.9	10.8	10.7	10.5	10.4
120	12.4	12.3	12.2	12.1	12.0	11.9	11.8	11.8	11.7	11.5	11.3
130	13.4	13.3	13.2	13.1	13.0	12.9	12.8	12.7	12.7	12.5	12.3
140	14.5	14.3	14.2	14.1	14.0	13.9	13.8	13.7	13.6	13.4	13.2
150	15.5	15.4	15.3	15.1	15.0	14.9	14.8	14.7	14.6	14.4	14.2
160	16.5	16.4	16.3	16.2	16.0	15.9	15.8	15.7	15.6	15.3	15.1
170	17.6	17.4	17.3	17.2	17.0	16.9	16.8	16.7	16.5	16.3	16.1
180	18.6	18.4	18.3	18.2	18.0	17.9	17.8	17.6	17.5	17.3	17.0
190	19.6	19.5	19.3	19.2	19.0	18.9	18.8	18.6	18.5	18.2	18.0
200	20.7	20.5	20.3	20.2	20.0	19.9	19.7	19.6	19.5	19.2	18.9
210	21.7	21.5	21.4	21.2	21.0	20.9	20.7	20.6	20.4	20.1	19.8
220	22.7	22.5	22.4	22.2	22.0	21.9	21.7	21.6	21.4	21.1	20.8
230	23.8	23.6	23.4	23.2	23.0	22.9	22.7	22.5	22.4	22.0	21.7
240	24.8	24.6	24.4	24.2	24.0	23.9	23.7	23.5	23.3	23.0	22.7
250	25.8	25.6	25.4	25.2	25.0	24.9	24.7	24.5	24.3	24.0	23.6
260	26.9	26.6	26.4	26.2	26.1	25.9	25.7	25.5	25.3	24.9	24.6
270	27.9	27.7	27.5	27.3	27.1	26.9	26.7	26.5	26.3	25.9	25.5
280	28.9	28.7	28.5	28.3	28.1	27.8	27.6	27.4	27.2	26.8	26.5
290	29.0	29.7	29.5	29.3	29.1	28.8	28.6	28.4	28.2	27.8	27.4
300	31.0	30.7	30.5	30.3	30.1	29.8	29.6	29.4	29.2	28.8	28.4
310	32.0	31.8	31.5	31.3	31.1	30.8	30.6	30.4	30.2	29.7	29.3
320	33.1	32.8	32.6	32.3	32.1	31.8	31.6	31.4	31.1	30.7	30.3
330	34.1	33.8	33.6	33.3	33.1	32.8	32.6	32.3	32.1	31.6	31.2
340	35.1	34.8	34.6	34.3	34.1	33.8	33.6	33.3	33.1	32.6	32.1
350	36.2	35.9	35.6	35.3	35.1	34.8	34.6	34.3	34.0	33.5	33.1
360	37.2	36.9	36.6	36.3	36.1	35.8	35.5	35.3	35.0	34.5	34.0
370	38.2	37.9	37.6	37.4	37.1	36.8	36.5	36.3	36.0	35.5	35.0
380	39.2	38.9	38.7	38.4	38.1	37.8	37.5	37.2	37.0	36.4	35.9
390	40.3	40.0	39.7	39.4	39.1	38.8	38.5	38.2	37.9	37.4	36.9
400	41.3	41.0	40.7	40.4	40.1	39.8	39.5	39.2	38.9	38.3	37.8
410	42.3	42.0	41.7	41.4	41.1	40.8	40.5	40.2	39.9	39.3	38.7
420	43.4	43.0	42.7	42.4	42.1	41.8	41.5	41.2	40.8	40.3	39.7
430	44.4	44.1	43.7	43.4	43.1	42.8	42.4	42.1	41.8	41.2	40.6
440	45.4	45.1	44.8	44.4	44.1	43.8	43.4	43.1	42.8	42.2	41.6
450	46.5	46.1	45.8	45.4	45.1	44.8	44.4	44.1	43.8	43.1	42.5
460	47.5	47.1	46.8	46.4	46.1	45.7	45.4	45.1	44.7	44.1	43.5
470	48.5	48.2	47.8	47.4	47.1	46.7	46.4	46.1	45.7	45.0	44.4
480	49.6	49.2	48.8	48.5	48.1	47.7	47.4	47.0	46.7	46.0	45.4
490	50.6	50.2	49.8	49.5	49.1	48.7	48.4	48.0	47.6	47.0	46.3
500	51.6	51.2	50.9	50.5	50.1	49.7	49.4	49.0	48.6	47.9	47.2

TABLE 17.
REDUCTION OF THE BAROMETER TO SEA LEVEL
METRIC MEASURES.

Values of $2000 \times m$.

$$m = \frac{Z(1-\beta)}{18444 + 67.53\theta + .003Z}$$

Altitude in metres. Z.	MEAN TEMPERATURE OF AIR COLUMN IN CENTIGRADE DEGREES (θ).										
	$-16°$	$-12°$	$-8°$	$-4°$	$-2°$	$0°$	$+2°$	$+4°$	$+6°$	$+8°$	$+10°$
500	57.5	56.6	55.8	54.9	54.5	54.1	53.7	53.3	52.9	52.4	52.0
510	58.7	57.8	56.9	56.0	55.6	55.2	54.8	54.3	53.9	53.5	53.1
520	59.8	58.9	58.0	57.1	56.7	56.3	55.8	55.4	55.0	54.5	54.1
530	61.0	60.0	59.1	58.2	57.8	57.3	56.9	56.5	56.0	55.6	55.2
540	62.1	61.2	60.2	59.3	58.9	58.4	58.0	57.5	57.1	56.6	56.2
550	63.3	62.3	61.4	60.4	60.0	59.5	59.0	58.6	58.1	57.7	57.2
560	64.4	63.4	62.5	61.5	61.1	60.6	60.1	59.7	59.2	58.7	58.3
570	65.6	64.6	63.6	62.6	62.1	61.7	61.2	60.7	60.3	59.8	59.3
580	66.7	65.7	64.7	63.7	63.2	62.7	62.3	61.8	61.3	60.8	60.4
590	67.9	66.8	65.8	64.8	64.3	63.8	63.3	62.9	62.4	61.9	61.4
600	69.0	68.0	66.9	65.9	65.4	64.9	64.4	63.9	63.4	62.9	62.4
610	70.2	69.1	68.0	67.0	66.5	66.0	65.5	65.0	64.5	64.0	63.5
620	71.4	70.2	69.2	68.1	67.6	67.1	66.6	66.0	65.5	65.0	64.5
630	72.5	71.4	70.3	69.2	68.7	68.2	67.6	67.1	66.6	66.1	65.6
640	73.7	72.5	71.4	70.3	69.8	69.2	68.7	68.2	67.7	67.1	66.6
650	74.8	73.6	72.5	71.4	70.9	70.3	69.8	69.2	68.7	68.2	67.6
660	76.0	74.8	73.6	72.5	72.0	71.4	70.9	70.3	69.8	69.2	68.7
670	77.1	75.9	74.7	73.6	73.0	72.5	71.9	71.4	70.8	70.3	69.7
680	78.3	77.0	75.9	74.7	74.1	73.6	73.0	72.4	71.9	71.3	70.8
690	79.4	78.2	77.0	75.8	75.2	74.6	74.1	73.5	72.9	72.4	71.8
700	80.6	79.3	78.1	76.9	76.3	75.7	75.1	74.6	74.0	73.4	72.9
710	81.7	80.4	79.2	78.0	77.4	76.8	76.2	75.6	75.1	74.5	73.9
720	82.9	81.6	80.3	79.1	78.5	77.9	77.3	76.7	76.1	75.5	74.9
730	83.0	82.7	81.4	80.2	79.6	79.0	78.4	77.8	77.2	76.6	76.0
740	85.2	83.8	82.5	81.3	80.7	80.1	79.4	78.8	78.2	77.6	77.0
750	86.3	85.0	83.7	82.4	81.8	81.1	80.5	79.9	79.3	78.7	78.1
760	87.5	86.1	84.8	83.5	82.9	82.2	81.6	81.0	80.3	79.7	79.1
770	88.6	87.2	85.9	84.6	83.9	83.3	82.7	82.0	81.4	80.8	80.1
780	89.8	88.4	87.0	85.7	85.0	84.4	83.7	83.1	82.5	81.8	81.2
790	90.9	89.5	88.1	86.8	86.1	85.5	84.8	84.2	83.5	82.9	82.2
800	92.1	90.6	89.2	87.9	87.2	86.5	85.9	85.2	84.6	83.9	83.3
810	93.2	91.8	90.4	89.0	88.3	87.6	87.0	86.3	85.6	85.0	84.3
820	94.4	92.9	91.5	90.1	89.4	88.7	88.0	87.4	86.7	86.0	85.3
830	95.5	94.0	92.6	91.2	90.5	89.8	89.1	88.4	87.7	87.1	86.4
840	96.7	95.2	93.7	92.3	91.6	90.9	90.2	89.5	88.8	88.1	87.4
850	97.8	96.3	94.8	93.4	92.7	92.0	91.2	90.5	89.8	89.2	88.5
860	99.0	97.4	95.9	94.5	93.8	93.0	92.3	91.6	90.9	90.2	89.5
870	100.1	98.6	97.0	95.6	94.8	94.1	93.4	92.7	92.0	91.3	90.5
880	101.3	99.7	98.2	96.7	95.9	95.2	94.5	93.7	93.0	92.3	91.6
890	102.4	100.8	99.3	97.8	97.0	96.3	95.5	94.8	94.1	93.3	92.6
900	103.6	102.0	100.4	98.9	98.1	97.4	96.6	95.9	95.1	94.4	93.7
910	104.7	103.1	101.5	100.0	99.2	98.4	97.7	96.9	96.2	95.4	94.7
920	105.9	104.2	102.6	101.1	100.3	99.5	98.8	98.0	97.2	96.5	95.7
930	107.0	105.4	103.7	102.2	101.4	100.6	99.8	99.1	98.3	97.5	96.8
940	108.2	106.5	104.9	103.3	102.5	101.7	100.9	100.1	99.4	98.6	97.8
950	109.3	107.6	106.0	104.4	103.6	102.8	102.0	101.2	100.4	99.6	98.9
960	110.5	108.8	107.1	105.5	104.7	103.9	103.1	102.3	101.5	100.7	99.9
970	111.6	109.9	108.2	106.6	105.7	104.9	104.1	103.3	102.5	101.7	100.9
980	112.8	111.0	109.3	107.6	106.8	106.0	105.2	104.4	103.6	102.8	102.0
990	113.9	112.1	110.4	108.7	107.9	107.1	106.3	105.5	104.6	103.8	103.0
1000	115.1	113.3	111.5	109.8	109.0	108.2	107.3	106.5	105.7	104.9	104.1

SMITHSONIAN TABLES.

TABLE 17.

REDUCTION OF THE BAROMETER TO SEA LEVEL.
METRIC MEASURES.

Values of 2000 × m.

$$m = \frac{Z(1-\beta)}{18444 + 67.53\theta + .003Z}$$

Altitude in metres. Z.	MEAN TEMPERATURE OF AIR COLUMN IN CENTIGRADE DEGREES (θ).									
	+12°	+14°	+16°	+18°	+20°	+22°	+24°	+26°	+28°	+32°	+36°
500	51.6	51.2	50.9	50.5	50.1	49.7	49.4	49.0	48.6	47.9	47.2
510	52.7	52.3	51.9	51.5	51.1	50.7	50.3	50.0	49.6	48.9	48.2
520	53.7	53.3	52.9	52.5	52.1	51.7	51.3	51.0	50.6	49.8	49.1
530	54.7	54.3	53.9	53.5	53.1	52.7	52.3	51.9	51.5	50.8	50.1
540	55.8	55.3	54.9	54.5	54.1	53.7	53.3	52.9	52.5	51.8	51.0
550	56.8	56.4	55.9	55.5	55.1	54.7	54.3	53.9	53.5	52.7	52.0
560	57.8	57.4	57.0	56.5	56.1	55.7	55.3	54.9	54.4	53.7	52.9
570	58.9	58.4	58.0	57.5	57.1	56.7	56.3	55.8	55.4	54.6	53.9
580	59.9	59.4	59.0	58.5	58.1	57.7	57.2	56.8	56.4	55.6	54.8
590	60.9	60.5	60.0	59.6	59.1	58.7	58.2	57.8	57.4	56.5	55.7
600	62.0	61.5	61.0	60.6	60.1	59.7	59.2	58.8	58.3	57.5	56.7
610	63.0	62.5	62.0	61.6	61.1	60.7	60.2	59.8	59.3	58.5	57.6
620	64.0	63.5	63.1	62.6	62.1	61.7	61.2	60.7	60.3	59.4	58.6
630	65.1	64.6	64.1	63.6	63.1	62.6	62.2	61.7	61.3	60.4	59.5
640	66.1	65.6	65.1	64.6	64.1	63.6	63.2	62.7	62.2	61.3	60.5
650	67.1	66.6	66.1	65.6	65.1	64.6	64.2	63.7	63.2	62.3	61.4
660	68.2	67.6	67.1	66.6	66.1	65.6	65.1	64.7	64.2	63.3	62.4
670	69.2	68.7	68.1	67.6	67.1	66.6	66.1	65.6	65.1	64.2	63.3
680	70.2	69.7	69.2	68.6	68.1	67.6	67.1	66.6	66.1	65.2	64.2
690	71.3	70.7	70.2	69.6	69.1	68.6	68.1	67.6	67.1	66.1	65.2
700	72.3	71.7	71.2	70.7	70.1	69.6	69.1	68.6	68.1	67.1	66.1
710	73.3	72.8	72.2	71.7	71.1	70.6	70.1	69.6	69.0	68.0	67.1
720	74.4	73.8	73.2	72.7	72.1	71.6	71.1	70.5	70.0	69.0	68.0
730	75.4	74.8	74.2	73.7	73.1	72.6	72.0	71.5	71.0	70.0	69.0
740	76.4	75.8	75.3	74.7	74.1	73.6	73.0	72.5	72.0	70.9	69.9
750	77.5	76.9	76.3	75.7	75.1	74.6	74.0	73.5	72.9	71.9	70.9
760	78.5	77.9	77.3	76.7	76.1	75.6	75.0	74.5	73.9	72.8	71.8
770	79.5	78.9	78.3	77.7	77.1	76.6	76.0	75.4	74.9	73.8	72.8
780	80.6	79.9	79.3	78.7	78.1	77.6	77.0	76.4	75.9	74.8	73.7
790	81.6	81.0	80.3	79.7	79.1	78.6	78.0	77.4	76.8	75.7	74.6
800	82.6	82.0	81.4	80.8	80.1	79.6	79.0	78.4	77.8	76.7	75.6
810	83.7	83.0	82.4	81.8	81.2	80.5	79.9	79.4	78.8	77.6	76.5
820	84.7	84.0	83.4	82.8	82.2	81.5	80.9	80.3	79.7	78.6	77.5
830	85.7	85.1	84.4	83.8	83.2	82.5	81.9	81.3	80.7	79.5	78.4
840	86.8	86.1	85.4	84.8	84.2	83.5	82.9	82.3	81.7	80.5	79.4
850	87.8	87.1	86.4	85.8	85.2	84.5	83.9	83.3	82.7	81.5	80.3
860	88.8	88.1	87.5	86.8	86.2	85.5	84.9	84.3	83.6	82.4	81.3
870	89.9	89.2	88.5	87.8	87.2	86.5	85.9	85.2	84.6	83.4	82.2
880	90.9	90.2	89.5	88.8	88.2	87.5	86.9	86.2	85.6	84.3	83.1
890	91.9	91.2	90.5	89.8	89.2	88.5	87.8	87.2	86.6	85.3	84.1
900	93.0	92.2	91.5	90.8	90.2	89.5	88.8	88.2	87.5	86.3	85.0
910	94.0	93.3	92.6	91.9	91.2	90.5	89.8	89.2	88.5	87.2	86.0
920	95.0	94.3	93.6	92.9	92.2	91.5	90.8	90.1	89.5	88.2	86.9
930	96.0	95.3	94.6	93.9	93.2	92.5	91.8	91.1	90.4	89.1	87.9
940	97.1	96.3	95.6	94.9	94.2	93.5	92.8	92.1	91.4	90.1	88.8
950	98.1	97.4	96.6	95.9	95.2	94.5	93.8	93.1	92.4	91.1	89.8
960	99.1	98.4	97.6	96.9	96.2	95.5	94.8	94.1	93.4	92.0	90.7
970	100.2	99.4	98.7	97.9	97.2	96.5	95.7	95.0	94.3	93.0	91.6
980	101.2	100.4	99.7	98.9	98.2	97.4	96.7	96.0	95.3	93.9	92.6
990	102.2	101.5	100.7	99.9	99.2	98.4	97.7	97.0	96.3	94.9	93.5
1000	103.3	102.5	101.7	100.9	100.2	99.4	98.7	98.0	97.3	95.9	94.5

TABLE 17.
REDUCTION OF THE BAROMETER TO SEA LEVEL.
METRIC MEASURES.

Values of $2000 \times m$.

$$m = \frac{Z(1-\beta)}{18444 + 67.53\theta + .003Z}$$

Altitude in metres. Z.	MEAN TEMPERATURE OF AIR COLUMN IN CENTIGRADE DEGREES (θ).										
	−16°	−12°	−8°	−4°	−2°	0°	+2°	+4°	+6°	+8°	+10°
1000	115.1	113.3	111.5	109.8	109.0	108.2	107.3	106.5	105.7	104.9	104.1
1010	116.2	114.4	112.7	110.9	110.1	109.3	108.4	107.6	106.8	105.9	105.1
1020	117.4	115.5	113.8	112.0	111.2	110.3	109.5	108.7	107.8	107.0	106.2
1030	118.5	116.7	114.9	113.1	112.3	111.4	110.6	109.7	108.9	108.0	107.2
1040	119.7	117.8	116.0	114.2	113.4	112.5	111.6	110.8	109.9	109.1	108.2
1050	120.8	118.9	117.1	115.3	114.5	113.6	112.7	111.8	111.0	110.1	109.3
1060	122.0	120.1	118.2	116.4	115.6	114.7	113.8	112.9	112.0	111.2	110.3
1070	123.1	121.2	119.3	117.5	116.6	115.7	114.9	114.0	113.1	112.2	111.4
1080	124.3	122.3	120.5	118.6	117.7	116.8	115.9	115.0	114.2	113.3	112.4
1090	125.4	123.5	121.6	119.7	118.8	117.9	117.0	116.1	115.2	114.3	113.4
1100	126.6	124.6	122.7	120.8	119.9	119.0	118.1	117.2	116.3	115.4	114.5
1110	127.7	125.7	123.8	121.9	121.0	120.1	119.2	118.2	117.3	116.4	115.5
1120	128.9	126.9	124.9	123.0	122.1	121.2	120.2	119.3	118.4	117.5	116.6
1130	130.0	128.0	126.0	124.1	123.2	122.2	121.3	120.4	119.4	118.5	117.6
1140	131.2	129.1	127.2	125.2	124.3	123.3	122.4	121.4	120.5	119.6	118.6
1150	132.3	130.3	128.3	126.3	125.4	124.4	123.4	122.5	121.6	120.6	119.7
1160	133.5	131.4	129.4	127.4	126.4	125.5	124.5	123.6	122.6	121.7	120.7
1170	134.6	132.5	130.5	128.5	127.5	126.6	125.6	124.6	123.7	122.7	121.8
1180	135.8	133.7	131.6	129.6	128.6	127.6	126.7	125.7	124.7	123.8	122.8
1190	136.9	134.8	132.7	130.7	129.7	128.7	127.7	126.8	125.8	124.8	123.8
1200	138.1	135.9	133.8	131.8	130.8	129.8	128.8	127.8	126.8	125.9	124.9
1210	139.2	137.1	135.0	132.9	131.9	130.9	129.9	128.9	127.9	126.9	125.9
1220	140.4	138.2	136.1	134.0	133.0	132.0	131.0	130.0	129.0	128.0	127.0
1230	141.5	139.3	137.2	135.1	134.1	133.1	132.0	131.0	130.0	129.0	128.0
1240	142.7	140.5	138.3	136.2	135.2	134.1	133.1	132.1	131.1	130.1	129.0
1250	143.8	141.6	139.4	137.3	136.3	135.2	134.2	133.1	132.1	131.1	130.1
1260	145.0	142.7	140.5	138.4	137.3	136.3	135.3	134.2	133.2	132.1	131.1
1270	146.1	143.9	141.7	139.5	138.4	137.4	136.3	135.3	134.2	133.2	132.2
1280	147.3	145.0	142.8	140.6	139.5	138.5	137.4	136.3	135.3	134.2	133.2
1290	148.4	146.1	143.9	141.7	140.6	139.5	138.5	137.4	136.3	135.3	134.2
1300	149.6	147.3	145.0	142.8	141.7	140.6	139.5	138.5	137.4	136.3	135.3
1310	150.7	148.4	146.1	143.9	142.8	141.7	140.6	139.5	138.5	137.4	136.3
1320	151.9	149.5	147.2	145.0	143.9	142.8	141.7	140.6	139.5	138.4	137.4
1330	153.0	150.7	148.3	146.1	145.0	143.9	142.8	141.7	140.6	139.5	138.4
1340	154.2	151.8	149.5	147.2	146.1	145.0	143.8	142.7	141.6	140.5	139.5
1350	155.3	152.9	150.6	148.3	147.2	146.0	144.9	143.8	142.7	141.6	140.5
1360	156.5	154.1	151.7	149.4	148.2	147.1	146.0	144.9	143.7	142.6	141.5
1370	157.6	155.2	152.8	150.5	149.3	148.2	147.1	145.9	144.8	143.7	142.6
1380	158.8	156.3	153.9	151.6	150.4	149.3	148.1	147.0	145.9	144.7	143.6
1390	159.9	157.5	155.0	152.7	151.5	150.4	149.2	148.1	146.9	145.8	144.7
1400	161.1	158.7	156.2	153.8	152.6	151.4	150.3	149.1	148.0	146.8	145.7
1410	162.2	159.7	157.3	154.9	153.7	152.5	151.4	150.2	149.0	147.9	146.7
1420	163.4	160.8	158.4	156.0	154.8	153.6	152.4	151.3	150.1	148.9	147.8
1430	164.5	162.0	159.5	157.1	155.9	154.7	153.5	152.3	151.1	150.0	148.8
1440	165.7	163.1	160.6	158.2	157.0	155.8	154.6	153.4	152.2	151.0	149.9
1450	166.8	164.2	161.7	159.3	158.1	156.8	155.7	154.5	153.3	152.1	150.9
1460	168.0	165.4	162.8	160.4	159.1	157.9	156.7	155.5	154.3	153.1	151.9
1470	169.1	166.5	164.0	161.5	160.2	159.0	157.8	156.6	155.4	154.2	153.0
1480	170.3	167.6	165.1	162.6	161.3	160.1	158.9	157.6	156.4	155.2	154.0
1490	171.4	168.8	166.2	163.7	162.4	161.2	159.9	158.7	157.5	156.3	155.1
1500	172.6	169.9	167.3	164.8	163.5	162.3	161.0	159.8	158.5	157.3	156.1

SMITHSONIAN TABLES.

TABLE 17.

REDUCTION OF THE BAROMETER TO SEA LEVEL.
METRIC MEASURES.

Values of 2000 × m.

$$m = \frac{Z(1-\beta)}{18444 + 67.53\theta + .003Z}$$

Altitude in metres. Z.	MEAN TEMPERATURE OF AIR COLUMN IN CENTIGRADE DEGREES (θ).									
	+12°	+14°	+16°	+18°	+20°	+22°	+24°	+26°	+28°	+32°	+36°
1000	103.3	102.5	101.7	100.9	100.2	99.4	98.7	98.0	97.3	95.9	94.5
1010	104.3	103.5	102.7	101.9	101.2	100.4	99.7	99.0	98.2	96.8	95.4
1020	105.3	104.5	103.7	103.0	102.2	101.4	100.7	99.9	99.2	97.8	96.4
1030	106.4	105.6	104.8	104.0	103.2	102.4	101.7	100.9	100.2	98.7	97.3
1040	107.4	106.6	105.8	105.0	104.2	103.4	102.6	101.9	101.1	99.7	98.3
1050	108.4	107.6	106.8	106.0	105.2	104.4	103.6	102.9	102.1	100.6	99.2
1060	109.5	108.6	107.8	107.0	106.2	105.4	104.6	103.9	103.1	101.6	100.1
1070	110.5	109.7	108.8	108.0	107.2	106.4	105.6	104.8	104.1	102.6	101.1
1080	111.5	110.7	109.8	109.0	108.2	107.4	106.6	105.8	105.0	103.5	102.0
1090	112.6	111.7	110.9	110.0	109.2	108.4	107.6	106.8	106.0	104.5	103.0
1100	113.6	112.7	111.9	111.0	110.2	109.4	108.6	107.8	107.0	105.4	103.9
1110	114.6	113.8	112.9	112.0	111.2	110.4	109.6	108.8	108.0	106.4	104.9
1120	115.7	114.8	113.9	113.1	112.2	111.4	110.5	109.7	108.9	107.4	105.8
1130	116.7	115.8	114.9	114.1	113.2	112.4	111.5	110.7	109.9	108.3	106.8
1140	117.7	116.8	115.9	115.1	114.2	113.4	112.5	111.7	110.9	109.3	107.7
1150	118.8	117.9	117.0	116.1	115.2	114.4	113.5	112.7	111.8	110.2	108.6
1160	119.8	118.9	118.0	117.1	116.2	115.3	114.5	113.6	112.8	111.2	109.6
1170	120.8	119.9	119.0	118.1	117.2	116.3	115.5	114.6	113.8	112.1	110.5
1180	121.9	120.9	120.0	119.1	118.2	117.3	116.5	115.6	114.8	113.1	111.5
1190	122.9	122.0	121.0	120.1	119.2	118.3	117.4	116.6	115.7	114.1	112.4
1200	123.9	123.0	122.0	121.1	120.2	119.3	118.4	117.6	116.7	115.0	113.4
1210	125.0	124.0	123.1	122.1	121.2	120.3	119.4	118.5	117.7	116.0	114.3
1220	126.0	125.0	124.1	123.1	122.2	121.3	120.4	119.5	118.6	116.9	115.3
1230	127.0	126.1	125.1	124.2	123.2	122.3	121.4	120.5	119.6	117.9	116.2
1240	128.1	127.1	126.1	125.2	124.2	123.3	122.4	121.5	120.6	118.9	117.2
1250	129.1	128.1	127.1	126.2	125.2	124.3	123.4	122.5	121.6	119.8	118.1
1260	130.1	129.1	128.1	127.2	126.2	125.3	124.4	123.4	122.5	120.8	119.0
1270	131.2	130.2	129.2	128.2	127.2	126.3	125.3	124.4	123.5	121.7	120.0
1280	132.2	131.2	130.2	129.2	128.2	127.3	126.3	125.4	124.5	122.7	120.9
1290	133.2	132.2	131.2	130.2	129.2	128.3	127.3	126.4	125.5	123.6	121.9
1300	134.3	133.2	132.2	131.2	130.2	129.3	128.3	127.4	126.4	124.6	122.8
1310	135.3	134.3	133.2	132.2	131.2	130.3	129.3	128.3	127.4	125.6	123.8
1320	136.3	135.3	134.2	133.2	132.2	131.3	130.3	129.3	128.4	126.5	124.7
1330	137.4	136.3	135.3	134.2	133.2	132.2	131.3	130.3	129.3	127.5	125.7
1340	138.4	137.3	136.3	135.3	134.2	133.2	132.2	131.3	130.3	128.4	126.6
1350	139.4	138.4	137.3	136.3	135.2	134.2	133.2	132.3	131.3	129.4	127.5
1360	140.5	139.4	138.3	137.3	136.2	135.2	134.2	133.2	132.3	130.3	128.5
1370	141.5	140.4	139.3	138.3	137.2	136.2	135.2	134.2	133.2	131.3	129.4
1380	142.5	141.4	140.3	139.3	138.2	137.2	136.2	135.2	134.2	132.3	130.4
1390	143.5	142.4	141.4	140.3	139.2	138.2	137.2	136.2	135.2	133.2	131.3
1400	144.6	143.5	142.4	141.3	140.2	139.2	138.2	137.2	136.2	134.2	132.3
1410	145.6	144.5	143.4	142.3	141.3	140.2	139.2	138.1	137.1	135.1	133.2
1420	146.6	145.5	144.4	143.3	142.3	141.2	140.1	139.1	138.1	136.1	134.2
1430	147.7	146.5	145.4	144.3	143.3	142.2	141.1	140.1	139.1	137.1	135.1
1440	148.7	147.6	146.4	145.3	144.3	143.2	142.1	141.1	140.0	138.0	136.0
1450	149.7	148.6	147.5	146.4	145.3	144.2	143.1	142.1	141.0	139.0	137.0
1460	150.8	149.6	148.5	147.4	146.3	145.2	144.1	143.0	142.0	139.9	137.9
1470	151.8	150.6	149.5	148.4	147.3	146.2	145.1	144.0	143.0	140.9	138.9
1480	152.8	151.7	150.5	149.4	148.3	147.2	146.1	145.0	143.9	141.8	139.8
1490	153.9	152.7	151.5	150.4	149.3	148.2	147.1	146.0	144.9	142.8	140.8
1500	154.9	153.7	152.5	151.4	150.3	149.1	148.0	147.0	145.9	143.8	141.7

SMITHSONIAN TABLES.

TABLE 17.
REDUCTION OF THE BAROMETER TO SEA LEVEL
METRIC MEASURES.

Values of $2000 \times m$.

$$m = \frac{Z(1-\beta)}{18444 + 67.53\theta + .003Z}$$

Altitude in metres. Z.	MEAN TEMPERATURE OF AIR COLUMN IN CENTIGRADE DEGREES (θ).									
	−16°	−12°	−8°	−4°	−2°	0°	+2°	+4°	+6°	+8°	+10°
1500	172.6	169.9	167.3	164.8	163.5	162.3	161.0	159.8	158.5	157.3	156.1
1510	173.7	171.0	168.4	165.9	164.6	163.3	162.1	160.8	159.6	158.4	157.1
1520	174.9	172.2	169.5	167.0	165.7	164.4	163.2	161.9	160.7	159.4	158.2
1530	176.0	173.3	170.7	168.1	166.8	165.5	164.2	163.0	161.7	160.5	159.2
1540	177.2	174.4	171.8	169.1	167.9	166.6	165.3	164.0	162.8	161.5	160.3
1550	178.3	175.6	172.9	170.2	169.0	167.7	166.4	165.1	163.8	162.6	161.3
1560	179.5	176.7	174.0	171.3	170.0	168.7	167.5	166.2	164.9	163.6	162.3
1570	180.6	177.8	175.1	172.4	171.1	169.8	168.5	167.2	165.9	164.7	163.4
1580	181.8	179.0	176.2	173.5	172.2	170.9	169.6	168.3	167.0	165.7	164.4
1590	182.9	180.1	177.3	174.6	173.3	172.0	170.7	169.4	168.1	166.8	165.5
1600	184.1	181.2	178.5	175.7	174.4	173.1	171.7	170.4	169.1	167.8	166.5
1610	185.2	182.4	179.6	176.8	175.5	174.2	172.8	171.5	170.2	168.9	167.5
1620	186.4	183.5	180.7	177.9	176.6	175.2	173.9	172.6	171.2	169.9	168.6
1630	187.5	184.6	181.8	179.0	177.7	176.3	175.0	173.6	172.3	170.9	169.6
1640	188.7	185.8	182.9	180.1	178.8	177.4	176.0	174.7	173.3	172.0	170.7
1650	189.8	186.9	184.0	181.2	179.8	178.5	177.1	175.7	174.4	173.0	171.7
1660	191.0	188.0	185.1	182.3	180.9	179.6	178.2	176.8	175.4	174.1	172.7
1670	192.2	189.2	186.3	183.4	182.0	180.6	179.3	177.9	176.5	175.1	173.8
1680	193.3	190.3	187.4	184.5	183.1	181.7	180.3	178.9	177.6	176.2	174.8
1690	194.5	191.4	188.5	185.6	184.2	182.8	181.4	180.0	178.6	177.2	175.9
1700	195.6	192.6	189.6	186.7	185.3	183.9	182.5	181.1	179.7	178.3	176.9
1710	196.8	193.7	190.7	187.8	186.4	185.0	183.5	182.1	180.7	179.3	177.9
1720	197.9	194.8	191.8	188.9	187.5	186.0	184.6	183.2	181.8	180.4	179.0
1730	199.1	196.0	193.0	190.0	188.6	187.1	185.7	184.3	182.8	181.4	180.0
1740	200.2	197.1	194.1	191.1	189.7	188.2	186.8	185.3	183.9	182.5	181.1
1750	201.4	198.2	195.2	192.2	190.7	189.3	187.8	186.4	185.0	183.5	182.1
1760	202.5	199.3	196.3	193.3	191.8	190.4	188.9	187.5	186.0	184.6	183.1
1770	203.7	200.5	197.4	194.4	192.9	191.5	190.0	188.5	187.1	185.6	184.2
1780	204.8	201.6	198.5	195.5	194.0	192.5	191.1	189.6	188.1	186.7	185.2
1790	206.0	202.7	199.6	196.6	195.1	193.6	192.1	190.7	189.2	187.7	186.3
1800	207.1	203.9	200.8	197.7	196.2	194.7	193.2	191.7	190.2	188.8	187.3
1810	208.3	205.0	201.9	198.8	197.3	195.8	194.3	192.8	191.3	189.8	188.3
1820	209.4	206.1	203.0	199.9	198.4	196.9	195.3	193.8	192.4	190.9	189.4
1830	210.6	207.3	204.1	201.0	199.5	197.9	196.4	194.9	193.4	191.9	190.4
1840	211.7	208.4	205.2	202.1	200.6	199.0	197.5	196.0	194.5	193.0	191.5
1850	212.9	209.5	206.3	203.2	201.6	200.1	198.6	197.0	195.5	194.0	192.5
1860	214.0	210.7	207.4	204.3	202.7	201.2	199.6	198.1	196.6	195.1	193.6
1870	215.2	211.8	208.6	205.4	203.8	202.3	200.7	199.2	197.6	196.1	194.6
1880	216.3	212.9	209.7	206.5	204.9	203.3	201.8	200.2	198.7	197.2	195.6
1890	217.5	214.1	210.8	207.6	206.0	204.4	202.9	201.3	199.7	198.2	196.7
1900	218.6	215.2	211.9	208.7	207.1	205.5	203.9	202.4	200.8	199.3	197.7
1910	219.8	216.3	213.0	209.8	208.2	206.6	205.0	203.4	201.9	200.3	198.8
1920	220.9	217.5	214.1	210.9	209.3	207.7	206.1	204.5	202.9	201.3	199.8
1930	222.1	218.6	215.2	212.0	210.4	208.8	207.2	205.6	204.0	202.4	200.8
1940	223.2	219.7	216.4	213.1	211.4	209.8	208.2	206.6	205.0	203.4	201.9
1950	224.4	220.9	217.5	214.2	212.5	210.9	209.3	207.7	206.1	204.5	202.9
1960	225.5	222.0	218.6	215.3	213.6	212.0	210.4	208.8	207.1	205.5	204.0
1970	226.7	223.1	219.7	216.4	214.7	213.1	211.4	209.8	208.2	206.6	205.0
1980	227.8	224.3	220.8	217.5	215.8	214.2	212.5	210.9	209.3	207.6	206.0
1990	229.0	225.4	221.9	218.6	216.9	215.2	213.6	211.9	210.3	208.7	207.1
2000	230.1	226.5	223.0	219.7	218.0	216.3	214.7	213.0	211.4	209.7	208.1

TABLE 17.
REDUCTION OF THE BAROMETER TO SEA LEVEL.
METRIC MEASURES.

Values of $2000 \times m$.

$$m = \frac{Z(1-\beta)}{18444 + 67.53\theta + .003Z}$$

Altitude in metres. Z.	MEAN TEMPERATURE OF AIR COLUMN IN CENTIGRADE DEGREES (θ).									
	+12°	+14°	+16°	+18°	+20°	+22°	+24°	+26°	+28°	+32°	+36°
1500	154.9	153.7	152.5	151.4	150.3	149.1	148.0	147.0	145.9	143.8	141.7
1510	155.9	154.7	153.6	152.4	151.3	150.1	149.0	147.9	146.8	144.7	142.7
1520	157.0	155.8	154.6	153.4	152.3	151.1	150.0	148.9	147.8	145.7	143.6
1530	158.0	156.8	155.6	154.4	153.3	152.1	151.0	149.9	148.8	146.6	144.5
1540	159.0	157.8	156.6	155.4	154.3	153.1	152.0	150.9	149.8	147.6	145.5
1550	160.1	158.8	157.6	156.4	155.3	154.1	153.0	151.8	150.7	148.6	146.4
1560	161.1	159.9	158.7	157.5	156.3	155.1	154.0	152.8	151.7	149.5	147.4
1570	162.1	160.9	159.7	158.5	157.3	156.1	154.9	153.8	152.7	150.5	148.3
1580	163.2	161.9	160.7	159.5	158.3	157.1	155.9	154.8	153.7	151.4	149.3
1590	164.2	162.9	161.7	160.5	159.3	158.1	156.9	155.8	154.6	152.4	150.2
1600	165.2	164.0	162.7	161.5	160.3	159.1	157.9	156.7	155.6	153.3	151.2
1610	166.3	165.0	163.7	162.5	161.3	160.1	158.9	157.7	156.6	154.3	152.1
1620	167.3	166.0	164.8	163.5	162.3	161.1	159.9	158.7	157.5	155.3	153.0
1630	168.3	167.0	165.8	164.5	163.3	162.1	160.9	159.7	158.5	156.2	154.0
1640	169.4	168.1	166.8	165.5	164.3	163.1	161.9	160.7	159.5	157.2	154.9
1650	170.4	169.1	167.8	166.5	165.3	164.1	162.8	161.6	160.5	158.1	155.9
1660	171.4	170.1	168.8	167.5	166.3	165.1	163.8	162.6	161.4	159.1	156.8
1670	172.5	171.1	169.8	168.6	167.3	166.0	164.8	163.6	162.4	160.1	157.8
1680	173.5	172.2	170.9	169.6	168.3	167.0	165.8	164.6	163.4	161.0	158.7
1690	174.5	173.2	171.9	170.6	169.3	168.0	166.8	165.6	164.3	162.0	159.7
1700	175.6	174.2	172.9	171.6	170.3	169.0	167.8	166.5	165.3	162.9	160.6
1710	176.6	175.2	173.9	172.6	171.3	170.0	168.8	167.5	166.3	163.9	161.5
1720	177.6	176.3	174.9	173.6	172.3	171.0	169.7	168.5	167.3	164.8	162.5
1730	178.7	177.3	175.9	174.6	173.3	172.0	170.7	169.5	168.2	165.8	163.4
1740	179.7	178.3	177.0	175.6	174.3	173.0	171.7	170.5	169.2	166.8	164.4
1750	180.7	179.3	178.0	176.6	175.3	174.0	172.7	171.4	170.2	167.7	165.3
1760	181.7	180.4	179.0	177.6	176.3	175.0	173.7	172.4	171.2	168.7	166.3
1770	182.8	181.4	180.0	178.6	177.3	176.0	174.7	173.4	172.1	169.6	167.2
1780	183.8	182.4	181.0	179.7	178.3	177.0	175.7	174.4	173.1	170.6	168.2
1790	184.8	183.4	182.0	180.7	179.3	178.0	176.7	175.4	174.1	171.6	169.1
1800	185.9	184.5	183.1	181.7	180.3	179.0	177.6	176.3	175.0	172.5	170.0
1810	186.9	185.5	184.1	182.7	181.3	180.0	178.6	177.3	176.0	173.5	171.0
1820	187.9	186.5	185.1	183.7	182.3	181.0	179.6	178.3	177.0	174.4	171.9
1830	189.0	187.5	186.1	184.7	183.3	181.9	180.6	179.3	178.0	175.4	172.9
1840	190.0	188.6	187.1	185.7	184.3	182.9	181.6	180.3	178.9	176.3	173.8
1850	191.0	189.6	188.1	186.7	185.3	183.9	182.6	181.2	179.9	177.3	174.8
1860	192.1	190.6	189.2	187.7	186.3	184.9	183.6	182.2	180.9	178.3	175.7
1870	193.1	191.6	190.2	188.7	187.3	185.9	184.5	183.2	181.8	179.2	176.7
1880	194.1	192.7	191.2	189.7	188.3	186.9	185.5	184.2	182.8	180.2	177.6
1890	195.2	193.7	192.2	190.8	189.3	187.9	186.5	185.1	183.8	181.1	178.5
1900	196.2	194.7	193.2	191.8	190.3	188.9	187.5	186.1	184.8	182.1	179.5
1910	197.2	195.7	194.2	192.8	191.3	189.9	188.5	187.1	185.7	183.1	180.4
1920	198.3	196.8	195.3	193.8	192.3	190.9	189.5	188.1	186.7	184.0	181.4
1930	199.3	197.8	196.3	194.8	193.3	191.9	190.5	189.1	187.7	185.0	182.3
1940	200.3	198.8	197.3	195.8	194.3	192.9	191.5	190.0	188.7	185.9	183.3
1950	201.4	199.8	198.3	196.8	195.3	193.9	192.4	191.0	189.6	186.9	184.2
1960	202.4	200.8	199.3	197.8	196.3	194.9	193.4	192.0	190.6	187.8	185.2
1970	203.4	201.9	200.3	198.8	197.3	195.9	194.4	193.0	191.6	188.8	186.1
1980	204.5	202.9	201.4	199.8	198.3	196.9	195.4	194.0	192.5	189.8	187.0
1990	205.5	203.9	202.4	200.8	199.3	197.9	196.4	194.9	193.5	190.7	188.0
2000	206.5	204.9	203.4	201.9	200.3	198.8	197.4	195.9	194.5	191.7	188.9

SMITHSONIAN TABLES.

TABLE 17.
REDUCTION OF THE BAROMETER TO SEA LEVEL.
METRIC MEASURES.

Values of $2000 \times m$.
$$m = \frac{Z(1-\beta)}{18444 + 67.53\theta + .003Z}$$

Altitude in metres. Z.	MEAN TEMPERATURE OF AIR COLUMN IN CENTIGRADE DEGREES (θ).									
	$-16°$	$-12°$	$-8°$	$-4°$	$-2°$	$0°$	$+2°$	$+4°$	$+6°$	$+8°$	$+10°$
2000	230.1	226.5	223.0	219.7	218.0	216.3	214.7	213.0	211.4	209.7	208.1
2010	231.3	227.7	224.2	220.8	219.1	217.4	215.7	214.1	212.4	210.8	209.2
2020	232.4	228.8	225.3	221.9	220.2	218.5	216.8	215.1	213.5	211.8	210.2
2030	233.6	229.9	226.4	223.0	221.3	219.6	217.9	216.2	214.5	212.9	211.2
2040	234.7	231.1	227.5	224.0	222.3	220.6	219.0	217.3	215.6	213.9	212.3
2050	235.9	232.2	228.6	225.1	223.4	221.7	220.0	218.3	216.7	215.0	213.3
2060	237.0	233.3	229.7	226.2	224.5	222.8	221.1	219.4	217.7	216.0	214.4
2070	238.2	234.4	230.9	227.3	225.6	223.9	222.2	220.5	218.8	217.1	215.4
2080	239.3	235.6	232.0	228.4	226.7	225.0	223.2	221.5	219.8	218.1	216.4
2090	240.5	236.7	233.1	229.5	227.8	226.1	224.3	222.6	220.9	219.2	217.5
2100	241.6	237.8	234.2	230.6	228.9	227.1	225.4	223.7	221.9	220.2	218.5
2110	242.8	239.0	235.3	231.7	230.0	228.2	226.5	224.7	223.0	221.3	219.6
2120	243.9	240.1	236.4	232.8	231.1	229.3	227.5	225.8	224.0	222.3	220.6
2130	245.1	241.2	237.5	233.9	232.2	230.4	228.6	226.9	225.1	223.4	221.6
2140	246.2	242.4	238.7	235.0	233.2	231.5	229.7	227.9	226.2	224.4	222.7
2150	247.4	243.5	239.8	236.1	234.3	232.5	230.8	229.0	227.2	225.5	223.7
2160	248.5	244.6	240.9	237.2	235.4	233.6	231.8	230.0	228.3	226.5	224.8
2170	249.7	245.8	242.0	238.3	236.5	234.7	232.9	231.1	229.3	227.6	225.8
2180	250.8	246.9	243.1	239.4	237.6	235.8	234.0	232.2	230.4	228.6	226.8
2190	252.0	248.0	244.2	240.5	238.7	236.9	235.1	233.2	231.4	229.7	227.9
2200	253.1	249.2	245.4	241.6	239.8	237.9	236.1	234.3	232.5	230.7	228.9
2210	254.3	250.3	246.5	242.7	240.9	239.0	237.2	235.4	233.6	231.7	230.0
2220	255.4	251.4	247.6	243.8	242.0	240.1	238.3	236.4	234.6	232.8	231.0
2230	256.6	252.6	248.7	244.9	243.0	241.2	239.3	237.5	235.7	233.8	232.0
2240	257.7	253.7	249.8	246.0	244.1	242.3	240.4	238.6	236.7	234.9	233.1
2250	258.9	254.8	250.9	247.1	245.2	243.4	241.5	239.6	237.8	235.9	234.1
2260	260.0	256.0	252.0	248.2	246.3	244.4	242.6	240.7	238.8	237.0	235.2
2270	261.2	257.1	253.2	249.3	247.4	245.5	243.6	241.8	239.9	238.0	236.2
2280	262.3	258.2	254.3	250.4	248.5	246.6	244.7	242.8	241.0	239.1	237.2
2290	263.5	259.4	255.4	251.5	249.6	247.7	245.8	243.9	242.0	240.1	238.3
2300	264.6	260.5	256.5	252.6	250.7	248.8	246.9	245.0	243.1	241.2	239.3
2310	265.8	261.6	257.6	253.7	251.8	249.8	247.9	246.0	244.1	242.2	240.4
2320	266.9	262.8	258.7	254.8	252.9	250.9	249.0	247.1	245.2	243.3	241.4
2330	268.1	263.9	259.8	255.9	253.9	252.0	250.1	248.1	246.2	244.3	242.4
2340	269.2	265.0	261.0	257.0	255.0	253.1	251.1	249.2	247.3	245.4	243.5
2350	270.4	266.1	262.1	258.1	256.1	254.2	252.2	250.3	248.3	246.4	244.5
2360	271.5	267.3	263.2	259.2	257.2	255.2	253.3	251.3	249.4	247.5	245.6
2370	272.7	268.4	264.3	260.3	258.3	256.3	254.4	252.4	250.5	248.5	246.6
2380	273.8	269.5	265.4	261.4	259.4	257.4	255.4	253.5	251.5	249.6	247.6
2390	275.0	270.7	266.5	262.5	260.5	258.5	256.5	254.5	252.6	250.6	248.7
2400	276.1	271.8	267.7	263.6	261.5	259.6	257.6	255.6	253.6	251.7	249.7
2410	277.3	272.9	268.8	264.7	262.7	260.7	258.7	256.7	254.7	252.7	250.8
2420	278.4	274.1	269.9	265.8	263.7	261.7	259.7	257.7	255.7	253.8	251.8
2430	279.6	275.2	271.0	266.9	264.8	262.8	260.8	258.8	256.8	254.8	252.8
2440	280.7	276.3	272.1	268.0	265.9	263.9	261.9	259.9	257.9	255.9	253.9
2450	281.9	277.5	273.2	269.1	267.0	265.0	262.9	260.9	258.9	256.9	254.9
2460	283.0	278.6	274.3	270.2	268.1	266.1	264.0	262.0	260.0	258.0	256.0
2470	284.2	279.7	275.5	271.3	269.2	267.1	265.1	263.1	261.0	259.0	257.0
2480	285.3	280.9	276.6	272.4	270.3	268.2	266.2	264.1	262.1	260.1	258.0
2490	286.5	282.0	277.7	273.5	271.4	269.3	267.2	265.2	263.1	261.1	259.1
2500	287.6	283.1	278.8	274.5	272.5	270.4	268.3	266.2	264.2	262.2	260.1

SMITHSONIAN TABLES.

TABLE 17.

REDUCTION OF THE BAROMETER TO SEA LEVEL.
METRIC MEASURES.

Values of $2000 \times m$.

$$m = \frac{Z(1-\beta)}{18444 + 67.53\theta + .003Z}$$

Altitude in metres. Z.	MEAN TEMPERATURE OF AIR COLUMN IN CENTIGRADE DEGREES (θ).									
	+12°	+14°	+16°	+18°	+20°	+22°	+24°	+26°	+28°	+32°	+36°
2000	206.5	204.9	203.4	201.9	200.3	198.8	197.4	195.9	194.5	191.7	188.9
2010	207.6	206.0	204.4	202.9	201.3	199.8	198.4	196.9	195.5	192.6	189.9
2020	208.6	207.0	205.4	203.9	202.3	200.8	199.3	197.9	196.4	193.6	190.8
2030	209.6	208.0	206.4	204.9	203.3	201.8	200.3	198.8	197.4	194.6	191.8
2040	210.7	209.0	207.5	205.9	204.3	202.8	201.3	199.8	198.4	195.5	192.7
2050	211.7	210.1	208.5	206.9	205.3	203.8	202.3	200.8	199.3	196.5	193.7
2060	212.7	211.1	209.5	207.9	206.3	204.8	203.3	201.8	200.3	197.4	194.6
2070	213.8	212.1	210.5	208.9	207.3	205.8	204.3	202.8	201.3	198.4	195.5
2080	214.8	213.1	211.5	209.9	208.3	206.8	205.3	203.7	202.3	199.3	196.5
2090	215.8	214.2	212.5	210.9	209.3	207.8	206.3	204.7	203.2	200.3	197.4
2100	216.8	215.2	213.5	211.9	210.4	208.8	207.2	205.7	204.2	201.3	198.4
2110	217.9	216.2	214.6	213.0	211.4	209.8	208.2	206.7	205.2	202.2	199.3
2120	218.9	217.2	215.6	214.0	212.4	210.8	209.2	207.7	206.2	203.2	200.3
2130	219.9	218.3	216.6	215.0	213.4	211.8	210.2	208.6	207.1	204.1	201.2
2140	221.0	219.3	217.6	216.0	214.4	212.8	211.2	209.6	208.1	205.1	202.2
2150	222.0	220.3	218.6	217.0	215.4	213.8	212.2	210.6	209.1	206.1	203.1
2160	223.0	221.3	219.6	218.0	216.4	214.7	213.2	211.6	210.0	207.0	204.0
2170	224.1	222.4	220.7	219.0	217.4	215.7	214.1	212.6	211.0	208.0	205.0
2180	225.1	223.4	221.7	220.0	218.4	216.7	215.1	213.5	212.0	208.9	205.9
2190	226.1	224.4	222.7	221.0	219.4	217.7	216.1	214.5	213.0	209.9	206.9
2200	227.2	225.4	223.7	222.0	220.4	218.7	217.1	215.5	213.9	210.8	207.8
2210	228.2	226.5	224.7	223.0	221.4	219.7	218.1	216.5	214.9	211.8	208.8
2220	229.2	227.5	225.7	224.0	222.4	220.7	219.1	217.5	215.9	212.8	209.7
2230	230.3	228.5	226.8	225.1	223.4	221.7	220.1	218.4	216.8	213.7	210.7
2240	231.3	229.5	227.8	226.1	224.4	222.7	221.0	219.4	217.8	214.7	211.6
2250	232.3	230.6	228.8	227.1	225.4	223.7	222.0	220.4	218.8	215.6	212.5
2260	233.4	231.6	229.8	228.1	226.4	224.7	223.0	221.4	219.8	216.6	213.5
2270	234.4	232.6	230.8	229.1	227.4	225.7	224.0	222.4	220.7	217.6	214.4
2280	235.4	233.6	231.8	230.1	228.4	226.7	225.0	223.3	221.7	218.5	215.4
2290	236.5	234.7	232.9	231.1	229.4	227.7	226.0	224.3	222.7	219.5	216.3
2300	237.5	235.7	233.9	232.1	230.4	228.7	227.0	225.3	223.6	220.4	217.3
2310	238.5	236.7	234.9	233.1	231.4	229.7	228.0	226.3	224.6	221.4	218.2
2320	239.6	237.7	235.9	234.1	232.4	230.7	228.9	227.3	225.6	222.3	219.2
2330	240.6	238.7	236.9	235.1	233.4	231.6	229.9	228.2	226.6	223.3	220.1
2340	241.6	239.8	237.9	236.2	234.4	232.6	230.9	229.2	227.5	224.3	221.0
2350	242.7	240.8	239.0	237.2	235.4	233.6	231.9	230.2	228.5	225.2	222.0
2360	243.7	241.8	240.0	238.2	236.4	234.6	232.9	231.2	229.5	226.2	222.9
2370	244.7	242.8	241.0	239.2	237.4	235.6	233.9	232.2	230.4	227.1	223.9
2380	245.7	243.9	242.0	240.2	238.4	236.6	234.9	233.1	231.4	228.1	224.8
2390	246.8	244.9	243.0	241.2	239.4	237.6	235.8	234.1	232.4	229.1	225.8
2400	247.8	245.9	244.0	242.2	240.4	238.6	236.8	235.1	233.4	230.0	226.7
2410	248.8	246.9	245.1	243.2	241.4	239.6	237.8	236.1	234.3	231.0	227.7
2420	249.9	248.0	246.1	244.2	242.4	240.6	238.8	237.1	235.3	231.9	228.6
2430	250.9	249.0	247.1	245.2	243.4	241.6	239.8	238.0	236.3	232.9	229.5
2440	251.9	250.0	248.1	246.2	244.4	242.6	240.8	239.0	237.3	233.8	230.5
2450	253.0	251.0	249.1	247.3	245.4	243.6	241.8	240.0	238.2	234.8	231.4
2460	254.0	252.1	250.1	248.3	246.4	244.6	242.8	241.0	239.2	235.8	232.4
2470	255.0	253.1	251.2	249.3	247.4	245.6	243.7	241.9	240.2	236.7	233.3
2480	256.1	254.1	252.2	250.3	248.4	246.6	244.7	242.9	241.1	237.7	234.3
2490	257.1	255.1	253.2	251.3	249.4	247.5	245.7	243.9	242.1	238.6	235.2
2500	258.1	256.2	254.2	252.3	250.4	248.5	246.7	244.9	243.1	239.6	236.2

TABLE 17.
REDUCTION OF THE BAROMETER TO SEA LEVEL.
METRIC MEASURES.

Values of $2000 \times m$.

$$m = \frac{Z(1-\beta)}{18444 + 67.53\theta + .003Z}$$

Altitude in metres. Z.	MEAN TEMPERATURE OF AIR COLUMN IN CENTIGRADE DEGREES (θ).									
	$-16°$	$-12°$	$-8°$	$-4°$	$-2°$	$0°$	$+2°$	$+4°$	$+6°$	$+8°$	$+10°$
2500	287.6	283.1	278.8	274.5	272.5	270.4	268.3	266.2	264.2	262.2	260.1
2510	288.8	284.3	279.9	275.6	273.6	271.5	269.4	267.3	265.2	263.2	261.2
2520	289.9	285.4	281.0	276.7	274.6	272.5	270.5	268.4	266.3	264.2	262.2
2530	291.1	286.5	282.1	277.8	275.7	273.6	271.5	269.4	267.4	265.3	263.2
2540	292.2	287.7	283.3	278.9	276.8	274.7	272.6	270.5	268.4	266.3	264.3
2550	293.4	288.8	284.4	280.0	277.9	275.8	273.7	271.6	269.5	267.4	265.3
2560	294.5	289.9	285.5	281.1	279.0	276.9	274.7	272.6	270.5	268.4	266.4
2570	295.7	291.1	286.6	282.2	280.1	277.9	275.8	273.7	271.6	269.5	267.4
2580	296.8	292.2	287.7	283.3	281.2	279.0	276.9	274.8	272.6	270.5	268.4
2590	298.0	293.3	288.8	284.4	282.3	280.1	278.0	275.8	273.7	271.6	269.5
2600	299.1	294.5	290.0	285.5	283.4	281.2	279.0	276.9	274.8	272.6	270.5
2610	300.3	295.6	291.1	286.6	284.4	282.3	280.1	278.0	275.8	273.7	271.6
2620	301.4	296.7	292.2	287.7	285.5	283.4	281.2	279.0	276.9	274.7	272.6
2630	302.6	297.8	293.3	288.8	286.6	284.4	282.3	280.1	277.9	275.8	273.6
2640	303.7	299.0	294.4	289.9	287.7	285.5	283.3	281.1	279.0	276.8	274.7
2650	304.9	300.1	295.5	291.0	288.8	286.6	284.4	282.2	280.0	277.9	275.7
2660	306.0	301.2	296.6	292.1	289.9	287.7	285.5	283.3	281.1	278.9	276.8
2670	307.2	302.4	297.8	293.2	291.0	288.8	286.5	284.3	282.1	280.0	277.8
2680	308.3	303.5	298.9	294.3	292.1	289.8	287.6	285.4	283.2	281.0	278.8
2690	309.5	304.6	300.0	295.4	293.2	290.9	288.7	286.5	284.3	282.1	279.9
2700	310.6	305.8	301.1	296.5	294.2	292.0	289.8	287.5	285.3	283.1	280.9
2710	311.8	306.9	302.2	297.6	295.3	293.1	290.8	288.6	286.4	284.2	282.0
2720	312.9	308.0	303.3	298.7	296.4	294.2	291.9	289.7	287.4	285.2	283.0
2730	314.1	309.2	304.5	299.8	297.5	295.2	293.0	290.7	288.5	286.3	284.0
2740	315.2	310.3	305.6	300.9	298.6	296.3	294.1	291.8	289.5	287.3	285.1
2750	316.4	311.4	306.7	302.0	299.7	297.4	295.1	292.9	290.6	288.4	286.1
2760	317.5	312.6	307.8	303.1	300.8	298.5	296.2	293.9	291.7	289.4	287.2
2770	318.7	313.7	308.9	304.2	301.9	299.6	297.3	295.0	292.7	290.5	288.2
2780	319.8	314.8	310.0	305.3	303.0	300.6	298.3	296.1	293.8	291.5	289.2
2790	321.0	316.0	311.1	306.4	304.1	301.7	299.4	297.1	294.8	292.5	290.3
2800	322.1	317.1	312.3	307.5	305.1	302.8	300.5	298.2	295.9	293.6	291.3
2810	323.3	318.2	313.4	308.6	306.2	303.9	301.6	299.2	296.9	294.6	292.4
2820	324.4	319.4	314.5	309.7	307.3	305.0	302.6	300.3	298.0	295.7	293.4
2830	325.6	320.5	315.6	310.8	308.4	306.1	303.7	301.4	299.0	296.7	294.4
2840	326.7	321.6	316.7	311.9	309.5	307.1	304.8	302.4	300.1	297.8	295.5
2850	327.9	322.8	317.8	313.0	310.6	308.2	305.9	303.5	301.2	298.8	296.5
2860	329.0	323.9	318.9	314.1	311.7	309.3	306.9	304.6	302.2	299.9	297.6
2870	330.2	325.0	320.1	315.2	312.8	310.4	308.0	305.6	303.3	300.9	298.6
2880	331.3	326.1	321.2	316.3	313.9	311.5	309.1	306.7	304.3	302.0	299.6
2890	332.5	327.3	322.3	317.4	314.9	312.5	310.1	307.8	305.4	303.0	300.7
2900	333.6	328.4	323.4	318.4	316.0	313.6	311.2	308.8	306.4	304.1	301.7
2910	334.8	329.5	324.5	319.5	317.1	314.7	312.3	309.9	307.5	305.1	302.8
2920	335.9	330.7	325.6	320.6	318.2	315.8	313.4	311.0	308.6	306.2	303.8
2930	337.1	331.8	326.7	321.7	319.3	316.9	314.4	312.0	309.6	307.2	304.8
2940	338.2	332.9	327.9	322.8	320.4	317.9	315.5	313.1	310.7	308.3	305.9
2950	339.4	334.1	329.0	323.9	321.5	319.0	316.6	314.2	311.7	309.3	306.9
2960	340.5	335.2	330.1	325.0	322.6	320.1	317.7	315.2	312.8	310.4	308.0
2970	341.7	336.3	331.2	326.1	323.7	321.2	318.7	316.3	313.8	311.4	309.0
2980	342.8	337.5	332.3	327.2	324.7	322.3	319.8	317.3	314.9	312.5	310.0
2990	344.0	338.6	333.4	328.3	325.8	323.3	320.9	318.4	315.9	313.5	311.1
3000	345.1	339.7	334.5	329.4	326.9	324.4	321.9	319.5	317.0	314.6	312.1

SMITHSONIAN TABLES.

TABLE 17.

REDUCTION OF THE BAROMETER TO SEA LEVEL
METRIC MEASURES.

Values of 2000 × m.

$$m = \frac{Z(1-\beta)}{18444 + 67.53\theta + .003Z}$$

Altitude in metres. Z.	MEAN TEMPERATURE OF AIR COLUMN IN CENTIGRADE DEGREES (θ).									
	+12°	+14°	+16°	+18°	+20°	+22°	+24°	+26°	+28°	+32°	+36°
2500	258.1	256.2	254.2	252.3	250.4	248.5	246.7	244.9	243.1	239.6	236.2
2510	259.2	257.2	255.2	253.3	251.4	249.5	247.7	245.9	244.1	240.6	237.1
2520	260.2	258.2	256.2	254.3	252.4	250.5	248.7	246.8	245.0	241.5	238.0
2530	261.2	259.2	257.3	255.3	253.4	251.5	249.7	247.8	246.0	242.5	239.0
2540	262.3	260.3	258.3	256.3	254.4	252.5	250.6	248.8	247.0	243.4	239.9
2550	263.3	261.3	259.3	257.3	255.4	253.5	251.6	249.8	247.9	244.4	240.9
2560	264.3	262.3	260.3	258.4	256.4	254.5	252.6	250.7	248.9	245.3	241.8
2570	265.4	263.3	261.3	259.4	257.4	255.5	253.6	251.7	249.9	246.3	242.8
2580	266.4	264.4	262.3	260.4	258.4	256.5	254.6	252.7	250.9	247.3	243.7
2590	267.4	265.4	263.4	261.4	259.4	257.5	255.6	253.7	251.8	248.2	244.7
2600	268.5	266.4	264.4	262.4	260.4	258.5	256.6	254.7	252.8	249.2	245.6
2610	269.5	267.4	265.4	263.4	261.4	259.5	257.5	255.6	253.8	250.1	246.5
2620	270.5	268.5	266.4	264.4	262.4	260.5	258.5	256.6	254.8	251.1	247.5
2630	271.6	269.5	267.4	265.4	263.4	261.5	259.5	257.6	255.7	252.0	248.4
2640	272.6	270.5	268.4	266.4	264.4	262.5	260.5	258.6	256.7	253.0	249.4
2650	273.6	271.5	269.5	267.4	265.4	263.4	261.5	259.6	257.7	254.0	250.3
2660	274.7	272.6	270.5	268.4	266.4	264.4	262.5	260.5	258.6	254.9	251.3
2670	275.7	273.6	271.5	269.4	267.4	265.4	263.5	261.5	259.6	255.9	252.2
2680	276.7	274.6	272.5	270.5	268.4	266.4	264.4	262.5	260.6	256.8	253.1
2690	277.7	275.6	273.5	271.5	269.4	267.4	265.4	263.5	261.6	257.8	254.1
2700	278.8	276.6	274.5	272.5	270.4	268.4	266.4	264.5	262.5	258.8	255.0
2710	279.8	277.7	275.6	273.5	271.4	269.4	267.4	265.4	263.5	259.7	256.0
2720	280.8	278.7	276.6	274.5	272.4	270.4	268.4	266.4	264.5	260.7	256.9
2730	281.9	279.7	277.6	275.5	273.4	271.4	269.4	267.4	265.4	261.6	257.9
2740	282.9	280.7	278.6	276.5	274.4	272.4	270.4	268.4	266.4	262.6	258.8
2750	283.9	281.8	279.6	277.5	275.4	273.4	271.4	269.4	267.4	263.5	259.8
2760	285.0	282.8	280.6	278.5	276.4	274.4	272.3	270.3	268.4	264.5	260.7
2770	286.0	283.8	281.7	279.5	277.4	275.4	273.3	271.3	269.3	265.5	261.6
2780	287.0	284.8	282.7	280.5	278.4	276.4	274.3	272.3	270.3	266.4	262.6
2790	288.1	285.9	283.7	281.5	279.4	277.4	275.3	273.3	271.3	267.4	263.5
2800	289.1	286.9	284.7	282.6	280.4	278.3	276.3	274.3	272.2	268.3	264.5
2810	290.1	287.9	285.7	283.6	281.4	279.3	277.3	275.2	273.2	269.3	265.4
2820	291.2	288.9	286.7	284.6	282.4	280.3	278.3	276.2	274.2	270.3	266.4
2830	292.2	290.0	287.8	285.6	283.4	281.3	279.2	277.2	275.2	271.2	267.3
2840	293.2	291.0	288.8	286.6	284.4	282.3	280.2	278.2	276.1	272.2	268.3
2850	294.3	292.0	289.8	287.6	285.4	283.3	281.2	279.2	277.1	273.1	269.2
2860	295.3	293.0	290.8	288.6	286.4	284.3	282.2	280.1	278.1	274.1	270.1
2870	296.3	294.1	291.8	289.6	287.4	285.3	283.2	281.1	279.0	275.0	271.1
2880	297.4	295.1	292.8	290.6	288.4	286.3	284.2	282.1	280.0	276.0	272.0
2890	298.4	296.1	293.8	291.6	289.4	287.3	285.2	283.1	281.0	277.0	273.0
2900	299.4	297.1	294.9	292.6	290.4	288.3	286.2	284.1	282.0	277.9	273.9
2910	300.4	298.1	295.9	293.7	291.5	289.3	287.1	285.0	282.9	278.9	274.9
2920	301.5	299.2	296.9	294.7	292.5	290.3	288.1	286.0	283.9	279.8	275.8
2930	302.5	300.2	297.9	295.7	293.5	291.3	289.1	287.0	284.9	280.8	276.8
2940	303.5	301.2	298.9	296.7	294.5	292.3	290.1	288.0	285.9	281.8	277.7
2950	304.6	302.2	299.9	297.7	295.5	293.3	291.1	289.0	286.8	282.7	278.6
2960	305.6	303.3	301.0	298.7	296.5	294.2	292.1	289.9	287.8	283.7	279.6
2970	306.6	304.3	302.0	299.7	297.5	295.2	293.1	290.9	288.8	284.6	280.5
2980	307.7	305.3	303.0	300.7	298.5	296.2	294.0	291.9	289.7	285.6	281.5
2990	308.7	306.3	304.0	301.7	299.5	297.2	295.0	292.9	290.7	286.5	282.4
3000	309.7	307.4	305.0	302.7	300.5	298.2	296.0	293.8	291.7	287.5	283.4

SMITHSONIAN TABLES.

TABLE 18.

REDUCTION OF THE BAROMETER TO SEA LEVEL.
METRIC MEASURES.

Correction of 2000 m for Latitude: $2000\,m \times 0.002662 \cos 2\phi$.

For latitudes 0° to 45°, the correction is to be subtracted.
For latitudes 45° to 90°, the correction is to be added.

2000 m.	LATITUDE.									
	0°	5°	10°	15°	20°	25°	30°	35°	40°	45°
10	0.0	0.0	0.0	0.0	0.0	0.0	0.0	0.0	0.0	0.0
20	0.1	0.1	0.1	0.0	0.0	0.0	0.0	0.0	0.0	0.0
30	0.1	0.1	0.1	0.1	0.1	0.1	0.0	0.0	0.0	0.0
40	0.1	0.1	0.1	0.1	0.1	0.1	0.1	0.0	0.0	0.0
50	0.1	0.1	0.1	0.1	0.1	0.1	0.1	0.0	0.0	0.0
60	0.2	0.2	0.2	0.1	0.1	0.1	0.1	0.1	0.0	0.0
70	0.2	0.2	0.2	0.2	0.1	0.1	0.1	0.1	0.0	0.0
80	0.2	0.2	0.2	0.2	0.2	0.1	0.1	0.1	0.0	0.0
90	0.2	0.2	0.2	0.2	0.2	0.2	0.1	0.1	0.0	0.0
100	0.3	0.3	0.3	0.2	0.2	0.2	0.1	0.1	0.0	0.0
110	0.3	0.3	0.3	0.3	0.2	0.2	0.1	0.1	0.1	0.0
120	0.3	0.3	0.3	0.3	0.2	0.2	0.2	0.1	0.1	0.0
130	0.3	0.3	0.3	0.3	0.3	0.2	0.2	0.1	0.1	0.0
140	0.4	0.4	0.4	0.3	0.3	0.2	0.2	0.1	0.1	0.0
150	0.4	0.4	0.4	0.3	0.3	0.3	0.2	0.1	0.1	0.0
160	0.4	0.4	0.4	0.4	0.3	0.3	0.2	0.1	0.1	0.0
170	0.5	0.4	0.4	0.4	0.3	0.3	0.2	0.2	0.1	0.0
180	0.5	0.5	0.5	0.4	0.4	0.3	0.2	0.2	0.1	0.0
190	0.5	0.5	0.5	0.4	0.4	0.3	0.3	0.2	0.1	0.0
200	0.5	0.5	0.5	0.5	0.4	0.3	0.3	0.2	0.1	0.0
210	0.6	0.6	0.5	0.5	0.4	0.4	0.3	0.2	0.1	0.0
220	0.6	0.6	0.6	0.5	0.4	0.4	0.3	0.2	0.1	0.0
230	0.6	0.6	0.6	0.5	0.5	0.4	0.3	0.2	0.1	0.0
240	0.6	0.6	0.6	0.6	0.5	0.4	0.3	0.2	0.1	0.0
250	0.7	0.7	0.6	0.6	0.5	0.4	0.3	0.2	0.1	0.0
260	0.7	0.7	0.7	0.6	0.5	0.4	0.3	0.2	0.1	0.0
270	0.7	0.7	0.7	0.6	0.6	0.5	0.4	0.2	0.1	0.0
280	0.7	0.7	0.7	0.6	0.6	0.5	0.4	0.3	0.1	0.0
290	0.8	0.8	0.7	0.7	0.6	0.5	0.4	0.3	0.1	0.0
300	0.8	0.8	0.8	0.7	0.6	0.5	0.4	0.3	0.1	0.0
310	0.8	0.8	0.8	0.7	0.6	0.5	0.4	0.3	0.1	0.0
320	0.9	0.8	0.8	0.7	0.7	0.5	0.4	0.3	0.1	0.0
330	0.9	0.9	0.8	0.8	0.7	0.6	0.4	0.3	0.2	0.0
340	0.9	0.9	0.9	0.8	0.7	0.6	0.5	0.3	0.2	0.0
350	0.9	0.9	0.9	0.8	0.7	0.6	0.5	0.3	0.2	0.0
	90°	85°	80°	75°	70°	65°	60°	55°	50°	45°

TABLE 19.

REDUCTION OF THE BAROMETER TO SEA LEVEL.
METRIC MEASURES.
$B_0 - B = B(10^m - 1)$.

Top argument: Height of the barometer (B).
Side argument: Values of 2000 m obtained from Table 17.

2000 m.	HEIGHT OF THE BAROMETER IN MILLIMETRES.										
	790	780	770	760	750	740	730	720	710	700	690
	mm.	mm.	mm.	mm.	mm.	mm.	mm.	mm.	mm.	mm.	mm.
1	0.9	0.9	0.9	0.9	0.9	0.9	0.8	0.8	0.8	0.8	0.8
2	1.8	1.8	1.8	1.8	1.7	1.7	1.7	1.7	1.6	1.6	1.6
3	2.7	2.7	2.7	2.6	2.6	2.6	2.5	2.5	2.5	2.4	2.4
4	3.6	3.6	3.6	3.5	3.5	3.4	3.4	3.3	3.3	3.2	3.2
5	4.6	4.5	4.4	4.4	4.3	4.3	4.2	4.2	4.1	4.0	4.0
6	5.5	5.4	5.3	5.3	5.2	5.1	5.1	5.0	4.9	4.9	4.8
7	6.4	6.3	6.2	6.1	6.1	6.0	5.9	5.8	5.7	5.7	5.6
8	7.3	7.2	7.1	7.0	6.9	6.8	6.8	6.7	6.6	6.5	6.4
9	8.2	8.1	8.0	7.9	7.8	7.7	7.6	7.5	7.4	7.3	7.2
10	9.1	9.0	8.9	8.8	8.7	8.6	8.5	8.3	8.2	8.1	8.0
11	10.1	9.9	9.8	9.7	9.6	9.4	9.3	9.2	9.0	8.9	8.8
12	11.0	10.9	10.7	10.6	10.4	10.3	10.2	10.0	9.9	9.7	9.6
13	11.9	11.8	11.6	11.5	11.3	11.2	11.0	10.9	10.7	10.6	10.4
14	12.8	12.7	12.5	12.3	12.2	12.0	11.9	11.7	11.5	11.4	11.2
15	13.8	13.6	13.4	13.2	13.1	12.9	12.7	12.5	12.4	12.2	12.0
16	14.7	14.5	14.3	14.1	13.9	13.8	13.6	13.4	13.2	13.0	12.8
17	15.6	15.4	15.2	15.0	14.8	14.6	14.4	14.2	14.0	13.8	13.6
18	16.5	16.3	16.1	15.9	15.7	15.5	15.3	15.1	14.9	14.7	14.4
19	17.5	17.3	17.0	16.8	16.6	16.4	16.1	15.9	15.7	15.5	15.3
20	18.4	18.2	17.9	17.7	17.5	17.2	17.0	16.8	16.5	16.3	16.1
21	19.3	19.1	18.8	18.6	18.4	18.1	17.9	17.6	17.4	17.1	16.9
22		20.0	19.8	19.5	19.2	19.0	18.7	18.5	18.2	18.0	17.7
23		20.9	20.7	20.4	20.1	19.9	19.6	19.3	19.1	18.8	18.5
24		21.9	21.6	21.3	21.0	20.7	20.5	20.2	19.9	19.6	19.3
25		22.8	22.5	22.2	21.9	21.6	21.3	21.0	20.7	20.4	20.1
26		23.7	23.4	23.1	22.8	22.5	22.2	21.9	21.6	21.3	21.0
27		24.6	24.3	24.0	23.7	23.4	23.0	22.7	22.4	22.1	21.8
28		25.6	25.2	24.9	24.6	24.2	23.9	23.6	23.3	22.9	22.6
29		26.5	26.1	25.8	25.5	25.1	24.8	24.4	24.1	23.8	23.4
30		27.4	27.1	26.7	26.4	26.0	25.7	25.3	25.0	24.6	24.2
31		28.3	28.0	27.6	27.3	26.9	26.5	26.2	25.8	25.4	25.1
32		29.3	28.9	28.5	28.1	27.8	27.4	27.0	26.6	26.3	25.9
33		30.2	29.8	29.4	29.0	28.7	28.3	27.9	27.5	27.1	26.7
34			30.7	30.3	29.9	29.5	29.1	28.7	28.3	27.9	27.5
35			31.7	31.2	30.8	30.4	30.0	29.6	29.2	28.8	28.4
36			32.6	32.2	31.7	31.3	30.9	30.5	30.0	29.6	29.2
37			33.5	33.1	32.6	32.2	31.8	31.3	30.9	30.5	30.0
38			34.4	34.0	33.5	33.1	32.6	32.2	31.8	31.3	30.9
39			35.4	34.9	34.4	34.0	33.5	33.1	32.6	32.1	31.7
40			36.3	35.8	35.3	34.9	34.4	33.9	33.5	33.0	32.5

TABLE 19.

REDUCTION OF THE BAROMETER TO SEA LEVEL.
METRIC MEASURES.

$B_0 - B = B (10^m - 1)$.

Top argument: Height of the barometer (B).
Side argument: Values of 2000 m obtained from Table 17.

2000 m.	HEIGHT OF THE BAROMETER IN MILLIMETRES.									
	760	750	740	730	720	710	700	690	680	670	660
	mm.	mm.	mm.	mm.	mm.	mm.	mm.	mm.	mm.	mm.	mm.
40	35.8	35.3	34.9	34.4	33.9	33.5	33.0	32.5	32.0	31.6	31.1
41	36.7	36.3	35.8	35.3	34.8	34.3	33.8	33.4	32.9	32.4	31.9
42	37.7	37.2	36.7	36.2	35.7	35.2	34.7	34.2	33.7	33.2	32.7
43	38.6	38.1	37.6	37.0	36.5	36.0	35.5	35.0	34.5	34.0	33.5
44	39.5	39.0	38.5	37.9	37.4	36.9	36.4	35.9	35.3	34.8	34.3
45	40.4	39.9	39.3	38.8	38.3	37.8	37.2	36.7	36.2	35.6	35.1
46	41.3	40.8	40.2	39.7	39.2	38.6	38.1	37.5	37.0	36.4	35.9
47	42.3	41.7	41.1	40.6	40.0	39.5	38.9	38.4	37.8	37.3	36.7
48	43.2	42.6	42.0	41.5	40.9	40.3	39.8	39.2	38.6	38.1	37.5
49	44.1	43.5	42.9	42.4	41.8	41.2	40.6	40.0	39.5	38.9	38.3
50	45.0	44.4	43.8	43.3	42.7	42.1	41.5	40.9	40.3	39.7	39.1
51	46.0	45.4	44.8	44.1	43.5	42.9	42.3	41.7	41.1	40.5	39.9
52	46.9	46.3	45.7	45.0	44.4	43.8	43.2	42.6	42.0	41.3	40.7
53	47.8	47.2	46.6	45.9	45.3	44.7	44.0	43.4	42.8	42.2	41.5
54	48.7	48.1	47.5	46.8	46.2	45.5	44.9	44.3	43.6	43.0	42.3
55	49.7	49.0	48.4	47.7	47.1	46.4	45.8	45.1	44.5	43.8	43.1
56	50.6	49.9	49.3	48.6	47.9	47.3	46.6	46.0	45.3	44.6	44.0
57	51.5	50.9	50.2	49.5	48.8	48.2	47.5	46.8	46.1	45.4	44.8
58	52.5	51.8	51.1	50.4	49.7	49.0	48.3	47.6	47.0	46.3	45.6
59	53.4	52.7	52.0	51.3	50.6	49.9	49.2	48.5	47.8	47.1	46.4
60		53.6	52.9	52.2	51.5	50.8	50.1	49.3	48.6	47.9	47.2
61		54.6	53.8	53.1	52.4	51.7	50.9	50.2	49.5	48.7	48.0
62		55.5	54.8	54.0	53.3	52.5	51.8	51.1	50.3	49.6	48.8
63		56.4	55.7	54.9	54.2	53.4	52.7	51.9	51.2	50.4	49.6
64		57.3	56.6	55.8	55.1	54.3	53.5	52.8	52.0	51.2	50.5
65		58.3	57.5	56.7	55.9	55.2	54.4	53.6	52.8	52.1	51.3
66		59.2	58.4	57.6	56.8	56.0	55.3	54.5	53.7	52.9	52.1
67		60.1	59.3	58.5	57.7	56.9	56.1	55.3	54.5	53.7	52.9
68			60.3	59.4	58.6	57.8	57.0	56.2	55.4	54.6	53.7
69			61.2	60.4	59.5	58.7	57.9	57.0	56.2	55.4	54.6
70			62.1	61.3	60.4	59.6	58.7	57.9	57.1	56.2	55.4
71			63.0	62.2	61.3	60.5	59.6	58.8	57.9	57.1	56.2
72			64.0	63.1	62.2	61.4	60.5	59.6	58.8	57.9	57.0
73			64.9	64.0	63.1	62.3	61.4	60.5	59.6	58.7	57.9
74			65.8	64.9	64.0	63.1	62.3	61.4	60.5	59.6	58.7
75			66.7	65.8	64.9	64.0	63.1	62.2	61.3	60.4	59.5
76			67.7	66.8	65.8	64.9	64.0	63.1	62.2	61.3	60.4
77			68.6	67.7	66.7	65.8	64.9	64.0	63.0	62.1	61.2
78			69.5	68.6	67.6	66.7	65.8	64.8	63.9	63.0	62.0
79			70.5	69.5	68.6	67.6	66.7	65.7	64.7	63.8	62.8
80				70.4	69.5	68.5	67.5	66.6	65.6	64.6	63.7
81				71.4	70.4	69.4	68.4	67.4	66.5	65.5	64.5
82				72.3	71.3	70.3	69.3	68.3	67.3	66.3	65.3
83				73.2	72.2	71.2	70.2	69.2	68.2	67.2	66.2
84				74.1	73.1	72.1	71.1	70.1	69.0	68.0	67.0
85				75.0	74.0	73.0	72.0	70.9	69.9	68.9	67.9

SMITHSONIAN TABLES.

TABLE 19.
REDUCTION OF THE BAROMETER TO SEA LEVEL.
METRIC MEASURES.
$B_0 - B = B(10^m - 1)$.

Top argument: Height of the barometer (B).
Side argument: Values of $2000\,m$ obtained from Table 17.

2000 m.	HEIGHT OF THE BAROMETER IN MILLIMETRES.								
	720	710	700	690	680	670	660	650	640	630
	mm.	mm.	mm.	mm.	mm.	mm.	mm.	mm.	mm.	mm.
80	69.5	68.5	67.5	66.6	65.6	64.6	63.7	62.7	61.7	60.8
81	70.4	69.4	68.4	67.4	66.5	65.5	64.5	63.5	62.6	61.6
82	71.3	70.3	69.3	68.3	67.3	66.3	65.3	64.4	63.4	62.4
83	72.2	71.2	70.2	69.2	68.2	67.2	66.2	65.2	64.2	63.2
84	73.1	72.1	71.1	70.1	69.0	68.0	67.0	66.0	65.0	64.0
85	74.0	73.0	72.0	70.9	69.9	68.9	67.9	66.8	65.8	64.8
86	74.9	73.9	72.9	71.8	70.8	69.7	68.7	67.7	66.6	65.6
87	75.9	74.8	73.7	72.7	71.6	70.6	69.5	68.5	67.4	66.4
88	76.8	75.7	74.6	73.6	72.5	71.4	70.4	69.3	68.2	67.2
89	77.7	76.6	75.5	74.5	73.4	72.3	71.2	70.1	69.1	68.0
90	78.6	77.5	76.4	75.3	74.2	73.1	72.1	71.0	69.9	68.8
91	79.5	78.4	77.3	76.2	75.1	74.0	72.9	71.8	70.7	69.6
92	80.4	79.3	78.2	77.1	76.0	74.9	73.7	72.6	71.5	70.4
93	81.4	80.2	79.1	78.0	76.8	75.7	74.6	73.5	72.3	71.2
94	82.3	81.1	80.0	78.9	77.7	76.6	75.4	74.3	73.1	72.0
95	83.2	82.1	80.9	79.7	78.6	77.4	76.3	75.1	74.0	72.8
96	84.1	83.0	81.8	80.6	79.5	78.3	77.1	76.0	74.8	73.6
97	85.1	83.9	82.7	81.5	80.3	79.2	78.0	76.8	75.6	74.4
98	86.0	84.8	83.6	82.4	81.2	80.0	78.8	77.6	76.4	75.2
99	86.9	85.7	84.5	83.3	82.1	80.9	79.7	78.5	77.3	76.1
100	87.9	86.6	85.4	84.2	83.0	81.8	80.5	79.3	78.1	76.9
101	88.8	87.6	86.3	85.1	83.9	82.6	81.4	80.2	78.9	77.7
102	89.7	88.5	87.2	86.0	84.7	83.5	82.2	81.0	79.7	78.5
103	90.6	89.4	88.1	86.9	85.6	84.4	83.1	81.8	80.6	79.3
104		90.3	89.0	87.8	86.5	85.2	84.0	82.7	81.4	80.1
105		91.2	89.9	88.7	87.4	86.1	84.8	83.5	82.2	81.0
106		92.2	90.9	89.6	88.3	87.0	85.7	84.4	83.1	81.8
107		93.1	91.8	90.5	89.1	87.8	86.5	85.2	83.9	82.6
108		94.0	92.7	91.4	90.0	88.7	87.4	86.1	84.7	83.4
109		94.9	93.6	92.3	90.9	89.6	88.2	86.9	85.6	84.2
110		95.9	94.5	93.2	91.8	90.5	89.1	87.8	86.4	85.1
111		96.8	95.4	94.1	92.7	91.3	90.0	88.6	87.2	85.9
112		97.7	96.3	95.0	93.6	92.2	90.8	89.5	88.1	86.7
113		98.6	97.3	95.9	94.5	93.1	91.7	90.3	88.9	87.5
114		99.6	98.2	96.8	95.4	94.0	92.6	91.2	89.8	88.4
115		100.5	99.1	97.7	96.3	94.8	93.4	92.0	90.6	89.2
116			100.0	98.6	97.2	95.7	94.3	92.9	91.4	90.0
117			100.9	99.5	98.1	96.6	95.2	93.7	92.3	90.8
118			101.9	100.4	98.9	97.5	96.0	94.6	93.1	91.7
119			102.8	101.3	99.8	98.4	96.9	95.4	94.0	92.5
120			103.7	102.2	100.7	99.3	97.8	96.3	94.8	93.3
121			104.6	103.1	101.6	100.1	98.7	97.2	95.7	94.2
122			105.6	104.1	102.5	101.0	99.5	98.0	96.5	95.0
123			106.5	105.0	103.4	101.9	100.4	98.9	97.4	95.8
124			107.4	105.9	104.3	102.8	101.3	99.7	98.2	96.7
125			108.3	106.8	105.3	103.7	102.2	100.6	99.1	97.5

SMITHSONIAN TABLES.

TABLE 19.
REDUCTION OF THE BAROMETER TO SEA LEVEL
METRIC MEASURES.
$B_o - B = B(10^m - 1)$.

Top argument: Height of the barometer (B).
Side argument: Values of 2000 m obtained from Table 17.

2000 m.	HEIGHT OF THE BAROMETER IN MILLIMETRES.									
	690	680	670	660	650	640	630	620	610	600
	mm.	mm.	mm.	mm.	mm.	mm.	mm.	mm.	mm.	mm.
125	106.8	105.3	103.7	102.2	100.6	99.1	97.5	96.0	94.4	92.9
126	107.7	106.2	104.6	103.0	101.5	99.9	98.4	96.8	95.2	93.7
127	108.6	107.1	105.5	103.9	102.3	100.8	99.2	97.6	96.0	94.5
128	109.6	108.0	106.4	104.8	103.2	101.6	100.0	98.4	96.9	95.3
129	110.5	108.9	107.3	105.7	104.1	102.5	100.9	99.3	97.7	96.1
130	111.4	109.8	108.2	106.6	104.9	103.3	101.7	100.1	98.5	96.9
131	112.3	110.7	109.1	107.4	105.8	104.2	102.6	100.9	99.3	97.7
132	113.2	111.6	110.0	108.3	106.7	105.0	103.4	101.8	100.1	98.5
133	114.2	112.5	110.9	109.2	107.6	105.9	104.2	102.6	100.9	99.3
134	115.1	113.4	111.8	110.1	108.4	106.8	105.1	103.4	101.8	100.1
135	116.0	114.3	112.7	111.0	109.3	107.6	105.9	104.3	102.6	100.9
136	117.0	115.3	113.6	111.9	110.2	108.5	106.8	105.1	103.4	101.7
137	117.9	116.2	114.5	112.8	111.1	109.3	107.6	105.9	104.2	102.5
138	118.8	117.1	115.4	113.6	111.9	110.2	108.5	106.8	105.0	103.3
139	119.7	118.0	116.3	114.5	112.8	111.1	109.3	107.6	105.9	104.1
140	120.7	118.9	117.2	115.4	113.7	111.9	110.2	108.4	106.7	104.9
141	121.6	119.9	118.1	116.3	114.6	112.8	111.0	109.3	107.5	105.8
142		120.8	119.0	117.2	115.4	113.7	111.9	110.1	108.3	106.6
143		121.7	119.9	118.1	116.3	114.5	112.7	111.0	109.2	107.4
144		122.6	120.8	119.0	117.2	115.4	113.6	111.8	110.0	108.2
145		123.5	121.7	119.9	118.1	116.3	114.5	112.6	110.8	109.0
146		124.5	122.6	120.8	119.0	117.1	115.3	113.5	111.7	109.8
147		125.4	123.6	121.7	119.9	118.0	116.2	114.3	112.5	110.6
148		126.3	124.5	122.6	120.7	118.9	117.0	115.2	113.3	111.5
149		127.3	125.4	123.5	121.6	119.8	117.9	116.0	114.2	112.3
150		128.2	126.3	124.4	122.5	120.6	118.8	116.9	115.0	113.1
151		129.1	127.2	125.3	123.4	121.5	119.6	117.7	115.8	113.9
152		130.0	128.1	126.2	124.3	122.4	120.5	118.6	116.7	114.7
153		131.0	129.1	127.1	125.2	123.3	121.3	119.4	117.5	115.6
154			130.0	128.0	126.1	124.2	122.2	120.3	118.3	116.4
155			130.9	128.9	127.0	125.0	123.1	121.1	119.2	117.2
156			131.8	129.8	127.9	125.9	123.9	122.0	120.0	118.0
157			132.7	130.8	128.8	126.8	124.8	122.8	120.9	118.9
158			133.7	131.7	129.7	127.7	125.7	123.7	121.7	119.7
159			134.6	132.6	130.6	128.6	126.6	124.5	122.5	120.5
160			135.5	133.5	131.5	129.4	127.4	125.4	123.4	121.4
161			136.4	134.4	132.4	130.3	128.3	126.3	124.2	122.2
162			137.4	135.3	133.3	131.2	129.2	127.1	125.1	123.0
163			138.3	136.2	134.2	132.1	130.0	128.0	125.9	123.9
164			139.2	137.2	135.1	133.0	130.9	128.8	126.8	124.7
165			140.2	138.1	136.0	133.9	131.8	129.7	127.6	125.5
166			141.1	139.0	136.9	134.8	132.7	130.6	128.5	126.4
167			142.0	139.9	137.8	135.7	133.6	131.4	129.3	127.2
168				140.8	138.7	136.6	134.4	132.3	130.2	128.0
169				141.8	139.6	137.5	135.3	133.2	131.0	128.9
170				142.7	140.5	138.4	136.2	134.0	131.9	129.7

TABLE 19.

REDUCTION OF THE BAROMETER TO SEA LEVEL.
METRIC MEASURES.
$B_o - B = B(10^m - 1)$.

Top argument: Height of the barometer (B).
Side argument: Values of 2000 m obtained from Table 17.

2000 m.	HEIGHT OF THE BAROMETER IN MILLIMETRES.								
	650	640	630	620	610	600	590	580	570	560
	mm.	mm.	mm.	mm.	mm.	mm.	mm.	mm.	mm.
170	140.5	138.4	136.2	134.0	131.9	129.7	127.5	125.4	123.2	121.1
171	141.4	139.3	137.1	134.9	132.7	130.6	128.4	126.2	124.0	121.8
172	142.3	140.2	138.0	135.8	133.6	131.4	129.2	127.0	124.8	122.6
173	143.3	141.1	138.8	136.6	134.4	132.2	130.0	127.8	125.6	123.4
174	144.2	142.0	139.7	137.5	135.3	133.1	130.9	128.6	126.4	124.2
175	145.1	142.9	140.6	138.4	136.2	133.9	131.7	129.5	127.2	125.0
176	146.0	143.8	141.5	139.3	137.0	134.8	132.5	130.3	128.0	125.8
177	146.9	144.7	142.4	140.1	137.9	135.6	133.4	131.1	128.8	126.6
178	147.8	145.6	143.3	141.0	138.7	136.5	134.2	131.9	129.6	127.4
179	148.8	146.5	144.2	141.9	139.6	137.3	135.0	132.7	130.4	128.2
180	149.7	147.4	145.1	142.8	140.5	138.2	135.9	133.6	131.3	129.0
181	150.6	148.3	146.0	143.6	141.3	139.0	136.7	134.4	132.1	129.7
182	151.5	149.2	146.9	144.5	142.2	139.9	137.5	135.2	132.9	130.5
183	152.4	150.1	147.8	145.4	143.1	140.7	138.4	136.0	133.7	131.3
184	153.4	151.0	148.6	146.3	143.9	141.6	139.2	136.8	134.5	132.1
185	154.3	151.9	149.5	147.2	144.8	142.4	140.0	137.7	135.3	132.9
186	155.2	152.8	150.4	148.1	145.7	143.3	140.9	138.5	136.1	133.7
187	156.1	153.7	151.3	148.9	146.5	144.1	141.7	139.3	136.9	134.5
188	157.1	154.7	152.2	149.8	147.4	145.0	142.6	140.2	137.7	135.3
189	158.0	155.6	153.1	150.7	148.3	145.8	143.4	141.0	138.6	136.1
190	158.9	156.5	154.0	151.6	149.2	146.7	144.3	141.8	139.4	136.9
191	159.9	157.4	154.9	152.5	150.0	147.6	145.1	142.7	140.2	137.7
192	160.8	158.3	155.9	153.4	150.9	148.4	146.0	143.5	141.0	138.5
193	161.7	159.2	156.8	154.3	151.8	149.3	146.8	144.3	141.8	139.3
194		160.2	157.7	155.2	152.7	150.2	147.7	145.2	142.6	140.1
195		161.1	158.6	156.1	153.5	151.0	148.5	146.0	143.5	141.0
196		162.0	159.5	156.9	154.4	151.9	149.4	146.8	144.3	141.8
197		162.9	160.4	157.8	155.3	152.8	150.2	147.7	145.1	142.6
198		163.9	161.3	158.7	156.2	153.6	151.1	148.5	145.9	143.4
199		164.8	162.2	159.6	157.1	154.5	151.9	149.3	146.8	144.2
200		165.7	163.1	160.5	157.9	155.4	152.8	150.2	147.6	145.0
201		166.6	164.0	161.4	158.8	156.2	153.6	151.0	148.4	145.8
202		167.6	165.0	162.3	159.7	157.1	154.5	151.9	149.2	146.6
203		168.5	165.9	163.2	160.6	158.0	155.3	152.7	150.1	147.4
204		169.4	166.8	164.1	161.5	158.8	156.2	153.5	150.9	148.3
205		170.4	167.7	165.0	162.4	159.7	157.1	154.4	151.7	149.1
206		171.3	168.6	165.9	163.3	160.6	157.9	155.2	152.6	149.9
207		172.2	169.5	166.8	164.2	161.5	158.8	156.1	153.4	150.7
208			170.5	167.8	165.1	162.3	159.6	156.9	154.2	151.5
209			171.4	168.7	165.9	163.2	160.5	157.8	155.1	152.3
210			172.3	169.6	166.8	164.1	161.4	158.6	155.9	153.2
211			173.2	170.5	167.7	165.0	162.2	159.5	156.7	154.0
212			174.2	171.4	168.6	165.9	163.1	160.3	157.6	154.8
213			175.1	172.3	169.5	166.7	164.0	161.2	158.4	155.6
214			176.0	173.2	170.4	167.6	164.8	162.0	159.2	156.5
215			176.9	174.1	171.3	168.5	165.7	162.9	160.1	157.3

SMITHSONIAN TABLES.

TABLE 19.
REDUCTION OF THE BAROMETER TO SEA LEVEL.
METRIC MEASURES.
$B_0 - B = B(10^m - 1)$.

Top argument: Height of the barometer (B).
Side argument: Values of 2000 m obtained from Table 17.

2000 m.	HEIGHT OF THE BAROMETER IN MILLIMETRES.									
	620	610	600	590	580	570	560	550	540	530
	mm.	mm.	mm.	mm.	mm.	mm.	mm.	mm.	mm.	mm.
215	174.1	171.3	168.5	165.7	162.9	160.1	157.3	154.5	151.7	148.9
216	175.0	172.2	169.4	166.6	163.8	160.9	158.1	155.3	152.5	149.6
217	176.0	173.1	170.3	167.4	164.6	161.8	158.9	156.1	153.3	150.4
218	176.9	174.0	171.2	168.3	165.5	162.6	159.8	156.9	154.1	151.2
219	177.8	174.9	172.1	169.2	166.3	163.5	160.6	157.7	154.9	152.0
220	178.7	175.8	172.9	170.1	167.2	164.3	161.4	158.5	155.7	152.8
221	179.6	176.7	173.8	170.9	168.0	165.1	162.3	159.4	156.5	153.6
222	180.6	177.6	174.7	171.8	168.9	166.0	163.1	160.2	157.3	154.3
223	181.5	178.6	175.6	172.7	169.8	166.8	163.9	161.0	158.1	155.1
224	182.4	179.5	176.5	173.6	170.6	167.7	164.7	161.8	158.9	155.9
225	183.3	180.4	177.4	174.5	171.5	168.5	165.6	162.6	159.7	156.7
226	184.3	181.3	178.3	175.3	172.4	169.4	166.4	163.4	160.5	157.5
227	185.2	182.2	179.2	176.2	173.2	170.2	167.3	164.3	161.3	158.3
228	186.1	183.1	180.1	177.1	174.1	171.1	168.1	165.1	162.1	159.1
229	187.0	184.0	181.0	178.0	175.0	172.0	168.9	165.9	162.9	159.9
230	188.0	184.9	181.9	178.9	175.8	172.8	169.8	166.7	163.7	160.7
231	188.9	185.8	182.8	179.8	176.7	173.7	170.6	167.6	164.5	161.5
232	189.8	186.8	183.7	180.6	177.6	174.5	171.5	168.4	165.3	162.3
233	190.8	187.7	184.6	181.5	178.5	175.4	172.3	169.2	166.1	163.1
234	191.7	188.6	185.5	182.4	179.3	176.2	173.1	170.0	167.0	163.9
235	192.6	189.5	186.4	183.3	180.2	177.1	174.0	170.9	167.8	164.7
236		190.4	187.3	184.2	181.1	178.0	174.8	171.7	168.6	165.5
237		191.4	188.2	185.1	182.0	178.8	175.7	172.5	169.4	166.3
238		192.3	189.1	186.0	182.8	179.7	176.5	173.4	170.2	167.1
239		193.2	190.0	186.9	183.7	180.5	177.4	174.2	171.0	167.9
240		194.1	191.0	187.8	184.6	181.4	178.2	175.0	171.9	168.7
241		195.1	191.9	188.7	185.5	182.3	179.1	175.9	172.7	169.5
242		196.0	192.8	189.6	186.4	183.1	179.9	176.7	173.5	170.3
243		196.9	193.7	190.5	187.2	184.0	180.8	177.5	174.3	171.1
244		197.8	194.6	191.4	188.1	184.9	181.6	178.4	175.1	171.9
245		198.8	195.5	192.3	189.0	185.7	182.5	179.2	176.0	172.7
246		199.7	196.4	193.2	189.9	186.6	183.3	180.1	176.8	173.5
247		200.6	197.4	194.1	190.8	187.5	184.2	180.9	177.6	174.3
248		201.6	198.3	195.0	191.7	188.4	185.1	181.7	178.4	175.1
249		202.5	199.2	195.9	192.6	189.2	185.9	182.6	179.3	176.0
250			200.1	196.8	193.4	190.1	186.8	183.4	180.1	176.8
251			201.0	197.7	194.3	191.0	187.6	184.3	180.9	177.6
252			202.0	198.6	195.2	191.9	188.5	185.1	181.8	178.4
253			202.9	199.5	196.1	192.7	189.4	186.0	182.6	179.2
254			203.8	200.4	197.0	193.6	190.2	186.8	183.4	180.0
255			204.7	201.3	197.9	194.5	191.1	187.7	184.3	180.8
256			205.7	202.2	198.8	195.4	191.9	188.5	185.1	181.7
257			206.6	203.1	199.7	196.3	192.8	189.4	185.9	182.5
258			207.5	204.1	200.6	197.1	193.7	190.2	186.8	183.3
259			208.4	205.0	201.5	198.0	194.6	191.1	187.6	184.1
260			209.4	205.9	202.4	198.9	195.4	191.9	188.4	185.0

TABLE 19.

REDUCTION OF THE BAROMETER TO SEA LEVEL.
METRIC MEASURES.
$B_0 - B = B(10^m - 1)$.

Top argument: Height of the barometer (B).
Side argument: Values of 2000 m obtained from Table 17.

2000 m.	HEIGHT OF THE BAROMETER IN MILLIMETRES.								
	590	580	570	560	550	540	530	520	510
	mm.	mm.	mm.	mm.	mm.	mm.	mm.	mm.	mm.
260	205.9	202.4	198.9	195.4	191.9	188.4	185.0	181.5	178.0
261	206.8	203.3	199.8	196.3	192.8	189.3	185.8	182.3	178.8
262	207.7	204.2	200.7	197.2	193.6	190.1	186.6	183.1	179.6
263	208.6	205.1	201.6	198.0	194.5	191.0	187.4	183.9	180.4
264	209.6	206.0	202.5	198.9	195.4	191.8	188.3	184.7	181.1
265	210.5	206.9	203.3	199.8	196.2	192.6	189.1	185.5	181.9
266	211.4	207.8	204.2	200.7	197.1	193.5	189.9	186.3	182.7
267	212.3	208.7	205.1	201.5	197.9	194.3	190.7	187.1	183.5
268	213.3	209.6	206.0	202.4	198.8	195.2	191.6	188.0	184.3
269	214.2	210.5	206.9	203.3	199.7	196.0	192.4	188.8	185.1
270	215.1	211.5	207.8	204.2	200.5	196.9	193.2	189.6	185.9
271	216.0	212.4	208.7	205.0	201.4	197.7	194.1	190.4	186.7
272	217.0	213.3	209.6	205.9	202.3	198.6	194.9	191.2	187.5
273	217.9	214.2	210.5	206.8	203.1	199.4	195.7	192.0	188.3
274	218.8	215.1	211.4	207.7	204.0	200.3	196.6	192.9	189.1
275	219.8	216.0	212.3	208.6	204.9	201.1	197.4	193.7	190.0
276	220.7	216.9	213.2	209.5	205.7	202.0	198.2	194.5	190.8
277	221.6	217.9	214.1	210.3	206.6	202.8	199.1	195.3	191.6
278	222.6	218.8	215.0	211.2	207.5	203.7	199.9	196.1	192.4
279	223.5	219.7	215.9	212.1	208.3	204.5	200.8	197.0	193.2
280		220.6	216.8	213.0	209.2	205.4	201.6	197.8	194.0
281		221.5	217.7	213.9	210.1	206.3	202.4	198.6	194.8
282		222.5	218.6	214.8	211.0	207.1	203.3	199.5	195.6
283		223.4	219.5	215.7	211.8	208.0	204.1	200.3	196.4
284		224.3	220.5	216.6	212.7	208.8	205.0	201.1	197.2
285		225.2	221.4	217.5	213.6	209.7	205.8	201.9	198.1
286		226.2	222.3	218.4	214.5	210.6	206.7	202.8	198.9
287		227.1	223.2	219.3	215.4	211.4	207.5	203.6	199.7
288		228.0	224.1	220.2	216.2	212.3	208.4	204.4	200.5
289		229.0	225.0	221.1	217.1	213.2	209.2	205.3	201.3
290		229.9	225.9	222.0	218.0	214.0	210.1	206.1	202.1
291		230.8	226.8	222.9	218.9	214.9	210.9	206.9	203.0
292		231.8	227.8	223.8	219.8	215.8	211.8	207.8	203.8
293		232.7	228.7	224.7	220.7	216.6	212.6	208.6	204.6
294		233.6	229.6	225.6	221.5	217.5	213.5	209.5	205.4
295			230.5	226.5	222.4	218.4	214.3	210.3	206.3
296			231.4	227.4	223.3	219.3	215.2	211.1	207.1
297			232.4	228.3	224.2	220.1	216.1	212.0	207.9
298			233.3	229.2	225.1	221.0	216.9	212.8	208.7
299			234.2	230.1	226.0	221.9	217.8	213.7	209.6
300			235.1	231.0	226.9	222.8	218.6	214.5	210.4
301			236.1	231.9	227.8	223.6	219.5	215.4	211.2
302			237.0	232.8	228.7	224.5	220.4	216.2	212.1
303			237.9	233.8	229.6	225.4	221.2	217.1	212.9
304			238.9	234.7	230.5	226.3	222.1	217.9	213.7
305			239.8	235.6	231.4	227.2	223.0	218.8	214.6

SMITHSONIAN TABLES.

TABLE 19.
REDUCTION OF THE BAROMETER TO SEA LEVEL.
METRIC MEASURES.
$B_0 - B = B(10^m - 1)$.

Top argument: Height of the barometer (B).
Side argument: Values of 2000 m obtained from Table 17.

2000 m.	HEIGHT OF THE BAROMETER IN MILLIMETRES.								
	560	550	540	530	520	510	500	490	480
	mm.	mm.	mm.	mm.	mm.	mm.	mm.	mm.	mm.
305	235.6	231.4	227.2	223.0	218.8	214.6	210.3	206.1	201.9
306	236.5	232.3	228.0	223.8	219.6	215.4	211.2	206.9	202.7
307	237.4	233.2	228.9	224.7	220.5	216.2	212.0	207.7	203.5
308	238.3	234.1	229.8	225.6	221.3	217.1	212.8	208.5	204.3
309	239.2	235.0	230.7	226.4	222.2	217.9	213.6	209.4	205.1
310	240.2	235.9	231.6	227.3	223.0	218.7	214.4	210.1	205.9
311	241.1	236.8	232.5	228.2	223.9	219.6	215.3	211.0	206.7
312	242.0	237.7	233.4	229.1	224.7	220.4	216.1	211.8	207.5
313	242.9	238.6	234.3	229.9	225.6	221.2	216.9	212.6	208.2
314	243.9	239.5	235.2	230.8	226.4	222.1	217.7	213.4	209.0
315	244.8	240.4	236.0	231.7	227.3	222.9	218.6	214.2	209.8
316	245.7	241.3	237.0	232.6	228.2	223.8	219.4	215.0	210.6
317	246.6	242.2	237.8	233.4	229.0	224.6	220.2	215.8	211.4
318	247.6	243.2	238.7	234.3	229.9	225.5	221.1	216.6	212.2
319	248.5	244.1	239.6	235.2	230.8	226.3	221.9	217.4	213.0
320	249.4	245.0	240.5	236.1	231.6	227.2	222.7	218.3	213.8
321	250.4	245.9	241.4	237.0	232.5	228.0	223.6	219.1	214.6
322	251.3	246.8	242.3	237.8	233.4	228.9	224.4	219.9	215.4
323	252.2	247.7	243.2	238.7	234.2	229.7	225.2	220.7	216.2
324	253.2	248.7	244.1	239.6	235.1	230.6	226.0	221.5	217.0
325	254.1	249.6	245.0	240.5	236.0	231.4	226.9	222.4	217.8
326		250.5	245.9	241.4	236.8	232.3	227.7	223.2	218.6
327		251.4	246.8	242.3	237.7	233.1	228.6	224.0	219.4
328		252.3	247.7	243.2	238.6	234.0	229.4	224.8	220.2
329		253.3	248.7	244.0	239.4	234.8	230.2	225.6	221.0
330		254.2	249.6	244.9	240.3	235.7	231.1	226.5	221.8
331		255.1	250.5	245.8	241.2	236.6	231.9	227.3	222.6
332		256.0	251.4	246.7	242.1	237.4	232.8	228.1	223.5
333		257.0	252.3	247.6	243.0	238.3	233.6	228.9	224.3
334		257.9	253.2	248.5	243.8	239.2	234.5	229.8	225.1
335		258.8	254.1	249.4	244.7	240.0	235.3	230.6	225.9
336		259.8	255.0	250.3	245.6	240.9	236.2	231.4	226.7
337		260.7	256.0	251.2	246.5	241.7	237.0	232.3	227.5
338		261.6	256.9	252.1	247.4	242.6	237.8	233.1	228.3
339		262.6	257.8	253.0	248.2	243.5	238.7	233.9	229.2
340		263.5	258.7	253.9	249.1	244.4	239.6	234.8	230.0
341		264.4	259.6	254.8	250.0	245.2	240.4	235.6	230.8
342			260.6	255.7	250.9	246.1	241.3	236.4	231.6
343			261.5	256.6	251.8	247.0	242.1	237.3	232.4
344			262.4	257.5	252.7	247.8	243.0	238.1	233.2
345			263.3	258.4	253.6	248.7	243.8	238.9	234.1

SMITHSONIAN TABLES.

TABLE 20.
DETERMINATION OF HEIGHTS BY THE BAROMETER.
ENGLISH MEASURES.

Values of $60368 [1 + 0.0010195 \times 36] \log \frac{29.90}{B}$.

Barometric Pressure. B.	.00	.01	.02	.03	.04	.05	.06	.07	.08	.09
Inches.	Feet.	Feet.	Feet.	Feet.	Feet.	Feet.	Feet.	Feet.	Feet.	Feet.
12.00	24814	24791	24769	24746	24723	24701	24678	24656	24633	24611
12.10	24588	24566	24543	24521	24499	24476	24454	24431	24409	24387
12.20	24365	24342	24320	24298	24276	24253	24231	24209	24187	24165
12.30	24143	24121	24098	24076	24054	24032	24010	23988	23966	23944
12.40	23923	23901	23879	23857	23835	23813	23791	23770	23748	23726
12.50	23704	23682	23661	23639	23617	23596	23574	23552	23531	23509
12.60	23488	23466	23445	23423	23402	23380	23359	23337	23316	23294
12.70	23273	23251	23230	23209	23187	23166	23145	23123	23102	23081
12.80	23060	23038	23017	22996	22975	22954	22933	22911	22890	22869
12.90	22848	22827	22806	22785	22764	22743	22722	22701	22680	22659
13.00	22638	22617	22596	22576	22555	22534	22513	22492	22471	22451
13.10	22430	22409	22388	22368	22347	22326	22306	22285	22264	22244
13.20	22223	22203	22182	22162	22141	22121	22100	22080	22059	22039
13.30	22018	21998	21977	21957	21937	21916	21896	21876	21855	21835
13.40	21815	21794	21774	21754	21734	21713	21693	21673	21653	21633
13.50	21612	21592	21572	21552	21532	21512	21492	21472	21452	21432
13.60	21412	21392	21372	21352	21332	21312	21292	21272	21252	21233
13.70	21213	21193	21173	21153	21134	21114	21094	21074	21054	21035
13.80	21015	20995	20976	20956	20936	20917	20897	20878	20858	20838
13.90	20819	20799	20780	20760	20741	20721	20702	20682	20663	20643
14.00	20624	20605	20585	20566	20546	20527	20508	20488	20469	20450
14.10	20431	20411	20392	20373	20354	20334	20315	20296	20277	20258
14.20	20238	20219	20200	20181	20162	20143	20124	20105	20086	20067
14.30	20048	20029	20010	19991	19972	19953	19934	19915	19896	19877
14.40	19858	19839	19821	19802	19783	19764	19745	19727	19708	19689
14.50	19670	19651	19633	19614	19595	19577	19558	19539	19521	19502
14.60	19483	19465	19446	19428	19409	19390	19372	19353	19335	19316
14.70	19298	19279	19261	19242	19224	19206	19187	19169	19150	19132
14.80	19114	19095	19077	19059	19040	19022	19004	18985	18967	18949
14.90	18931	18912	18894	18876	18858	18840	18821	18803	18785	18767
15.00	18749	18731	18713	18694	18676	18658	18640	18622	18604	18586
15.10	18568	18550	18532	18514	18496	18478	18460	18442	18425	18407
15.20	18389	18371	18353	18335	18317	18300	18282	18264	18246	18228
15.30	18211	18193	18175	18157	18140	18122	18104	18086	18069	18051
15.40	18033	18016	17998	17981	17963	17945	17928	17910	17893	17875
15.50	17858	17840	17823	17805	17788	17770	17753	17735	17718	17700
15.60	17683	17665	17648	17631	17613	17596	17578	17561	17544	17526
15.70	17509	17492	17474	17457	17440	17423	17405	17388	17371	17354
15.80	17337	17319	17302	17285	17268	17251	17234	17216	17199	17182
15.90	17165	17148	17131	17114	17097	17080	17063	17046	17029	17012
16.00	16995	16978	16961	16944	16927	16910	16893	16876	16859	16842
16.10	16825	16808	16792	16775	16758	16741	16724	16707	16691	16674
16.20	16657	16640	16623	16607	16590	16573	16557	16540	16523	16506
16.30	16490	16473	16456	16440	16423	16406	16390	16373	16357	16340
16.40	16324	16307	16290	16274	16257	16241	16224	16208	16191	16175
16.50	16158	16142	16125	16109	16092	16076	16060	16043	16027	16010
16.60	15994	15978	15961	15945	15929	15912	15896	15880	15863	15847
16.70	15831	15815	15798	15782	15766	15750	15733	15717	15701	15685
16.80	15669	15652	15636	15620	15604	15588	15572	15556	15539	15523
16.90	15507	15491	15475	15459	15443	15427	15411	15395	15379	15363
17.00	15347	15331	15315	15299	15283	15267	15251	15235	15219	15203

TABLE 20.

DETERMINATION OF HEIGHTS BY THE BAROMETER.
ENGLISH MEASURES.

Values of $60368 [1 + 0.0010195 \times 36] \log \frac{29.90}{B}$.

Barometric Pressure. B.	.00	.01	.02	.03	.04	.05	.06	.07	.08	.09
Inches.	Feet.	Feet.	Feet.	Feet.	Feet.	Feet.	Feet.	Feet.	Feet.	Feet.
17.00	15347	15331	15315	15299	15283	15267	15251	15235	15219	15203
17.10	15187	15172	15156	15140	15124	15108	15092	15076	15061	15045
17.20	15029	15013	14997	14982	14966	14950	14934	14919	14903	14887
17.30	14871	14856	14840	14824	14809	14793	14777	14762	14746	14730
17.40	14715	14699	14684	14668	14652	14637	14621	14606	14590	14575
17.50	14559	14544	14528	14512	14497	14481	14466	14451	14435	14420
17.60	14404	14389	14373	14358	14342	14327	14312	14296	14281	14266
17.70	14250	14235	14219	14204	14189	14173	14158	14143	14128	14112
17.80	14097	14082	14067	14051	14036	14021	14006	13990	13975	13960
17.90	13945	13930	13914	13899	13884	13869	13854	13839	13824	13808
18.00	13793	13778	13763	13748	13733	13718	13703	13688	13673	13658
18.10	13643	13628	13613	13598	13583	13568	13553	13538	13523	13508
18.20	13493	13478	13463	13448	13433	13418	13404	13389	13374	13359
18.30	13344	13329	13314	13300	13285	13270	13255	13240	13226	13211
18.40	13196	13181	13166	13152	13137	13122	13107	13093	13078	13063
18.50	13049	13034	13019	13005	12990	12975	12961	12946	12931	12917
18.60	12902	12888	12873	12858	12844	12829	12815	12800	12785	12771
18.70	12756	12742	12727	12713	12698	12684	12669	12655	12640	12626
18.80	12611	12597	12583	12568	12554	12539	12525	12510	12496	12482
18.90	12467	12453	12438	12424	12410	12395	12381	12367	12352	12338
19.00	12324	12310	12295	12281	12267	12252	12238	12224	12210	12195
19.10	12181	12167	12153	12138	12124	12110	12096	12082	12068	12053
19.20	12039	12025	12011	11997	11983	11969	11954	11940	11926	11912
19.30	11898	11884	11870	11856	11842	11828	11814	11800	11786	11772
19.40	11758	11744	11730	11716	11702	11688	11674	11660	11646	11632
19.50	11618	11604	11590	11576	11562	11548	11534	11520	11507	11493
19.60	11479	11465	11451	11437	11423	11410	11396	11382	11368	11354
19.70	11340	11327	11313	11299	11285	11272	11258	11244	11230	11217
19.80	11203	11189	11175	11162	11148	11134	11121	11107	11093	11080
19.90	11066	11052	11039	11025	11011	10998	10984	10970	10957	10943
20.00	10930	10916	10903	10889	10875	10862	10848	10835	10821	10808
20.10	10794	10781	10767	10754	10740	10727	10713	10700	10686	10673
20.20	10659	10646	10632	10619	10605	10592	10579	10565	10552	10538
20.30	10525	10512	10498	10485	10472	10458	10445	10431	10418	10405
20.40	10391	10378	10365	10352	10338	10325	10312	10298	10285	10272
20.50	10259	10245	10232	10219	10206	10192	10179	10166	10153	10139
20.60	10126	10113	10100	10087	10074	10060	10047	10034	10021	10008
20.70	9995	9982	9968	9955	9942	9929	9916	9903	9890	9877
20.80	9864	9851	9838	9825	9812	9799	9786	9772	9759	9746
20.90	9733	9720	9707	9694	9681	9668	9655	9642	9629	9617
21.00	9604	9591	9578	9565	9552	9539	9526	9513	9500	9487
21.10	9474	9462	9449	9436	9423	9410	9397	9384	9372	9359
21.20	9346	9333	9320	9307	9295	9282	9269	9256	9244	9231
21.30	9218	9205	9193	9180	9167	9154	9142	9129	9116	9103
21.40	9091	9078	9065	9053	9040	9027	9015	9002	8989	8977
21.50	8964	8951	8939	8926	8913	8901	8888	8876	8863	8850
21.60	8838	8825	8813	8800	8788	8775	8762	8750	8737	8725
21.70	8712	8700	8687	8675	8662	8650	8637	8625	8612	8600
21.80	8587	8575	8562	8550	8538	8525	8513	8500	8488	8475
21.90	8463	8451	8438	8426	8413	8401	8389	8376	8364	8352
22.00	8339	8327	8314	8302	8290	8277	8265	8253	8240	8228

SMITHSONIAN TABLES.

TABLE 20.
DETERMINATION OF HEIGHTS BY THE BAROMETER.
ENGLISH MEASURES.

Values of $60368 [1 + 0.0010195 \times 36] \log \frac{29.90}{B}$.

Barometric Pressure. B.	.00	.01	.02	.03	.04	.05	.06	.07	.08	.09
Inches.	Feet.	Feet.	Feet.	Feet.	Feet.	Feet.	Feet.	Feet.	Feet.	Feet.
22.00	8339	8327	8314	8302	8290	8277	8265	8253	8240	8228
22.10	8216	8204	8191	8179	8167	8154	8142	8130	8118	8105
22.20	8093	8081	8069	8056	8044	8032	8020	8008	7995	7983
22.30	7971	7959	7947	7935	7922	7910	7898	7886	7874	7862
22.40	7849	7837	7825	7813	7801	7789	7777	7765	7753	7740
22.50	7728	7716	7704	7692	7680	7668	7656	7644	7632	7620
22.60	7608	7596	7584	7572	7560	7548	7536	7524	7512	7500
22.70	7488	7476	7464	7452	7440	7428	7416	7404	7392	7380
22.80	7368	7356	7345	7333	7321	7309	7297	7285	7273	7261
22.90	7249	7238	7226	7214	7202	7190	7178	7166	7155	7143
23.00	7131	7119	7107	7096	7084	7072	7060	7048	7037	7025
23.10	7013	7001	6990	6978	6966	6954	6943	6931	6919	6907
23.20	6896	6884	6872	6861	6849	6837	6825	6814	6802	6790
23.30	6779	6767	6755	6744	6732	6721	6709	6697	6686	6674
23.40	6662	6651	6639	6628	6616	6604	6593	6581	6570	6558
23.50	6546	6535	6523	6512	6500	6489	6477	6466	6454	6443
23.60	6431	6420	6408	6397	6385	6374	6362	6351	6339	6328
23.70	6316	6305	6293	6282	6270	6259	6247	6236	6225	6213
23.80	6202	6190	6179	6167	6156	6145	6133	6122	6110	6099
23.90	6088	6076	6065	6054	6042	6031	6020	6008	5997	5986
24.00	5974	5963	5952	5940	5929	5918	5906	5895	5884	5872
24.10	5861	5850	5839	5827	5816	5805	5794	5782	5771	5760
24.20	5749	5737	5726	5715	5704	5693	5681	5670	5659	5648
24.30	5637	5625	5614	5603	5592	5581	5570	5558	5547	5536
24.40	5525	5514	5503	5492	5480	5469	5458	5447	5436	5425
24.50	5414	5403	5392	5381	5369	5358	5347	5336	5325	5314
24.60	5303	5292	5281	5270	5259	5248	5237	5226	5215	5204
24.70	5193	5182	5171	5160	5149	5138	5127	5116	5105	5094
24.80	5083	5072	5061	5050	5039	5028	5017	5006	4995	4985
24.90	4974	4963	4952	4941	4930	4919	4908	4897	4886	4876
25.00	4865	4854	4843	4832	4821	4810	4800	4789	4778	4767
25.10	4756	4745	4735	4724	4713	4702	4691	4681	4670	4659
25.20	4648	4637	4627	4616	4605	4594	4584	4573	4562	4551
25.30	4540	4530	4519	4508	4498	4487	4476	4465	4455	4444
25.40	4433	4423	4412	4401	4391	4380	4369	4358	4348	4337
25.50	4326	4316	4305	4295	4284	4273	4263	4252	4241	4231
25.60	4220	4209	4199	4188	4178	4167	4156	4146	4135	4125
25.70	4114	4104	4093	4082	4072	4061	4051	4040	4030	4019
25.80	4009	3998	3988	3977	3966	3956	3945	3935	3924	3914
25.90	3903	3893	3882	3872	3861	3851	3841	3830	3820	3809
26.00	3799	3788	3778	3767	3757	3746	3736	3726	3715	3705
26.10	3694	3684	3674	3663	3653	3642	3632	3622	3611	3601
26.20	3590	3580	3570	3559	3549	3539	3528	3518	3508	3497
26.30	3487	3477	3466	3456	3446	3435	3425	3415	3404	3394
26.40	3384	3373	3363	3353	3343	3332	3322	3312	3301	3291
26.50	3281	3270	3260	3250	3240	3230	3219	3209	3199	3189

TABLE 20.

DETERMINATION OF HEIGHTS BY THE BAROMETER.
ENGLISH MEASURES.

Values of $60368 [1 + 0.0010195 \times 36] \log \frac{29.90}{B}$.

Barometric Pressure. B.	.00	.01	.02	.03	.04	.05	.06	.07	.08	.09
Inches.	Feet.	Feet.	Feet.	Feet.	Feet.	Feet.	Feet.	Feet.	Feet.	Feet.
26.50	3281	3270	3260	3250	3240	3230	3219	3209	3199	3189
26.60	3179	3168	3158	3148	3138	3128	3117	3107	3097	3087
26.70	3077	3066	3056	3046	3036	3026	3016	3005	2995	2985
26.80	2975	2965	2955	2945	2934	2924	2914	2904	2894	2884
26.90	2874	2864	2854	2843	2833	2823	2813	2803	2793	2783
27.00	2773	2763	2753	2743	2733	2723	2713	2703	2692	2682
27.10	2672	2662	2652	2642	2632	2622	2612	2602	2592	2582
27.20	2572	2562	2552	2542	2532	2522	2512	2502	2493	2483
27.30	2473	2463	2453	2443	2433	2423	2413	2403	2393	2383
27.40	2373	2363	2353	2343	2334	2324	2314	2304	2294	2284
27.50	2274	2264	2254	2245	2235	2225	2215	2205	2195	2185
27.60	2176	2166	2156	2146	2136	2126	2116	2107	2097	2087
27.70	2077	2067	2058	2048	2038	2028	2018	2009	1999	1989
27.80	1979	1970	1960	1950	1940	1930	1921	1911	1901	1891
27.90	1882	1872	1862	1852	1843	1833	1823	1814	1804	1794
28.00	1784	1775	1765	1755	1746	1736	1726	1717	1707	1697
28.10	1688	1678	1668	1659	1649	1639	1630	1620	1610	1601
28.20	1591	1581	1572	1562	1552	1543	1533	1524	1514	1504
28.30	1495	1485	1476	1466	1456	1447	1437	1428	1418	1408
28.40	1399	1389	1380	1370	1361	1351	1342	1332	1322	1313
28.50	1303	1294	1284	1275	1265	1256	1246	1237	1227	1218
28.60	1208	1199	1189	1180	1170	1161	1151	1142	1132	1123
28.70	1113	1104	1094	1085	1075	1066	1057	1047	1038	1028
28.80	1019	1009	1000	990	981	972	962	953	943	934
28.90	925	915	906	896	887	878	868	859	849	840
29.00	831	821	812	803	793	784	775	765	756	746
29.10	737	728	718	709	700	690	681	672	663	653
29.20	644	635	625	616	607	597	588	579	570	560
29.30	551	542	532	523	514	505	495	486	477	468
29.40	458	449	440	431	421	412	403	394	384	375
29.50	366	357	348	338	329	320	311	302	292	283
29.60	274	265	256	247	237	228	219	210	201	192
29.70	182	173	164	155	146	137	128	118	109	100
29.80	+ 91	+ 82	+ 73	+ 64	+ 55	+ 45	+ 36	+ 27	+ 18	+ 9
29.90	0	− 9	− 18	− 27	− 36	− 45	− 55	− 64	− 73	− 82
30.00	− 91	− 100	− 109	− 118	− 127	− 136	− 145	− 154	− 163	− 172
30.10	− 181	− 190	− 199	− 208	− 217	− 226	− 235	− 244	− 253	− 262
30.20	− 271	− 280	− 289	− 298	− 307	− 316	− 325	− 334	− 343	− 352
30.30	− 361	− 370	− 379	− 388	− 397	− 406	− 415	− 424	− 433	− 442
30.40	− 451	− 460	− 469	− 478	− 486	− 495	− 504	− 513	− 522	− 531
30.50	− 540	− 549	− 558	− 567	− 576	− 585	− 593	− 602	− 611	− 620
30.60	− 629	− 638	− 647	− 656	− 665	− 673	− 682	− 691	− 700	− 709
30.70	− 718	− 727	− 735	− 744	− 753	− 762	− 771	− 780	− 788	− 797
30.80	− 806	− 815	− 824	− 833	− 841	− 850	− 859	− 868	− 877	− 885

TABLE 21.
DETERMINATION OF HEIGHTS BY THE BAROMETER.
ENGLISH MEASURES.

Term for Temperature: $0.002039 \, (\theta - 50°) \, z$.

For temperatures { above 50° F. } the values are to be { added.
{ below 50° F. } { subtracted.

Mean Temperature. θ.		APPROXIMATE DIFFERENCE OF HEIGHT OBTAINED FROM TABLE 20.										
		20	40	60	80	100	200	300	400	500	600	700	800	900
F.	F.	Feet.	Feet.	Feet.	Feet.	Feet.	Feet.	Feet.	Feet.	Feet.	Feet.	Feet.	Feet.	Feet.
49°	51°	0	0	0	0	0	0	1	1	1	1	1	2	2
48	52	0	0	0	0	0	1	1	2	2	2	3	3	4
47	53	0	0	0	0	1	1	2	2	3	4	4	5	6
46	54	0	0	0	1	1	2	2	3	4	5	6	7	7
45	55	0	0	1	1	1	2	3	4	5	6	7	8	9
44	56	0	0	1	1	1	2	4	5	6	7	9	10	11
43	57	0	1	1	1	1	3	4	6	7	9	10	11	13
42	58	0	1	1	1	2	3	5	7	8	10	11	13	15
41	59	0	1	1	1	2	4	6	7	9	11	13	15	17
40	60	0	1	1	2	2	4	6	8	10	12	14	16	18
39	61	0	1	1	2	2	4	7	9	11	13	16	18	20
38	62	0	1	1	2	2	5	7	10	12	15	17	20	22
37	63	1	1	2	2	3	5	8	11	13	16	19	21	24
36	64	1	1	2	2	3	6	9	11	14	17	20	23	26
35	65	1	1	2	2	3	6	9	12	15	18	21	24	28
34	66	1	1	2	3	3	7	10	13	16	20	23	26	29
33	67	1	1	2	3	3	7	10	14	17	21	24	28	31
32	68	1	1	2	3	4	7	11	15	18	22	26	29	33
31	69	1	2	2	3	4	8	12	15	19	23	27	31	35
30	70	1	2	2	3	4	8	12	16	20	24	29	33	37
29	71	1	2	3	3	4	9	13	17	21	26	30	34	39
28	72	1	2	3	4	4	9	13	18	22	27	31	36	40
27	73	1	2	3	4	5	9	14	19	23	28	33	38	42
26	74	1	2	3	4	5	10	15	20	24	29	34	39	44
25	75	1	2	3	4	5	10	15	20	25	31	36	41	46
24	76	1	2	3	4	5	11	16	21	27	32	37	42	48
23	77	1	2	3	4	6	11	17	22	28	33	39	44	50
22	78	1	2	3	5	6	11	17	23	29	34	40	46	51
21	79	1	2	4	5	6	12	18	24	30	35	41	47	53
20	80	1	2	4	5	6	12	18	24	31	37	43	49	55
19	81	1	3	4	5	6	13	19	25	32	38	44	51	57
18	82	1	3	4	5	7	13	20	26	33	39	46	52	59
17	83	1	3	4	5	7	13	20	27	34	40	47	54	61
16	84	1	3	4	6	7	14	21	28	35	42	49	55	62
15	85	1	3	4	6	7	14	21	29	36	43	50	57	64
14	86	1	3	4	6	7	15	22	29	37	44	51	59	66
13	87	2	3	5	6	8	15	23	30	38	45	53	60	68
12	88	2	3	5	6	8	15	23	31	39	46	54	62	70
11	89	2	3	5	6	8	16	24	32	40	48	56	64	72
10	90	2	3	5	7	8	16	24	33	41	49	57	65	73
9	91	2	3	5	7	8	17	25	33	42	50	59	67	75
8	92	2	3	5	7	9	17	26	34	43	51	60	69	77
7	93	2	4	5	7	9	18	26	35	44	53	61	70	79
6	94	2	4	5	7	9	18	27	36	45	54	63	72	81
5	95	2	4	6	7	9	18	28	37	46	55	64	73	83
4	96	2	4	6	8	9	19	28	38	47	56	66	75	84
3	97	2	4	6	8	10	19	29	38	48	57	67	77	86
2	98	2	4	6	8	10	20	29	39	49	59	69	78	88
1	99	2	4	6	8	10	20	30	40	50	60	70	80	90
0	100	2	4	6	8	10	20	31	41	51	61	71	82	92

SMITHSONIAN TABLES.

TABLE 21.
DETERMINATION OF HEIGHTS BY THE BAROMETER.
ENGLISH MEASURES.

Term for Temperature: $0.002039 \, (\theta - 50°) \, z$.

For temperatures {above 50° F.} {below 50° F.} the values are to be {added.} {subtracted.}

Mean Temperature. θ.		APPROXIMATE DIFFERENCE OF HEIGHT OBTAINED FROM TABLE 20.									
		1000	2000	3000	4000	5000	6000	7000	8000	9000	10000	20000
F.	F.	Feet.	Feet.	Feet.	Feet.	Feet.	Feet.	Feet.	Feet.	Feet.	Feet.	Feet.
49°	51°	2	4	6	8	10	12	14	16	18	20	41
48	52	4	8	12	16	20	24	29	33	37	41	82
47	53	6	12	18	24	31	37	43	49	55	61	122
46	54	8	16	24	33	41	49	57	65	73	82	163
45	55	10	20	31	41	51	61	71	82	92	102	204
44	56	12	24	37	49	61	73	86	98	110	122	245
43	57	14	29	43	57	71	86	100	114	128	143	285
42	58	16	33	49	65	82	98	114	130	147	163	326
41	59	18	37	55	73	92	110	128	147	165	184	367
40	60	20	41	61	82	102	122	143	163	184	204	408
39	61	22	45	67	90	112	135	157	179	202	224	449
38	62	24	49	73	98	122	147	171	196	220	245	489
37	63	27	53	80	106	133	159	186	212	239	265	530
36	64	29	57	86	114	143	171	200	228	257	285	571
35	65	31	61	92	122	153	184	214	245	275	306	612
34	66	33	65	98	130	163	196	228	261	294	326	652
33	67	35	69	104	139	173	208	243	277	312	347	693
32	68	37	73	110	147	184	220	257	294	330	367	734
31	69	39	77	116	155	194	232	271	310	349	387	775
30	70	41	82	122	163	204	245	285	326	367	408	816
29	71	43	86	128	171	214	257	300	343	385	428	856
28	72	45	90	135	179	224	269	314	359	404	449	897
27	73	47	94	141	188	234	281	328	375	422	469	938
26	74	49	98	147	196	245	294	343	391	440	489	979
25	75	51	102	153	204	255	306	357	408	459	510	1020
24	76	53	106	159	212	265	318	371	424	477	530	1060
23	77	55	110	165	220	275	330	385	440	495	551	1101
22	78	57	114	171	228	285	343	400	457	514	571	1142
21	79	59	118	177	236	296	355	414	473	532	591	1183
20	80	61	122	184	245	306	367	428	489	551	612	1223
19	81	63	126	190	253	316	379	442	506	569	632	1264
18	82	65	130	196	261	326	391	457	522	587	652	1305
17	83	67	135	202	269	336	404	471	538	606	673	1346
16	84	69	139	208	277	347	416	485	555	624	693	1387
15	85	71	143	214	285	357	428	500	571	642	714	1427
14	86	73	147	220	294	367	440	514	587	661	734	1468
13	87	75	151	226	302	377	453	528	604	679	754	1509
12	88	77	155	232	310	387	465	542	620	697	775	1550
11	89	80	159	239	318	398	477	557	636	716	795	1590
10	90	82	163	245	326	408	489	571	652	734	816	1631
9	91	84	167	251	334	418	502	585	669	752	836	1672
8	92	86	171	257	343	428	514	599	685	771	856	1713
7	93	88	175	263	351	438	526	614	701	789	877	1754
6	94	90	179	269	359	449	538	628	718	807	897	1794
5	95	92	184	275	367	459	551	642	734	826	918	1835
4	96	94	188	281	375	469	563	657	750	844	938	1876
3	97	96	192	287	383	479	575	671	767	862	958	1917
2	98	98	196	294	391	489	587	685	783	881	979	1957
1	99	100	200	300	400	500	599	699	799	899	999	1998
0	100	102	204	306	408	510	612	714	816	918	1020	2039

SMITHSONIAN TABLES.

TABLE 22.
DETERMINATION OF HEIGHTS BY THE BAROMETER.
ENGLISH MEASURES.

Correction for Latitude and Weight of Mercury: $z(0.002662 \cos 2\phi + 0.00239)$.

Latitude. ϕ.	APPROXIMATE DIFFERENCE OF HEIGHT OBTAINED FROM TABLES 20-21.										
	500	1000	1500	2000	2500	3000	3500	4000	4500	5000	5500
	Feet.	Feet.	Feet.	Feet.	Feet.	Feet.	Feet.	Feet.	Feet.	Feet.	Feet.
0°	+3	+5	+8	+10	+13	+15	+18	+20	+23	+25	+28
2	3	5	8	10	13	15	18	20	23	25	28
4	3	5	8	10	13	15	18	20	23	25	28
6	2	5	7	10	12	15	17	20	22	25	27
8	2	5	7	10	12	15	17	20	22	25	27
10	+2	+5	+7	+10	+12	+15	+17	+20	+22	+24	+27
12	2	5	7	10	12	14	17	19	22	24	27
14	2	5	7	9	12	14	17	19	21	24	26
16	2	5	7	9	12	14	16	19	21	23	26
18	2	5	7	9	11	14	16	18	20	23	25
20	+2	+4	+7	+9	+11	+13	+16	+17	+20	+22	+24
22	2	4	6	9	11	13	15	17	19	22	24
24	2	4	6	8	10	13	15	17	19	21	23
26	2	4	6	8	10	12	14	16	18	20	22
28	2	4	6	8	10	12	14	16	17	19	21
30	+2	+4	+6	+7	+9	+11	+13	+15	+17	+19	+20
32	2	4	5	7	9	11	12	14	16	18	20
34	2	3	5	7	8	10	12	14	15	17	19
36	2	3	5	6	8	10	11	13	14	16	18
38	2	3	5	6	8	9	11	12	14	15	17
40	+1	+3	+4	+6	+7	+9	+10	+11	+13	+14	+16
42	1	3	4	5	7	8	9	11	12	13	15
44	1	2	4	5	6	7	9	10	11	12	14
45	+1	+2	+4	+5	+6	+7	+8	+10	+11	+12	+13
46	+1	+2	+3	+5	+6	+7	+8	+9	+10	+11	+13
48	1	2	3	4	5	6	7	8	10	11	12
50	1	2	3	4	5	6	7	8	9	10	11
52	+1	+2	+3	+3	+4	+5	+6	+7	+8	+9	+10
54	1	2	2	3	4	5	5	6	7	8	9
56	1	1	2	3	3	4	5	6	7	7	8
58	1	1	2	2	3	4	4	5	6	6	7
60	1	1	2	2	3	3	4	4	5	5	6
62	0	+1	+1	+2	+2	+3	+3	+4	+4	+5	+5
64	0	1	1	2	2	2	3	3	3	4	4
66	0	1	1	1	2	2	2	2	3	3	3
68	0	0	1	1	1	1	2	2	2	2	3
70	0	0	1	1	1	1	1	1	2	2	2
72	0	0	0	0	+1	+1	+1	+1	+1	+1	+1
74	0	0	0	0	0	0	1	1	1	+1	+1
76	0	0	0	0	0	0	0	0	0	0	0
78	0	0	0	0	0	0	0	0	0	0	0
80	0	0	0	0	0	0	0	0	0	−1	−1

SMITHSONIAN TABLES.

TABLE 22.
DETERMINATION OF HEIGHTS BY THE BAROMETER.
ENGLISH MEASURES.

Correction for Latitude and Weight of Mercury: $z(0.002662 \cos 2\phi + 0.00239)$.

Latitude. ϕ.	APPROXIMATE DIFFERENCE OF HEIGHT OBTAINED FROM TABLES 20-21.									
	6000	7000	8000	9000	10000	11000	12000	13000	14000	15000	20000
	Feet.	Feet.	Feet.	Feet.	Feet.	Feet.	Feet.	Feet.	Feet.	Feet.	Feet.
0°	+30	+35	+40	+45	+51	+56	+61	+66	+71	+76	+101
2	30	35	40	45	50	56	61	66	71	76	101
4	30	35	40	45	50	56	60	65	70	75	101
6	30	35	40	45	50	55	60	65	70	75	100
8	30	35	40	45	49	54	59	64	69	74	99
10	+29	+34	+39	+44	+49	+54	+59	+64	+68	+73	+98
12	29	34	39	43	48	53	58	63	68	72	96
14	28	33	38	43	47	52	57	62	66	71	95
16	28	33	37	42	46	51	56	60	65	70	93
18	27	32	36	41	45	50	55	59	64	68	91
20	+27	+31	+35	+40	+44	+49	+53	+58	+62	+66	+89
22	26	30	34	39	43	47	52	56	60	65	86
24	25	29	33	38	42	46	50	54	58	63	83
26	24	28	32	36	40	44	48	52	56	60	81
28	23	27	31	35	39	43	47	50	54	58	78
30	+22	+26	+30	+33	+37	+41	+45	+48	+52	+56	+74
32	21	25	28	32	36	39	43	46	50	53	71
34	20	24	27	30	34	37	41	44	47	51	68
36	19	22	26	29	32	35	39	42	45	48	64
38	18	21	24	27	30	33	36	39	42	46	61
40	+17	+20	+23	+26	+29	+31	+34	+37	+40	+43	+57
42	16	19	21	24	27	29	32	35	37	40	53
44	15	17	20	22	25	27	30	32	35	37	50
45	+14	+17	+19	+22	+24	+26	+29	+31	+33	+36	+48
46	+14	+16	+18	+21	+23	+25	+28	+30	+32	+35	+46
48	13	15	17	19	21	23	25	27	30	32	42
50	12	13	15	17	19	21	23	25	27	29	39
52	+10	+12	+14	+16	+17	+19	+21	+23	+24	+26	+35
54	9	11	13	14	16	17	19	20	22	24	31
56	8	10	11	13	14	15	17	18	20	21	28
58	7	9	10	11	12	13	15	16	17	18	24
60	6	7	8	10	11	12	13	14	15	16	21
62	+5	+6	+7	+8	+9	+10	+11	+12	+13	+14	+18
64	5	5	6	7	8	8	9	10	11	11	15
66	4	4	5	5	6	7	7	8	9	9	12
68	3	3	4	4	5	5	6	6	7	7	10
70	2	2	3	3	4	4	4	5	5	5	7
72	+1	+2	+2	+2	+2						
74	+1	+1	+1	+1	+1						
76	0	0	0	0	0						
78	0	0	0	0	0						
80	−1	−1	−1	−1	−1						

SMITHSONIAN TABLES.

TABLE 23.
DETERMINATION OF HEIGHTS BY THE BAROMETER.
ENGLISH MEASURES.
Correction for an Average Degree of Humidity.

Mean Temperature.	APPROXIMATE DIFFERENCE OF HEIGHT OBTAINED FROM TABLES 20-21.											
	500	1000	2000	3000	4000	5000	6000	7000	8000	9000	10000	20000
F.	Feet.	Feet.	Feet.	Feet.	Feet.	Feet.	Feet.	Feet.	Feet.	Feet.	Feet.	Feet.
−20°	0	0	0	0	0	0	0	+1	+1	+1	+1	+2
−16	0	0	0	+1	+1	+1	+1	1	2	2	2	4
−12	0	0	+1	1	1	2	2	2	3	3	3	6
−8	0	0	1	1	2	2	3	3	4	4	4	9
−6	0	0	1	1	2	2	3	3	4	4	5	10
−4	0	+1	1	2	2	3	3	4	4	5	6	11
−2	0	1	1	2	2	3	4	4	5	6	6	12
0	0	1	1	2	3	3	4	5	5	6	7	14
+2	0	1	1	2	3	4	4	5	6	7	7	15
4	0	1	2	2	3	4	5	6	7	7	8	16
6	0	1	2	3	4	4	5	6	7	8	9	18
8	0	1	2	3	4	5	6	7	8	9	10	19
10	+1	1	2	3	4	5	6	7	8	9	10	21
12	1	1	2	3	4	6	7	8	9	10	11	22
14	1	1	2	4	5	6	7	8	9	11	12	24
16	1	1	3	4	5	6	8	9	10	11	13	25
18	1	1	3	4	5	7	8	9	11	12	13	27
20	1	1	3	4	6	7	9	10	11	13	14	29
22	1	2	3	5	6	8	9	11	12	14	15	31
24	1	2	3	5	7	8	10	11	13	15	16	33
26	1	2	3	5	7	9	10	12	14	16	17	35
28	1	2	4	6	7	9	11	13	15	17	19	37
30	1	2	4	6	8	10	12	14	16	18	20	41
32	1	2	4	7	9	11	13	16	18	20	22	44
34	1	2	5	7	10	12	15	17	19	22	24	49
36	1	3	5	8	11	13	16	19	21	24	27	53
38	1	3	6	9	12	15	18	21	23	26	29	59
40	2	3	6	10	13	16	19	23	26	29	32	64
42	2	4	7	11	14	18	21	25	28	32	35	71
44	2	4	8	12	15	19	23	27	31	35	39	77
46	2	4	8	13	17	21	25	29	34	38	42	84
48	2	5	9	14	18	23	27	32	37	41	46	92
50	2	5	10	15	20	25	30	35	40	45	50	99
52	3	5	11	16	21	27	32	37	43	48	53	107
54	3	6	11	17	23	29	34	40	46	51	57	114
56	3	6	12	18	24	30	37	43	49	55	61	122
58	3	6	13	19	26	32	39	45	52	58	65	130
60	3	7	14	21	27	34	41	48	55	62	69	137
62	4	7	14	22	29	36	43	51	58	65	72	145
64	4	8	15	23	30	38	46	53	61	69	76	152
66	4	8	16	24	32	40	48	56	64	72	80	160
68	4	8	17	25	34	42	50	59	67	76	84	168
70	4	9	18	26	35	44	53	61	70	79	88	175
72	5	9	18	27	37	46	55	64	73	82	91	183
76	5	10	20	30	40	49	59	69	79	89	99	198
80	5	11	21	32	43	53	64	75	85	96	106	213
84	6	11	23	34	46	57	68	80	91	103	114	228
88	6	12	24	37	49	61	73	85	97	110	122	243
92	6	13	26	39	52	65	78	91	103	116	129	259
96	7	14	27	41	55	68	82	96	110	123	137	274

SMITHSONIAN TABLES.

TABLE 24.

DETERMINATION OF HEIGHTS BY THE BAROMETER.
ENGLISH MEASURES.

Correction for the Variation of Gravity with Altitude: $\dfrac{z(z+2h_0)}{R}$.

Approximate difference of height. Z.	HEIGHT OF LOWER STATION IN FEET (h_0).										
	0	1000	2000	3000	4000	5000	6000	7000	8000	9000	10000	12000
Feet.	Feet.	Feet.	Feet.	Feet.	Feet.	Feet.	Feet.	Feet.	Feet.	Feet.	Feet.	Feet.
500	0	0	0	0	0	0	0	0	0	0	0	+1
1000	0	0	0	0	0	+1	+1	+1	+1	+1	+1	1
1500	0	0	0	+1	+1	1	1	1	1	1	2	2
2000	0	0	+1	1	1	1	1	2	2	2	2	2
2500	0	+1	1	1	1	1	2	2	2	2	3	3
3000	0	1	1	1	2	2	2	2	3	3	3	4
3500	+1	1	1	2	2	2	3	3	3	4	4	5
4000	1	1	2	2	2	3	3	3	4	4	5	5
4500	1	1	2	2	3	3	4	4	4	5	5	6
5000	1	2	2	3	3	4	4	5	5	6	6	7
5500	1	2	3	3	4	4	5	5	6	6	7	8
6000	2	2	3	3	4	5	5	6	6	7	7	9
6500	2	3	3	4	5	5	6	6	7	8	8	9
7000	2	3	4	4	5	6	6	7	8	8	9	10
7500	3	3	4	5	6	6	7	8	8	9	10	11
8000	3	4	5	5	6	7	8	8	9	10	11	12
8500	3	4	5	6	7	8	8	9	10	11	12	13
9000	4	5	6	6	7	8	9	10	11	12	12	14
9500	4	5	6	7	8	9	10	11	12	13	13	15
10000	5	6	7	8	9	10	11	12	13	14	16
11000	6	7	8	9	10	11	12	13	14	15	16	18
12000	7	8	9	10	11	13	14	15	16	17	18	21
13000	8	9	11	12	13	14	16	17	18	19	21	23
14000	9	11	12	13	15	16	17	19	20	21	23	25
15000	11	12	14	15	17	18	19	21	22	24	25	28
16000	12	14	15	17	18	20	21	23	25	26	28	31
17000	14	15	17	19	20	22	24	25	27	28	30	
18000	16	17	19	21	22	24	26	28	30	31		
19000	17	19	21	23	25	26	28	30	32			
20000	19	21	23	25	27	29	31					

TABLE 25.
DETERMINATION OF HEIGHTS BY THE BAROMETER.
METRIC MEASURES.
Values of $18400 \log \frac{760}{B}$.

Barometric Pressure.	0	1	2	3	4	5	6	7	8	9
mm.	m.	m.	m.	m.	m.	m.	m.	m.	m.	m.
300	7428	7401	7375	7348	7322	7296	7270	7244	7218	7192
310	7166	7140	7115	7089	7064	7038	7013	6987	6962	6937
320	6912	6887	6862	6838	6813	6789	6764	6740	6715	6691
330	6666	6642	6618	6594	6570	6546	6522	6498	6475	6451
340	6428	6405	6381	6358	6334	6311	6288	6265	6242	6219
350	6196	6173	6151	6128	6106	6083	6061	6038	6016	5993
360	5971	5949	5927	5905	5883	5861	5839	5817	5795	5773
370	5752	5730	5709	5687	5666	5644	5623	5602	5581	5560
380	5539	5518	5497	5476	5455	5434	5414	5393	5373	5352
390	5332	5311	5291	5270	5250	5229	5209	5189	5169	5149
400	5129	5109	5089	5069	5049	5029	5010	4990	4971	4951
410	4932	4912	4893	4873	4854	4834	4815	4796	4777	4758
420	4739	4720	4701	4682	4663	4644	4625	4606	4588	4569
430	4551	4532	4514	4495	4477	4458	4440	4422	4404	4386
440	4368	4350	4332	4314	4296	4278	4260	4242	4224	4206
450	4188	4170	4152	4134	4117	4099	4082	4064	4047	4029
460	4012	3994	3977	3959	3942	3925	3908	3791	3774	3757
470	3840	3823	3806	3789	3772	3755	3738	3721	3705	3688
480	3672	3655	3639	3622	3606	3589	3573	3556	3540	3523
490	3507	3490	3474	3458	3442	3426	3410	3394	3378	3362
500	3346	3330	3314	3298	3282	3266	3250	3235	3219	3203
510	3188	3172	3157	3141	3126	3110	3095	3079	3064	3048
520	3033	3017	3002	2986	2971	2955	2940	2925	2910	2895
530	2880	2865	2850	2835	2820	2805	2790	2775	2760	2745
540	2731	2716	2701	2687	2672	2657	2643	2628	2613	2599
550	2584	2570	2555	2541	2526	2512	2497	2483	2468	2454
560	2440	2426	2411	2397	2383	2369	2355	2341	2327	2313
570	2299	2285	2271	2257	2243	2229	2215	2201	2188	2174
580	2160	2146	2133	2119	2105	2092	2078	2064	2051	2037
590	2023	2010	1996	1983	1969	1956	1942	1929	1915	1902
600	1889	1875	1862	1848	1835	1822	1809	1796	1783	1770
610	1757	1744	1731	1718	1705	1692	1679	1666	1653	1640
620	1627	1614	1601	1588	1576	1563	1550	1537	1525	1512
630	1499	1486	1474	1461	1448	1436	1423	1411	1398	1386
640	1373	1361	1348	1336	1323	1311	1298	1286	1273	1261
650	1249	1236	1224	1212	1199	1187	1175	1163	1151	1139
660	1127	1115	1103	1091	1079	1067	1055	1043	1031	1019
670	1007	995	983	971	960	948	936	924	913	901
680	889	877	866	854	842	831	819	807	796	784
690	772	761	749	738	726	715	703	692	680	669
700	657	646	635	623	612	601	589	578	567	555
710	544	533	521	510	499	487	476	465	454	443
720	432	421	410	399	388	377	366	355	344	333
730	322	311	300	289	278	267	256	245	234	224
740	213	202	192	181	170	160	149	138	128	117
750	+ 106	+ 95	+ 85	+ 74	+ 64	+ 53	+ 43	+ 32	+ 22	+ 11
760	0	− 10	− 21	− 31	− 42	− 52	− 63	− 73	− 83	− 94
770	− 104	− 115	− 125	− 136	− 146	− 156	− 166	− 177	− 187	− 197

TABLE 26.

DETERMINATION OF HEIGHTS BY THE BAROMETER.
METRIC MEASURES.

Term for Temperature: $0.00367 \, \theta \times z$.

For temperatures { above 0° C. / below 0° C. } the values are to be { added. / subtracted. }

Approximate difference of height. z.	MEAN TEMPERATURE OF AIR COLUMN IN CENTIGRADE DEGREES (θ).											
	1°	2°	3°	4°	5°	6°	7°	8°	9°	10°	20°	30°	40°
m.	m.	m.	m.	m.	m.	m.	m.	m.	m.	m.	m.	m.	m.
100	0	1	1	1	2	2	3	3	3	4	7	11	15
200	1	1	2	3	4	4	5	6	7	7	15	22	29
300	1	2	3	4	6	7	8	9	10	11	22	33	44
400	1	3	4	6	7	9	10	12	13	15	29	44	59
500	2	4	6	7	9	11	13	15	17	18	37	55	73
600	2	4	7	9	11	13	15	18	20	22	44	66	88
700	3	5	8	10	13	15	18	21	23	26	51	77	103
800	3	6	9	12	15	18	21	23	26	29	59	88	117
900	3	7	10	13	17	20	23	26	30	33	66	99	132
1000	4	7	11	15	18	22	26	29	33	37	73	110	147
1100	4	8	12	16	20	24	28	32	36	40	81	121	161
1200	4	9	13	18	22	26	31	35	40	44	88	132	176
1300	5	10	14	19	24	29	33	38	43	48	95	143	191
1400	5	10	15	21	26	31	36	41	46	51	103	154	206
1500	6	11	17	22	28	33	39	44	50	55	110	165	220
1600	6	12	18	23	29	35	41	47	53	59	117	176	235
1700	6	12	19	25	31	37	44	50	56	62	125	187	250
1800	7	13	20	26	33	40	46	53	59	66	132	198	264
1900	7	14	21	28	35	42	49	56	63	70	139	209	279
2000	7	15	22	29	37	44	51	59	66	73	147	220	294
2100	8	15	23	31	39	46	54	62	69	77	154	231	308
2200	8	16	24	32	40	48	57	65	73	81	161	242	323
2300	8	17	25	34	42	51	59	68	76	84	169	253	338
2400	9	18	26	35	44	53	62	70	79	88	176	264	352
2500	9	18	28	37	46	55	64	73	83	92	184	275	367
2600	10	19	29	38	48	57	67	76	86	95	191	286	382
2700	10	20	30	40	50	59	69	79	89	99	198	297	396
2800	10	21	31	41	51	62	72	82	92	103	206	308	411
2900	11	21	32	43	53	64	75	85	96	106	213	319	426
3000	11	22	33	44	55	66	77	88	99	110	220	330	440
3100	11	23	34	46	57	68	80	91	102	114	228	341	455
3200	12	23	35	47	59	70	82	94	106	117	235	352	470
3300	12	24	36	48	61	73	85	97	109	121	242	363	484
3400	12	25	37	50	62	75	87	100	112	125	250	374	499
3500	13	26	39	51	64	77	90	103	116	128	257	385	514
3600	13	26	40	53	66	79	92	106	119	132	264	396	528
3700	14	27	41	54	68	81	95	109	122	136	272	407	543
3800	14	28	42	56	70	84	98	112	126	139	279	418	558
3900	14	29	43	57	72	86	100	115	129	143	286	429	573
4000	15	29	44	59	73	88	103	117	132	147	294	440	587
5000	18	37	55	73	92	110	128	147	165	183	367	551	734
6000	22	44	66	88	110	132	154	176	198	220	440	661	881
7000	26	51	77	103	128	154	180	206	231	257	514	771	1028

SMITHSONIAN TABLES.

III

TABLE 27.
DETERMINATION OF HEIGHTS BY THE BAROMETER.
METRIC MEASURES.
Correction for Humidity: Values of 10000 β.

$$\beta = 0.378 \frac{e}{b} = 0.378 \frac{f+f_o}{B+B_o}.$$

Mean Vapor Pressure. $e = \frac{f+f_o}{2}$	MEAN BAROMETRIC PRESSURE IN MILLIMETRES $\left(\frac{B+B_o}{2}\right)$.												
	500	520	540	560	580	600	620	640	660	680	700	720	740	760
mm.	mm.	mm.	mm.	mm.	mm.	mm.	mm.	mm.	mm.	mm.	mm.	mm.	mm.	mm.
1	8	7	7	7	7	6	6	6	6	6	5	5	5	5
2	15	15	14	14	13	13	12	12	11	11	11	11	10	10
3	23	22	21	20	20	19	18	18	17	17	16	16	15	15
4	30	29	28	27	26	25	24	24	23	22	22	21	20	20
5	38	36	35	34	33	31	30	30	29	28	27	26	26	25
6	45	44	42	41	39	38	37	35	34	33	32	32	31	30
7	53	51	49	47	46	44	43	41	40	39	38	37	36	35
8	60	58	56	54	52	50	49	47	46	44	43	42	41	40
9	68	65	63	61	59	57	55	53	52	50	49	47	46	45
10	76	73	70	68	65	63	61	59	57	56	54	53	51	50
11	83	80	77	74	72	69	67	65	63	61	59	58	56	55
12	91	87	84	81	78	76	73	71	69	67	65	63	61	60
13	98	95	91	88	85	82	79	77	74	72	70	68	66	65
14	106	102	98	95	91	88	85	83	80	78	76	74	72	70
15	113	109	105	101	98	95	91	89	86	83	81	79	77	75
16	121	116	112	108	104	101	98	94	92	89	86	84	82	80
17	129	124	119	115	111	107	104	100	97	94	92	89	87	85
18	136	131	126	122	117	113	110	106	103	100	97	95	92	90
19	144	138	133	128	124	120	116	112	109	106	103	100	97	95
20	151	145	140	135	130	126	122	118	115	111	108	105	102	99
21	159	153	147	142	137	132	128	124	120	117	113	110	107	104
22	166	160	154	149	143	139	134	130	126	122	119	116	112	109
23	174	167	161	155	150	145	140	136	132	128	124	121	117	114
24	181	174	168	162	156	151	146	142	137	133	130	126	123	119
25	189	182	175	169	163	157	152	148	143	139	135	131	128	124
26	197	189	182	175	169	164	159	154	149	145	140	137	133	129
27	204	196	189	182	176	170	165	159	155	150	146	142	138	134
28	212	204	196	189	182	176	171	165	160	156	151	147	143	139
29	219	211	203	196	189	183	177	171	166	161	157	152	148	144
30	227	218	210	203	196	189	183	177	172	167	162	158	153	149
31	234	225	217	209	202	195	189	183	178	172	167	163	158	154
32	242	233	224	216	209	202	195	189	183	178	173	168	163	159
33	249	240	231	223	215	208	201	195	189	183	178	173	169	164
34	257	247	238	230	222	214	207	201	195	189	184	179	174	169
35	265	254	245	236	228	220	213	207	200	195	189	184	179	174
36	272	262	252	243	235	227	219	213	206	200	194	189	184	179
37	280	269	259	250	241	233	226	219	212	206	200	194	189	184
38	287	276	266	257	248	239	232	224	218	211	205	200	194	189
39	295	283	273	263	254	246	238	230	223	217	211	205	199	194
40	302	291	280	270	261	252	244	236	229	222	216	210	204	199

SMITHSONIAN TABLES.

TABLE 27.
DETERMINATION OF HEIGHTS BY THE BAROMETER.
METRIC MEASURES.

Correction for Humidity: $10000 \, \beta \times z$.

Top argument: Values of $10000 \, \beta$ obtained from page 112.
Side argument: Approximate difference of height (z).

Approximate Difference of Height. z.	$10000 \, \beta$.									
	25	50	75	100	125	150	175	200	225	250	275	300
m.	m.	m.	m.	m.	m.	m.	m.	m.	m.	m.	m.	m.
100	0.3	0.5	0.8	1.0	1.3	1.5	1.8	2.0	2.3	2.5	2.8	3.0
200	0.5	1.0	1.5	2.0	2.5	3.0	3.5	4.0	4.5	5.0	5.5	6.0
300	0.8	1.5	2.3	3.0	3.8	4.5	5.3	6.0	6.8	7.5	8.3	9.0
400	1.0	2.0	3.0	4.0	5.0	6.0	7.0	8.0	9.0	10.0	11.0	12.0
500	1.3	2.5	3.8	5.0	6.3	7.5	8.8	10.0	11.3	12.5	13.8	15.0
600	1.5	3.0	4.5	6.0	7.5	9.0	10.5	12.0	13.5	15.0	16.5	18.0
700	1.8	3.5	5.3	7.0	8.8	10.5	12.3	14.0	15.8	17.5	19.3	21.0
800	2.0	4.0	6.0	8.0	10.0	12.0	14.0	16.0	18.0	20.0	22.0	24.0
900	2.3	4.5	6.8	9.0	11.3	13.5	15.8	18.0	20.3	22.5	24.8	27.0
1000	2.5	5.0	7.5	10.0	12.5	15.0	17.5	20.0	22.5	25.0	27.5	30.0
1100	2.8	5.5	8.3	11.0	13.8	16.5	19.3	22.0	24.8	27.5	30.3	33.0
1200	3.0	6.0	9.0	12.0	15.0	18.0	21.0	24.0	27.0	30.0	33.0	36.0
1300	3.3	6.5	9.8	13.0	16.3	19.5	22.8	26.0	29.3	32.5	35.8	39.0
1400	3.5	7.0	10.5	14.0	17.5	21.0	24.5	28.0	31.5	35.0	38.5	42.0
1500	3.8	7.5	11.3	15.0	18.8	22.5	26.3	30.0	33.8	37.5	41.3	45.0
1600	4.0	8.0	12.0	16.0	20.0	24.0	28.0	32.0	36.0	40.0	44.0	48.0
1700	4.3	8.5	12.8	17.0	21.3	25.5	29.8	34.0	38.3	42.5	46.8	51.0
1800	4.5	9.0	13.5	18.0	22.5	27.0	31.5	36.0	40.5	45.0	49.5	54.0
1900	4.8	9.5	14.3	19.0	23.8	28.5	33.3	38.0	42.8	47.5	52.3	57.0
2000	5.0	10.0	15.0	20.0	25.0	30.0	35.0	40.0	45.0	50.0	55.0	60.0
2100	5.3	10.5	15.8	21.0	26.3	31.5	36.8	42.0	47.3	52.5	57.8	63.0
2200	5.5	11.0	16.5	22.0	27.5	33.0	38.5	44.0	49.5	55.0	60.5	66.0
2300	5.8	11.5	17.3	23.0	28.8	34.5	40.3	46.0	51.8	57.5	63.3	69.0
2400	6.0	12.0	18.0	24.0	30.0	36.0	42.0	48.0	54.0	60.0	66.0	72.0
2500	6.3	12.5	18.8	25.0	31.3	37.5	43.8	50.0	56.3	62.5	68.8	75.0
2600	6.5	13.0	19.5	26.0	32.5	39.0	45.5	52.0	58.5	65.0	71.5	78.0
2700	6.8	13.5	20.3	27.0	33.8	40.5	47.3	54.0	60.8	67.5	74.3	81.0
2800	7.0	14.0	21.0	28.0	35.0	42.0	49.0	56.0	63.0	70.0	77.0	84.0
2900	7.3	14.5	21.8	29.0	36.3	43.5	50.8	58.0	65.3	72.5	79.8	87.0
3000	7.5	15.0	22.5	30.0	37.5	45.0	52.5	60.0	67.5	75.0	82.5	90.0
3100	7.8	15.5	23.3	31.0	38.8	46.5	54.3	62.0	69.8	77.5	85.3	93.0
3200	8.0	16.0	24.0	32.0	40.0	48.0	56.0	64.0	72.0	80.0	88.0	96.0
3300	8.3	16.5	24.8	33.0	41.3	49.5	57.8	66.0	74.3	82.5	90.8	99.0
3400	8.5	17.0	25.5	34.0	42.5	51.0	59.5	68.0	76.5	85.0	93.5	102.0
3500	8.8	17.5	26.3	35.0	43.8	52.5	61.3	70.0	78.8	87.5	96.3	105.0
3600	9.0	18.0	27.0	36.0	45.0	54.0	63.0	72.0	81.0	90.0	99.0	108.0
3700	9.3	18.5	27.8	37.0	46.3	55.5	64.8	74.0	83.3	92.5	101.8	111.0
3800	9.5	19.0	28.5	38.0	47.5	57.0	66.5	76.0	85.5	95.0	104.5	114.0
3900	9.8	19.5	29.3	39.0	48.8	58.5	68.3	78.0	87.8	97.5	107.3	117.0
4000	10.0	20.0	30.0	40.0	50.0	60.0	70.0	80.0	90.0	100.0	110.0	120.0
5000	12.5	25.0	37.5	50.0	62.5	75.0	87.5	100.0	112.5	125.0	137.5	150.0
6000	15.0	30.0	45.0	60.0	75.0	90.0	105.0	120.0	135.0	150.0	165.0	180.0
7000	17.5	35.0	52.5	70.0	87.5	105.0	122.5	140.0	157.5	175.0	192.5	210.0

SMITHSONIAN TABLES.

TABLE 28.
DETERMINATION OF HEIGHTS BY THE BAROMETER
METRIC MEASURES.

Correction for Latitude and Weight of Mercury: $z(0.002662 \cos 2\phi + 0.00239)$.

Approximate difference of Height. Z.	LATITUDE (ϕ).															
	0°	5°	10°	15°	20°	25°	30°	35°	40°	45°	50°	55°	60°	65°	70°	75°
metres.	m.	m.	m.	m.	m.	m.	m.	m.	m.	m.	m.	m.	m.	m.	m.	m.
100	1	1	0	0	0	0	0	0	0	0	0	0	0	0	0	0
200	1	1	1	1	1	1	1	1	1	0	0	0	0	0	0	0
300	2	2	1	1	1	1	1	1	1	1	1	0	0	0	0	0
400	2	2	2	2	2	2	1	1	1	1	1	1	0	0	0	0
500	3	3	2	2	2	2	2	2	1	1	1	1	1	0	0	0
600	3	3	3	3	3	2	2	2	2	1	1	1	1	0	0	0
700	4	4	3	3	3	3	3	2	2	2	1	1	1	0	0	0
800	4	4	4	4	4	3	3	3	2	2	2	1	1	1	0	0
900	5	5	4	4	4	4	3	3	3	2	2	2	1	1	1	0
1000	5	5	5	5	4	4	4	3	3	2	2	1	1	1	0	0
1100	6	6	5	5	5	5	4	4	3	3	2	2	1	1	0	0
1200	6	6	6	6	5	5	4	4	3	3	2	2	1	1	0	0
1300	7	7	6	6	6	5	5	4	4	3	3	2	1	1	0	0
1400	7	7	7	7	6	6	5	5	4	3	3	2	1	1	0	0
1500	8	8	7	7	7	6	6	5	4	4	3	2	2	1	1	0
1600	8	8	8	8	7	7	6	5	5	4	3	2	2	1	1	0
1700	9	9	8	8	8	7	6	6	5	4	3	3	2	1	1	0
1800	9	9	9	8	8	7	7	6	5	4	3	3	2	1	1	0
1900	10	10	9	9	8	8	7	6	5	5	4	3	2	1	1	0
2000	10	10	10	9	9	8	7	7	6	5	4	3	2	1	1	0
2100	11	11	10	10	9	9	8	7	6	5	4	3	2	1	1	0
2200	11	11	11	10	10	9	8	7	6	5	4	3	2	1	1	0
2300	12	12	11	11	10	9	9	8	7	5	4	3	2	2	1	0
2400	12	12	12	11	11	10	9	8	7	6	5	4	3	2	1	0
2500	13	13	12	12	11	10	9	8	7	6	5	4	3	2	1	0
2600	13	13	13	12	12	11	10	9	7	6	5	4	3	2	1	0
2700	14	14	13	13	12	11	10	9	8	6	5	4	3	2	1	0
2800	14	14	14	13	12	11	10	9	8	7	5	4	3	2	1	0
2900	15	15	14	14	13	12	11	10	8	7	6	4	3	2	1	0
3000	15	15	15	14	13	12	11	10	9	7	6	4	3	2	1	0
3100	16	16	15	15	14	13	12	10	9	7	6	5	3	2	1	0
3200	16	16	16	15	14	13	12	11	9	8	6	5	3	2	1	0
3300	17	17	16	15	15	14	12	11	9	8	6	5	3	2	1	0
3400	17	17	17	16	15	14	13	11	10	8	7	5	4	2	1	0
3500	18	18	17	16	16	14	13	12	10	8	7	5	4	2	1	0
3600	18	18	18	17	16	15	13	12	10	9	7	5	4	2	1	0
3700	19	19	18	17	16	15	14	12	11	9	7	5	4	3	1	0
3800	19	19	19	18	17	16	14	13	11	9	7	6	4	3	1	0
3900	20	20	19	18	17	16	15	13	11	9	8	6	4	3	1	0
4000	20	20	20	19	18	16	15	13	11	10	8	6	4	3	1	0
4500	23	23	22	21	20	18	17	15	13	11	9	7	5	3	2	0
5000	25	25	24	23	22	21	19	17	14	12	10	7	5	3	2	0
5500	28	28	27	26	24	23	20	18	16	13	11	8	6	4	2	0
6000	30	30	29	28	27	25	22	20	17	14	12	9	6	4	2	1
6500	33	33	32	31	29	27	24	21	19	16	13	10	7	4	2	1
7000	35	35	34	33	31	29	26	23	20	17	13	10	7	5	2	1

SMITHSONIAN TABLES.

TABLE 29.

DETERMINATION OF HEIGHTS BY THE BAROMETER.
METRIC MEASURES.

Correction for the variation of gravity with altitude: $\dfrac{z(z+2h_0)}{R}$

Approximate difference of height. Z.	HEIGHT OF LOWER STATION IN METRES (h_0).												
	0	200	400	600	800	1000	1200	1400	1600	1800	2000	2500	3000	4000
metres.	m.	m.	m.	m.	m.	m.	m.	m.	m.	m.	m.	m.	m.	m.
200	0	0	0	0	0	0	0	0	0	0	0	0	0	0
300	0	0	0	0	0	0	0	0	0	0	0	0	0	0
400	0	0	0	0	0	0	0	0	0	0	0	0	0	1
500	0	0	0	0	0	0	0	0	0	0	0	0	1	1
600	0	0	0	0	0	0	0	0	0	0	0	1	1	1
700	0	0	0	0	0	0	0	0	0	0	1	1	1	1
800	0	0	0	0	0	0	0	0	1	1	1	1	1	1
900	0	0	0	0	0	0	0	1	1	1	1	1	1	1
1000	0	0	0	0	0	1	1	1	1	1	1	1	1	1
1100	0	0	0	0	0	1	1	1	1	1	1	1	1	2
1200	0	0	0	0	1	1	1	1	1	1	1	1	1	2
1300	0	0	0	1	1	1	1	1	1	1	1	1	1	2
1400	0	0	0	1	1	1	1	1	1	1	1	1	2	2
1500	0	0	1	1	1	1	1	1	1	1	1	2	2	2
1600	0	1	1	1	1	1	1	1	1	1	1	2	2	2
1700	0	1	1	1	1	1	1	1	1	1	2	2	2	3
1800	1	1	1	1	1	1	1	1	1	2	2	2	2	3
1900	1	1	1	1	1	1	1	1	2	2	2	2	2	3
2000	1	1	1	1	1	1	1	2	2	2	2	2	3	3
2100	1	1	1	1	1	1	1	2	2	2	2	2	3	3
2200	1	1	1	1	1	1	2	2	2	2	2	2	3	4
2300	1	1	1	1	1	2	2	2	2	2	2	3	3	4
2400	1	1	1	1	2	2	2	2	2	2	2	3	3	4
2500	1	1	1	1	2	2	2	2	2	2	3	3	3	4
2600	1	1	1	2	2	2	2	2	2	3	3	3	4	4
2700	1	1	1	2	2	2	2	2	3	3	3	3	4	5
2800	1	1	2	2	2	2	2	2	3	3	3	3	4	5
2900	1	2	2	2	2	2	2	3	3	3	3	4	4	5
3000	1	2	2	2	2	2	3	3	3	3	3	4	4	5
3100	2	2	2	2	2	2	3	3	3	3	3	4	4	5
3200	2	2	2	2	2	3	3	3	3	3	4	4	5	6
3300	2	2	2	2	3	3	3	3	3	4	4	4	5	6
3400	2	2	2	2	3	3	3	3	4	4	4	4	5	6
3500	2	2	2	3	3	3	3	3	4	4	4	5	5	6
3600	2	2	2	3	3	3	3	4	4	4	4	5	5	7
3700	2	2	3	3	3	3	4	4	4	4	4	5	6	7
3800	2	3	3	3	3	3	4	4	4	4	5	5	6	7
3900	2	3	3	3	3	4	4	4	4	5	5	5	6	7
4000	3	3	3	3	4	4	4	5	5	5	5	6	6	8
4500	3	3	4	4	5	5	5	5	6	6	7	7	9
5000	4	4	5	5	5	6	6	6	7	7	8	9	10
5500	5	5	5	6	6	7	7	8	8	8	9	10	12
6000	6	6	6	7	7	8	8	9	9	9	10	11	13
6500	7	7	7	8	8	9	9	9	10	10	11	12	13	15
7000	8	8	9	9	9	10	10	11	11	12	12	13	14	16

SMITHSONIAN TABLES.

TABLE 30.
DIFFERENCE OF HEIGHT CORRESPONDING TO A CHANGE OF 0.1 INCH IN THE BAROMETER.
ENGLISH MEASURES.

Barometric Pressure.	MEAN TEMPERATURE OF THE AIR IN FAHRENHEIT DEGREES.											
	30°	35°	40°	45°	50°	55°	60°	65°	70°	75°	80°	85°
Inches	Feet.	Feet.	Feet.	Feet.	Feet.	Feet.	Feet.	Feet.	Feet.	Feet.	Feet.	Feet.
22.0	119.2	120.5	121.8	123.1	124.4	125.8	127.1	128.5	129.8	131.2	132.5	133.9
.2	118.2	119.4	120.7	122.0	123.3	124.7	126.0	127.3	128.7	130.0	131.3	132.7
.4	117.1	118.3	119.6	120.9	122.2	123.6	124.9	126.2	127.5	128.8	130.2	131.5
.6	116.1	117.3	118.6	119.8	121.1	122.5	123.8	125.1	126.4	127.7	129.0	130.3
.8	115.0	116.3	117.5	118.8	120.1	121.4	122.7	124.0	125.3	126.6	127.9	129.2
23.0	114.0	115.3	116.5	117.8	119.0	120.3	121.6	122.9	124.2	125.5	126.8	128.1
.2	113.1	114.3	115.5	116.8	118.0	119.3	120.6	121.8	123.1	124.4	125.7	127.0
.4	112.1	113.3	114.5	115.8	117.0	118.3	119.5	120.8	122.1	123.3	124.6	125.9
.6	111.1	112.3	113.5	114.8	116.0	117.3	118.5	119.8	121.0	122.3	123.5	124.8
.8	110.2	111.4	112.6	113.8	115.1	116.3	117.5	118.8	120.0	121.3	122.5	123.8
24.0	109.3	110.5	111.7	112.9	114.1	115.3	116.5	117.8	119.0	120.2	121.5	122.7
.2	108.4	109.5	110.7	111.9	113.1	114.4	115.6	116.8	118.0	119.2	120.5	121.7
.4	107.5	108.6	109.8	111.0	112.2	113.4	114.6	115.9	117.1	118.3	119.5	120.7
.6	106.6	107.8	108.9	110.1	111.3	112.5	113.7	114.9	116.1	117.3	118.5	119.7
.8	105.8	106.9	108.1	109.2	110.4	111.6	112.8	114.0	115.2	116.4	117.6	118.8
25.0	104.9	106.0	107.2	108.3	109.5	110.7	111.9	113.1	114.2	115.4	116.6	117.8
.2	104.1	105.2	106.3	107.5	108.7	109.8	111.0	112.2	113.3	114.5	115.7	116.9
.4	103.3	104.4	105.5	106.6	107.8	109.0	110.1	111.3	112.4	113.6	114.8	116.0
.6	102.5	103.6	104.7	105.8	107.0	108.1	109.3	110.4	111.6	112.7	113.9	115.1
.8	101.7	102.8	103.9	105.0	106.1	107.3	108.4	109.6	110.7	111.9	113.0	114.2
26.0	100.9	102.0	103.1	104.2	105.3	106.4	107.6	108.7	109.9	111.0	112.1	113.3
.2	100.1	101.2	102.3	103.4	104.5	105.6	106.8	107.9	109.0	110.1	111.3	112.4
.4	99.4	100.4	101.5	102.6	103.7	104.8	106.0	107.1	108.2	109.3	110.4	111.6
.6	98.6	99.7	100.7	101.8	102.9	104.0	105.2	106.3	107.4	108.5	109.6	110.7
.8	97.9	98.9	100.0	101.1	102.2	103.3	104.4	105.5	106.6	107.7	108.8	109.9
27.0	97.1	98.2	99.2	100.3	101.4	102.5	103.6	104.7	105.8	106.9	108.0	109.1
.2	96.4	97.5	98.5	99.6	100.7	101.8	102.8	103.9	105.0	106.1	107.2	108.3
.4	95.7	96.8	97.8	98.9	99.9	101.0	102.1	103.2	104.2	105.3	106.4	107.5
.6	95.0	96.1	97.1	98.1	99.2	100.3	101.3	102.4	103.5	104.6	105.6	106.7
.8	94.3	95.4	96.4	97.4	98.5	99.6	100.6	101.7	102.7	103.8	104.9	105.9
28.0	93.7	94.7	95.7	96.7	97.8	98.8	99.9	101.0	102.0	103.1	104.1	105.2
.2	93.0	94.0	95.0	96.1	97.1	98.1	99.2	100.2	101.3	102.3	103.4	104.4
.4	92.4	93.4	94.4	95.4	96.4	97.5	98.5	99.5	100.6	101.6	102.7	103.7
.6	91.7	92.7	93.7	94.7	95.7	96.8	97.8	98.8	99.9	100.9	101.9	103.0
.8	91.1	92.1	93.1	94.1	95.1	96.1	97.1	98.2	99.2	100.2	101.2	102.3
29.0	90.4	91.4	92.4	93.4	94.4	95.4	96.5	97.5	98.5	99.5	100.5	101.6
.2	89.8	90.8	91.8	92.8	93.8	94.8	95.8	96.8	97.8	98.8	99.9	100.9
.4	89.2	90.2	91.1	92.1	93.1	94.1	95.1	96.1	97.1	98.2	99.2	100.2
.6	88.6	89.6	90.5	91.5	92.5	93.5	94.5	95.5	96.5	97.5	98.5	99.5
.8	88.0	89.0	89.9	90.9	91.9	92.9	93.9	94.9	95.8	96.8	97.8	98.8
30.0	87.4	88.4	89.3	90.3	91.3	92.3	93.2	94.2	95.2	96.2	97.2	98.2
.2	86.8	87.8	88.7	89.7	90.7	91.7	92.6	93.6	94.6	95.6	96.5	97.5
.4	86.3	87.2	88.2	89.1	90.1	91.1	92.0	93.0	94.0	94.9	95.9	96.9
.6	85.7	86.7	87.6	88.5	89.5	90.5	91.4	92.4	93.3	94.3	95.3	96.2
.8	85.2	86.1	87.0	88.0	88.9	89.9	90.8	91.8	92.7	93.7	94.7	95.6

TABLE 31.
DIFFERENCE OF HEIGHT CORRESPONDING TO A CHANGE OF 1 MILLIMETRE IN THE BAROMETER.
METRIC MEASURES.

Barometric Pressure.	MEAN TEMPERATURE OF THE AIR IN CENTIGRADE DEGREES.									
	−2°	0°	2°	4°	6°	8°	10°	12°	14°	16°
mm.	Metres.	Metres.	Metres.	Metres.	Metres.	Metres.	Metres.	Metres.	Metres.	Metres.
760	10.48	10.57	10.65	10.73	10.81	10.89	10.98	11.06	11.15	11.23
750	10.62	10.71	10.79	10.87	10.95	11.04	11.13	11.21	11.30	11.38
740	10.77	10.85	10.93	11.02	11.10	11.19	11.28	11.36	11.45	11.54
730	10.91	11.00	11.08	11.17	11.26	11.35	11.43	11.52	11.61	11.70
720	11.06	11.15	11.24	11.32	11.42	11.51	11.59	11.68	11.77	11.86
710	11.22	11.31	11.40	11.48	11.58	11.67	11.75	11.85	11.94	12.03
700	11.38	11.47	11.56	11.65	11.74	11.83	11.92	12.02	12.11	12.20
690	11.55	11.63	11.72	11.82	11.91	12.00	12.09	12.19	12.28	12.38
680	11.72	11.80	11.89	11.99	12.08	12.18	12.27	12.37	12.46	12.56
670	11.89	11.98	12.07	12.17	12.26	12.36	12.46	12.55	12.65	12.75
660	12.07	12.16	12.26	12.35	12.45	12.55	12.65	12.74	12.84	12.94
650	12.26	12.35	12.45	12.54	12.64	12.74	12.84	12.94	13.04	13.14
640	12.45	12.55	12.64	12.74	12.84	12.94	13.04	13.14	13.24	13.35
630	12.65	12.75	12.84	12.94	13.04	13.15	13.25	13.35	13.45	13.56
620	12.85	12.96	13.05	13.15	13.25	13.36	13.46	13.57	13.67	13.78
610	13.06	13.17	13.27	13.37	13.47	13.58	13.68	13.79	13.89	14.01
600	13.28	13.39	13.49	13.59	13.70	13.80	13.91	14.02	14.13	14.24
590	13.51	13.62	13.72	13.82	13.93	14.03	14.15	14.26	14.37	14.48
580	13.74	13.85	13.96	14.06	14.17	14.28	14.39	14.51	14.62	14.73
570	13.98	14.09	14.20	14.31	14.42	14.53	14.64	14.76	14.88	14.99
560	14.23	14.34	14.45	14.57	14.68	14.79	14.90	15.02	15.14	15.25

Barometric Pressure.	MEAN TEMPERATURE OF THE AIR IN CENTIGRADE DEGREES.									
	18°	20°	22°	24°	26°	28°	30°	32°	34°	36°
mm.	Metres.	Metres.	Metres.	Metres.	Metres.	Metres.	Metres.	Metres.	Metres.	Metres.
760	11.32	11.41	11.49	11.58	11.66	11.75	11.84	11.92	12.01	12.10
750	11.47	11.56	11.64	11.73	11.82	11.91	12.00	12.08	12.17	12.26
740	11.63	11.72	11.80	11.89	11.98	12.07	12.16	12.24	12.33	12.42
730	11.79	11.88	11.96	12.05	12.15	12.23	12.32	12.41	12.50	12.59
720	11.95	12.04	12.13	12.22	12.32	12.40	12.49	12.58	12.68	12.77
710	12.12	12.21	12.30	12.39	12.49	12.58	12.67	12.76	12.86	12.95
700	12.29	12.39	12.48	12.57	12.67	12.76	12.85	12.94	13.04	13.13
690	12.47	12.57	12.66	12.75	12.85	12.94	13.04	13.13	13.23	13.32
680	12.66	12.75	12.85	12.94	13.04	13.13	13.23	13.32	13.42	13.52
670	12.85	12.94	13.04	13.14	13.23	13.33	13.43	13.52	13.62	13.72
660	13.04	13.14	13.24	13.34	13.43	13.53	13.63	13.73	13.83	13.93
650	13.24	13.34	13.44	13.54	13.64	13.74	13.84	13.94	14.04	14.15
640	13.45	13.55	13.65	13.75	13.85	13.96	14.06	14.15	14.26	14.37
630	13.66	13.76	13.87	13.97	14.07	14.18	14.28	14.38	14.49	14.60
620	13.88	13.98	14.09	14.20	14.30	14.41	14.51	14.62	14.72	14.83
610	14.11	14.21	14.32	14.43	14.54	14.64	14.75	14.86	14.96	15.07
600	14.35	14.45	14.56	14.67	14.78	14.89	15.00	15.11	15.21	15.32
590	14.59	14.70	14.81	14.92	15.03	15.14	15.25	15.36	15.47	15.59
580	14.84	14.95	15.07	15.17	15.29	15.40	15.52	15.63	15.74	15.86
570	15.10	15.21	15.33	15.44	15.56	15.67	15.79	15.91	16.02	16.14
560	15.37	15.48	15.60	15.72	15.84	15.95	16.07	16.19	16.30	16.42

SMITHSONIAN TABLES.

TABLE 32.

DETERMINATION OF HEIGHTS BY THE BAROMETER.

Formula of Babinet.

$$Z = C \frac{B_0 - B}{B_0 + B}$$

C (in feet) $= 52494 \left[1 + \frac{t_0 + t - 64}{900} \right]$ —English Measures.

C (in metres) $= 16000 \left[1 + \frac{2(t_0 + t)}{1000} \right]$ —Metric Measures.

In which $Z =$ Difference of height of two stations in feet or metres.

$B_0, B =$ Barometric readings at the lower and upper stations respectively, corrected for all sources of instrumental error.

$t_0, t =$ Air temperatures at the lower and upper stations respectively.

Values of C.

ENGLISH MEASURES.			METRIC MEASURES.		
½ ($t_0 + t$).	log C.	C.	½ ($t_0 + t$).	log C.	C.
F.		Feet.	C.		Metres.
10°	4.69834	49928	−10°	4.18639	15360
15	.70339	50511	−8	.19000	15488
20	.70837	51094	−6	.19357	15616
25	.71330	51677	−4	.19712	15744
30	.71818	52261	−2	.20063	15872
			0	4.20412	16000
			+2	.20758	16128
35	4.72300	52844	4	.21101	16256
40	.72777	53428	6	.21442	16384
45	.73248	54011	8	.21780	16512
50	.73715	54595			
55	.74177	55178	10	4.22115	16640
			12	.22448	16768
			14	.22778	16896
60	4.74633	55761	16	.23106	17024
65	.75085	56344	18	.23431	17152
70	.75532	56927	20	4.23754	17280
75	.75975	57511	22	.24075	17408
80	.76413	58094	24	.24393	17536
			26	.24709	17664
			28	.25022	17792
85	4.76847	58677			
90	.77276	59260	30	4.25334	17920
95	.77702	59844	32	.25643	18048
100	.78123	60427	34	.25950	18176
			36	.26255	18304

TABLE 33.
BAROMETRIC PRESSURES CORRESPONDING TO THE TEMPERATURE OF THE BOILING POINT OF WATER.
ENGLISH MEASURES.

Temperature. F.	0°.0 Inches.	0°.1 Inches.	0°.2 Inches.	0°.3 Inches.	0°.4 Inches.	0°.5 Inches.	0°.6 Inches.	0°.7 Inches.	0°.8 Inches.	0°.9 Inches.
185°	17.05	17.08	17.12	17.16	17.20	17.23	17.27	17.31	17.35	17.39
186	17.42	17.46	17.50	17.54	17.58	17.61	17.65	17.69	17.73	17.77
187	17.81	17.84	17.88	17.92	17.96	18.00	18.04	18.08	18.12	18.16
188	18.20	18.24	18.27	18.31	18.35	18.39	18.43	18.47	18.51	18.55
189	18.59	18.63	18.67	18.71	18.75	18.79	18.83	18.87	18.91	18.95
190	19.00	19.04	19.08	19.12	19.16	19.20	19.24	19.28	19.32	19.36
191	19.41	19.45	19.49	19.53	19.57	19.61	19.66	19.70	19.74	19.78
192	19.82	19.87	19.91	19.95	19.99	20.04	20.08	20.12	20.17	20.21
193	20.25	20.29	20.34	20.38	20.42	20.47	20.51	20.55	20.60	20.64
194	20.68	20.73	20.77	20.82	20.86	20.90	20.95	20.99	21.04	21.08
195	21.13	21.17	21.22	21.26	21.30	21.35	21.39	21.44	21.48	21.53
196	21.58	21.62	21.67	21.71	21.76	21.80	21.85	21.89	21.94	21.99
197	22.03	22.08	22.12	22.17	22.22	22.26	22.31	22.36	22.40	22.45
198	22.50	22.54	22.59	22.64	22.69	22.73	22.78	22.83	22.88	22.92
199	22.97	23.02	23.07	23.11	23.16	23.21	23.26	23.31	23.36	23.40
200	23.45	23.50	23.55	23.60	23.65	23.70	23.75	23.80	23.85	23.89
201	23.94	23.99	24.04	24.09	24.14	24.19	24.24	24.29	24.34	24.39
202	24.44	24.49	24.54	24.59	24.64	24.69	24.74	24.80	24.85	24.90
203	24.95	25.00	25.05	25.10	25.15	25.21	25.26	25.31	25.36	25.41
204	25.46	25.52	25.57	25.62	25.67	25.73	25.78	25.83	25.88	25.94
205	25.99	26.04	26.10	26.15	26.20	26.25	26.31	26.36	26.42	26.47
206	26.52	26.58	26.63	26.68	26.74	26.79	26.85	26.90	26.96	27.01
207	27.07	27.12	27.18	27.23	27.29	27.34	27.40	27.45	27.51	27.56
208	27.62	27.67	27.73	27.79	27.84	27.90	27.95	28.01	28.07	28.12
209	28.18	28.24	28.29	28.35	28.41	28.46	28.52	28.58	28.64	28.69
210	28.75	28.81	28.87	28.92	28.98	29.04	29.10	29.16	29.21	29.27
211	29.33	29.39	29.45	29.51	29.57	29.62	29.68	29.74	29.80	29.86
212	29.92	29.98	30.04	30.10	30.16	30.22	30.28	30.34	30.40	30.46

METRIC MEASURES.
TABLE 34.

Temperature. C.	0°.0 mm.	0°.1 mm.	0°.2 mm.	0°.3 mm.	0°.4 mm.	0°.5 mm.	0°.6 mm.	0°.7 mm.	0°.8 mm.	0°.9 mm.
80°	354.6	356.1	357.5	359.0	360.4	361.9	363.3	364.8	366.3	367.8
81	369.3	370.8	372.3	373.8	375.3	376.8	378.3	379.8	381.3	382.9
82	384.4	385.9	387.5	389.0	390.6	392.2	393.7	395.3	396.9	398.5
83	400.1	401.7	403.3	404.9	406.5	408.1	409.7	411.3	413.0	414.6
84	416.3	417.9	419.6	421.2	422.9	424.6	426.2	427.9	429.6	431.3
85	433.0	434.7	436.4	438.1	439.9	441.6	443.3	445.1	446.8	448.6
86	450.3	452.1	453.8	455.6	457.4	459.2	461.0	462.8	464.6	466.4
87	468.2	470.0	471.8	473.7	475.5	477.3	479.2	481.0	482.9	484.8
88	486.6	488.5	490.4	492.3	494.2	496.1	498.0	499.9	501.8	503.8
89	505.7	507.6	509.6	511.5	513.5	515.5	517.4	519.4	521.4	523.4
90	525.4	527.4	529.4	531.4	533.4	535.5	537.5	539.6	541.6	543.7
91	545.7	547.8	549.9	551.9	554.0	556.1	558.2	560.3	562.4	564.6
92	566.7	568.8	571.0	573.1	575.3	577.4	579.6	581.8	584.0	586.1
93	588.3	590.5	592.7	595.0	597.2	599.4	601.6	603.9	606.1	608.4
94	610.7	612.9	615.2	617.5	619.8	622.1	624.4	626.7	629.0	631.4
95	633.7	636.0	638.4	640.7	643.1	645.5	647.9	650.2	652.6	655.0
96	657.4	659.9	662.3	664.7	667.1	669.6	672.0	674.5	677.0	679.4
97	681.9	684.4	686.9	689.4	691.9	694.5	697.0	699.5	702.1	704.6
98	707.2	709.7	712.3	714.9	717.5	720.1	722.7	725.3	727.9	730.5
99	733.2	735.8	738.5	741.2	743.8	746.5	749.2	751.9	754.6	757.3
100	760.0	762.7	765.5	768.2	770.9	773.7	776.5	779.2	782.0	784.8

SMITHSONIAN TABLES.

HYGROMETRICAL TABLES.

Pressure of aqueous vapor (*Broch*) —
    English measures . . . . TABLE 35
    Metric measures . . . . $\begin{cases} 36 \\ 43 \end{cases}$

Pressure of aqueous vapor at low temperatures (*C. F. Marvin*) —
    English and Metric measures . . . . TABLE 37

Weight of aqueous vapor in a cubic foot of saturated air —
    English measures . . . . TABLE 38

Weight of aqueous vapor in a cubic metre of saturated air —
    Metric measures . . . . TABLE 39

Reduction of psychrometric observations — English measures.
    Pressure of aqueous vapor . . . . TABLE 40
    Values of $0.000367\, B\, (t - t_1)\left(1 + \dfrac{t - t_1}{1571}\right)$ 41

Relative humidity — Temperature Fahrenheit . . . . . . TABLE 42

Reduction of psychrometric observations — Metric measures.
    Pressure of aqueous vapor . . . . TABLE 43
    Values of $0.000660\, B\, (t - t_1)\left(1 + \dfrac{t - t_1}{873}\right)$ 44

Relative humidity — Temperature Centigrade . . . . . . TABLE 45

Reduction of snowfall measurements.
    Depth of water corresponding to the weight of snow
        (or rain) collected in an 8-inch gage . . . . . . TABLE 46

Rate of decrease of vapor pressure with altitude . . . . TABLE 47

TABLE 35.

PRESSURE OF AQUEOUS VAPOR.
(*Broch.*)
ENGLISH MEASURES.

Temperature.	Vapor Pressure.	Diff. for 0°.1	Temperature.	Vapor Pressure.	Diff. for 0°.1	Temperature.	Vapor Pressure.	Diff. for 0°.1	Temperature.	Vapor Pressure.	Diff. for 0°.1
F.	Inch.		F.	Inch.		F.	Inch.		F.	Inch.	
−20°.0	0.0167		−10°.0	0.0277		0°.0	0.0449	2	10°.0	0.0710	3
−19.8	0.0168	1	− 9.8	0.0280	1	+0.2	.0454	2	10.2	.0716	3
19.6	.0170	1	9.6	.0283	1	0.4	.0458	2	10.4	.0723	3
19.4	.0172	1	9.4	.0286	1	0.6	.0462	2	10.6	.0729	3
19.2	.0174	1	9.2	.0289	1	0.8	.0467	2	10.8	.0736	3
19.0	.0176	1	9.0	.0292	1	1.0	.0471	2	11.0	0.0742	3
		1			1	1.2	.0475	2	11.2	.0749	3
−18.8	0.0177	1	− 8.8	0.0294	1	1.4	.0480	2	11.4	.0756	3
18.6	.0179	1	8.6	.0297	1	1.6	.0484	2	11.6	.0762	3
18.4	.0181	1	8.4	.0300	2	1.8	.0489	2	11.8	.0769	3
18.2	.0183	1	8.2	.0303	2			2			3
18.0	.0185	1	8.0	.0306	2	2.0	0.0493	2	12.0	0.0776	3
					2	2.2	.0498	2	12.2	.0783	3
−17.8	0.0187	1	− 7.8	0.0309	2	2.4	.0503	2	12.4	.0790	3
17.6	.0189	1	7.6	.0312	2	2.6	.0507	2	12.6	.0797	3
17.4	.0191	1	7.4	.0315	2	2.8	.0512	2	12.8	.0804	4
17.2	.0193	1	7.2	.0318	2			2			4
17.0	.0195	1	7.0	.0322	2	3.0	0.0517	2	13.0	0.0811	4
					2	3.2	.0522	2	13.2	.0818	4
−16.8	0.0197	1	− 6.8	0.0325	2	3.4	.0526	2	13.4	.0825	4
16.6	.0199	1	6.6	.0328	2	3.6	.0531	2	13.6	.0832	4
16.4	.0201	1	6.4	.0331	2	3.8	.0536	2	13.8	.0840	4
16.2	.0203	1	6.2	.0334	2			2			4
16.0	.0205	1	6.0	.0338	2	4.0	0.0541	3	14.0	0.0847	4
					2	4.2	.0546	3	14.2	.0854	4
−15.8	0.0207	1	− 5.8	0.0341	2	4.4	.0551	3	14.4	.0862	4
15.6	.0209	1	5.6	.0344	2	4.6	.0556	3	14.6	.0869	4
15.4	.0211	1	5.4	.0347	2	4.8	.0561	3	14.8	.0877	4
15.2	.0213	1	5.2	.0351	2			3			4
15.0	.0216	1	5.0	.0354	2	5.0	0.0567	3	15.0	0.0885	4
					2	5.2	.0572	3	15.2	.0892	4
−14.8	0.0218	1	− 4.8	0.0358	2	5.4	.0577	3	15.4	.0900	4
14.6	.0220	1	4.6	.0361	2	5.6	.0582	3	15.6	.0908	4
14.4	.0222	1	4.4	.0365	2	5.8	.0588	3	15.8	.0916	4
14.2	.0225	1	4.2	.0368	2			3			4
14.0	.0227	1	4.0	.0372	2	6.0	0.0593	3	16.0	0.0924	4
					2	6.2	.0598	3	16.2	.0932	4
−13.8	0.0229	1	− 3.8	0.0375	2	6.4	.0604	3	16.4	.0940	4
13.6	.0232	1	3.6	.0379	2	6.6	.0609	3	16.6	.0948	4
13.4	.0234	1	3.4	.0383	2	6.8	.0615	3	16.8	.0956	4
13.2	.0236	1	3.2	.0386	2			3			4
13.0	.0239	1	3.0	.0390	2	7.0	0.0620	3	17.0	0.0965	4
					2	7.2	.0626	3	17.2	.0973	4
−12.8	0.0241	1	− 2.8	0.0394	2	7.4	.0632	3	17.4	.0981	4
12.6	.0244	1	2.6	.0397	2	7.6	.0637	3	17.6	.0990	4
12.4	.0246	1	2.4	.0401	2	7.8	.0643	3	17.8	.0999	4
12.2	.0249	1	2.2	.0405	2			3			4
12.0	.0251	1	2.0	.0409	2	8.0	0.0649	3	18.0	0.1007	4
					2	8.2	.0655	3	18.2	.1016	4
−11.8	0.0254	1	− 1.8	0.0413	2	8.4	.0661	3	18.4	.1024	4
11.6	.0256	1	1.6	.0417	2	8.6	.0667	3	18.6	.1033	4
11.4	.0259	1	1.4	.0421	2	8.8	.0673	3	18.8	.1042	4
11.2	.0261	1	1.2	.0425	2			3			4
11.0	.0264	1	1.0	.0429	2	9.0	0.0679	3	19.0	0.1051	5
					2	9.2	.0685	3	19.2	.1060	5
−10.8	0.0267	1	− 0.8	0.0433	2	9.4	.0691	3	19.4	.1069	5
10.6	.0269	1	0.6	.0437	2	9.6	.0697	3	19.6	.1078	5
10.4	.0272	1	0.4	.0441	2	9.8	.0704	3	19.8	.1087	5
10.2	.0275	1	0.2	.0445	2			3			5

SMITHSONIAN TABLES.

TABLE 35.

PRESSURE OF AQUEOUS VAPOR.

(Broch.)

ENGLISH MEASURES.

Temperature.	Vapor Pressure.	Diff. for 0°.1	Temperature.	Vapor Pressure.	Diff. for 0°.1	Temperature.	Vapor Pressure.	Diff. for 0°.1	Temperature.	Vapor Pressure.	Diff. for 0°.1
F.	Inch.		F.	Inch.		F.	Inch.		F.	Inch.	
20°.0	0.1097		30°.0	0.1660		40°.0	0.2465		50°.0	0.3598	
20.2	.1106	5	30.2	.1673	7	40.2	.2484	10	50.2	.3625	13
20.4	.1115	5	30.4	.1687	7	40.4	.2503	10	50.4	.3652	13
20.6	.1125	5	30.6	.1700	7	40.6	.2523	10	50.6	.3679	14
20.8	.1134	5	30.8	.1714	7	40.8	.2542	10	50.8	.3706	14
21.0	0.1144	5	31.0	0.1728	7	41.0	0.2562	10	51.0	0.3734	14
21.2	.1154	5	31.2	.1742	7	41.2	.2582	10	51.2	.3761	14
21.4	.1163	5	31.4	.1756	7	41.4	.2601	10	51.4	.3789	14
21.6	.1173	5	31.6	.1770	7	41.6	.2621	10	51.6	.3817	14
21.8	.1183	5	31.8	.1784	7	41.8	.2642	10	51.8	.3845	14
22.0	0.1193	5	32.0	0.1799	7	42.0	0.2662	10	52.0	0.3874	14
22.2	.1203	5	32.2	.1813	7	42.2	.2683	10	52.2	.3902	14
22.4	.1213	5	32.4	.1828	7	42.4	.2703	10	52.4	.3931	14
22.6	.1223	5	32.6	.1842	7	42.6	.2724	11	52.6	.3960	14
22.8	.1234	5	32.8	.1857	7	42.8	.2745	11	52.8	.3989	15
23.0	0.1244	5	33.0	0.1872	8	43.0	0.2766	11	53.0	0.4018	15
23.2	.1255	5	33.2	.1887	8	43.2	.2787	11	53.2	.4048	15
23.4	.1265	5	33.4	.1902	8	43.4	.2808	11	53.4	.4077	15
23.6	.1276	5	33.6	.1917	8	43.6	.2830	11	53.6	.4107	15
23.8	.1287	5	33.8	.1933	8	43.8	.2851	11	53.8	.4137	15
24.0	0.1297	5	34.0	0.1948	8	44.0	0.2873	11	54.0	0.4168	15
24.2	.1308	5	34.2	.1964	8	44.2	.2895	11	54.2	.4198	15
24.4	.1319	5	34.4	.1979	8	44.4	.2917	11	54.4	.4229	15
24.6	.1330	6	34.6	.1995	8	44.6	.2939	11	54.6	.4259	16
24.8	.1341	6	34.8	.2011	8	44.8	.2962	11	54.8	.4290	16
25.0	0.1352	6	35.0	0.2027	8	45.0	0.2984	11	55.0	0.4322	16
25.2	.1364	6	35.2	.2043	8	45.2	.3007	11	55.2	.4353	16
25.4	.1375	6	35.4	.2059	8	45.4	.3030	12	55.4	.4385	16
25.6	.1386	6	35.6	.2076	8	45.6	.3053	12	55.6	.4417	16
25.8	.1398	6	35.8	.2092	8	45.8	.3076	12	55.8	.4449	16
26.0	0.1409	6	36.0	0.2109	8	46.0	0.3099	12	56.0	0.4481	16
26.2	.1421	6	36.2	.2125	8	46.2	.3123	12	56.2	.4513	16
26.4	.1433	6	36.4	.2142	9	46.4	.3146	12	56.4	.4546	16
26.6	.1445	6	36.6	.2159	9	46.6	.3170	12	56.6	.4579	17
26.8	.1457	6	36.8	.2176	9	46.8	.3194	12	56.8	.4612	17
27.0	0.1469	6	37.0	0.2193	9	47.0	0.3218	12	57.0	0.4645	17
27.2	.1481	6	37.2	.2210	9	47.2	.3242	12	57.2	.4679	17
27.4	.1493	6	37.4	.2228	9	47.4	.3267	12	57.4	.4712	17
27.6	.1505	6	37.6	.2245	9	47.6	.3291	12	57.6	.4746	17
27.8	.1518	6	37.8	.2263	9	47.8	.3316	12	57.8	.4780	17
28.0	0.1530	6	38.0	0.2281	9	48.0	0.3341	13	58.0	0.4815	17
28.2	.1543	6	38.2	.2298	9	48.2	.3366	13	58.2	.4849	17
28.4	.1555	6	38.4	.2316	9	48.4	.3391	13	58.4	.4884	18
28.6	.1568	6	38.6	.2334	9	48.6	.3416	13	58.6	.4919	18
28.8	.1581	6	38.8	.2353	9	48.8	.3442	13	58.8	.4954	18
29.0	0.1594	7	39.0	0.2371	9	49.0	0.3467	13	59.0	0.4990	18
29.2	.1607	7	39.2	.2390	9	49.2	.3493	13	59.2	.5025	18
29.4	.1620	7	39.4	.2408	9	49.4	.3519	13	59.4	.5061	18
29.6	.1633	7	39.6	.2427	9	49.6	.3546	13	59.6	.5097	18
29.8	.1646	7	39.8	.2446	10	49.8	.3572	13	59.8	.5134	18

SMITHSONIAN TABLES.

TABLE 35.

PRESSURE OF AQUEOUS VAPOR.
(*Broch.*)
ENGLISH MEASURES.

Temperature.	Vapor Pressure.	Diff. for 0°.1	Temperature.	Vapor Pressure.	Diff. for 0°.1	Temperature.	Vapor Pressure.	Diff. for 0°.1	Temperature.	Vapor Pressure.	Diff. for 0°.1
F.	Inch.		F.	Inch.		F.	Inches.		F.	Inches.	
60.0	0.5170	18	70.0	0.7320	25	80.0	1.0219	34	90.0	1.4081	44
60.2	.5207	19	70.2	.7370	25	80.2	.0286	34	90.2	.4170	45
60.4	.5244	19	70.4	.7420	25	80.4	.0354	34	90.4	.4259	45
60.6	.5282	19	70.6	.7471	26	80.6	.0422	34	90.6	.4349	45
60.8	.5319	19	70.8	.7522	26	80.8	.0490	34	90.8	.4439	45
61.0	0.5357	19	71.0	0.7573	26	81.0	1.0558	35	91.0	1.4530	46
61.2	.5395	19	71.2	.7625	26	81.2	.0627	35	91.2	.4621	46
61.4	.5433	19	71.4	.7676	26	81.4	.0697	35	91.4	.4712	46
61.6	.5471	19	71.6	.7728	26	81.6	.0767	35	91.6	.4805	46
61.8	.5510	19	71.8	.7781	26	81.8	.0837	35	91.8	.4897	47
62.0	0.5549	20	72.0	0.7834	27	82.0	1.0907	36	92.0	1.4990	47
62.2	.5588	20	72.2	.7887	27	82.2	.0978	36	92.2	.5084	47
62.4	.5628	20	72.4	.7940	27	82.4	.1050	36	92.4	.5178	47
62.6	.5667	20	72.6	.7994	27	82.6	.1121	36	92.6	.5273	48
62.8	.5707	20	72.8	.8048	27	82.8	.1194	36	92.8	.5368	48
63.0	0.5748	20	73.0	0.8102	27	83.0	1.1266	37	93.0	1.5464	48
63.2	.5788	20	73.2	.8157	28	83.2	.1339	37	93.2	.5560	48
63.4	.5829	21	73.4	.8212	28	83.4	.1413	37	93.4	.5657	49
63.6	.5870	21	73.6	.8267	28	83.6	.1487	37	93.6	.5755	49
63.8	.5911	21	73.8	.8323	28	83.8	.1561	37	93.8	.5853	49
64.0	0.5952	21	74.0	0.8379	28	84.0	1.1635	38	94.0	1.5951	49
64.2	.5994	21	74.2	.8435	28	84.2	.1710	38	94.2	.6050	50
64.4	.6036	21	74.4	.8492	29	84.4	.1786	38	94.4	.6149	50
64.6	.6078	21	74.6	.8549	29	84.6	.1862	38	94.6	.6249	50
64.8	.6120	21	74.8	.8606	29	84.8	.1938	38	94.8	.6350	51
65.0	0.6163	22	75.0	0.8664	29	85.0	1.2015	39	95.0	1.6451	51
65.2	.6206	22	75.2	.8722	29	85.2	.2093	39	95.2	.6552	51
65.4	.6249	22	75.4	.8780	29	85.4	.2170	39	95.4	.6655	51
65.6	.6293	22	75.6	.8839	30	85.6	.2248	39	95.6	.6758	52
65.8	.6337	22	75.8	.8898	30	85.8	.2327	39	95.8	.6861	52
66.0	0.6381	22	76.0	0.8957	30	86.0	1.2406	40	96.0	1.6964	52
66.2	.6425	22	76.2	.9017	30	86.2	.2485	40	96.2	.7069	52
66.4	.6470	22	76.4	.9077	30	86.4	.2565	40	96.4	.7174	53
66.6	.6514	23	76.6	.9137	30	86.6	.2645	40	96.6	.7279	53
66.8	.6560	23	76.8	.9198	31	86.8	.2726	41	96.8	.7385	53
67.0	0.6605	23	77.0	0.9259	31	87.0	1.2807	41	97.0	1.7492	54
67.2	.6651	23	77.2	.9321	31	87.2	.2889	41	97.2	.7599	54
67.4	.6697	23	77.4	.9383	31	87.4	.2971	41	97.4	.7707	54
67.6	.6743	23	77.6	.9445	31	87.6	.3054	42	97.6	.7815	54
67.8	.6789	23	77.8	.9507	31	87.8	.3137	42	97.8	.7924	55
68.0	0.6836	24	78.0	0.9570	32	88.0	1.3220	42	98.0	1.8034	55
68.2	.6883	24	78.2	.9633	32	88.2	.3304	42	98.2	.8144	55
68.4	.6930	24	78.4	.9697	32	88.4	.3388	42	98.4	.8254	56
68.6	.6978	24	78.6	.9761	32	88.6	.3473	43	98.6	.8366	56
68.8	.7026	24	78.8	.9825	32	88.8	.3558	43	98.8	.8477	56
69.0	0.7074	24	79.0	0.9890	33	89.0	1.3644	43	99.0	1.8590	57
69.2	.7123	24	79.2	.9955	33	89.2	.3731	43	99.2	.8703	57
69.4	.7172	25	79.4	1.0021	33	89.4	.3818	44	99.4	.8817	57
69.6	.7221	25	79.6	.0087	33	89.6	.3905	44	99.6	.8931	57
69.8	.7270	25	79.8	.0153	33	89.8	.3993	44	99.8	.9046	58

TABLE 35.
PRESSURE OF AQUEOUS VAPOR
(*Broch.*)
ENGLISH MEASURES.

Temperature.	Vapor Pressure.	Diff. for 0°.1	Temperature.	Vapor Pressure.	Diff. for 0°.1	Temperature.	Vapor Pressure.	Diff. for 0°.1	Temperature.	Vapor Pressure.	Diff. for 0°.1
F.	Inches.		F.	Inches.		F.	Inches.		F.	Inches.	
100°.0	1.9161		110°.0	2.5765		120°.0	3.4253		130°.0	4.5044	
100.2	.9277	58	110.2	.5915	75	120.2	.4445	96	130.2	.5286	121
100.4	.9394	58	110.4	.6066	75	120.4	.4637	96	130.4	.5530	122
100.6	.9511	59	110.6	.6217	76	120.6	.4831	97	130.6	.5775	122
100.8	.9629	59	110.8	.6369	76	120.8	.5026	97	130.8	.6020	123
		59			77			98			123
101.0	1.9747	60	111.0	2.6522	77	121.0	3.5221	98	131.0	4.6267	124
101.2	.9867	60	111.2	.6676	77	121.2	.5417	99	131.2	.6515	125
101.4	.9986	60	111.4	.6831	78	121.4	.5615	99	131.4	.6764	125
101.6	2.0107	60	111.6	.6986	78	121.6	.5813	100	131.6	.7015	126
101.8	.0228	61	111.8	.7142	78	121.8	.6012	100	131.8	.7266	126
102.0	2.0349	61	112.0	2.7299	79	122.0	3.6213	101	132.0	4.7519	127
102.2	.0471	61	112.2	.7457	79	122.2	.6414	101	132.2	.7773	127
102.4	.0594	62	112.4	.7616	80	122.4	.6616	102	132.4	.8028	128
102.6	.0718	62	112.6	.7775	80	122.6	.6819	102	132.6	.8284	129
102.8	.0842	62	112.8	.7935	80	122.8	.7023	102	132.8	.8541	129
103.0	2.0967	63	113.0	2.8096	81	123.0	3.7228	103	133.0	4.8800	130
103.2	.1092	63	113.2	.8257	81	123.2	.7434	103	133.2	.9059	130
103.4	.1218	63	113.4	.8420	82	123.4	.7641	104	133.4	.9320	131
103.6	.1345	64	113.6	.8583	82	123.6	.7849	104	133.6	.9582	132
103.8	.1473	64	113.8	.8747	82	123.8	.8058	105	133.8	.9845	132
104.0	2.1601	64	114.0	2.8912	83	124.0	3.8267	105	134.0	5.0110	133
104.2	.1730	65	114.2	.9078	83	124.2	.8478	106	134.2	.0375	133
104.4	.1859	65	114.4	.9244	84	124.4	.8690	106	134.4	.0642	134
104.6	.1989	65	114.6	.9412	84	124.6	.8903	107	134.6	.0910	135
104.8	.2120	66	114.8	.9580	85	124.8	.9117	107	134.8	.1179	135
105.0	2.2251	66	115.0	2.9749	85	125.0	3.9332	108	135.0	5.1450	136
105.2	.2384	66	115.2	.9919	85	125.2	.9548	109	135.2	.1722	136
105.4	.2516	67	115.4	3.0089	86	125.4	.9765	109	135.4	.1994	137
105.6	.2650	67	115.6	.0261	86	125.6	.9983	110	135.6	.2269	138
105.8	.2784	67	115.8	.0433	87	125.8	4.0202	110	135.8	.2544	138
106.0	2.2919	68	116.0	3.0606	87	126.0	4.0422	111	136.0	5.2820	139
106.2	.3054	68	116.2	.0780	87	126.2	.0643	111	136.2	.3098	139
106.4	.3190	68	116.4	.0955	88	126.4	.0865	112	136.4	.3377	140
106.6	.3327	69	116.6	.1131	88	126.6	.1088	112	136.6	.3657	141
106.8	.3465	69	116.8	.1308	89	126.8	.1312	113	136.8	.3939	141
107.0	2.3603	70	117.0	3.1485	89	127.0	4.1537	113	137.0	5.4221	142
107.2	.3742	70	117.2	.1663	90	127.2	.1764	114	137.2	.4505	143
107.4	.3882	70	117.4	.1842	90	127.4	.1991	114	137.4	.4791	143
107.6	.4023	71	117.6	.2023	91	127.6	.2219	115	137.6	.5077	144
107.8	.4164	71	117.8	.2204	91	127.8	.2448	115	137.8	.5365	145
108.0	2.4306	71	118.0	3.2386	91	128.0	4.2679	116	138.0	5.5654	145
108.2	.4449	72	118.2	.2568	92	128.2	.2910	116	138.2	.5945	146
108.4	.4592	72	118.4	.2752	92	128.4	.3143	117	138.4	.6237	147
108.6	.4736	72	118.6	.2937	93	128.6	.3377	117	138.6	.6530	147
108.8	.4881	73	118.8	.3122	93	128.8	.3612	118	138.8	.6824	148
109.0	2.5026	73	119.0	3.3308	94	129.0	4.3848	119	139.0	5.7120	149
109.2	.5172	73	119.2	.3495	94	129.2	.4085	119	139.2	.7417	149
109.4	.5319	74	119.4	.3683	95	129.4	.4323	120	139.4	.7715	150
109.6	.5467	74	119.6	.3872	95	129.6	.4562	120	139.6	.8014	150
109.8	.5616	75	119.8	.4062	95	129.8	.4802	121	139.8	.8315	151

TABLE 35.

PRESSURE OF AQUEOUS VAPOR.
(*Broch.*)

ENGLISH MEASURES.

Temperature.	Vapor Pressure.	Diff. for 0°.1	Temperature.	Vapor Pressure.	Diff. for 0°.1	Temperature.	Vapor Pressure.	Diff. for 0°.1	Temperature.	Vapor Pressure.	Diff. for 0°.1
F.	Inches.		F.	Inches.		F.	Inches.		F.	Inches.	
140.0	5.8617	152	150.0	7.5521	188	160.0	9.6374	231	170.0	12.1870	281
140.2	.8921	152	150.2	.5897	189	160.2	.6836	232	170.2	.2432	282
140.4	.9226	153	150.4	.6275	190	160.4	.7300	233	170.4	.2997	283
140.6	.9532	154	150.6	.6654	191	160.6	.7765	234	170.6	.3564	285
140.8	.9839	154	150.8	.7035	191	160.8	.8233	235	170.8	.4133	286
141.0	6.0148	155	151.0	7.7418	192	161.0	9.8702	236	171.0	12.4704	287
141.2	.0458	156	151.2	.7802	193	161.2	.9173	237	171.2	.5278	288
141.4	.0770	157	151.4	.8188	194	161.4	.9647	238	171.4	.5853	289
141.6	.1083	157	151.6	.8575	195	161.6	10.0122	239	171.6	.6431	290
141.8	.1397	158	151.8	.8964	195	161.8	.0599	240	171.8	.7011	291
142.0	6.1713	159	152.0	7.9355	196	162.0	10.1078	241	172.0	12.7593	292
142.2	.2030	159	152.2	.9747	197	162.2	.1559	241	172.2	.8177	293
142.4	.2348	160	152.4	8.0141	198	162.4	.2042	242	172.4	.8764	295
142.6	.2668	161	152.6	.0536	199	162.6	.2526	243	172.6	.9353	296
142.8	.2989	161	152.8	.0934	199	162.8	.3013	244	172.8	.9945	297
143.0	6.3312	162	153.0	8.1332	200	163.0	10.3501	245	173.0	13.0538	298
143.2	.3636	163	153.2	.1733	201	163.2	.3992	246	173.2	.1134	299
143.4	.3961	164	153.4	.2135	202	163.4	.4484	247	173.4	.1732	300
143.6	.4288	164	153.6	.2539	203	163.6	.4979	248	173.6	.2332	301
143.8	.4616	165	153.8	.2944	204	163.8	.5475	249	173.8	.2935	303
144.0	6.4946	166	154.0	8.3351	205	164.0	10.5974	250	174.0	13.3540	304
144.2	.5277	166	154.2	.3760	205	164.2	.6474	251	174.2	.4147	305
144.4	.5610	167	154.4	.4171	206	164.4	.6976	252	174.4	.4756	306
144.6	.5944	168	154.6	.4583	207	164.6	.7480	253	174.6	.5368	307
144.8	.6279	168	154.8	.4997	208	164.8	.7986	254	174.8	.5982	308
145.0	6.6616	169	155.0	8.5413	209	165.0	10.8495	255	175.0	13.6599	309
145.2	.6954	170	155.2	.5830	210	165.2	.9005	256	175.2	.7218	311
145.4	.7294	171	155.4	.6249	210	165.4	.9517	257	175.4	.7839	312
145.6	.7635	171	155.6	.6670	211	165.6	11.0032	258	175.6	.8462	313
145.8	.7978	172	155.8	.7092	212	165.8	.0548	259	175.8	.9088	314
146.0	6.8322	173	156.0	8.7516	213	166.0	11.1067	260	176.0	13.9716	315
146.2	.8668	174	156.2	.7942	214	166.2	.1587	261	176.2	14.0347	317
146.4	.9015	174	156.4	.8370	215	166.4	.2109	262	176.4	.0980	318
146.6	.9363	175	156.6	.8799	216	166.6	.2634	263	176.6	.1616	319
146.8	.9713	176	156.8	.9231	217	166.8	.3160	264	176.8	.2253	320
147.0	7.0065	177	157.0	8.9664	217	167.0	11.3689	265	177.0	14.2894	321
147.2	.0418	177	157.2	9.0098	218	167.2	.4220	266	177.2	.3536	323
147.4	.0773	178	157.4	.0535	219	167.4	.4752	267	177.4	.4181	324
147.6	.1129	179	157.6	.0973	220	167.6	.5287	268	177.6	.4823	325
147.8	.1486	180	157.8	.1413	221	167.8	.5824	270	177.8	.5478	326
148.0	7.1845	180	158.0	9.1855	222	168.0	11.6363	271	178.0	14.6131	327
148.2	.2206	181	158.2	.2299	223	168.2	.6904	272	178.2	.6785	329
148.4	.2568	182	158.4	.2745	224	168.4	.7447	273	178.4	.7443	330
148.6	.2932	183	158.6	.3192	225	168.6	.7993	274	178.6	.8102	331
148.8	.3297	183	158.8	.3641	226	168.8	.8540	275	178.8	.8764	332
149.0	7.3664	184	159.0	9.4092	226	169.0	11.9090	276	179.0	14.9429	334
149.2	.4032	185	159.2	.4545	227	169.2	.9641	277	179.2	15.0096	335
149.4	.4402	186	159.4	.4999	228	169.4	12.0195	278	179.4	.0765	336
149.6	.4774	187	159.6	.5456	229	169.6	.0751	279	179.6	.1437	337
149.8	.5147	187	159.8	.5914	230	169.8	.1309	280	179.8	.2112	339

SMITHSONIAN TABLES.

TABLE 35.

PRESSURE OF AQUEOUS VAPOR.
(*Broch.*)

ENGLISH MEASURES.

Temperature.	Vapor Pressure.	Diff. for 0°.1	Temperature.	Vapor Pressure.	Diff. for 0°.1	Temperature.	Vapor Pressure.	Diff. for 0°.1	Temperature.	Vapor Pressure.	Diff. for 0°.1
F.	Inches.		F.	Inches.		F.	Inches.		F.	Inches.	
180.0	15.2789		190.0	19.0009		200.0	23.4530		210.0	28.7497	
180.2	.3468	340	190.2	.0825	408	200.2	.5502	486	210.2	.8651	577
180.4	.4150	341	190.4	.1643	409	200.4	.6478	488	210.4	.9809	579
180.6	.4835	342	190.6	.2464	411	200.6	.7457	490	210.6	29.0972	581
180.8	.5522	344	190.8	.3288	412	200.8	.8440	491	210.8	.2138	583
		345			414			493			585
181.0	15.6212	346	191.0	19.4115	415	201.0	23.9426	495	211.0	29.3308	587
181.2	.6904	347	191.2	.4945	416	201.2	24.0415	497	211.2	.4482	589
181.4	.7599	349	191.4	.5778	418	201.4	.1408	498	211.4	.5660	591
181.6	.8296	350	191.6	.6614	419	201.6	.2404	500	211.6	.6842	593
181.8	.8996	351	191.8	.7453	421	201.8	.3404	502	211.8	.8028	595
182.0	15.9699	353	192.0	19.8295	422	202.0	24.4407	503	212.0	29.9218	597
182.2	16.0404	354	192.2	.9140	424	202.2	.5414	505	212.2	30.0412	599
182.4	.1112	355	192.4	.9988	426	202.4	.6424	507	212.4	.1610	601
182.6	.1822	357	192.6	20.0839	427	202.6	.7438	509	212.6	.2813	603
182.8	.2535	358	192.8	.1693	429	202.8	.8455	510	212.8	.4019	605
183.0	16.3250	359	193.0	20.2550	430	203.0	24.9476	512	213.0	30.5229	607
183.2	.3968	361	193.2	.3410	432	203.2	25.0500	514	213.2	.6444	609
183.4	.4689	362	193.4	.4273	433	203.4	.1528	516	213.4	.7662	611
183.6	.5413	363	193.6	.5139	435	203.6	.2559	518	213.6	.8885	613
183.8	.6139	364	193.8	.6008	436	203.8	.3594	519	213.8	31.0111	615
184.0	16.6868	366	194.0	20.6881	438	204.0	25.4633	521	214.0	31.1342	
184.2	.7599	367	194.2	.7756	439	204.2	.5675	523			
184.4	.8334	368	194.4	.8635	441	204.4	.6720	525			
184.6	.9071	370	194.6	.9517	442	204.6	.7769	527			
184.8	.9810	371	194.8	21.0402	444	204.8	.8822	528			
185.0	17.0552	373	195.0	21.1289	446	205.0	25.9878	530			
185.2	.1297	374	195.2	.2180	447	205.2	26.0939	532			
185.4	.2045	375	195.4	.3074	449	205.4	.2002	534			
185.6	.2795	377	195.6	.3971	450	205.6	.3070	536			
185.8	.3548	378	195.8	.4872	452	205.8	.4141	537			
186.0	17.4304	379	196.0	21.5776	454	206.0	26.5215	539			
186.2	.5063	381	196.2	.6683	455	206.2	.6294	541			
186.4	.5824	382	196.4	.7593	457	206.4	.7376	543			
186.6	.6588	384	196.6	.8506	458	206.6	.8461	545			
186.8	.7355	385	196.8	.9422	460	206.8	.9551	547			
187.0	17.8125	386	197.0	22.0342	462	207.0	27.0644	549			
187.2	.8897	388	197.2	.1265	463	207.2	.1741	550			
187.4	.9672	389	197.4	.2191	465	207.4	.2842	552			
187.6	18.0451	391	197.6	.3120	466	207.6	.3946	554			
187.8	.1231	392	197.8	.4053	468	207.8	.5054	556			
188.0	18.2015	393	198.0	22.4989	470	208.0	27.6166	558			
188.2	.2802	395	198.2	.5928	471	208.2	.7282	560			
188.4	.3591	396	198.4	.6871	473	208.4	.8402	562			
188.6	.4383	398	198.6	.7816	475	208.6	.9525	564			
188.8	.5178	399	198.8	.8765	476	208.8	28.0652	566			
189.0	18.5976	400	199.0	22.9718	478	209.0	28.1784	568			
189.2	.6777	402	199.2	23.0673	480	209.2	.2919	569			
189.4	.7581	403	199.4	.1632	481	209.4	.4057	571			
189.6	.8388	405	199.6	.2595	483	209.6	.5200	573			
189.8	.9197	406	199.8	.3560	485	209.8	.6346	575			

SMITHSONIAN TABLES.

TABLE 36.

PRESSURE OF AQUEOUS VAPOR.
(Broch.)

METRIC MEASURES.

Temperature.	0°.0	0°.1	0°.2	0°.3	0°.4	0°.5	0°.6	0°.7	0°.8	0°.9
C.	mm.	mm.	mm.	mm.	mm.	mm.	mm.	mm.	mm.	mm.
−29°	0.42	0.41	0.41	0.41	0.40	0.40	0.40	0.39	0.39	0.38
28	0.46	0.46	0.45	0.45	0.44	0.44	0.43	0.43	0.43	0.42
27	0.50	0.50	0.50	0.49	0.49	0.48	0.48	0.47	0.47	0.46
26	0.55	0.55	0.54	0.54	0.53	0.53	0.52	0.52	0.51	0.51
−25	0.61	0.60	0.60	0.59	0.58	0.58	0.57	0.57	0.56	0.56
24	0.66	0.66	0.65	0.65	0.64	0.63	0.63	0.62	0.62	0.61
23	0.73	0.72	0.71	0.71	0.70	0.69	0.69	0.68	0.68	0.67
22	0.79	0.79	0.78	0.77	0.77	0.76	0.75	0.75	0.74	0.73
21	0.87	0.86	0.85	0.84	0.84	0.83	0.82	0.81	0.81	0.80
−20	0.94	0.94	0.93	0.92	0.91	0.90	0.90	0.89	0.88	0.87
19	1.03	1.02	1.01	1.00	0.99	0.99	0.98	0.97	0.96	0.95
18	1.12	1.11	1.10	1.09	1.08	1.07	1.06	1.06	1.05	1.04
17	1.22	1.21	1.20	1.19	1.18	1.17	1.16	1.15	1.14	1.13
16	1.32	1.31	1.30	1.29	1.28	1.27	1.26	1.25	1.24	1.23
−15	1.44	1.43	1.42	1.40	1.39	1.38	1.37	1.36	1.35	1.34
14	1.56	1.55	1.54	1.52	1.51	1.50	1.49	1.48	1.46	1.45
13	1.69	1.68	1.67	1.65	1.64	1.63	1.61	1.60	1.59	1.57
12	1.84	1.82	1.81	1.79	1.78	1.76	1.75	1.74	1.72	1.71
11	1.99	1.97	1.96	1.94	1.93	1.91	1.90	1.88	1.87	1.85
−10	2.15	2.13	2.12	2.10	2.08	2.07	2.05	2.04	2.02	2.00
9	2.33	2.31	2.29	2.27	2.26	2.24	2.22	2.20	2.19	2.17
8	2.51	2.50	2.48	2.46	2.44	2.42	2.40	2.38	2.36	2.34
7	2.72	2.69	2.67	2.65	2.63	2.61	2.59	2.57	2.55	2.53
6	2.93	2.91	2.89	2.86	2.84	2.81	2.80	2.78	2.76	2.74
−5	3.16	3.14	3.11	3.09	3.07	3.04	3.02	3.00	2.98	2.95
4	3.41	3.38	3.36	3.33	3.31	3.28	3.26	3.23	3.21	3.18
3	3.67	3.64	3.62	3.59	3.56	3.54	3.51	3.48	3.46	3.43
2	3.95	3.92	3.89	3.86	3.84	3.81	3.78	3.75	3.72	3.70
1	4.25	4.22	4.19	4.16	4.13	4.10	4.07	4.04	4.01	3.98
−0	4.57	4.54	4.50	4.47	4.44	4.41	4.37	4.34	4.31	4.28
Values for temperatures between 0° and 45° are given in Table 43.										
+45°	71.36	71.73	72.10	72.48	72.85	72.23	73.60	73.98	74.36	74.75
46	75.13	75.52	75.91	76.30	76.69	77.08	77.47	77.87	78.27	78.67
47	79.07	79.47	79.88	80.29	80.70	81.11	81.52	81.93	82.35	82.77
48	83.19	83.61	84.03	84.46	84.89	84.32	85.75	86.18	86.61	87.05
49	87.49	87.93	88.37	88.81	89.26	89.71	90.16	90.61	91.06	91.52
50	91.98	92.44	92.90	93.36	93.83	94.30	94.77	95.24	95.71	96.19
51	96.66	97.14	97.63	98.11	98.60	99.08	99.57	100.07	100.56	101.06
52	101.55	102.05	102.56	103.06	103.57	104.08	104.59	105.10	105.62	106.14
53	106.65	107.18	107.70	108.23	108.76	109.29	109.82	110.35	110.89	111.43
54	111.97	112.52	113.06	113.61	114.16	114.72	115.27	115.83	116.39	116.95
55	117.52	118.08	118.65	119.22	119.80	120.37	120.95	121.53	122.12	122.70
56	123.29	123.88	124.48	125.07	125.67	126.27	126.87	127.48	128.09	128.70
57	129.31	129.92	130.54	131.16	131.79	132.41	133.04	133.67	134.30	134.94
58	135.58	136.22	136.86	137.50	138.15	138.80	139.46	140.11	140.77	141.43
59	142.10	142.76	143.43	144.11	144.78	145.46	146.14	146.82	147.51	148.19
60	148.88	149.58	150.27	150.97	151.68	152.38	153.09	153.80	154.51	155.23

TABLE 36.

PRESSURE OF AQUEOUS VAPOR.
(*Broch.*)
METRIC MEASURES.

Temperature.	0°.0	0°.1	0°.2	0°.3	0°.4	0°.5	0°.6	0°.7	0°.8	0°.9
C.	mm.	mm.	mm.	mm.	mm.	mm.	mm.	mm.	mm.	mm.
60°	148.88	149.58	150.27	150.97	151.68	152.38	153.09	153.80	154.51	155.23
61	155.95	156.67	157.39	158.12	158.85	159.58	160.32	161.06	161.80	162.54
62	163.29	164.04	164.79	165.55	166.31	167.07	167.83	168.60	169.37	170.15
63	170.92	171.70	172.49	173.27	174.06	174.85	175.65	176.45	177.25	178.05
64	178.86	179.67	180.48	181.30	182.12	182.94	183.77	184.60	185.43	186.26
65	187.10	187.94	188.79	189.64	190.49	191.34	192.20	193.06	193.93	194.80
66	195.67	196.54	197.42	198.30	199.18	200.07	200.96	201.86	202.75	203.65
67	204.56	205.47	206.38	207.29	208.21	209.13	210.06	210.98	211.92	212.85
68	213.79	214.73	215.68	216.63	217.58	218.54	219.50	220.46	221.43	222.40
69	223.37	224.35	225.33	226.31	227.30	228.29	229.29	230.29	231.29	232.30
70	233.31	234.32	235.34	236.36	237.39	238.42	239.45	240.48	241.52	242.57
71	243.62	244.67	245.72	246.78	247.85	248.91	249.98	251.06	252.14	253.22
72	254.30	255.40	256.49	257.59	258.69	259.80	260.91	262.02	263.14	264.26
73	265.38	266.51	267.65	268.79	269.93	271.08	272.23	273.38	274.54	275.70
74	276.87	278.04	279.21	280.39	281.58	282.76	283.95	285.15	286.35	287.56
75	288.76	289.98	291.19	292.42	293.64	294.87	296.11	297.34	298.59	299.83
76	301.09	302.34	303.60	304.87	306.14	307.41	308.69	309.97	311.26	312.55
77	313.85	315.15	316.45	317.76	319.07	320.39	321.72	323.04	324.38	325.71
78	327.05	328.40	329.75	331.11	332.47	333.83	335.20	336.58	337.95	339.34
79	340.73	342.12	343.52	344.92	346.33	347.74	349.16	350.58	352.01	353.44
80	354.87	356.31	357.76	359.21	360.67	362.13	363.59	365.07	366.54	368.02
81	369.51	371.00	372.49	374.00	375.50	377.01	378.53	380.05	381.58	383.11
82	384.64	386.18	387.73	389.28	390.84	392.40	393.97	395.54	397.12	398.70
83	400.29	401.89	403.49	405.09	406.70	408.32	409.94	411.56	413.19	414.83
84	416.47	418.12	419.77	421.43	423.09	424.76	426.44	428.12	429.81	431.50
85	433.19	434.90	436.60	438.32	440.04	441.76	443.49	445.23	446.97	448.72
86	450.47	452.23	454.00	455.77	457.54	459.33	461.11	462.91	464.71	466.51
87	468.32	470.14	471.96	473.79	475.63	477.47	479.32	481.17	483.03	484.89
88	486.76	488.64	490.52	492.41	494.31	496.21	498.12	500.03	501.95	503.87
89	505.81	507.74	509.69	511.64	513.60	515.56	517.53	519.50	521.48	523.47
90	525.47	527.47	529.48	531.49	533.51	535.54	537.57	539.61	541.65	543.71
91	545.77	547.83	549.90	551.98	554.07	556.16	558.26	560.36	562.47	564.59
92	566.71	568.85	570.98	573.13	575.28	577.44	579.61	581.78	583.96	586.14
93	588.33	590.53	592.74	594.95	597.17	599.40	601.64	603.88	606.13	608.38
94	610.64	612.91	615.19	617.47	619.76	622.06	624.37	626.68	629.00	631.32
95	633.66	636.00	638.35	640.70	643.06	645.43	647.81	650.20	652.59	654.99
96	657.40	659.81	662.23	664.66	667.10	669.54	672.00	674.45	676.92	679.40
97	681.88	684.37	686.87	689.37	691.89	694.41	696.93	699.47	702.02	704.57
98	707.13	709.69	712.27	714.85	717.44	720.04	722.65	725.27	727.89	730.52
99	733.16	735.81	738.46	741.13	743.80	746.48	749.17	751.86	754.57	757.28
100	760.00	762.73	765.47	768.21	770.97	773.73	776.50	779.28	782.07	784.86
101	787.67									

SMITHSONIAN TABLES.

TABLE 37.
PRESSURE OF AQUEOUS VAPOR AT LOW TEMPERATURES.
(*C. F. Marvin.*)
ENGLISH AND METRIC MEASURES.

Temperatures.	0°.0		0°.2		0°.4		0°.6		0°.8	
F.	Inch.	mm.	Inch.	mm.	Inch.	mm.	Inch.	mm.	Inch.	mm.
−60°	0.0010	0.026								
59	.0011	.028	0.0011	0.028	0.0011	0.027	0.0011	0.027	0.0010	0.026
58	.0012	.030	.0012	.030	.0011	.029	.0011	.029	.0011	.028
57	.0013	.032	.0013	.032	.0012	.031	.0012	.031	.0012	.030
56	.0013	.034	.0013	.034	.0013	.033	.0013	.033	.0013	.032
−55°	0.0015	0.037	0.0014	0.036	0.0014	0.036	0.0014	0.035	0.0014	0.035
54	.0016	.040	.0015	.039	.0015	.039	.0015	.038	.0015	.037
53	.0017	.043	.0017	.042	.0017	.042	.0016	.041	.0016	.040
52	.0018	.046	.0018	.045	.0018	.045	.0017	.044	.0017	.043
51	.0019	.049	.0019	.048	.0019	.048	.0019	.047	.0018	.046
−50°	0.0021	0.053	0.0020	0.052	0.0020	0.051	0.0020	0.051	0.0020	0.050
49	.0022	.057	.0022	.056	.0022	.055	.0022	.055	.0021	.054
48	.0024	.061	.0024	.060	.0023	.059	.0023	.059	.0023	.058
47	.0026	.065	.0025	.064	.0025	.063	.0025	.063	.0024	.062
46	.0027	.069	.0027	.068	.0027	.068	.0026	.067	.0026	.066
−45°	0.0029	0.074	0.0029	0.073	0.0028	0.072	0.0028	0.071	0.0028	0.070
44	.0031	.079	.0031	.078	.0030	.077	.0030	.076	.0030	.075
43	.0033	.084	.0033	.083	.0032	.082	.0032	.081	.0031	.080
42	.0035	.089	.0035	.088	.0034	.087	.0034	.086	.0033	.085
41	.0037	.094	.0037	.093	.0036	.092	.0036	.091	.0035	.090
−40°	0.0039	0.100	0.0039	0.098	0.0038	0.097	0.0038	0.096	0.0037	0.095
39	.0041	.105	.0041	.104	.0041	.103	.0040	.102	.0040	.101
38	.0044	.111	.0043	.109	.0043	.108	.0042	.107	.0042	.106
37	.0046	.117	.0045	.115	.0045	.114	.0044	.113	.0044	.112
36	.0048	.123	.0048	.121	.0047	.120	.0047	.119	.0046	.118
−35°	0.0051	0.130	0.0051	0.129	0.0050	0.127	0.0050	0.126	0.0049	0.124
34	.0054	.138	.0054	.136	.0053	.135	.0052	.133	.0052	.132
33	.0057	.146	.0057	.144	.0056	.142	.0056	.141	.0055	.139
32	.0061	.155	.0060	.153	.0059	.151	.0059	.149	.0058	.147
31	.0065	.165	.0064	.163	.0063	.161	.0063	.159	.0062	.157
−30°	0.0069	0.176	0.0069	0.174	0.0067	0.171	0.0067	0.169	0.0066	0.167
29	.0074	.187	.0073	.185	.0072	.183	.0071	.180	.0070	.178
28	.0078	.199	.0078	.197	.0077	.195	.0076	.192	.0075	.190
27	.0083	.212	.0083	.210	.0081	.207	.0080	.204	.0080	.202
26	.0089	.225	.0088	.223	.0087	.220	.0085	.217	.0085	.215
−25°	0.0094	0.239	0.0093	0.236	0.0092	0.233	0.0091	0.230	0.0089	0.227
24	.0100	.253	.0098	.250	.0097	.247	.0096	.244	.0095	.242
23	.0106	.268	.0104	.265	.0103	.262	.0102	.259	.0101	.256
22	.0112	.284	.0111	.281	.0109	.278	.0108	.274	.0107	.271
21	.0119	.301	.0117	.297	.0116	.294	.0115	.291	.0113	.287
−20°	0.0126	0.319	0.0124	0.315	0.0122	0.311	0.0121	0.308	0.0120	0.304
19	.0133	.338	.0131	.334	.0130	.330	.0128	.326	.0127	.322
18	.0141	.358	.0139	.354	.0138	.350	.0136	.346	.0135	.342
17	.0150	.380	.0148	.375	.0146	.371	.0144	.366	.0143	.362
16	.0159	.403	.0157	.398	.0155	.393	.0153	.389	.0151	.384
−15°	0.0168	0.427	0.0166	0.422	0.0164	0.417	0.0162	0.412	0.0160	0.407

TABLE 37.
PRESSURE OF AQUEOUS VAPOR AT LOW TEMPERATURES.
(C. F. Marvin.)
ENGLISH AND METRIC MEASURES.

Tempera-ture.	0°.0		0°.2		0°.4		0°.6		0°.8	
F.	Inch.	mm.	Inch.	mm.	Inch.	mm.	Inch.	mm.	Inch.	mm.
−15°	0.0168	0.427	0.0166	0.422	0.0164	0.417	0.0162	0.412	0.0160	0.407
14	.0178	.452	.0176	.447	.0174	.442	.0172	.437	.0170	.432
13	.0188	.478	.0186	.473	.0184	.468	.0182	.462	.0180	.457
12	.0199	.505	.0196	.499	.0194	.494	.0192	.488	.0190	.483
11	.0210	.534	.0208	.528	.0206	.522	.0203	.516	.0201	.510
−10	0.0222	0.564	0.0220	0.558	0.0217	0.552	0.0215	0.546	0.0213	0.540
9	.0234	.595	.0232	.588	.0229	.582	.0227	.576	.0224	.570
8	.0247	.627	.0244	.620	.0242	.614	.0239	.607	.0237	.601
7	.0260	.661	.0257	.654	.0255	.647	.0252	.640	.0249	.633
6	.0275	.698	.0272	.691	.0269	.683	.0266	.676	.0263	.669
−5	0.0291	0.738	0.0287	0.730	0.0284	0.722	0.0281	0.714	0.0278	0.706
4	.0307	.781	.0304	.772	.0301	.764	.0297	.755	.0294	.747
3	.0325	.826	.0322	.817	.0318	.808	.0315	.799	.0311	.790
2	.0344	.873	.0340	.863	.0336	.854	.0332	.844	.0329	.835
1	.0363	.922	.0359	.912	.0355	.902	.0351	.892	.0347	.882
−0	.0383	.972	.0379	.962	.0375	.952	.0371	.942	.0367	.932
+0	0.0383	0.972	0.0387	0.982	0.0391	0.992	0.0394	1.002	0.0398	1.012
1	.0403	1.023	.0407	1.033	.0411	1.043	.0415	1.054	.0419	1.064
2	.0423	1.075	.0428	1.086	.0431	1.096	.0436	1.107	.0440	1.118
3	.0444	1.129	.0449	1.140	.0453	1.151	.0458	1.163	.0462	1.174
4	.0467	1.186	.0472	1.198	.0476	1.210	.0481	1.222	.0486	1.234
5	0.0491	1.246	0.0495	1.258	0.0500	1.271	0.0505	1.283	0.0510	1.296
6	.0515	1.309	.0520	1.322	.0526	1.335	.0531	1.349	.0536	1.362
7	.0542	1.376	.0547	1.390	.0553	1.404	.0558	1.418	.0564	1.433
8	.0570	1.447	.0576	1.462	.0582	1.477	.0587	1.492	.0594	1.508
9	.0600	1.523	.0606	1.539	.0612	1.555	.0618	1.571	.0625	1.587
10	0.0631	1.603	0.0638	1.620	0.0644	1.636	0.0651	1.653	0.0657	1.670
11	.0665	1.688	.0671	1.705	.0678	1.722	.0685	1.740	.0692	1.758
12	.0699	1.776	.0706	1.794	.0713	1.812	.0720	1.830	.0728	1.848
13	.0735	1.867	.0742	1.885	.0750	1.904	.0757	1.923	.0765	1.942
14	.0772	1.961	.0780	1.980	.0787	1.999	.0794	2.018	.0802	2.038
15	0.0810	2.058	0.0818	2.078	0.0826	2.098	0.0834	2.118	0.0842	2.138
16	.0850	2.158	.0857	2.178	.0866	2.199	.0874	2.220	.0882	2.241
17	.0891	2.262	.0899	2.283	.0907	2.305	.0916	2.327	.0925	2.349
18	.0933	2.371	.0942	2.393	.0951	2.416	.0960	2.439	.0969	2.462
19	.0979	2.486	.0988	2.510	.0998	2.534	.1007	2.558	.1017	2.582
20	0.1026	2.607	0.1036	2.632	0.1046	2.657	0.1056	2.683	0.1067	2.709
21	.1077	2.735	.1087	2.761	.1098	2.788	.1108	2.815	.1119	2.842
22	.1130	2.869	.1141	2.897	.1152	2.925	.1163	2.953	.1174	2.981
23	.1185	3.009	.1196	3.037	.1207	3.066	.1219	3.095	.1230	3.125
24	.1242	3.155	.1254	3.185	.1266	3.215	.1278	3.245	.1290	3.276
25	0.1302	3.307	0.1314	3.338	0.1327	3.370	0.1339	3.402	0.1352	3.434
26	.1365	3.466	.1377	3.498	.1390	3.531	.1403	3.564	.1416	3.597
27	.1430	3.631	.1443	3.665	.1456	3.699	.1470	3.733	.1483	3.768
28	.1497	3.803	.1511	3.838	.1525	3.874	.1539	3.910	.1554	3.946
29	.1568	3.982	.1582	4.018	.1596	4.055	.1611	4.093	.1626	4.131
30	0.1641	4.169	0.1656	4.207	0.1671	4.245	0.1687	4.284	0.1702	4.324
31	.1718	4.364	.1734	4.404	.1750	4.444	.1766	4.485	.1782	4.526
32	.1798	4.568								

SMITHSONIAN TABLES.

TABLE 38.

WEIGHT OF AQUEOUS VAPOR IN A CUBIC FOOT OF SATURATED AIR.

ENGLISH MEASURES.

Temperature.	0°.0	0°.5	Diff. for 0°.1	Temperature.	0°.0	0°.5	Diff. for 0°.1	Temperature.	0°.0	0°.5	Diff. for 0°.1
F.	Grains troy.	Grains troy.		F.	Grains troy.	Grains troy.		F.	Grains troy.	Grains troy.	
−19°	0.230	0.224	1	26°	1.675	1.709	7	71°	8.240	8.372	27
−18	.242	.236	1	27	1.743	1.777	7	72	8.508	8.644	27
−17	.254	.248	1	28	1.812	1.847	7	73	8.782	8.923	28
−16	.267	.260	1	29	1.882	1.919	7	74	9.066	9.210	29
−15	0.280	0.273	1	30	1.956	1.995	8	75	9.356	9.504	30
−14	.294	.286	1	31	2.034	2.073	8	76	9.655	9.807	31
−13	.309	.301	1	32	2.113	2.153	8	77	9.962	10.118	32
−12	.324	.316	2	33	2.194	2.236	8	78	10.277	10.438	33
−11	.340	.332	2	34	2.279	2.322	9	79	10.601	10.766	33
−10	0.356	0.348	2	35	2.366	2.411	9	80	10.934	11.103	34
−9	.373	.365	2	36	2.457	2.503	9	81	11.275	11.450	35
−8	.391	.382	2	37	2.550	2.598	10	82	11.626	11.805	36
−7	.411	.400	2	38	2.646	2.695	10	83	11.987	12.170	37
−6	.430	.420	2	39	2.746	2.797	10	84	12.356	12.545	38
−5	0.450	0.439	2	40	2.849	2.901	11	85	12.736	12.930	39
−4	.471	.460	2	41	2.955	3.009	11	86	13.127	13.325	40
−3	.493	.482	2	42	3.064	3.120	11	87	13.526	13.730	41
−2	.516	.504	2	43	3.177	3.235	12	88	13.937	14.146	42
−1	.540	.528	2	44	3.294	3.354	12	89	14.359	14.573	43
−0	.564	.552	2								
+0	0.564	0.577	3	45	3.414	3.477	12	90	14.790	15.011	44
1	.590	.603	3	46	3.539	3.603	13	91	15.234	15.460	45
2	.617	.630	3	47	3.667	3.733	13	92	15.689	15.920	47
3	.645	.659	3	48	3.800	3.868	14	93	16.155	16.393	48
4	.674	.689	3	49	3.936	4.006	14	94	16.634	16.877	49
5	0.705	0.719	3	50	4.076	4.148	15	95	17.124	17.374	50
6	.735	.751	3	51	4.222	4.296	15	96	17.626	17.883	52
7	.767	.784	3	52	4.372	4.448	15	97	18.142	18.404	53
8	.801	.819	4	53	4.526	4.604	16	98	18.671	18.940	54
9	.837	.854	4	54	4.685	4.766	16	99	19.212	19.487	55
10	0.873	0.891	4	55	4.849	4.933	17	100	19.766	20.049	57
11	.910	.930	4	56	5.016	5.103	17	101	20.335	20.624	58
12	.950	.970	4	57	5.191	5.280	18	102	20.917	21.214	60
13	.991	1.011	4	58	5.370	5.462	18	103	21.514	21.817	61
14	1.033	1.054	4	59	5.555	5.649	19	104	22.125	22.436	62
15	1.077	1.098	5	60	5.745	5.842	20	105	22.750	23.070	64
16	1.122	1.144	5	61	5.941	6.040	20	106	23.392	23.718	66
17	1.169	1.193	5	62	6.142	6.245	21	107	24.048	24.382	67
18	1.217	1.242	5	63	6.349	6.456	21	108	24.720	25.062	69
19	1.268	1.294	5	64	6.563	6.672	22	109	25.408	25.758	70
20	1.321	1.347	5	65	6.782	6.895	23	110	26.112	26.470	72
21	1.374	1.402	6	66	7.009	7.124	23	111	26.832	27.199	74
22	1.430	1.459	6	67	7.241	7.360	24	112	27.570	27.946	75
23	1.488	1.518	6	68	7.480	7.602	25	113	28.325	28.708	77
24	1.549	1.580	6	69	7.726	7.852	25	114	29.096	29.489	79
25	1.611	1.643	6	70	7.980	8.109	26	115	29.887		

SMITHSONIAN TABLES.

TABLE 39.

WEIGHT OF AQUEOUS VAPOR IN A CUBIC METRE OF SATURATED AIR.

METRIC MEASURES.

Temperature.		Temperature.	0°.0	0°.5	Temperature.	0°.0	0°.2	0°.4	0°.6	0°.8
C.	Gram's.	C.	Gram's.	Gram's.	C.	Gram's.	Gram's.	Gram's.	Gram's.	Gram's.
−29°	0.496	−17°	1.375	1.321	−5°	3.407	3.359	3.311	3.263	3.217
28	.542	16	1.489	1.432	4	3.659	3.607	3.556	3.506	3.456
27	.593	15	1.611	1.549	3	3.926	3.871	3.817	3.763	3.711
26	.647	14	1.742	1.676	2	4.211	4.152	4.095	4.038	3.982
25	.706	13	1.882	1.811	1	4.513	4.451	4.390	4.329	4.270
24	.770	12	2.032	1.956	−0	4.835	4.769	4.704	4.640	4.576
−23	0.839	−11	2.192	2.111	+0	4.835	4.901	4.969	5.037	5.106
22	.913	10	2.363	2.276	1	5.176	5.247	5.318	5.391	5.464
21	.992	9	2.546	2.453	2	5.538	5.613	5.689	5.766	5.844
20	1.078	8	2.741	2.642	3	5.922	6.002	6.082	6.164	6.246
19	1.170	7	2.949	2.843	4	6.430	6.414	6.499	6.585	6.673
−18	1.269	−6	3.171	3.058	+5	6.761	6.851	6.941	7.033	7.125

Temperature.	0°.0	0°.1	0°.2	0°.3	0°.4	0°.5	0°.6	0°.7	0°.8	0°.9
C.	Gram's.	Gram's.	Gram's.	Gram's.	Gram's.	Gram's.	Gram's.	Gram's.	Gram's.	Gram's.
+6°	7.219	7.266	7.313	7.361	7.409	7.457	7.506	7.555	7.614	7.653
7	7.703	7.753	7.803	7.853	7.904	7.955	8.007	8.058	8.110	8.162
8	8.215	8.268	8.321	8.374	8.428	8.482	8.536	8.591	8.646	8.701
9	8.757	8.813	8.869	8.926	8.982	9.039	9.097	9.155	9.213	9.271
10	9.330	9.389	9.448	9.508	9.568	9.628	9.689	9.750	9.811	9.873
11	9.935	9.997	10.060	10.123	10.186	10.250	10.314	10.378	10.443	10.508
12	10.574	10.640	10.706	10.773	10.840	10.907	10.975	11.043	11.111	11.180
13	11.249	11.318	11.388	11.458	11.529	11.600	11.672	11.744	11.816	11.888
14	11.961	12.035	12.108	12.182	12.257	12.332	12.407	12.483	12.559	12.635
15	12.712	12.790	12.867	12.945	13.024	13.103	13.182	13.262	13.342	13.423
16	13.505	13.586	13.668	13.750	13.833	13.916	14.000	14.085	14.169	14.254
17	14.339	14.425	14.511	14.598	14.685	14.773	14.861	14.950	15.039	15.128
18	15.218	15.308	15.399	15.491	15.583	15.675	15.768	15.861	15.955	16.049
19	16.144	16.239	16.335	16.431	16.528	16.625	16.723	16.821	16.920	17.019
20	17.118	17.218	17.319	17.420	17.522	17.624	17.727	17.830	17.934	18.039
21	18.143	18.248	18.353	18.460	18.568	18.676	18.784	18.893	19.002	19.111
22	19.222	19.332	19.444	19.556	19.668	19.781	19.895	20.009	20.124	20.239
23	20.355	20.471	20.588	20.706	20.824	20.943	21.062	21.182	21.303	21.424
24	21.546	21.668	21.791	21.914	22.038	22.163	22.287	22.414	22.541	22.668
25	22.796	22.925	23.054	23.184	23.314	23.445	23.577	23.709	23.842	23.975
26	24.109	24.244	24.380	24.516	24.653	24.790	24.928	25.067	25.207	25.347
27	25.487	25.629	25.771	25.914	26.058	26.202	26.347	26.492	26.639	26.786
28	26.933	27.082	27.231	27.381	27.531	27.682	27.834	27.988	28.142	28.295
29	28.450	28.605	28.762	28.919	29.077	29.235	29.394	29.555	29.715	29.877
30	30.039	30.202	30.366	30.530	30.696	30.862	31.029	31.197	31.365	31.534
31	31.704	31.875	32.047	32.219	32.392	32.567	32.741	32.917	33.094	33.271
32	33.449	33.628	33.807	33.988	34.169	34.351	34.534	34.718	34.903	35.089
33	35.275	35.462	35.651	35.840	36.030	36.220	36.412	36.604	36.798	36.992
34	37.187	37.383	37.580	37.777	37.976	38.176	38.376	38.577	38.780	38.983
35	39.187	39.390	39.598	39.805	40.012	40.221	40.431	40.641	40.853	41.065
36	41.279	41.493	41.708	41.924	42.142	42.360	42.579	42.799	43.020	43.242
37	43.465	43.690	43.914	44.140	44.367	44.596	44.825	45.054	45.286	45.518
38	45.751	45.985	46.220	46.456	46.693	46.931	47.171	47.411	47.653	47.895
39	48.138	48.385	48.628	48.875	49.123	49.372	49.621	49.872	50.124	50.377

SMITHSONIAN TABLES.

TABLE 40.

REDUCTION OF PSYCHROMETRIC OBSERVATIONS.

ENGLISH MEASURES.

Pressure of Aqueous Vapor.

Temperature.	0°	1°	2°	3°	4°	5°	6°	7°	8°	9°
F.	Inch.	Inch.	Inch.	Inch.	Inch.	Inch.	Inch.	Inch.	Inch.	Inch.
−30°	0.007	0.006	0.006	0.006	0.005	0.005	0.005	0.005	0.004	0.004
−20	.013	.012	.011	.011	.010	.009	.009	.008	.008	.007
−10	.022	.021	.020	.019	.018	.017	.016	.015	.014	.013
−0	.038	.036	.034	.033	.031	.029	.027	.026	.025	.023
+0	0.038	0.040	0.042	0.044	0.047	0.049	0.052	0.054	0.057	0.060
10	.063	.066	.070	.074	.077	.081	.085	.089	.093	.098
20	.103	.108	.113	.118	.124	.130	.136	.143	.150	.157

Temperature.	0°.0	0°.1	0°.2	0°.3	0°.4	0°.5	0°.6	0°.7	0°.8	0°.9
F.	Inch.	Inch.	Inch.	Inch.	Inch.	Inch.	Inch.	Inch.	Inch.	Inch.
+30°	0.164	0.165	0.166	0.166	0.167	0.168	0.169	0.169	0.170	0.171
31	.172	.173	.173	.174	.175	.176	.177	.177	.178	.179
32	.180	.181	.181	.182	.183	.184	.184	.185	.186	.186
33	.187	.188	.189	.190	.190	.191	.192	.193	.193	.194
34	.195	.196	.196	.197	.198	.199	.200	.200	.201	.202
35	0.203	0.204	0.204	0.205	0.206	0.207	0.208	0.208	0.209	0.210
36	.211	.212	.213	.213	.214	.215	.216	.217	.218	.219
37	.219	.220	.221	.222	.223	.224	.225	.225	.226	.227
38	.228	.229	.230	.231	.232	.233	.233	.234	.235	.236
39	.237	.238	.239	.240	.241	.242	.243	.244	.245	.246
40	0.247	0.247	0.248	0.249	0.250	0.251	0.252	0.253	0.254	0.255
41	.256	.257	.258	.259	.260	.261	.262	.263	.264	.265
42	.266	.267	.268	.269	.270	.271	.272	.273	.274	.276
43	.277	.278	.279	.280	.281	.282	.283	.284	.285	.286
44	.287	.288	.289	.291	.292	.293	.294	.295	.296	.297
45	0.298	0.300	0.301	0.302	0.303	0.304	0.305	0.306	0.308	0.309
46	.310	.311	.312	.313	.315	.316	.317	.318	.319	.321
47	.322	.323	.324	.325	.327	.328	.329	.330	.332	.333
48	.334	.335	.337	.338	.339	.340	.342	.343	.344	.345
49	.347	.348	.349	.351	.352	.353	.355	.356	.357	.358
50	0.360	0.361	0.362	0.364	0.365	0.367	0.368	0.369	0.371	0.372
51	.373	.375	.376	.377	.379	.380	.382	.383	.384	.386
52	.387	.389	.390	.392	.393	.394	.396	.397	.399	.400
53	.402	.403	.405	.406	.408	.409	.411	.412	.414	.415
54	.417	.418	.420	.421	.423	.424	.426	.427	.429	.431
55	0.432	0.434	0.435	0.437	0.438	0.440	0.442	0.443	0.445	0.446
56	.448	.450	.451	.453	.455	.456	.458	.460	.461	.463
57	.465	.466	.468	.470	.471	.473	.475	.476	.478	.480
58	.482	.483	.485	.487	.488	.490	.492	.494	.495	.497
59	.499	.501	.503	.504	.506	.508	.510	.512	.513	.515
60	0.517	0.519	0.521	0.523	0.524	0.526	0.528	0.530	0.532	0.534
61	.536	.538	.539	.541	.543	.545	.547	.549	.551	.553
62	.555	.557	.559	.561	.563	.565	.567	.569	.571	.573
63	.575	.577	.579	.581	.583	.585	.587	.589	.591	.593
64	.595	.597	.599	.601	.604	.606	.608	.610	.612	.614
65	0.616	0.618	0.621	0.623	0.625	0.627	0.629	0.631	0.634	0.636
66	.638	.640	.643	.645	.647	.649	.651	.654	.656	.658
67	.661	.663	.665	.667	.670	.672	.674	.677	.679	.681
68	.684	.686	.688	.691	.693	.695	.698	.700	.703	.705
69	.707	.710	.712	.715	.717	.720	.722	.725	.727	.729

SMITHSONIAN TABLES.

TABLE 40.

REDUCTION OF PSYCHROMETRIC OBSERVATIONS.
ENGLISH MEASURES.
Pressure of Aqueous Vapor.

Temperature.	0°.0	0°.1	0°.2	0°.3	0°.4	0°.5	0°.6	0°.7	0°.8	0°.9
F.	Inch.	Inch.	Inch.	Inch.	Inch.	Inch.	Inch.	Inch.	Inch.	Inch.
70°	0.732	0.734	0.737	0.739	0.742	0.744	0.747	0.750	0.752	0.755
71	0.757	0.760	0.762	0.765	0.768	0.770	0.773	0.775	0.778	0.781
72	0.783	0.786	0.789	0.791	0.794	0.797	0.799	0.802	0.805	0.807
73	0.810	0.813	0.816	0.818	0.821	0.824	0.827	0.830	0.832	0.835
74	0.838	0.841	0.843	0.846	0.849	0.852	0.855	0.858	0.861	0.863
75	0.866	0.869	0.872	0.875	0.878	0.881	0.884	0.887	0.890	0.893
76	0.896	0.899	0.902	0.905	0.908	0.911	0.914	0.917	0.920	0.923
77	0.926	0.929	0.932	0.935	0.938	0.941	0.944	0.948	0.951	0.954
78	0.957	0.960	0.963	0.966	0.970	0.973	0.976	0.979	0.982	0.986
79	0.989	0.992	0.995	0.999	1.002	1.005	1.009	1.012	1.015	1.019
80	1.022	1.025	1.029	1.032	1.035	1.039	1.042	1.046	1.049	1.052
81	1.056	1.059	1.063	1.066	1.070	1.073	1.077	1.080	1.084	1.087
82	1.091	1.094	1.098	1.101	1.105	1.109	1.112	1.116	1.119	1.123
83	1.127	1.130	1.134	1.138	1.141	1.145	1.149	1.152	1.156	1.160
84	1.163	1.167	1.171	1.175	1.179	1.182	1.186	1.190	1.194	1.198
85	1.201	1.205	1.209	1.213	1.217	1.221	1.225	1.229	1.233	1.237
86	1.241	1.245	1.248	1.253	1.256	1.260	1.264	1.269	1.273	1.277
87	1.281	1.285	1.289	1.293	1.297	1.301	1.305	1.310	1.314	1.318
88	1.322	1.326	1.330	1.335	1.339	1.343	1.347	1.352	1.356	1.360
89	1.364	1.369	1.373	1.377	1.382	1.386	1.390	1.395	1.399	1.404
90	1.408	1.413	1.417	1.421	1.426	1.430	1.435	1.439	1.444	1.448
91	1.453	1.458	1.462	1.467	1.471	1.476	1.480	1.485	1.490	1.494
92	1.499	1.504	1.508	1.513	1.518	1.523	1.527	1.532	1.537	1.542
93	1.546	1.551	1.556	1.561	1.566	1.571	1.576	1.580	1.585	1.590
94	1.595	1.600	1.605	1.610	1.615	1.620	1.625	1.630	1.635	1.640
95	1.645	1.650	1.655	1.660	1.665	1.671	1.676	1.681	1.686	1.691
96	1.696	1.702	1.707	1.712	1.717	1.723	1.728	1.733	1.738	1.744
97	1.749	1.755	1.760	1.765	1.771	1.776	1.781	1.787	1.792	1.798
98	1.803	1.809	1.814	1.820	1.825	1.831	1.837	1.842	1.848	1.853
99	1.859	1.865	1.870	1.876	1.882	1.887	1.893	1.899	1.905	1.910
100	1.916	1.922	1.928	1.934	1.939	1.945	1.951	1.957	1.963	1.969
101	1.975	1.981	1.987	1.993	1.999	2.005	2.011	2.017	2.023	2.029
102	2.035	2.041	2.047	2.053	2.059	2.066	2.072	2.078	2.084	2.090
103	2.097	2.103	2.109	2.116	2.122	2.128	2.134	2.141	2.147	2.154
104	2.160	2.166	2.173	2.179	2.186	2.192	2.199	2.205	2.212	2.219
105	2.225	2.232	2.238	2.245	2.252	2.258	2.265	2.272	2.278	2.285
106	2.292	2.299	2.305	2.312	2.319	2.326	2.333	2.340	2.346	2.353
107	2.360	2.367	2.374	2.381	2.388	2.395	2.402	2.409	2.416	2.423
108	2.431	2.438	2.445	2.452	2.459	2.466	2.474	2.481	2.488	2.495
109	2.503	2.510	2.517	2.525	2.532	2.539	2.547	2.554	2.562	2.569
110	2.576	2.584	2.591	2.599	2.607	2.614	2.622	2.629	2.637	2.645
111	2.652	2.660	2.668	2.675	2.683	2.691	2.699	2.706	2.714	2.722
112	2.730	2.738	2.746	2.754	2.762	2.770	2.777	2.785	2.793	2.801
113	2.810	2.818	2.826	2.834	2.842	2.850	2.858	2.866	2.875	2.883
114	2.891	2.899	2.908	2.916	2.924	2.933	2.941	2.950	2.958	2.966
115	2.975	2.983	2.992	3.000	3.009	3.017	3.026	3.035	3.043	3.052
116	3.061	3.069	3.078	3.087	3.095	3.104	3.113	3.122	3.131	3.140
117	3.148	3.157	3.166	3.175	3.184	3.193	3.202	3.211	3.220	3.229
118	3.239	3.248	3.257	3.266	3.275	3.284	3.294	3.303	3.312	3.321
119	3.331	3.340	3.349	3.359	3.368	3.378	3.387	3.397	3.406	3.416

TABLE 41.
REDUCTION OF PSYCHROMETRIC OBSERVATIONS
ENGLISH MEASURES.

Values of $0.000367 \, B \, (t-t_1)\left(1+\dfrac{t-t_1}{1571}\right)$.

$B =$ Barometric pressure.
$t =$ Temperature of the dry-bulb thermometer.
$t_1 =$ Temperature of the wet-bulb thermometer.

$t - t_1$	BAROMETRIC PRESSURE IN INCHES (B).										
	30.5	30.0	29.5	29.0	28.5	28.0	27.5	27.0	26.5	26.0	25.5	25.0
F.	Inch.	Inch.	Inch.	Inch.	Inch.	Inch.	Inch.	Inch.	Inch.	Inch.	Inch.	Inch.
1°	0.011	0.011	0.011	0.011	0.010	0.010	0.010	0.010	0.010	0.010	0.009	0.009
2	.022	.022	.022	.021	.021	.021	.020	.020	.019	.019	.019	.018
3	.034	.033	.033	.032	.031	.031	.030	.030	.029	.029	.028	.027
4	.045	.044	.043	.043	.042	.041	.040	.040	.039	.038	.038	.037
5	0.056	0.055	0.054	0.053	0.052	0.052	0.051	0.050	0.049	0.048	0.047	0.046
6	.067	.066	.065	.064	.063	.062	.061	.060	.059	.057	.056	.055
7	.079	.077	.076	.075	.073	.072	.071	.070	.068	.067	.066	.064
8	.090	.088	.087	.086	.084	.083	.081	.080	.078	.077	.075	.074
9	.101	.099	.098	.096	.095	.093	.091	.090	.088	.086	.085	.083
10	0.113	0.111	0.109	0.107	0.105	0.103	0.102	0.100	0.098	0.096	0.094	0.092
11	.124	.122	.120	.118	.116	.114	.112	.110	.108	.106	.104	.102
12	.135	.133	.131	.129	.126	.124	.122	.120	.118	.115	.113	.111
13	.147	.144	.142	.140	.137	.135	.132	.130	.127	.125	.123	.120
14	.158	.156	.153	.150	.148	.145	.143	.140	.137	.135	.132	.130
15	0.170	0.167	0.164	0.161	0.158	0.156	0.153	0.150	0.147	0.144	0.142	0.139
16	.181	.178	.175	.172	.169	.166	.163	.160	.157	.154	.151	.148
17	.192	.189	.186	.183	.180	.177	.173	.170	.167	.164	.161	.158
18	.204	.200	.197	.194	.190	.187	.184	.180	.177	.174	.170	.167
19	.215	.212	.208	.205	.201	.198	.194	.191	.187	.183	.180	.176
20	0.227	0.223	0.219	0.216	0.212	0.208	0.204	0.201	0.197	0.193	0.190	0.186
21	.238	.234	.230	.226	.223	.219	.215	.211	.207	.203	.199	.195
22	.250	.246	.242	.237	.233	.229	.225	.221	.217	.213	.209	.205
23	.261	.257	.253	.248	.244	.240	.236	.231	.227	.223	.218	.214
24	.273	.268	.264	.259	.255	.250	.246	.241	.237	.233	.228	.224
25	0.284	0.280	0.275	0.270	0.266	0.261	0.256	0.252	0.247	0.242	0.238	0.233
26	.296	.291	.286	.281	.277	.272	.267	.262	.257	.252	.247	.243
27	.307	.302	.297	.292	.287	.282	.277	.272	.267	.262	.257	.252
28	.319	.314	.309	.303	.298	.293	.288	.282	.277	.272	.267	.261
29	.331	.325	.320	.314	.309	.304	.298	.293	.287	.282	.276	.271
30	0.342	0.337	0.331	0.325	0.320	0.314	0.309	0.303	0.297	0.292	0.286	0.281
31	.354	.348	.342	.336	.331	.325	.319	.313	.307	.302	.296	.290
32	.365	.359	.354	.348	.342	.336	.330	.324	.318	.312	.306	.300
33	.377	.371	.365	.359	.352	.346	.340	.334	.328	.322	.315	.309
34	.389	.382	.376	.370	.363	.357	.351	.344	.338	.331	.325	.319
35	0.401	0.394	0.387	0.381	0.374	0.368	0.361	0.355	0.348	0.341	0.335	0.328
36	.412	.405	.399	.392	.385	.378	.372	.365	.358	.351	.345	.338
37	.424	.417	.410	.403	.396	.389	.382	.375	.368	.361	.354	.347
38	.436	.428	.421	.414	.407	.400	.393	.386	.379	.371	.364	.357
39	.447	.440	.433	.425	.418	.411	.403	.396	.389	.381	.374	.367
40	0.459	0.452	0.444	0.437	0.429	0.422	0.414	0.406	0.399	0.391	0.384	0.376

SMITHSONIAN TABLES.

TABLE 41.
REDUCTION OF PSYCHROMETRIC OBSERVATIONS
ENGLISH MEASURES.

Values of $0.000367 \, B \, (t-t_1)\left(1+\dfrac{t-t_1}{1571}\right)$.

$B =$ Barometric pressure.
$t =$ Temperature of the dry-bulb thermometer.
$t_1 =$ Temperature of the wet-bulb thermometer.

$t-t_1$	BAROMETRIC PRESSURE IN INCHES (B).												
	24.5	24.0	23.5	23.0	22.5	22.0	21.5	21.0	20.5	20.0	19.5	19.0	18.5
F.	Inch.	Inch.	Inch.	Inch.	Inch.	Inch.	Inch.	Inch.	Inch.	Inch.	Inch.	Inch.	Inch.
1°	0.009	0.009	0.009	0.008	0.008	0.008	0.008	0.008	0.008	0.007	0.007	0.007	0.007
2	.018	.018	.017	.017	.016	.016	.016	.015	.015	.015	.014	.014	.014
3	.028	.026	.026	.025	.025	.024	.024	.023	.023	.022	.021	.021	.020
4	.036	.035	.035	.034	.033	.032	.032	.031	.030	.029	.029	.028	.027
5	0.045	0.044	0.043	0.042	0.041	0.040	0.040	0.039	0.038	0.037	0.036	0.035	0.034
6	.054	.053	.052	.051	.050	.049	.048	.046	.045	.044	.043	.042	.041
7	.063	.062	.061	.059	.058	.057	.055	.054	.053	.052	.050	.049	.048
8	.072	.071	.070	.068	.066	.065	.063	.062	.060	.059	.057	.056	.055
9	.081	.080	.078	.076	.075	.073	.071	.070	.068	.066	.064	.063	.061
10	0.090	0.089	0.087	0.085	0.083	0.081	0.079	0.077	0.076	0.074	0.072	0.070	0.068
11	.100	.097	.095	.093	.091	.089	.087	.085	.083	.081	.079	.077	.075
12	.109	.106	.104	.102	.100	.097	.095	.093	.091	.089	.086	.084	.082
13	.118	.115	.113	.110	.108	.106	.103	.101	.098	.096	.093	.091	.089
14	.127	.124	.122	.119	.117	.114	.111	.109	.106	.104	.101	.098	.095
15	0.136	0.133	0.131	0.128	0.125	0.122	0.119	0.117	0.114	0.111	0.108	0.105	0.102
16	.145	.142	.139	.136	.133	.130	.127	.124	.121	.118	.116	.113	.110
17	.155	.151	.148	.145	.142	.139	.135	.132	.129	.126	.123	.120	.117
18	.164	.160	.157	.154	.151	.147	.143	.140	.137	.134	.130	.127	.124
19	.173	.169	.166	.162	.159	.155	.152	.148	.144	.141	.137	.134	.131
20	0.182	0.178	0.175	0.171	0.167	0.163	0.160	0.156	0.152	0.148	0.144	0.141	0.137
21	.191	.187	.183	.180	.176	.172	.168	.164	.160	.156	.152	.148	.144
22	.201	.196	.192	.188	.184	.180	.176	.172	.168	.164	.160	.155	.151
23	.210	.205	.201	.197	.193	.188	.184	.180	.175	.171	.167	.163	.158
24	.219	.214	.210	.205	.201	.196	.192	.188	.183	.179	.174	.170	.165
25	0.228	0.223	0.219	0.214	0.210	0.205	0.200	0.196	0.191	0.186	0.181	0.177	0.172
26	.238	.233	.228	.223	.218	.213	.208	.203	.199	.194	.189	.184	.179
27	.247	.242	.237	.232	.227	.222	.216	.211	.206	.201	.196	.191	.186
28	.256	.251	.245	.240	.235	.230	.225	.219	.214	.209	.203	.198	.193
29	.266	.260	.254	.249	.244	.238	.233	.227	.222	.216	.211	.206	.200
30	0.275	0.269	0.263	0.258	0.252	0.247	0.241	0.235	0.230	0.224	0.218	0.213	0.207
31	.284	.278	.272	.267	.261	.255	.249	.243	.238	.232	.226	.220	.214
32	.294	.287	.281	.275	.269	.263	.257	.251	.245	.239	.233	.227	.221
33	.303	.296	.290	.284	.278	.272	.266	.259	.253	.247	.241	.235	.229
34	.312	.306	.299	.293	.286	.280	.274	.267	.261	.255	.248	.242	.236
35	0.322	0.315	0.308	0.302	0.295	0.289	0.282	0.275	0.268	0.262	0.255	0.249	0.243
36	.331	.324	.317	.311	.304	.297	.290	.284	.277	.270	.263	.257	.250
37	.341	.333	.326	.319	.312	.305	.299	.292	.285	.278	.271	.264	.257
38	.350	.342	.335	.328	.321	.314	.307	.300	.293	.285	.278	.271	.264
39	.359	.352	.344	.337	.330	.322	.315	.308	.300	.293	.285	.278	.271
40	0.369	0.361	0.353	0.346	0.338	0.331	0.323	0.316	0.308	0.301	0.293	0.286	0.278

SMITHSONIAN TABLES.

RELATIVE HUMIDITY.

TEMPERATURES FAHRENHEIT.

DEPRESSION OF THE DEW-POINT $(t-d)$.

1°.5	2°.0	2°.5	3°.0	3°.5	4°.0	4°.5	5°.0	5°.5	6°.0	6°.5	7°.0	7°.5
92	89											
92	89	86	83	81	78	75	73	71	69	67	65	63
92	89	87	84	81	79	76	74	72	70	67	65	63
92	89	87	84	82	79	77	75	73	71	69	66	65
92	90	87	85	82	80	78	75	73	71	69	67	65
92	90	87	85	82	80	78	75	73	71	69	67	65
92	90	87	85	83	80	78	76	74	72	70	68	66
92	90	87	85	83	80	78	76	74	72	70	68	66
92	90	87	85	83	80	78	76	74	72	70	68	66
93	90	88	86	84	82	80	78	76	74	71	70	68
93	90	88	86	84	82	80	78	76	74	72	71	69
93	90	88	86	84	82	80	78	76	74	72	71	69
93	91	89	86	84	82	80	78	76	74	72	71	69
93	91	89	87	85	83	81	79	77	75	73	72	70
93	91	89	87	85	83	81	79	77	75	73	72	70
93	91	89	87	85	83	81	79	77	76	74	72	70
94	92	90	87	85	83	81	79	78	76	74	72	71
94	92	91	89	87	85	83	81	80	78	76	74	73
94	93	91	89	87	86	84	82	81	79	78	76	74
94	93	91	89	87	86	84	83	81	79	78	76	75
95	93	91	89	88	86	84	83	81	80	78	77	75
95	93	91	89	88	86	85	83	82	80	79	77	76
95	93	91	90	88	86	85	83	82	80	79	77	76
95	93	91	90	88	87	85	84	82	81	79	78	76
95	93	92	90	88	87	85	84	82	81	79	78	77
95	93	92	90	89	87	86	84	83	81	80	78	77
95	93	92	90	89	87	86	84	83	81	80	79	77
95	94	92	90	89	87	86	85	83	82	80	79	77
95	94	92	91	89	88	86	85	83	82	81	79	78
95	94	92	91	89	88	86	85	84	82	81	80	78
95	94	92	91	89	88	87	85	84	83	81	80	79
95	94	92	91	90	88	87	85	84	83	81	80	79
95	94	93	91	90	88	87	86	84	83	82	80	79
96	94	93	91	90	89	87	86	85	83	82	81	79
96	94	93	91	90	89	87	86	85	83	82	81	80
96	94	93	92	90	89	88	86	85	84	82	81	80
96	94	93	92	90	89	88	86	85	84	83	82	80
96	94	93	92	90	89	88	87	85	84	83	82	81
96	95	93	92	91	89	88	87	86	84	83	82	81

RELATIVE HUMIDITY.

TEMPERATURES FAHRENHEIT.

Air Temperature. t.	DEPRESSION OF THE DEW-POINT $(t-d)$.														
	8°0	8°5	9°0	9°5	10°0	10°5	11°0	11°5	12°0	12°5	13°0	13°5	14°0	14°5	15°0
F. −24°	61	60	58	56	55	53	51	50	49						
−20	62	61	59	57	55	53	52	50	49	47	46	45	43	42	41
−16	63	61	59	58	56	54	53	51	49	48	46	45	44	42	41
−12	63	61	60	58	56	55	53	52	50	49	47	46	45	43	42
−8	64	62	61	59	57	56	54	52	51	49	48	47	45	44	43
−4	65	63	61	60	58	56	55	53	52	50	49	47	46	45	43
0	65	63	61	60	58	57	55	53	52	51	49	48	47	45	44
+4	66	64	62	61	59	57	56	54	52	51	50	49	48	46	45
8	67	65	64	62	60	59	57	55	54	52	51	50	48	47	46
12	67	65	64	62	61	59	58	56	55	53	52	51	49	48	47
16	67	65	64	62	61	60	58	56	55	54	52	51	50	49	47
20	68	66	65	63	61	60	58	57	56	54	53	51	50	49	48
24	68	67	65	64	62	61	59	58	56	55	54	52	51	50	48
28	69	67	65	64	62	61	59	58	57	55	54	53	52	50	49
32	69	68	66	64	63	61	60	58	57	56	54	53	52	51	50
36	71	69	68	66	65	63	62	60	59	58	56	55	54	52	51
40	73	71	70	68	67	65	64	62	61	59	58	57	55	54	53
44	73	72	71	69	68	66	65	64	63	61	60	58	57	56	55
48	74	72	71	70	68	67	66	64	63	62	61	59	58	57	56
52	74	73	71	70	69	67	66	65	64	62	61	60	59	58	57
56	75	73	72	70	69	68	67	65	64	63	62	61	59	58	57
60	75	74	72	71	70	68	67	66	65	63	62	61	60	59	58
64	75	74	73	71	70	69	68	66	65	64	63	62	60	59	58
68	76	74	73	72	70	69	68	67	66	64	63	62	61	60	59
72	76	75	73	72	71	70	68	67	66	65	64	63	61	60	59
76	76	75	74	72	71	70	69	68	66	65	64	63	62	61	60
80	77	75	74	73	72	70	69	68	67	66	65	64	62	61	60
84	77	76	74	73	72	71	70	68	67	66	65	64	63	62	61
88	77	76	75	74	72	71	70	69	68	67	66	64	63	62	61
92	78	76	75	74	73	72	70	69	68	67	66	65	64	63	62
96	78	77	75	74	73	72	71	70	69	67	66	65	64	63	62
100	78	77	76	75	73	72	71	70	69	68	67	66	65	64	63
104	79	77	76	75	74	73	72	70	69	68	67	66	65	64	63
108	79	78	76	75	74	73	72	71	70	69	68	67	66	65	64
112	79	78	77	76	75	73	72	71	70	69	68	67	66	65	64
116	79	78	77	76	75	74	73	72	71	70	69	67	66	65	65
120	80	79	77	76	75	74	73	72	71	70	69	68	67	66	65

SMITHSONIAN TABLES.

TABLE 42.

RELATIVE HUMIDITY.

TEMPERATURES FAHRENHEIT.

Air Temperature. t.	DEPRESSION OF THE DEW-POINT $(t-d)$.															
	15°	16°	17°	18°	19°	20°	21°	22°	23°	24°	25°	26°	27°	28°	29°	30°
F.																
−20°	41	39														
−16	41	39	36	34	32	31	29									
−12	42	39	37	35	33	31	29	27	26	24	23	22	21			
−8	43	40	38	36	34	32	30	28	26	25	23	22	21	20	19	18
−4	43	41	39	36	34	32	31	29	27	25	24	23	21	20	19	18
0	44	42	39	37	35	33	31	29	28	26	25	23	22	21	19	18
+4	45	43	40	38	36	34	32	30	29	27	25	24	23	21	20	19
8	46	43	41	39	37	35	33	31	30	28	26	25	23	22	21	20
12	47	44	42	39	37	35	34	32	30	28	27	25	24	23	21	20
16	47	45	43	40	38	36	34	32	31	29	28	26	25	23	22	21
20	48	46	43	41	39	37	35	33	32	30	28	27	25	24	23	22
24	48	46	44	42	40	38	36	34	32	31	29	28	26	25	23	22
28	49	47	44	42	40	38	36	34	33	31	30	28	27	26	24	23
32	50	47	45	43	41	39	37	35	33	32	30	29	27	26	25	24
36	51	49	46	44	42	40	38	37	35	33	32	30	28	27	26	25
40	53	50	48	46	44	42	40	38	36	34	33	31	30	28	27	26
44	55	52	50	48	45	43	41	39	38	36	34	32	31	30	28	27
48	56	54	51	49	47	45	43	41	39	37	35	34	32	31	29	28
52	57	54	52	50	48	46	44	42	41	39	37	35	34	32	31	29
56	57	55	53	51	49	47	45	43	42	40	38	37	35	33	32	30
60	58	56	54	51	50	48	46	44	42	41	39	38	36	35	33	32
64	58	56	54	52	50	48	46	45	43	41	40	38	37	35	34	33
68	59	57	55	53	51	49	47	45	44	42	40	39	37	36	35	33
72	59	57	55	53	51	49	48	46	44	43	41	40	38	37	35	34
76	60	58	56	54	52	50	48	47	45	43	42	40	39	37	36	35
80	60	58	56	54	52	51	49	47	45	44	42	41	39	38	37	35
84	61	59	57	55	53	51	49	48	46	44	43	41	40	39	37	36
88	61	59	57	55	54	52	50	48	47	45	43	42	41	39	38	36
92	62	60	58	56	54	52	51	49	47	46	44	43	41	40	38	37
96	62	60	58	56	55	53	51	49	48	46	45	43	42	40	39	38
100	63	61	59	57	55	53	52	50	48	47	45	44	42	41	40	38
104	63	61	59	57	56	54	52	50	49	47	46	44	43	41	40	39
108	64	62	60	58	56	54	53	51	49	48	46	45	43	42	41	39
112	64	62	60	58	57	55	53	52	50	48	47	45	44	43	41	40
116	65	63	61	59	57	55	54	52	51	49	48	46	45	43	42	41
120	65	63	61	59	58	56	54	53	51	50	48	47	45	44	42	41

TABLE 42.

RELATIVE HUMIDITY.

TEMPERATURES FAHRENHEIT.

Air Temperature, t.	DEPRESSION OF THE DEW-POINT $(t-d)$.													
	33°	36°	39°	42°	45°	48°	51°	54°	57°	60°	63°	66°	69°	72°	75°
F.															
−4	15	13													
0	15	13	11												
+4	16	13	11	9	8										
8	17	14	11	10	8	7									
12	17	14	12	10	8	7	6								
16	18	15	12	10	9	7	6	5	4						
20	18	15	13	11	9	8	6	5	4	4					
24	19	16	14	11	10	8	7	6	5	4	3				
28	19	16	14	12	10	8	7	6	5	4	3	3			
32	20	17	14	12	10	9	7	6	5	4	4	3	3	2	
36	21	18	15	13	11	9	8	7	6	5	4	3	3	2	2
40	22	19	16	14	12	10	9	7	6	5	4	4	3	2	2
44	23	20	17	15	13	11	9	8	7	6	5	4	3	3	2
48	24	21	18	15	13	11	10	8	7	6	5	4	4	3	2
52	25	22	19	16	14	12	10	9	8	6	5	5	4	3	3
56	26	23	20	17	15	13	11	9	8	7	6	5	4	4	3
60	28	24	21	18	16	13	12	10	9	7	6	5	5	4	3
64	29	25	22	19	16	14	12	11	9	8	7	6	5	4	4
68	30	26	23	20	17	15	13	11	10	8	7	6	5	5	4
72	30	27	24	21	18	16	14	12	10	9	8	7	6	5	4
76	31	28	24	22	19	17	15	13	11	9	8	7	6	5	4
80	31	28	25	22	20	18	15	13	12	10	9	8	7	6	5
84	32	29	26	23	20	18	16	14	12	11	9	8	7	6	5
88	33	29	26	23	21	19	17	15	13	11	10	9	8	7	6
92	33	30	27	24	21	19	17	15	14	12	10	9	8	7	6
96	34	30	27	25	22	20	18	16	14	12	11	10	8	7	6
100	34	31	28	25	23	20	18	16	14	13	11	10	9	8	7
104	35	32	29	26	23	21	19	17	15	13	12	11	9	8	7
108	36	32	29	26	24	21	19	17	15	14	12	11	10	9	8
112	36	33	30	27	24	22	20	18	16	14	13	11	10	9	8
116	37	33	30	27	25	22	20	18	16	15	13	12	11	9	8
120	37	34	31	28	25	23	21	19	17	15	14	12	11	10	9

SMITHSONIAN TABLES.

TABLE 43.
REDUCTION OF PSYCHROMETRIC OBSERVATIONS.
METRIC MEASURES.
Pressure of Aqueous Vapor.
(*Broch.*)

Temperature.	0°	1°	2°	3°	4°	5°	6°	7°	8°	9°
C.	mm.	mm.	mm.	mm.	mm.	mm.	mm.	mm.	mm.	mm.
−30°	0.38	0.35	0.32	0.29	0.26	0.23	0.21	0.19	0.17	0.15
−20	0.94	0.87	0.79	0.73	0.66	0.61	0.55	0.50	0.46	0.42
−10	2.15	1.99	1.84	1.69	1.56	1.44	1.32	1.22	1.12	1.03
−0	4.57	4.25	3.95	3.67	3.41	3.16	2.93	2.72	2.51	2.33

Temperature.	0°.0	0°.1	0°.2	0°.3	0°.4	0°.5	0°.6	0°.7	0°.8	0°.9
C.	mm.	mm.	mm.	mm.	mm.	mm.	mm.	mm.	mm.	mm.
+0°	4.57	4.60	4.64	4.67	4.70	4.74	4.77	4.80	4.84	4.87
1	4.91	4.94	4.98	5.02	5.05	5.09	5.12	5.16	5.20	5.23
2	5.27	5.31	5.35	5.39	5.42	5.46	5.50	5.54	5.58	5.62
3	5.66	5.70	5.74	5.78	5.82	5.86	5.90	5.94	5.99	6.03
4	6.07	6.11	6.15	6.20	6.24	6.28	6.33	6.37	6.42	6.46
5	6.51	6.55	6.60	6.64	6.69	6.74	6.78	6.83	6.88	6.92
6	6.97	7.02	7.07	7.12	7.17	7.22	7.26	7.31	7.36	7.42
7	7.47	7.52	7.57	7.62	7.67	7.72	7.78	7.83	7.88	7.94
8	7.99	8.05	8.10	8.15	8.21	8.27	8.32	8.38	8.43	8.49
9	8.55	8.61	8.66	8.72	8.78	8.84	8.90	8.96	9.02	9.08
10	9.14	9.20	9.26	9.32	9.39	9.45	9.51	9.58	9.64	9.70
11	9.77	9.83	9.90	9.96	10.03	10.09	10.16	10.23	10.30	10.36
12	10.43	10.50	10.57	10.64	10.71	10.78	10.85	10.92	10.99	11.07
13	11.14	11.21	11.28	11.36	11.43	11.50	11.58	11.66	11.73	11.81
14	11.88	11.96	12.04	12.12	12.19	12.27	12.35	12.43	12.51	12.59
15	12.67	12.76	12.84	12.92	13.00	13.09	13.17	13.25	13.34	13.42
16	13.51	13.60	13.68	13.77	13.86	13.95	14.04	14.12	14.21	14.30
17	14.40	14.49	14.58	14.67	14.76	14.86	14.95	15.04	15.14	15.23
18	15.33	15.43	15.52	15.62	15.72	15.82	15.92	16.02	16.12	16.22
19	16.32	16.42	16.52	16.63	16.73	16.83	16.94	17.04	17.15	17.26
20	17.36	17.47	17.58	17.69	17.80	17.91	18.02	18.13	18.24	18.35
21	18.47	18.58	18.69	18.81	18.92	19.04	19.16	19.27	19.39	19.51
22	19.63	19.75	19.87	19.99	20.11	20.24	20.36	20.48	20.61	20.73
23	20.86	20.98	21.11	21.24	21.37	21.50	21.63	21.76	21.89	22.02
24	22.15	22.29	22.42	22.55	22.69	22.83	22.96	23.10	23.24	23.38
25	23.52	23.66	23.80	23.94	24.08	24.23	24.37	24.52	24.66	24.81
26	24.96	25.10	25.25	25.40	25.55	25.70	25.86	26.01	26.16	26.32
27	26.47	26.63	26.78	26.94	27.10	27.26	27.42	27.58	27.74	27.90
28	28.07	28.23	28.39	28.56	28.73	28.89	29.06	29.23	29.40	29.57
29	29.74	29.92	30.09	30.26	30.44	30.62	30.79	30.97	31.15	31.33
30	31.51	31.69	31.87	32.06	32.24	32.43	32.61	32.80	32.99	33.18
31	33.37	33.56	33.75	33.94	34.14	34.33	34.53	34.72	34.92	35.12
32	35.32	35.52	35.72	35.92	36.13	36.33	36.54	36.74	36.95	37.16
33	37.37	37.58	37.79	38.00	38.22	38.43	38.65	38.87	39.08	39.30
34	39.52	39.74	39.97	40.19	40.41	40.64	40.87	41.09	41.32	41.55
35	41.78	42.02	42.25	42.48	42.72	42.96	43.19	43.43	43.67	43.92
36	44.16	44.40	44.65	44.89	45.14	45.39	45.64	45.89	46.14	46.39
37	46.65	46.90	47.16	47.42	47.68	47.94	48.20	48.46	48.73	48.99
38	49.26	49.53	49.80	50.07	50.34	50.61	50.89	51.16	51.44	51.72
39	52.00	52.28	52.56	52.84	53.13	53.41	53.70	53.99	54.28	54.57
40	54.87	55.16	55.46	55.75	56.05	56.35	56.65	56.95	57.26	57.56
41	57.87	58.18	58.49	58.80	59.11	59.43	59.74	60.06	60.38	60.70
42	61.02	61.34	61.66	61.99	62.32	62.65	62.98	63.31	63.64	63.97
43	64.31	64.65	64.99	65.33	65.67	66.01	66.36	66.71	67.05	67.41
44	67.76	68.11	68.47	68.82	69.18	69.54	69.90	70.26	70.63	70.99
45	71.36	71.73	72.10	72.48	72.85	73.23	73.60	73.98	74.36	74.75

Smithsonian Tables.

TABLE 44.
REDUCTION OF PSYCHROMETRIC OBSERVATIONS.
METRIC MEASURES.

Values of $0.000660 \, B \, (t-t_1)\left(1+\dfrac{t-t_1}{873}\right)$.

$t =$ Temperature of the dry-bulb thermometer.
$t_1 =$ Temperature of the wet-bulb thermometer.

$t-t_1$	BAROMETRIC PRESSURE IN MILLIMETRES (B).													
	770	760	750	740	730	720	710	700	690	680	670	660	650	640	630
C.	mm.	mm.	mm.	mm.	mm.	mm.	mm.	mm.	mm.	mm.	mm.	mm.	mm.	mm.	mm.
1°	0.52	0.51	0.50	0.50	0.49	0.48	0.48	0.47	0.46	0.46	0.45	0.44	0.44	0.43	0.42
2	1.03	1.01	1.00	0.98	0.97	0.96	0.94	0.93	0.92	0.90	0.89	0.88	0.87	0.85	0.84
3	1.54	1.52	1.49	1.47	1.45	1.43	1.41	1.39	1.37	1.35	1.33	1.32	1.30	1.28	1.26
4	2.04	2.02	1.99	1.97	1.94	1.91	1.89	1.86	1.83	1.81	1.78	1.75	1.73	1.70	1.67
5	2.56	2.52	2.49	2.46	2.43	2.39	2.36	2.32	2.29	2.26	2.23	2.19	2.17	2.13	2.09
6	3.07	3.03	2.99	2.95	2.91	2.87	2.83	2.79	2.75	2.71	2.67	2.63	2.59	2.55	2.51
7	3.59	3.54	3.50	3.45	3.40	3.36	3.31	3.26	3.22	3.17	3.12	3.08	3.04	2.99	2.94
8	4.11	4.05	4.00	3.95	3.89	3.84	3.79	3.73	3.68	3.63	3.58	3.53	3.48	3.42	3.36
9	4.62	4.56	4.50	4.44	4.38	4.32	4.27	4.21	4.15	4.09	4.03	3.97	3.91	3.85	3.79
10	5.15	5.08	5.01	4.94	4.88	4.81	4.74	4.68	4.61	4.54	4.47	4.41	4.35	4.28	4.21
11	5.66	5.59	5.51	5.44	5.37	5.30	5.22	5.15	5.08	5.00	4.93	4.86	4.79	4.71	4.63
12	6.19	6.11	6.02	5.94	5.86	5.78	5.70	5.62	5.54	5.46	5.38	5.30	5.22	5.14	5.06
13	6.71	6.62	6.53	6.45	6.36	6.27	6.18	6.10	6.01	5.92	5.83	5.75	5.66	5.57	5.49
14	7.23	7.14	7.05	6.95	6.86	6.76	6.67	6.58	6.48	6.39	6.29	6.20	6.11	6.01	5.92
15	7.76	7.66	7.56	7.46	7.36	7.26	7.16	7.06	6.95	6.85	6.75	6.65	6.55	6.45	6.35
16	8.29	8.18	8.07	7.96	7.86	7.75	7.64	7.54	7.43	7.32	7.21	7.11	7.00	6.89	6.78
17	8.82	8.70	8.59	8.47	8.36	8.24	8.13	8.02	7.90	7.79	7.67	7.56	7.45	7.33	7.21
18	9.35	9.22	9.10	8.98	8.86	8.74	8.62	8.50	8.37	8.25	8.13	8.01	7.89	7.77	7.65
19	9.87	9.75	9.62	9.49	9.36	9.23	9.11	8.98	8.85	8.72	8.59	8.47	8.34	8.21	8.08
20	10.41	10.27	10.14	10.00	9.87	9.73	9.60	9.46	9.32	9.19	9.05	8.92	8.78	8.65	8.51

$t-t_1$	BAROMETRIC PRESSURE IN MILLIMETRES (B).													
	620	610	600	590	580	570	560	550	540	530	520	510	500	490	480
C.	mm.	mm.	mm.	mm.	mm.	mm.	mm.	mm.	mm.	mm.	mm.	mm.	mm.	mm.	mm.
1°	0.42	0.41	0.40	0.40	0.39	0.38	0.38	0.37	0.36	0.36	0.35	0.34	0.34	0.33	0.32
2	0.82	0.81	0.80	0.78	0.77	0.76	0.75	0.73	0.72	0.70	0.69	0.68	0.67	0.65	0.64
3	1.24	1.22	1.20	1.17	1.15	1.13	1.12	1.10	1.08	1.06	1.04	1.02	1.00	0.98	0.96
4	1.65	1.62	1.60	1.57	1.54	1.51	1.49	1.46	1.44	1.41	1.38	1.36	1.33	1.30	1.28
5	2.06	2.03	1.99	1.96	1.93	1.90	1.86	1.83	1.80	1.76	1.73	1.70	1.66	1.63	1.60
6	2.47	2.43	2.39	2.35	2.32	2.28	2.24	2.20	2.16	2.12	2.08	2.04	2.00	1.96	1.92
7	2.89	2.84	2.80	2.75	2.71	2.66	2.61	2.56	2.52	2.47	2.43	2.38	2.33	2.28	2.24
8	3.31	3.26	3.20	3.15	3.10	3.04	2.99	2.94	2.88	2.83	2.78	2.72	2.67	2.62	2.56
9	3.73	3.67	3.61	3.55	3.49	3.43	3.37	3.31	3.25	3.19	3.13	3.06	3.00	2.94	2.88
10	4.14	4.07	4.01	3.94	3.88	3.81	3.74	3.67	3.61	3.54	3.48	3.41	3.34	3.27	3.21
11	4.56	4.49	4.42	4.34	4.27	4.19	4.12	4.05	3.97	3.90	3.83	3.75	3.68	3.60	3.53
12	4.98	4.90	4.82	4.74	4.66	4.58	4.50	4.42	4.34	4.26	4.18	4.10	4.02	3.93	3.85
13	5.40	5.31	5.23	5.14	5.05	4.96	4.88	4.79	4.70	4.62	4.53	4.44	4.36	4.27	4.18
14	5.83	5.73	5.64	5.54	5.45	5.35	5.26	5.17	5.07	4.98	4.88	4.79	4.70	4.60	4.51
15	6.25	6.15	6.05	5.95	5.85	5.74	5.64	5.54	5.44	5.34	5.24	5.14	5.04	4.94	4.84
16	6.68	6.57	6.46	6.35	6.24	6.14	6.03	5.92	5.81	5.71	5.60	5.49	5.38	5.27	5.17
17	7.10	6.98	6.87	6.75	6.64	6.53	6.41	6.30	6.18	6.07	5.95	5.84	5.72	5.61	5.50
18	7.52	7.40	7.28	7.16	7.04	6.92	6.80	6.67	6.55	6.43	6.31	6.19	6.07	5.95	5.83
19	7.95	7.82	7.70	7.57	7.44	7.31	7.18	7.05	6.93	6.80	6.67	6.54	6.42	6.29	6.16
20	8.38	8.24	8.11	7.97	7.84	7.70	7.57	7.43	7.30	7.16	7.03	6.90	6.76	6.62	6.49

TABLE 45.

RELATIVE HUMIDITY.

TEMPERATURE CENTIGRADE.

Depression of the dew-point. $t-d$.	DEW-POINT (d).									
	$-15°$	$-10°$	$-5°$	$0°$	$+5°$	$+10°$	$+15°$	$+20°$	$+25°$	$+30°$
C.										
0°.0	100	100	100	100	100	100	100	100	100	100
0.2	98	98	99	99	99	99	99	99	99	99
0.4	97	97	97	97	97	97	97	98	98	98
0.6	95	95	96	96	96	96	96	96	97	97
0.8	94	94	94	94	95	95	95	95	95	96
1.0	92	92	93	93	93	94	94	94	94	94
1.2	91	91	91	92	92	92	93	93	93	93
1.4	89	90	90	90	91	91	91	92	92	92
1.6	88	88	89	89	90	90	90	91	91	91
1.8	86	87	87	88	88	89	89	90	90	90
2.0	85	86	86	87	87	88	88	88	89	89
2.2	84	84	85	85	86	86	87	87	88	88
2.4	83	83	84	84	85	85	86	86	87	87
2.6	82	82	82	83	84	84	85	85	86	86
2.8	80	80	81	82	83	83	84	84	85	85
3.0	78	79	80	81	81	82	83	83	84	84
3.2	77	78	79	80	80	81	82	82	83	83
3.4	76	77	78	79	79	80	81	81	82	82
3.6	75	76	77	77	78	79	80	80	81	82
3.8	74	75	75	76	77	78	79	79	80	81
4.0	72	73	74	75	76	77	78	78	79	80
4.2	71	72	73	74	75	76	77	77	78	79
4.4	70	71	72	73	74	75	76	77	77	78
4.6	69	70	71	72	73	74	75	76	76	77
4.8	68	69	70	71	72	73	74	75	75	76
5.0	67	68	69	70	71	72	73	74	75	75
5.2	66	67	68	69	70	71	72	73	74	75
5.4	65	66	67	68	69	70	71	72	73	74
5.6	64	65	66	67	68	69	70	71	72	73
5.8	63	64	65	66	68	69	69	70	71	72
6.0	62	63	64	66	67	68	69	70	70	71
6.2	61	62	63	65	66	67	68	69	70	71
6.4	60	61	63	64	65	66	67	68	69	70
6.6	59	60	62	63	64	65	66	67	68	69
6.8	58	60	61	62	63	64	65	66	67	68
7.0	57	59	60	61	62	63	65	66	67	68
7.2	56	58	59	60	62	63	64	65	66	67
7.4	55	57	58	60	61	62	63	64	65	66
7.6	55	56	57	59	60	61	62	63	64	65
7.8	54	55	57	58	59	60	62	63	64	65
8.0	53	54	56	57	58	60	61	62	63	64

TABLE 45

RELATIVE HUMIDITY.
TEMPERATURE CENTIGRADE.

Depression of the dew-point. $t-d$.	DEW-POINT (d).									
	−15°	−10°	−5°	0°	+5°	+10°	+15°	+20°	+25°	+30°
C.										
8.0	53	54	56	57	58	60	61	62	63	64
8.2	52	54	55	56	57	59	60	61	62	63
8.4	51	53	54	56	57	58	59	60	62	63
8.6	51	52	54	55	56	57	58	60	61	62
8.8	50	51	53	54	55	57	58	59	60	61
9.0	49	51	52	53	55	56	57	58	60	61
9.2	48	50	51	53	54	55	57	58	59	60
9.4	48	49	51	52	53	55	56	57	58	59
9.6	47	48	50	51	53	54	55	56	58	59
9.8	46	48	49	51	52	53	55	56	57	58
10.0	46	47	49	50	51	53	54	55	56	57
10.5	44	45	47	48	50	51	52	54	55	
11.0	42	44	45	47	48	49	51	52	53	
11.5	41	42	44	45	·47	48	49	51	52	
12.0	39	41	42	44	45	47	48	49	50	
12.5	38	39	41	42	44	45	46	48	49	
13.0	36	38	40	41	43	44	45	46	48	
13.5	35	37	38	40	42	43	44	45	46	
14.0	34	35	37	38	40	41	43	44	45	
14.5	33	34	36	37	39	40	41	43	44	
15.0	31	33	35	36	37	39	40	42		
15.5	30	32	33	35	36	38	39	40		
16.0	29	31	32	34	35	37	38	39		
16.5	28	30	31	33	34	36	37	38		
17.0	27	29	30	32	33	35	36	37		
17.5	26	28	29	31	32	34	35	36		
18.0	25	27	28	30	31	33	34	35		
18.5	25	26	27	29	30	32	33	34		
19.0	24	25	27	28	29	31	32	33		
19.5	23	24	26	27	29	30	31	33		
20.0	22	24	25	26	28	29	30	32		
21.0	21	22	23	25	26	27	29			
22.0	19	21	22	23	25	26	27			
23.0	18	19	21	22	23	24	26			
24.0	17	18	19	21	22	23	24			
25.0	16	17	18	19	21	22	23			
26.0	15	16	17	18	20	21				
27.0	14	15	16	17	18	20				
28.0	13	14	15	16	17	19				
29.0	12	13	14	15	16	18				
30.0	11	12	13	14	16	17				

SMITHSONIAN TABLES.

TABLE 46.
REDUCTION OF SNOWFALL MEASUREMENTS.

Depth of water corresponding to the weight of snow (or rain) collected in an 8-inch gage.

Weight of Snow.	Oz. 0	Oz. $\frac{1}{4}$	Oz. $\frac{1}{2}$	Oz. $\frac{3}{4}$	Weight of Snow.	Oz. 0	Oz. $\frac{1}{4}$	Oz. $\frac{1}{2}$	Oz. $\frac{3}{4}$	Weight of Snow.	Oz. 0	Oz. $\frac{1}{4}$	Oz. $\frac{1}{2}$	Oz. $\frac{3}{4}$
Lb.Oz.	Inch.	Inch.	Inch.	Inch.	Lb.Oz.	Inch's	Inch's	Inch's	Inch's	Lb.Oz.	Inch's	Inch's	Inch's	Inch's
0	0.00	0.01	0.02	0.03	1 8	0.83	0.83	0.84	0.85	2 13	1.55	1.56	1.57	1.57
1	.03	.04	.05	.06	1 9	.86	.87	.88	.89	2 14	1.58	1.59	1.60	1.61
2	.07	.08	.09	.09	1 10	.89	.90	.91	.92	2 15	1.62	1.63	1.63	1.64
3	.10	.11	.12	.13	1 11	.93	.94	.94	.95					
4	.14	.15	.15	.16	1 12	.96	.97	.98	.99					
5	0.17	0.18	0.19	0.20	1 13	1.00	1.01	1.01	1.02	3 0	1.65	1.66	1.67	1.68
6	.21	.22	.22	.23	1 14	1.03	1.04	1.05	1.06	3 1	1.69	1.69	1.70	1.71
7	.24	.25	.26	.27	1 15	1.07	1.08	1.08	1.09	3 2	1.72	1.73	1.74	1.75
8	.28	.28	.29	.30						3 3	1.75	1.76	1.77	1.78
9	.31	.32	.33	.34						3 4	1.79	1.80	1.81	1.81
10	0.34	0.35	0.36	0.37	2 0	1.10	1.11	1.12	1.13	3 5	1.82	1.83	1.84	1.85
11	.38	.39	.40	.41	2 1	1.14	1.14	1.15	1.16	3 6	1.86	1.87	1.87	1.88
12	.41	.42	.43	.44	2 2	1.17	1.18	1.19	1.20	3 7	1.89	1.90	1.91	1.92
13	.45	.46	.46	.47	2 3	1.20	1.21	1.22	1.23	3 8	1.93	1.94	1.94	1.95
14	.48	.49	.50	.51	2 4	1.24	1.25	1.26	1.26	3 9	1.96	1.97	1.98	1.99
15	.52	.52	.53	.54										
1 0	0.55	0.56	0.57	0.58	2 5	1.27	1.28	1.29	1.30	3 10	2.00	2.00	2.01	2.02
1 1	.58	.59	.60	.61	2 6	1.31	1.32	1.32	1.33	3 11	2.03	2.04	2.05	2.06
1 2	.62	.63	.64	.65	2 7	1.34	1.35	1.36	1.37	3 12	2.06	2.07	2.08	2.09
1 3	.65	.66	.67	.68	2 8	1.38	1.38	1.39	1.40	3 13	2.10	2.11	2.12	2.12
1 4	.69	.70	.71	.71	2 9	1.41	1.42	1.43	1.44	3 14	2.13	2.14	2.15	2.16
										3 15	2.17	2.18	2.18	2.19
1 5	0.72	0.73	0.74	0.75	2 10	1.44	1.45	1.46	1.47					
1 6	.76	.77	.77	.78	2 11	1.48	1.49	1.50	1.51	4 0	2.20			
1 7	.79	.80	.81	.82	2 12	1.51	1.52	1.53	1.54	5 0	2.75			

TABLE 47.
RATE OF DECREASE OF VAPOR PRESSURE WITH ALTITUDE.

(*According to the empirical formula of Dr. J. Hann*).

$$\frac{f}{f_0} = 10^{-\frac{h}{6517}}.$$

$f, f_0 =$ Vapor pressures at an upper and a lower station respectively.

$h =$ Difference of altitude in metres.

Difference of Altitude.		$\frac{f}{f_0}$.	Difference of Altitude.		$\frac{f}{f_0}$.	Difference of Altitude.		$\frac{f}{f_0}$.
metres.	Feet.		metres.	Feet.		metres.	Feet.	
200	656	0.93	1800	5905	0.53	3400	11155	0.30
400	1312	.87	2000	6562	.49	3600	11811	.28
600	1968	.81	2200	7218	.46	3800	12467	.26
800	2625	.75	2400	7874	.43	4000	13123	.24
1000	3281	0.70	2600	8530	0.40	4500	14764	0.20
1200	3937	.65	2800	9186	.37	5000	16404	.17
1400	4593	.61	3000	9842	.35	5500	18045	.14
1600	5249	.57	3200	10499	.32	6000	19685	.12

WIND TABLES.

Mean direction of the wind by Lambert's formula —
    Multiples of cos 45°; form and example of computation . TABLE 48
    Values of the mean direction (a) or its complement ($90 - a$)   49

Synoptic conversion of velocities . . . . TABLE 50

Miles per hour into feet per second . . . . TABLE 51

Feet per second into miles per hour . . . . TABLE 52

Metres per second into miles per hour . . . . TABLE 53

Miles per hour into metres per second . . . . . . - . TABLE 54

Metres per second into kilometres per hour . . . . . . TABLE 55

Kilometres per hour into metres per second . . . . . . TABLE 56

Beaufort wind scale and its conversion into velocity . . . TABLE 57

TABLE 48.
MEAN DIRECTION OF THE WIND BY LAMBERT'S FORMULA

$$\tan a = \frac{E-W+(NE+SE-NW-SW)\cos 45°}{N-S+(NE+NW-SE-SW)\cos 45°}$$

Multiples of $\cos 45°$.

Number.	0	1	2	3	4	5	6	7	8	9
0	0.0	0.7	1.4	2.1	2.8	3.5	4.2	4.9	5.7	6.4
10	7.1	7.8	8.5	9.2	9.9	10.6	11.3	12.0	12.7	13.4
20	14.1	14.8	15.6	16.3	17.0	17.7	18.4	19.1	19.8	20.5
30	21.2	21.9	22.6	23.3	24.0	24.7	25.5	26.2	26.9	27.6
40	28.3	29.0	29.7	30.4	31.1	31.8	32.5	33.2	33.9	34.6
50	35.4	36.1	36.8	37.5	38.2	38.9	39.6	40.3	41.0	41.7
60	42.4	43.1	43.8	44.5	45.3	46.0	46.7	47.4	48.1	48.8
70	49.5	50.2	50.9	51.6	52.3	53.0	53.7	54.4	55.2	55.9
80	56.6	57.3	58.0	58.7	59.4	60.1	60.8	61.5	62.2	62.9
90	63.6	64.3	65.1	65.8	66.5	67.2	67.9	68.6	69.3	70.0
100	70.7	71.4	72.1	72.8	73.5	74.2	75.0	75.7	76.4	77.1
110	77.8	78.5	79.2	79.9	80.6	81.3	82.0	82.7	83.4	84.1
120	84.9	85.6	86.3	87.0	87.7	88.4	89.1	89.8	90.5	91.2
130	91.9	92.6	93.3	94.0	94.8	95.5	96.2	96.9	97.6	98.3
140	99.0	99.7	100.4	101.1	101.8	102.5	103.2	103.9	104.7	105.4
150	106.1	106.8	107.5	108.2	108.9	109.6	110.3	111.0	111.7	112.4
160	113.1	113.8	114.6	115.3	116.0	116.7	117.4	118.1	118.8	119.5
170	120.2	120.9	121.6	122.3	123.0	123.7	124.5	125.2	125.9	126.6
180	127.3	128.0	128.7	129.4	130.1	130.8	131.5	132.2	132.9	133.6
190	134.4	135.1	135.8	136.5	137.2	137.9	138.6	139.3	140.0	140.7
200	141.4	142.1	142.8	143.5	144.2	145.0	145.7	146.4	147.1	147.8

Form for Computing the Numerator and Denominator.

Directions.	E	W	N	S	NE	SW	SE	NW
Observed values.	7	12	6	26	13	45	2	24
	$E-W$		$N-S$		$NE-SW$		$SE-NW$	
	[-5]		[-20]		[-32]$\times \cos 45°$		[-22]$\times \cos 45°$	
Numerator (n).	[-5]		$+$		[-22.6] $+$		[-15.6] $=$ [-43.2]	
Denominator (d).			[-20] $+$		[-22.6] $-$		[-15.6] $=$ [-27.0]	

a is the angle between the mean wind direction and the meridian.

The signs of the numerator (n) and denominator (d) determine the quadrant in which a lies.

When n and d are positive, a lies between N and E: $\frac{+}{+} = NE$.

When n is positive and d negative, a lies between S and E: $\frac{+}{-} = SE$.

When n and d are negative, a lies between S and W: $\frac{-}{-} = SW$.

When n is negative and d positive, a lies between N and W: $\frac{-}{+} = NW$.

TABLE 49.

MEAN DIRECTION OF THE WIND BY LAMBERT'S FORMULA.

Values of the mean direction (a) or its complement ($90°-a$).

$$a = \tan^{-1} n/d$$

n or d.	DENOMINATOR OR NUMERATOR (d OR n).																		
	10	15	20	25	30	35	40	45	50	55	60	65	70	75	80	85	90	95	100
1	6°	4°	3°	2°	2°	2°	1°	1°	1°	1°	1°	1°	1°	1°	1°	1°	1°	1°	1°
2	11	8	6	5	4	3	3	3	2	2	2	2	2	2	1	1	1	1	1
3	17	11	9	7	6	5	4	4	3	3	3	3	2	2	2	2	2	2	2
4	22	15	11	9	8	7	6	5	5	4	4	4	3	3	3	3	3	2	2
5	27	18	14	11	9	8	7	6	6	5	5	4	4	4	4	3	3	3	3
6	31	22	17	13	11	10	9	8	7	6	6	5	5	5	4	4	4	4	3
7	35	25	19	16	13	11	10	9	8	7	7	6	6	5	5	5	4	4	4
8	39	28	22	18	15	13	11	10	9	8	8	7	7	6	6	5	5	5	5
9	42	31	24	20	17	14	13	11	10	9	9	8	7	7	6	6	6	5	5
10	45	34	27	22	18	16	14	13	11	10	9	9	8	8	7	7	6	6	6
11		36	29	24	20	17	15	14	12	11	10	10	9	8	8	7	7	7	6
12		39	31	26	22	19	17	15	13	12	11	10	10	9	9	8	8	7	7
13		41	33	27	23	20	18	16	15	13	12	11	11	10	9	9	8	8	7
14		43	35	29	25	22	19	17	16	14	13	12	11	11	10	9	9	8	8
15		45	37	31	27	23	21	18	17	15	14	13	12	11	11	10	9	9	9
16			39	33	28	25	22	20	18	16	15	14	13	12	11	11	10	10	9
17			40	34	30	26	23	21	19	17	16	15	14	13	12	11	11	10	10
18			42	36	31	27	24	22	20	18	17	15	14	13	13	12	11	11	10
19			44	37	32	28	25	23	21	19	18	16	15	14	13	13	12	11	11
20			45	39	34	30	27	24	22	20	18	17	16	15	14	13	13	12	11
21				40	35	31	28	25	23	21	19	18	17	16	15	14	13	12	12
22				41	36	32	29	26	24	22	20	19	17	16	15	15	14	13	12
23				43	37	33	30	27	25	23	21	19	18	17	16	15	14	14	13
24				44	39	34	31	28	26	24	22	20	19	18	17	16	15	14	13
25				45	40	36	32	29	27	24	23	21	20	18	17	16	16	15	14
26					41	37	33	30	27	25	23	22	20	19	18	17	16	15	15
27					42	38	34	31	28	26	24	22	21	20	19	18	17	16	15
28					43	39	35	32	29	27	25	23	22	20	19	18	17	16	16
29					44	40	36	33	30	28	26	24	23	21	20	19	18	17	16
30					45	41	37	34	31	29	27	25	23	22	21	19	18	18	17
31						42	38	35	32	29	27	25	24	22	21	20	19	18	17
32						42	39	35	33	30	28	26	25	23	22	21	20	19	18
33						43	40	36	33	31	29	27	25	24	22	21	20	19	18
34						44	40	37	34	32	30	28	26	24	23	22	21	20	19
35						45	41	38	35	32	30	28	27	25	24	22	21	20	19
36							42	39	36	33	31	29	27	26	24	23	22	21	20
37							43	39	37	34	32	30	28	26	25	24	22	21	20
38							44	40	37	35	32	30	28	27	25	24	23	22	21
39							44	41	38	35	33	31	29	27	26	25	23	22	21
40							45	42	39	36	34	32	30	28	27	25	24	23	22
41								42	39	37	34	32	30	29	27	26	24	23	22
42								43	40	37	35	33	31	29	28	26	25	24	23
43								44	41	38	36	33	32	30	28	27	26	24	23
44								44	41	39	36	34	32	30	29	27	26	25	24
45								45	42	39	37	35	33	31	29	28	27	25	24
46									43	40	37	35	33	32	30	28	27	26	25
47									43	41	38	36	34	32	30	29	28	26	25
48									44	41	39	36	34	33	31	29	28	27	26
49									44	42	39	37	35	33	31	30	29	27	26
50									45	42	40	38	36	34	32	30	29	28	27

SMITHSONIAN TABLES

TABLE 49.

MEAN DIRECTION OF THE WIND BY LAMBERT'S FORMULA.

Values of the mean direction (a) or its complement ($90°-a$).

n or d.	DENOMINATOR OR NUMERATOR (d OR n).									
	105	110	115	120	125	130	135	140	145	150
1	1°	1°	0°	0°	0°	0°	0°	0°	0°	0°
2	1	1	1	1	1	1	1	1	1	1
3	2	2	1	1	1	1	1	1	1	1
4	2	2	2	2	2	2	2	2	2	2
5	3	3	2	2	2	2	2	2	2	2
6	3	3	3	3	3	3	3	2	2	2
7	4	4	3	3	3	3	3	3	3	3
8	4	4	4	4	4	4	3	3	3	3
9	4	4	4	4	4	4	4	4	4	3
10	5	5	5	5	5	4	4	4	4	4
11	6	6	5	5	5	5	5	4	4	4
12	7	6	6	6	5	5	5	5	5	5
13	7	7	6	6	6	6	6	5	5	5
14	8	7	7	7	6	6	6	6	6	5
15	8	8	7	7	7	7	6	6	6	6
16	9	8	8	8	7	7	7	7	6	6
17	9	9	8	8	8	7	7	7	7	6
18	10	9	9	9	8	8	8	7	7	7
19	10	10	9	9	9	8	8	8	7	7
20	11	10	10	9	9	9	8	8	8	8
21	11	11	10	10	10	9	9	9	8	8
22	12	11	11	10	10	10	9	9	9	8
23	12	12	11	11	10	10	10	9	9	9
24	13	12	12	11	11	10	10	10	9	9
25	13	13	12	12	11	11	10	10	10	9
26	14	13	13	12	12	11	11	11	10	10
27	14	14	13	13	12	12	11	11	11	10
28	15	14	14	13	13	12	12	11	11	11
29	15	15	14	14	13	13	12	12	11	11
30	16	15	15	14	13	13	13	12	12	11
31	16	16	15	14	14	13	13	12	12	12
32	17	16	16	15	14	14	13	13	12	12
33	17	17	16	15	15	14	14	13	13	12
34	18	17	16	16	15	15	14	14	13	13
35	18	18	17	16	16	15	15	14	14	13
36	19	18	17	17	16	15	15	15	14	13
37	19	19	18	17	16	16	15	15	14	14
38	20	19	18	18	17	16	16	15	15	14
39	20	20	19	18	17	17	16	16	15	15
40	21	20	19	18	18	17	17	16	15	15
41	21	20	20	19	18	18	17	16	16	15
42	22	21	20	19	19	18	17	17	16	16
43	22	21	21	20	19	18	18	17	17	16
44	23	22	21	20	19	19	18	17	17	16
45	23	22	21	21	20	19	18	18	17	17
46	24	23	22	21	20	19	19	18	18	17
47	24	23	22	21	21	20	19	19	18	17
48	25	24	23	22	21	20	20	19	18	18
49	25	24	23	22	21	21	20	19	19	18
50	25	24	23	23	22	21	20	20	19	18

TABLE 49.

MEAN DIRECTION OF THE WIND BY LAMBERT'S FORMULA.

Values of the mean direction (a) or its complement ($90° - a$).

n or d.	DENOMINATOR OR NUMERATOR (d OR n).									
	155	160	165	170	175	180	185	190	195	200
1	0°	0°	0°	0°	0°	0°	0°	0°	0°	0°
2	1	1	1	1	1	1	1	1	1	1
3	1	1	1	1	1	1	1	1	1	1
4	1	1	1	1	1	1	1	1	1	1
5	2	2	2	2	2	2	2	2	1	1
6	2	2	2	2	2	2	2	2	2	2
7	3	3	2	2	2	2	2	2	2	2
8	3	3	3	3	3	3	2	2	2	2
9	3	3	3	3	3	3	3	3	3	3
10	4	4	3	3	3	3	3	3	3	3
11	4	4	4	4	4	3	3	3	3	3
12	4	4	4	4	4	4	4	4	4	3
13	5	5	5	4	4	4	4	4	4	4
14	5	5	5	5	5	4	4	4	4	4
15	6	5	5	5	5	5	5	5	4	4
16	6	6	6	5	5	5	5	5	5	5
17	6	6	6	6	6	5	5	5	5	5
18	7	7	6	6	6	6	6	6	5	5
19	7	7	7	6	6	6	6	6	6	5
20	7	7	7	7	7	6	6	6	6	6
21	8	7	7	7	7	7	6	6	6	6
22	8	8	8	7	7	7	7	7	6	6
23	8	8	8	8	7	7	7	7	7	7
24	9	9	8	8	8	8	7	7	7	7
25	9	9	9	8	8	8	8	7	7	7
26	10	9	9	9	8	8	8	8	8	7
27	10	10	9	9	9	9	8	8	8	8
28	10	10	10	9	9	9	9	8	8	8
29	11	10	10	10	9	9	9	9	8	8
30	11	11	10	10	10	9	9	9	9	9
31	11	11	11	10	10	10	10	9	9	9
32	12	11	11	11	10	10	10	10	9	9
33	12	12	11	11	11	10	10	10	10	9
34	12	12	12	11	11	11	10	10	10	10
35	13	12	12	12	11	11	11	10	10	10
36	13	13	12	12	12	11	11	11	10	10
37	13	13	13	12	12	12	11	11	11	10
38	14	13	13	13	12	12	12	11	11	11
39	14	14	13	13	13	12	12	12	11	11
40	14	14	14	13	13	13	12	12	12	11
41	15	14	14	14	13	13	12	12	12	12
42	15	15	14	14	13	13	13	12	12	12
43	16	15	15	14	14	13	13	13	12	12
44	16	15	15	15	14	14	13	13	13	12
45	16	16	15	15	14	14	14	13	13	13
46	17	16	16	15	15	14	14	14	13	13
47	17	16	16	15	15	15	14	14	14	13
48	17	17	16	16	15	15	15	14	14	13
49	18	17	17	16	16	15	15	14	14	14
50	18	17	17	16	16	16	15	15	14	14

SMITHSONIAN TABLES.

TABLE 49.

MEAN DIRECTION OF THE WIND BY LAMBERT'S FORMULA.

Values of the mean direction (a) or its complement ($90° - a$).

$$a = \tan^{-1} \frac{n}{d}$$

n or d.	\multicolumn{16}{c}{DENOMINATOR OR NUMERATOR (d OR n).}
	55	60	65	70	75	80	85	90	95	100	105	110	115	120	125	130
50	42°	40°	38°	36°	34°	32°	30°	29°	28°	27°	25°	24°	23°	23°	22°	21°
52	43	41	39	37	35	33	31	30	29	27	26	25	24	23	23	22
54	44	42	40	38	36	34	32	31	30	28	27	26	25	24	23	22
56		43	41	39	37	35	33	32	31	29	28	27	26	25	24	23
58		44	42	40	38	36	34	33	31	30	29	28	27	26	25	24
60		45	43	41	39	37	35	34	32	31	30	29	28	27	26	25
62			44	42	40	38	36	35	33	32	31	29	28	27	26	25
64			45	42	40	39	37	35	34	33	31	30	29	28	27	26
66				43	41	40	38	36	35	33	32	31	30	29	28	27
68				44	42	40	39	37	36	34	33	32	31	30	29	28
70				45	43	41	39	38	36	35	34	32	31	30	29	28
72					44	42	40	39	37	36	34	33	32	31	30	29
74					45	43	41	39	38	37	35	34	33	32	31	30
76						44	42	40	39	37	36	35	33	32	31	30
78						44	43	41	39	38	37	35	34	33	32	31
80						45	43	42	40	39	37	36	35	34	33	32
82							44	42	41	39	38	37	35	34	33	32
84							45	43	41	40	39	37	36	35	34	33
86								44	42	41	39	38	37	36	35	33
88								44	43	41	40	39	37	36	35	34
90								45	43	42	41	39	38	37	36	35
92									44	43	41	40	39	37	36	35
94									45	43	42	41	39	38	37	36
96										44	42	41	40	39	38	36
98										44	43	42	40	39	38	37
100										45	44	42	41	40	39	38
102											44	43	42	40	39	38
104											45	43	42	41	40	39
106												44	43	41	40	39
108												44	43	42	41	40
110												45	44	43	41	40
112													44	43	42	41
114													45	44	42	41
116														44	43	42
118														45	43	42
120														45	44	43
122															44	43
124															45	44
126																44
128																45
130																45

TABLE 49.

MEAN DIRECTION OF THE WIND BY LAMBERT'S FORMULA.

Values of the mean direction (a) or its complement ($90° - a$).

n or $d.$	DENOMINATOR OR NUMERATOR (d OR n).													
	130	135	140	145	150	155	160	165	170	175	180	185	190	195	200
50	21°	20°	20°	19°	18°	18°	17°	17°	16°	16°	16°	15°	15°	14°	14°
52	22	21	20	20	19	19	18	17	17	17	16	16	15	15	15
54	22	22	21	20	20	19	19	18	18	17	17	16	16	15	15
56	23	23	22	21	20	20	19	19	18	18	17	17	16	16	16
58	24	23	23	22	21	21	20	19	19	18	18	17	17	17	16
60	25	24	23	22	22	21	21	20	19	19	18	18	18	17	17
62	25	25	24	23	22	22	21	21	20	20	19	19	18	18	17
64	26	25	25	24	23	22	22	21	21	20	20	19	19	18	18
66	27	26	25	24	24	23	22	22	21	21	20	20	19	19	18
68	28	27	26	25	24	24	23	22	22	21	21	20	20	19	19
70	28	27	27	26	25	24	24	23	22	22	21	21	20	20	19
72	29	28	27	26	26	25	24	24	23	22	22	21	21	20	20
74	30	29	28	27	26	26	25	24	24	23	22	22	21	21	20
76	30	29	28	28	27	26	25	25	24	23	23	22	22	21	21
78	31	30	29	28	27	27	26	25	25	24	23	23	22	22	21
80	32	31	30	29	28	27	27	26	25	25	24	23	23	22	22
82	32	31	30	29	29	28	27	26	26	25	24	24	23	23	22
84	33	32	31	30	29	28	28	27	26	26	25	24	24	23	23
86	33	32	32	31	30	29	28	28	27	26	26	25	24	24	23
88	34	33	32	31	30	30	29	28	27	27	26	25	25	24	24
90	35	34	33	32	31	30	29	29	28	27	27	26	25	25	24
92	35	34	33	32	32	31	30	29	28	28	27	26	26	25	25
94	36	35	34	33	32	31	30	30	29	28	28	27	26	26	25
96	36	35	34	34	33	32	31	30	29	29	28	27	27	26	26
98	37	36	35	34	33	32	31	31	30	29	29	28	27	27	26
100	38	37	36	35	34	33	32	31	30	30	29	28	28	27	27
102	38	37	36	35	34	33	33	32	31	30	30	29	28	28	27
104	39	38	37	36	35	34	33	32	31	31	30	29	29	28	27
106	39	38	37	36	35	34	34	33	32	31	30	30	29	29	28
108	40	39	38	37	36	35	34	33	32	32	31	30	30	29	28
110	40	39	38	37	36	35	35	34	33	32	31	31	30	29	29
112	41	40	39	38	37	36	35	34	33	33	32	31	31	30	29
114	41	40	39	38	37	36	35	35	34	33	32	32	31	30	30
116	42	41	40	39	38	37	36	35	34	34	33	32	31	31	30
118	42	41	40	39	38	37	36	36	35	34	33	33	32	31	31
120	43	42	41	40	39	38	37	36	35	34	34	33	32	32	31
122	43	42	41	40	39	38	37	36	36	35	34	33	33	32	31
124	44	43	42	41	40	39	38	37	36	35	35	34	33	32	32
126	44	43	42	41	40	39	38	37	36	36	35	34	34	33	32
128	45	43	42	41	40	40	39	38	37	36	35	35	34	33	33
130	45	44	43	42	41	40	39	38	37	37	36	35	34	34	33
132		44	43	42	41	40	40	39	38	37	36	35	35	34	33
134		45	44	43	42	41	40	39	38	37	37	36	35	34	34
136			44	43	42	41	40	39	39	38	37	36	36	35	34
138			45	44	43	42	41	40	39	38	38	37	36	35	35
140			45	44	43	42	41	40	39	39	38	37	36	36	35
142				44	43	42	42	41	40	39	38	38	37	36	35
144				45	44	43	42	41	40	39	39	38	37	36	36
146					44	43	42	41	41	40	39	38	38	37	36
148					45	44	43	42	41	40	39	39	38	37	37
150						45	44	43	42	41	41	40	39	38	38	37

SMITHSONIAN TABLES.

TABLE 50.

SYNOPTIC CONVERSION OF VELOCITIES.

Miles per hour into metres per second, feet per second and kilometres per hour.

Miles per hour.	Metres per second.	Feet per second.	Kilometres per hour.	Miles per hour.	Metres per second.	Feet per second.	Kilometres per hour.	Miles per hour.	Metres per second.	Feet per second.	Kilometres per hour.
0.0	0.0	0.0	0.0	26.0	11.6	38.1	41.8	52.0	23.2	76.3	83.7
0.5	0.2	0.7	0.8	26.5	11.8	38.9	42.6	52.5	23.5	77.0	84.5
1.0	0.4	1.5	1.6	27.0	12.1	39.6	43.5	53.0	23.7	77.7	85.3
1.5	0.7	2.2	2.4	27.5	12.3	40.3	44.3	53.5	23.9	78.5	86.1
2.0	0.9	2.9	3.2	28.0	12.5	41.1	45.1	54.0	24.1	79.2	86.9
2.5	1.1	3.7	4.0	28.5	12.7	41.8	45.9	54.5	24.4	79.9	87.7
3.0	1.3	4.4	4.8	29.0	13.0	42.5	46.7	55.0	24.6	80.7	88.5
3.5	1.6	5.1	5.6	29.5	13.2	43.3	47.5	55.5	24.8	81.4	89.3
4.0	1.8	5.9	6.4	30.0	13.4	44.0	48.3	56.0	25.0	82.1	90.1
4.5	2.0	6.6	7.2	30.5	13.6	44.7	49.1	56.5	25.3	82.9	90.9
5.0	2.2	7.3	8.0	31.0	13.9	45.5	49.9	57.0	25.5	83.6	91.7
5.5	2.5	8.1	8.9	31.5	14.1	46.2	50.7	57.5	25.7	84.3	92.5
6.0	2.7	8.8	9.7	32.0	14.3	46.9	51.5	58.0	25.9	85.1	93.3
6.5	2.9	9.5	10.5	32.5	14.5	47.7	52.3	58.5	26.2	85.8	94.1
7.0	3.1	10.3	11.3	33.0	14.8	48.4	53.1	59.0	26.4	86.5	95.0
7.5	3.4	11.0	12.1	33.5	15.0	49.1	53.9	59.5	26.6	87.3	95.8
8.0	3.6	11.7	12.9	34.0	15.2	49.9	54.7	60.0	26.8	88.0	96.6
8.5	3.8	12.5	13.7	34.5	15.4	50.6	55.5	60.5	27.0	88.7	97.4
9.0	4.0	13.2	14.5	35.0	15.6	51.3	56.3	61.0	27.3	89.5	98.2
9.5	4.2	13.9	15.3	35.5	15.9	52.1	57.1	61.5	27.5	90.2	99.0
10.0	4.5	14.7	16.1	36.0	16.1	52.8	57.9	62.0	27.7	90.9	99.8
10.5	4.7	15.4	16.9	36.5	16.3	53.5	58.7	62.5	27.9	91.7	100.6
11.0	4.9	16.1	17.7	37.0	16.5	54.3	59.5	63.0	28.2	92.4	101.4
11.5	5.1	16.9	18.5	37.5	16.8	55.0	60.4	63.5	28.4	93.1	102.2
12.0	5.4	17.6	19.3	38.0	17.0	55.7	61.2	64.0	28.6	93.9	103.0
12.5	5.6	18.3	20.1	38.5	17.2	56.5	62.0	64.5	28.8	94.6	103.8
13.0	5.8	19.1	20.9	39.0	17.4	57.2	62.8	65.0	29.1	95.3	104.6
13.5	6.0	19.8	21.7	39.5	17.7	57.9	63.6	65.5	29.3	96.1	105.4
14.0	6.3	20.5	22.5	40.0	17.9	58.7	64.4	66.0	29.5	96.8	106.2
14.5	6.5	21.3	23.3	40.5	18.1	59.4	65.2	66.5	29.7	97.5	107.0
15.0	6.7	22.0	24.1	41.0	18.3	60.1	66.0	67.0	30.0	98.3	107.8
15.5	6.9	22.7	24.9	41.5	18.6	60.9	66.8	67.5	30.2	99.0	108.6
16.0	7.2	23.5	25.7	42.0	18.8	61.6	67.6	68.0	30.4	99.7	109.4
16.5	7.4	24.2	26.6	42.5	19.0	62.3	68.4	68.5	30.6	100.5	110.2
17.0	7.6	24.9	27.4	43.0	19.2	63.1	69.2	69.0	30.8	101.2	111.0
17.5	7.8	25.7	28.2	43.5	19.4	63.8	70.0	69.5	31.1	101.9	111.8
18.0	8.0	26.4	29.0	44.0	19.7	64.5	70.8	70.0	31.3	102.7	112.7
18.5	8.3	27.1	29.8	44.5	19.9	65.3	71.6	70.5	31.5	103.4	113.5
19.0	8.5	27.9	30.6	45.0	20.1	66.0	72.4	71.0	31.7	104.1	114.3
19.5	8.7	28.6	31.4	45.5	20.3	66.7	73.2	71.5	32.0	104.9	115.1
20.0	8.9	29.3	32.2	46.0	20.6	67.5	74.0	72.0	32.2	105.6	115.9
20.5	9.2	30.1	33.0	46.5	20.8	68.2	74.8	72.5	32.4	106.3	116.7
21.0	9.4	30.8	33.8	47.0	21.0	68.9	75.6	73.0	32.6	107.1	117.5
21.5	9.6	31.5	34.6	47.5	21.2	69.7	76.4	73.5	32.9	107.8	118.3
22.0	9.8	32.3	35.4	48.0	21.5	70.4	77.2	74.0	33.1	108.5	119.1
22.5	10.1	33.0	36.2	48.5	21.7	71.1	78.1	74.5	33.3	109.3	119.9
23.0	10.3	33.7	37.0	49.0	21.9	71.9	78.9	75.0	33.5	110.0	120.7
23.5	10.5	34.5	37.8	49.5	22.1	72.6	79.7	75.5	33.8	110.7	121.5
24.0	10.7	35.2	38.6	50.0	22.4	73.3	80.5	76.0	34.0	111.5	122.3
24.5	11.0	35.9	39.4	50.5	22.6	74.1	81.3	76.5	34.2	112.2	123.1
25.0	11.2	36.7	40.2	51.0	22.8	74.8	82.1	77.0	34.4	112.9	123.9
25.5	11.4	37.4	41.0	51.5	23.0	75.5	82.9	77.5	34.6	113.7	124.7
26.0	11.6	38.1	41.8	52.0	23.2	76.3	83.7	78.0	34.9	114.4	125.5

TABLE 51.
MILES PER HOUR INTO FEET PER SECOND.

1 mile per hour = $\frac{44}{30}$ feet per second.

Miles per hour.	0	1	2	3	4	5	6	7	8	9
	Feet per sec.	Feet per sec.	Feet per sec.	Feet per sec.	Feet per sec.	Feet per sec.	Feet per sec.	Feet per sec.	Feet per sec.	Feet per sec.
0	0.0	1.5	2.9	4.4	5.9	7.3	8.8	10.3	11.7	13.2
10	14.7	16.1	17.6	19.1	20.5	22.0	23.5	24.9	26.4	27.9
20	29.3	30.8	32.3	33.7	35.2	36.7	38.1	39.6	41.1	42.5
30	44.0	45.5	46.9	48.4	49.9	51.3	52.8	54.3	55.7	57.2
40	58.7	60.1	61.6	63.1	64.5	66.0	67.5	68.9	70.4	71.9
50	73.3	74.8	76.3	77.7	79.2	80.7	82.1	83.6	85.1	86.5
60	88.0	89.5	90.9	92.4	93.9	95.3	96.8	98.3	99.7	101.2
70	102.7	104.1	105.6	107.1	108.5	110.0	111.5	112.9	114.4	115.9
80	117.3	118.8	120.3	121.7	123.2	124.7	126.1	127.6	129.1	130.5
90	132.0	133.5	134.9	136.4	137.9	139.3	140.8	142.3	143.7	145.2
100	146.7	148.1	149.6	151.1	152.5	154.0	155.5	156.9	158.4	159.9
110	161.3	162.8	164.3	165.7	167.2	168.7	170.1	171.6	173.1	174.5
120	176.0	177.5	178.9	180.4	181.9	183.3	184.8	186.3	187.7	189.2
130	190.7	192.1	193.6	195.1	196.5	198.0	199.5	200.9	202.4	203.9
140	205.3	206.8	208.3	209.7	211.2	212.7	214.1	215.6	217.1	218.5

TABLE 52.
FEET PER SECOND INTO MILES PER HOUR.

1 foot per second = $\frac{30}{44}$ miles per hour.

Feet per sec.	0	1	2	3	4	5	6	7	8	9
	Miles per hr.	Miles per hr.	Miles per hr.	Miles per hr.	Miles per hr.	Miles per hr.	Miles per hr.	Miles per hr.	Miles per hr.	Miles per hr.
0	0.0	0.7	1.4	2.0	2.7	3.4	4.1	4.8	5.5	6.1
10	6.8	7.5	8.2	8.9	9.5	10.2	10.9	11.6	12.3	13.0
20	13.6	14.3	15.0	15.7	16.4	17.0	17.7	18.4	19.1	19.8
30	20.5	21.1	21.8	22.5	23.2	23.9	24.5	25.2	25.9	26.6
40	27.3	28.0	28.6	29.3	30.0	30.7	31.4	32.0	32.7	33.4
50	34.1	34.8	35.5	36.1	36.8	37.5	38.2	38.9	39.5	40.2
60	40.9	41.6	42.3	43.0	43.6	44.3	45.0	45.7	46.4	47.0
70	47.7	48.4	49.1	49.8	50.5	51.1	51.8	52.5	53.2	53.9
80	54.5	55.2	55.9	56.6	57.3	58.0	58.6	59.3	60.0	60.7
90	61.4	62.0	62.7	63.4	64.1	64.8	65.5	66.1	66.8	67.5
100	68.2	68.9	69.5	70.2	70.9	71.6	72.3	73.0	73.6	74.3
110	75.0	75.7	76.4	77.0	77.7	78.4	79.1	79.8	80.5	81.1
120	81.8	82.5	83.2	83.9	84.5	85.2	85.9	86.6	87.3	88.0
130	88.6	89.3	90.0	90.7	91.4	92.0	92.7	93.4	94.1	94.8
140	95.5	96.1	96.8	97.5	98.2	98.9	99.5	100.2	100.9	101.6
150	102.3	103.0	103.6	104.3	105.0	105.7	106.4	107.0	107.7	108.4
160	109.1	109.8	110.5	111.1	111.8	112.5	113.2	113.9	114.5	115.2
170	115.9	116.6	117.3	118.0	118.6	119.3	120.0	120.7	121.4	120.0
180	122.7	123.4	124.1	124.8	125.5	126.1	126.8	127.5	128.2	128.9
190	129.5	130.2	130.9	131.6	132.3	133.0	133.6	134.3	135.0	135.7

TABLE 53.

METRES PER SECOND INTO MILES PER HOUR.

1 metre per second = 2.236932 miles per hour.

Metres per second.	0.0	0.1	0.2	0.3	0.4	0.5	0.6	0.7	0.8	0.9
	Miles per hr.	Miles per hr.	Miles per hr.	Miles per hr.	Miles per hr.	Miles per hr.	Miles per hr.	Miles per hr.	Miles per hr.	Miles per hr.
0	0.0	0.2	0.4	0.7	0.9	1.1	1.3	1.6	1.8	2.0
1	2.2	2.5	2.7	2.9	3.1	3.4	3.6	3.8	4.0	4.3
2	4.5	4.7	4.9	5.1	5.4	5.6	5.8	6.0	6.3	6.5
3	6.7	6.9	7.2	7.4	7.6	7.8	8.1	8.3	8.5	8.7
4	8.9	9.2	9.4	9.6	9.8	10.1	10.3	10.5	10.7	11.0
5	11.2	11.4	11.6	11.9	12.1	12.3	12.5	12.8	13.0	13.2
6	13.4	13.6	13.9	14.1	14.3	14.5	14.8	15.0	15.2	15.4
7	15.7	15.9	16.1	16.3	16.6	16.8	17.0	17.2	17.4	17.7
8	17.9	18.1	18.3	18.6	18.8	19.0	19.2	19.5	19.7	19.9
9	20.1	20.4	20.6	20.8	21.0	21.3	21.5	21.7	21.9	22.1
10	22.4	22.6	22.8	23.0	23.3	23.5	23.7	23.9	24.2	24.4
11	24.6	24.8	25.1	25.3	25.5	25.7	25.9	26.2	26.4	26.6
12	26.8	27.1	27.3	27.5	27.7	28.0	28.2	28.4	28.6	28.9
13	29.1	29.3	29.5	29.8	30.0	30.2	30.4	30.6	30.9	31.1
14	31.3	31.5	31.8	32.0	32.2	32.4	32.7	32.9	33.1	33.3
15	33.6	33.8	34.0	34.2	34.4	34.7	34.9	35.1	35.3	35.6
16	35.8	36.0	36.2	36.5	36.7	36.9	37.1	37.4	37.6	37.8
17	38.0	38.3	38.5	38.7	38.9	39.1	39.4	39.6	39.8	40.0
18	40.3	40.5	40.7	40.9	41.2	41.4	41.6	41.8	42.1	42.3
19	42.5	42.7	43.0	43.2	43.4	43.6	43.8	44.1	44.3	44.5
20	44.7	45.0	45.2	45.4	45.6	45.9	46.1	46.3	46.5	46.8
21	47.0	47.2	47.4	47.6	47.9	48.1	48.3	48.5	48.8	49.0
22	49.2	49.4	49.7	49.9	50.1	50.3	50.6	50.8	51.0	51.2
23	51.5	51.7	51.9	52.1	52.3	52.6	52.8	53.0	53.2	53.5
24	53.7	53.9	54.1	54.4	54.6	54.8	55.0	55.3	55.5	55.7
25	55.9	56.1	56.4	56.6	56.8	57.0	57.3	57.5	57.7	57.9
26	58.2	58.4	58.6	58.8	59.1	59.3	59.5	59.7	60.0	60.2
27	60.4	60.6	60.8	61.1	61.3	61.5	61.7	62.0	62.2	62.4
28	62.6	62.9	63.1	63.3	63.5	63.8	64.0	64.2	64.4	64.6
29	64.9	65.1	65.3	65.5	65.8	66.0	66.2	66.4	66.7	66.9
30	67.1	67.3	67.6	67.8	68.0	68.2	68.5	68.7	68.9	69.1
31	69.3	69.6	69.8	70.0	70.2	70.5	70.7	70.9	71.1	71.4
32	71.6	71.8	72.0	72.3	72.5	72.7	72.9	73.1	73.4	73.6
33	73.8	74.0	74.3	74.5	74.7	74.9	75.2	75.4	75.6	75.8
34	76.1	76.3	76.5	76.7	77.0	77.2	77.4	77.6	77.8	78.1
35	78.3	78.5	78.7	79.0	79.2	79.4	79.6	79.9	80.1	80.3
36	80.5	80.8	81.0	81.2	81.4	81.6	81.9	82.1	82.3	82.5
37	82.8	83.0	83.2	83.4	83.7	84.0	84.1	84.3	84.6	84.8
38	85.0	85.2	85.5	85.7	85.9	86.1	86.3	86.6	86.8	87.0
39	87.2	87.5	87.7	87.9	88.1	88.4	88.6	88.8	89.0	89.3
40	89.5	89.7	89.9	90.2	90.4	90.6	90.8	91.0	91.3	91.5
41	91.7	91.9	92.2	92.4	92.6	92.8	93.1	93.3	93.5	93.7
42	94.0	94.2	94.4	94.6	94.8	95.1	95.3	95.5	95.7	96.0
43	96.2	96.4	96.6	96.9	97.1	97.3	97.5	97.8	98.0	98.2
44	98.4	98.7	98.9	99.1	99.3	99.5	99.8	100.0	100.2	100.4

METRES PER SECOND INTO MILES PER HOUR.

Metres per second.	0.0	0.1	0.2	0.3	0.4	0.5	0.6	0.7	0.8	0.9
	Miles per hr.	Miles per hr.	Miles per hr.	Miles per hr.	Miles per hr.	Miles per hr.	Miles per hr.	Miles per hr.	Miles per hr.	Miles per hr.
45	100.7	100.9	101.1	101.3	101.6	101.8	102.0	102.2	102.5	102.7
46	102.9	103.1	103.3	103.6	103.8	104.0	104.2	104.5	104.7	104.9
47	105.1	105.4	105.6	105.8	106.0	106.3	106.5	106.7	106.9	107.2
48	107.4	107.6	107.8	108.0	108.3	108.5	108.7	108.9	109.2	109.4
49	109.6	109.8	110.1	110.3	110.5	110.7	111.0	111.2	111.4	111.6
50	111.8	112.1	112.3	112.5	112.7	113.0	113.2	113.4	113.6	113.9
51	114.1	114.3	114.5	114.8	115.0	115.2	115.4	115.7	115.9	116.1
52	116.3	116.6	116.8	117.0	117.2	117.4	117.7	117.9	118.1	118.3
53	118.6	118.8	119.0	119.2	119.5	119.7	119.9	120.1	120.4	120.6
54	120.8	121.0	121.3	121.5	121.7	121.9	122.1	122.4	122.6	122.8
55	123.0	123.3	123.5	123.7	123.9	124.2	124.4	124.6	124.8	125.1
56	125.3	125.5	125.7	126.0	126.2	126.4	126.6	126.8	127.1	127.3
57	127.5	127.8	128.0	128.2	128.4	128.6	128.9	129.1	129.3	129.5
58	129.8	130.0	130.2	130.4	130.7	130.9	131.1	131.3	131.6	131.8
59	132.0	132.2	132.5	132.7	132.9	133.1	133.3	133.6	133.8	134.0

TABLE 54.

MILES PER HOUR INTO METRES PER SECOND.

1 mile per hour = 0.4470409 metres per second.

Miles per hour.	0	1	2	3	4	5	6	7	8	9
	metres per sec.	metres per sec.	metres per sec.	metres per sec.	metres per sec.	metres per sec.	metres per sec.	metres per sec.	metres per sec.	metres per sec.
0	0.00	0.45	0.89	1.34	1.79	2.24	2.68	3.13	3.58	4.02
10	4.47	4.92	5.36	5.81	6.26	6.71	7.15	7.60	8.05	8.49
20	8.94	9.39	9.83	10.28	10.73	11.18	11.62	12.07	12.52	12.96
30	13.41	13.86	14.31	14.75	15.20	15.65	16.09	16.54	16.99	17.43
40	17.88	18.33	18.78	19.22	19.67	20.12	20.56	21.01	21.46	21.91
50	22.35	22.80	23.25	23.69	24.14	24.59	25.03	25.48	25.93	26.38
60	26.82	27.27	27.72	28.16	28.61	29.06	29.50	29.95	30.40	30.85
70	31.29	31.74	32.19	32.63	33.08	33.53	33.98	34.42	34.87	35.32
80	35.76	36.21	36.66	37.10	37.55	38.00	38.45	38.89	39.34	39.79
90	40.23	40.68	41.13	41.57	42.02	42.47	42.92	43.36	43.81	44.26
100	44.70	45.15	45.60	46.05	46.49	46.94	47.39	47.83	48.28	48.73
110	49.17	49.62	50.07	50.51	50.96	51.41	51.86	52.30	52.75	53.20
120	53.64	54.09	54.54	54.99	55.43	55.88	56.33	56.77	57.22	57.67
130	58.12	58.56	59.01	59.46	59.90	60.35	60.80	61.24	61.69	62.14
140	62.59	63.03	63.48	63.93	64.37	64.82	65.27	65.72	66.16	66.61

SMITHSONIAN TABLES.

TABLE 55.

METRES PER SECOND INTO KILOMETRES PER HOUR.

1 metre per second = 3.6 kilometres per hour.

Metres per second.	0.0	0.1	0.2	0.3	0.4	0.5	0.6	0.7	0.8	0.9
	km. per hr.	km. per hr.	km. per hr.	km. per hr.	km. per hr.	km. per hr.	km. per hr.	km. per hr.	km. per hr.	km. per hr.
0	0.0	0.4	0.7	1.1	1.4	1.8	2.2	2.5	2.9	3.2
1	3.6	4.0	4.3	4.7	5.0	5.4	5.8	6.1	6.5	6.8
2	7.2	7.6	7.9	8.3	8.6	9.0	9.4	9.7	10.1	10.4
3	10.8	11.2	11.5	11.9	12.2	12.6	13.0	13.3	13.7	14.0
4	14.4	14.8	15.1	15.5	15.8	16.2	16.6	16.9	17.3	17.6
5	18.0	18.4	18.7	19.1	19.4	19.8	20.2	20.5	20.9	21.2
6	21.6	22.0	22.3	22.7	23.0	23.4	23.8	24.1	24.5	24.8
7	25.2	25.6	25.9	26.3	26.6	27.0	27.4	27.7	28.1	28.4
8	28.8	29.2	29.5	29.9	30.2	30.6	31.0	31.3	31.7	32.0
9	32.4	32.8	33.1	33.5	33.8	34.2	34.6	34.9	35.3	35.6
10	36.0	36.4	36.7	37.1	37.4	37.8	38.2	38.5	38.9	39.2
11	39.6	40.0	40.3	40.7	41.0	41.4	41.8	42.1	42.5	42.8
12	43.2	43.6	43.9	44.3	44.6	45.0	45.4	45.7	46.1	46.4
13	46.8	47.2	47.5	47.9	48.2	48.6	49.0	49.3	49.7	50.0
14	50.4	50.8	51.1	51.5	51.8	52.2	52.6	52.9	53.3	53.6
15	54.0	54.4	54.7	55.1	55.4	55.8	56.2	56.5	56.9	57.2
16	57.6	58.0	58.3	58.7	59.0	59.4	59.8	60.1	60.5	60.8
17	61.2	61.6	61.9	62.3	62.6	63.0	63.4	63.7	64.1	64.4
18	64.8	65.2	65.5	65.9	66.2	66.6	67.0	67.3	67.7	68.0
19	68.4	68.8	69.1	69.5	69.8	70.2	70.6	70.9	71.3	71.6
20	72.0	72.4	72.7	73.1	73.4	73.8	74.2	74.5	74.9	75.2
21	75.6	76.0	76.3	76.7	77.0	77.4	77.8	78.1	78.5	78.8
22	79.2	79.6	79.9	80.3	80.6	81.0	81.4	81.7	82.1	82.4
23	82.8	83.2	83.5	83.9	84.2	84.6	85.0	85.3	85.7	86.0
24	86.4	86.8	87.1	87.5	87.8	88.2	88.6	88.9	89.3	89.6
25	90.0	90.4	90.7	91.1	91.4	91.8	92.2	92.5	92.9	93.2
26	93.6	94.0	94.3	94.7	95.0	95.4	95.8	96.1	96.5	96.8
27	97.2	97.6	97.9	98.3	98.6	99.0	99.4	99.7	100.1	100.4
28	100.8	101.2	101.5	101.9	102.2	102.6	103.0	103.3	103.7	104.0
29	104.4	104.8	105.1	105.5	105.8	106.2	106.6	106.9	107.3	107.6
30	108.0	108.4	108.7	109.1	109.4	109.8	110.2	110.5	110.9	111.2
31	111.6	112.0	112.3	112.7	113.0	113.4	113.8	114.1	114.5	114.8
32	115.2	115.6	115.9	116.3	116.6	117.0	117.4	117.7	118.1	118.4
33	118.8	119.2	119.5	119.9	120.2	120.6	121.0	121.3	121.7	122.0
34	122.4	122.8	123.1	123.5	123.8	124.2	124.6	124.9	125.3	125.6
35	126.0	126.4	126.7	127.1	127.4	127.8	128.2	128.5	128.9	129.2
36	129.6	130.0	130.3	130.7	131.0	131.4	131.8	132.1	132.5	132.8
37	133.2	133.6	133.9	134.3	134.6	135.0	135.4	135.7	136.1	136.4
38	136.8	137.2	137.5	137.9	138.2	138.6	139.0	139.3	139.7	140.0
39	140.4	140.8	141.1	141.5	141.8	142.2	142.6	142.9	143.3	143.6
40	144.0	144.4	144.7	145.1	145.4	145.8	146.2	146.5	146.9	147.2
41	147.6	148.0	148.3	148.7	149.0	149.4	149.8	150.1	150.5	150.8
42	151.2	151.6	151.9	152.3	152.6	153.0	153.4	153.7	154.1	154.4
43	154.8	155.2	155.5	155.9	156.2	156.6	157.0	157.3	157.7	158.0
44	158.4	158.8	159.1	159.5	159.8	160.2	160.6	160.9	161.3	161.6

SMITHSONIAN TABLES.

TABLE 55.

METRES PER SECOND INTO KILOMETRES PER HOUR.

Metres per second.	0.0	0.1	0.2	0.3	0.4	0.5	0.6	0.7	0.8	0.9
	km. per hr.	km. per hr.	km. per hr.	km. per hr.	km. per hr.	km. per hr.	km. per hr.	km. per hr.	km. per hr.	km. per hr.
45	162.0	162.4	162.7	163.1	163.4	163.8	164.2	164.5	164.9	165.2
46	165.6	166.0	166.3	166.7	167.0	167.4	167.8	168.1	168.5	168.8
47	169.2	169.6	169.9	170.3	170.6	171.0	171.4	171.7	172.1	172.4
48	172.8	173.2	173.5	173.9	174.2	174.6	175.0	175.3	175.7	176.0
49	176.4	176.8	177.1	177.5	177.8	178.2	178.6	178.9	179.3	179.6
50	180.0	180.4	180.7	181.1	181.4	181.8	182.2	182.5	182.9	183.2
51	183.6	184.0	184.3	184.7	185.0	185.4	185.8	186.1	186.5	186.8
52	187.2	187.6	187.9	188.3	188.6	189.0	189.4	189.7	190.1	190.4
53	190.8	191.2	191.5	191.9	192.2	192.6	193.0	193.3	193.7	194.0
54	194.4	194.8	195.1	195.5	195.8	196.2	196.6	196.9	197.3	197.6
55	198.0	198.4	198.7	199.1	199.4	199.8	200.2	200.5	200.9	201.2
56	201.6	202.0	202.3	202.7	203.0	203.4	203.8	204.1	204.5	204.8
57	205.2	205.6	205.9	206.3	206.6	207.0	207.4	207.7	208.1	208.4
58	208.8	209.2	209.5	209.9	210.2	210.6	211.0	211.3	211.7	212.0
59	212.4	212.8	213.1	213.5	213.8	214.2	214.6	214.9	215.3	215.6

TABLE 56.

KILOMETRES PER HOUR INTO METRES PER SECOND.

1 kilometre per hour $= \dfrac{10}{36}$ metres per second.

Kilometres per hour.	0	1	2	3	4	5	6	7	8	9
	metres per sec.	metres per sec.	metres per sec.	metres per sec.	metres per sec.	metres per sec.	metres per sec.	metres per sec.	metres per sec.	metres per sec.
0	0.00	0.28	0.56	0.83	1.11	1.39	1.67	1.94	2.22	2.50
10	2.78	3.06	3.33	3.61	3.89	4.17	4.44	4.72	5.00	5.28
20	5.56	5.83	6.11	6.39	6.67	6.94	7.22	7.50	7.78	8.06
30	8.33	8.61	8.89	9.17	9.44	9.72	10.00	10.28	10.56	10.83
40	11.11	11.39	11.67	11.94	12.22	12.50	12.78	13.06	13.33	13.61
50	13.89	14.17	14.44	14.72	15.00	15.28	15.56	15.83	16.11	16.39
60	16.67	16.94	17.22	17.50	17.78	18.06	18.33	18.61	18.89	19.17
70	19.44	19.72	20.00	20.28	20.56	20.83	21.11	21.39	21.67	21.94
80	22.22	22.50	22.78	23.06	23.33	23.61	23.89	24.17	24.44	24.72
90	25.00	25.28	25.56	25.83	26.11	26.39	26.67	26.94	27.22	27.50
100	27.78	28.06	28.33	28.61	28.89	29.17	29.44	29.72	30.00	30.28
110	30.56	30.83	31.11	31.39	31.67	31.94	32.22	32.50	32.78	33.06
120	33.33	33.61	33.89	34.17	34.44	34.72	35.00	35.28	35.56	35.83
130	36.11	36.39	36.67	36.94	37.22	37.50	37.78	38.06	38.33	38.61
140	38.89	39.17	39.44	39.72	40.00	40.28	40.56	40.83	41.11	41.39
150	41.67	41.94	42.22	42.50	42.78	43.06	43.33	43.61	43.89	44.17
160	44.44	44.72	45.00	45.28	45.56	45.83	46.11	46.39	46.67	46.94
170	47.22	47.50	47.78	48.06	48.33	48.61	48.89	49.17	49.44	49.72
180	50.00	50.28	50.56	50.83	51.11	51.39	51.67	51.94	52.22	52.50
190	52.78	53.06	53.33	53.61	53.89	54.17	54.44	54.72	55.00	55.28

TABLE 57.

BEAUFORT WIND SCALE AND ITS CONVERSION INTO VELOCITY.

Grade.	Designation.	Velocity in miles per hour.				
		a	b	c	d	e
0	Calm.	0	3.3*	0	0	3
1	Light air.	7	6.6	2	1	8
2	Light breeze.	14	10.0	4	4	13
3	Gentle breeze.	21	17.5	8	10	18
4	Moderate breeze.	28	25.0	16	17	23
5	Fresh breeze.	35	32.5	24	24	28
6	Strong breeze.	42	40.0	32	32	34
7	Moderate gale.	49	47.5	40	40	40
8	Fresh gale.	56	55.0	50	48	48
9	Strong gale.	63	62.5	62	56	56
10	Whole gale.	70	70.0	78	67	65
11	Storm.	77	77.5	96	82	75
12	Hurricane.	84	85.0	120	100	90

* Velocity 3.3 is assigned to 0.5 grade.

(*a.*) COLONEL SIR HENRY JAMES: Instructions for taking meteorological observations; with tables for their correction and notes on meteorological phenomena. 8vo. *Lond.*, 1860.

(*b.*) GEORGE NEUMAYER: Discussion of the meteorological and magnetical observations made at the Flagstaff Observatory, Melbourne, during the years 1858 to 1863. 4to. *Mannheim*, 1867.

(*c.*) J. K. LAUGHTON: Physical geography and its relation to the prevailing winds and currents. 8vo. *Lond.*, 1870. 2d ed., 8vo. *Lond.*, 1873.

(*d.*) C. A. SCHOTT: Meteorological observations in the Arctic seas. By Sir Francis Leopold McClintock, R. N. Made on board the Arctic searching yacht "Fox," in Baffin Bay and Prince Regent's Inlet, in 1857, 1858 and 1859. Reduced and discussed by Charles A. Schott. *Smithsonian Contributions to Knowledge*, 146. *Washington*, 1862.

(*e.*) ROBERT H. SCOTT: An attempt to establish a relation between the velocity of the wind and its force (Beaufort scale). *Quarterly Journal Meteorological Society, Lond.*, 1874-'75, ii, p. 109-123.

Instructions in the use of meteorological instruments. Compiled by direction of the Meteorological Committee. 8vo. *Lond.*, 1877.

GEODETICAL TABLES.

Relative acceleration of gravity at different latitudes . . . TABLE 58

Length of one degree of the meridian at different latitudes . . TABLE 59

Length of one degree of the parallel at different latitudes . . TABLE 60

Duration of sunshine at different latitudes . . . . TABLE 61

Declination of the sun for the year 1894 . . . . TABLE 62

Relative intensity of solar radiation at different latitudes for the first and sixteenth day of each month . . . . . . TABLE 63

TABLE 58.

RELATIVE ACCELERATION OF GRAVITY AT DIFFERENT LATITUDES.

Ratio of the acceleration of gravity at sea level for each 10′ of latitude, to its acceleration at latitude 45°.

$$\frac{g_\phi}{g_{45}} = 1 - 0.002662 \cos 2\phi$$

Latitude. ϕ.	0′	10′	20′	30′	40′	50′
0°	0.997 338	0.997 338	0.997 338	0.997 338	0.997 339	0.997 339
1	340	340	341	342	343	344
2	345	346	347	348	350	351
3	353	354	356	358	360	362
4	364	366	368	371	373	376
5	0.997 378	0.997 381	0.997 384	0.997 387	0.997 390	0.997 393
6	396	399	403	406	410	413
7	417	421	425	429	433	437
8	441	445	450	454	459	564
9	468	473	478	483	488	493
10	0.997 499	0.997 504	0.997 509	0.997 515	0.997 520	0.997 526
11	532	538	544	550	556	562
12	568	574	581	587	594	601
13	607	614	621	628	635	642
14	650	657	664	672	679	687
15	0.997 695	0.997 702	0.997 710	0.997 718	0.997 726	0.997 734
16	742	751	759	767	776	786
17	793	802	811	819	828	837
18	846	856	865	874	883	893
19	902	912	922	931	941	951
20	0.997 961	0.997 971	0.997 981	0.997 991	0.998 001	0.998 011
21	0.998 022	0.998 032	0.998 043	0.998 053	064	074
22	085	096	107	118	129	140
23	151	162	173	185	196	207
24	219	230	242	254	265	277
25	0.998 289	0.998 301	0.998 313	0.998 325	0.998 337	0.998 349
26	361	373	386	398	410	423
27	435	448	460	473	486	499
28	511	524	537	550	563	576
29	589	603	616	629	642	656
30	0.998 669	0.998 682	9.998 696	0.998 709	0.998 723	0.998 737
31	750	764	778	791	805	819
32	833	847	861	875	889	903
33	917	931	946	960	974	988
34	0.999 003	0.999 017	0.999 032	0.999 046	0.999 060	0.999 075
35	0.999 090	0.999 104	0.999 119	0.999 133	0.999 148	0.999 163
36	177	192	207	222	237	251
37	266	281	296	311	326	341
38	356	371	386	401	416	431
39	447	462	477	492	507	523
40	0.999 538	0.999 553	0.999 568	0.999 584	0.999 599	0.999 614
41	630	645	660	676	691	706
42	722	737	753	768	783	799
43	814	830	845	861	876	892
44	907	923	938	954	970	985
45	1.000 000	1.000 015	1.000 030	1.000 046	1.000 062	1.000 077

SMITHSONIAN TABLES.

TABLE 58.

RELATIVE ACCELERATION OF GRAVITY AT DIFFERENT LATITUDES.

Ratio of the acceleration of gravity at sea level for each 10′ of latitude, to its acceleration at latitude 45°.

$$\frac{g_\phi}{g_{45}} = 1 - 0.002662 \cos 2\phi$$

Latitude. ϕ.	0′	10′	20′	30′	40′	50′
45°	1.000 000	1.000 015	1.000 030	1.000 046	1.000 062	1.000 077
46	093	108	124	139	155	170
47	186	201	217	232	247	263
48	278	294	309	324	340	355
49	370	386	401	416	432	447
50	1.000 462	1.000 477	1.000 493	1.000 508	1.000 523	1.000 538
51	553	569	584	599	614	629
52	644	659	674	689	704	719
53	734	749	763	778	793	808
54	823	837	852	867	881	896
55	1.000 910	1.000 925	1.000 940	1.000 954	1.000 968	1.000 983
56	0 997	1 012	1 026	1 040	1 054	1 069
57	1 083	1 097	1 111	1 125	1 139	1 153
58	1 167	1 181	1 195	1 209	1 222	1 236
59	1 250	1 263	1 277	1 291	1 304	1 318
60	1.001 331	1.001 344	1.001 358	1.001 371	1.001 384	1.001 397
61	1 411	1 424	1 437	1 450	1 463	1 476
62	1 489	1 501	1 514	1 527	1 540	1 552
63	1 565	1 577	1 590	1 602	1 614	1 627
64	1 639	1 651	1 663	1 675	1 687	1 699
65	1.001 711	1.001 723	1.001 735	1.001 746	1.001 758	1.001 770
66	1 781	1 793	1 804	1 815	1 827	1 838
67	1 849	1 860	1 871	1 882	1 893	1 904
68	1 915	1 926	1 936	1 947	1 957	1 968
69	1 978	1 989	1 999	2 009	2 019	2 029
70	1.002 039	1.002 049	1.002 059	1.002 069	1.002 078	1.002 088
71	2 098	2 107	2 117	2 126	2 135	2 144
72	2 154	2 163	2 172	2 181	2 189	2 198
73	2 207	2 216	2 224	2 233	2 241	2 249
74	2 258	2 266	2 274	2 282	2 290	2 298
75	1.002 305	1.002 313	1.002 321	1.002 328	1.002 336	1.002 343
76	2 350	2 358	2 365	2 372	2 379	2 386
77	2 393	2 399	2 406	2 413	2 419	2 426
78	2 432	2 438	2 444	2 450	2 456	2 462
79	2 468	2 474	2 480	2 485	2 491	2 496
80	1.002 501	1.002 507	1.002 512	1.002 517	1.002 522	1.002 527
81	2 532	2 536	2 541	2 546	2 550	2 555
82	2 559	2 563	2 567	2 571	2 575	2 579
83	2 583	2 587	2 590	2 594	2 597	2 601
84	2 604	2 607	2 610	2 613	2 616	2 619
85	1.002 622	1.002 624	1.002 627	1.002 629	1.002 632	1.002 634
86	2 636	2 638	2 640	2 642	2 644	2 646
87	2 647	2 649	2 650	2 652	2 653	2 654
88	2 655	2 656	2 657	2 658	2 659	2 660
89	2 660	2 661	2 661	2 662	2 662	2 662

SMITHSONIAN TABLES.

TABLE 59.

LENGTH OF ONE DEGREE OF THE MERIDIAN AT DIFFERENT LATITUDES. (N-S)

Latitude.	Metres.	Statute Miles.	Geographic Miles. 1' of the Eq.	Latitude.	Metres.	Statute Miles.	Geographic Miles. 1' of the Eq.
0°	110 568.5	68.703	59.594	45°	111 132.1	69.054	59.898
1	110 568.8	68.704	59.594	46	111 151.9	69.067	59.908
2	110 569.8	68.705	59.595	47	111 171.6	69.079	59.919
3	110 571.5	68.706	59.596	48	111 191.3	69.091	59.929
4	110 573.9	68.707	59.597	49	111 210.9	69.103	59.940
5	110 577.0	68.709	59.598	50	111 230.5	69.115	59.951
6	110 580.7	68.711	59.600	51	111 249.9	69.127	59.961
7	110 585.1	68.714	59.603	52	111 269.2	69.139	59.972
8	110 590.2	68.717	59.606	53	111 288.3	69.151	59.982
9	110 595.9	68.721	59.609	54	111 307.3	69.163	59.992
10	110 602.3	68.725	59.612	55	111 326.0	69.175	60.002
11	110 609.3	68.729	59.616	56	111 344.5	69.186	60.012
12	110 617.0	68.734	59.620	57	111 362.7	69.198	60.022
13	110 625.3	68.739	59.625	58	111 380.7	69.209	60.032
14	110 634.2	68.745	59.629	59	111 398.4	69.220	60.041
15	110 643.7	68.751	59.634	60	111 415.7	69.230	60.051
16	110 653.8	68.757	59.640	61	111 432.7	69.241	60.060
17	110 664.5	68.763	59.646	62	111 449.4	69.251	60.069
18	110 675.7	68.770	59.652	63	111 465.7	69.261	60.077
19	110 687.5	68.778	59.658	64	111 481.5	69.271	60.086
20	110 699.9	68.786	59.665	65	111 497.0	69.281	60.094
21	110 712.8	68.794	59.672	66	111 512.0	69.290	60.102
22	110 726.2	68.802	59.679	67	111 526.5	69.299	60.110
23	110 740.1	68.810	59.686	68	111 540.5	69.308	60.118
24	110 754.4	68.819	59.694	69	111 554.1	69.316	60.125
25	110 769.2	68.829	59.702	70	111 567.1	69.324	60.132
26	110 784.5	68.838	59.710	71	111 579.7	69.332	60.139
27	110 800.2	68.848	59.719	72	111 591.6	69.340	60.145
28	110 816.3	68.858	59.727	73	111 603.0	69.347	60.151
29	110 832.8	68.868	59.736	74	111 613.9	69.354	60.157
30	110 849.7	68.879	59.745	75	111 624.1	69.360	60.163
31	110 866.9	68.889	59.755	76	111 633.8	69.366	60.168
32	110 884.4	68.900	59.764	77	111 642.8	69.372	60.173
33	110 902.3	68.911	59.774	78	111 651.2	69.377	60.177
34	110 920.4	68.923	59.784	79	111 659.0	69.382	60.182
35	110 938.8	68.934	59.794	80	111 666.2	69.386	60.186
36	110 957.4	68.946	59.804	81	111 672.6	69.390	60.189
37	110 976.3	68.957	59.814	82	111 678.5	69.394	60.192
38	110 995.3	68.969	59.824	83	111 683.6	69.397	60.195
39	111 014.5	68.981	59.834	84	111 688.1	69.400	60.197
40	111 033.9	68.993	59.845	85	111 691.9	69.402	60.199
41	111 053.4	69.005	59.855	86	111 695.0	69.404	60.201
42	111 073.0	69.017	59.866	87	111 697.4	69.405	60.202
43	111 092.6	69.029	59.876	88	111 699.2	69.407	60.203
44	111 112.4	69.042	59.887	89	111 700.2	69.407	60.204
45	111 132.1	69.054	59.898	90	111 700.6	69.407	60.204

SMITHSONIAN TABLES.

TABLE 60.

LENGTH OF ONE DEGREE OF THE PARALLEL AT DIFFERENT LATITUDES. (E-W)

Latitude.	Metres.	Statute Miles.	Geographic Miles. 1' of the Eq.	Latitude.	Metres.	Statute Miles.	Geographic Miles. 1' of the Eq.
0°	111 321.9	69.171	60.000	45°	78 850.0	48.995	42.498
1	111 305.2	69.162	59.991	46	77 466.5	48.135	41.753
2	111 254.6	69.130	59.964	47	76 059.2	47.261	40.994
3	111 170.4	69.078	59.918	48	74 628.5	46.372	40.223
4	111 052.6	69.005	59.855	49	73 174.9	45.469	39.440
5	110 901.2	68.911	59.773	50	71 698.9	44.552	38.644
6	110 716.2	68.796	59.673	51	70 200.8	43.621	37.837
7	110 497.7	68.660	59.556	52	68 681.1	42.676	37.018
8	110 245.8	68.503	59.420	53	67 140.3	41.719	36.187
9	109 960.5	68.326	59.266	54	65 578.8	40.749	35.346
10	109 641.9	68.128	59.095	55	63 997.1	39.766	34.493
11	109 290.1	67.909	58.905	56	62 395.7	38.771	33.630
12	108 905.2	67.670	58.697	57	60 775.1	37.764	32.757
13	108 487.3	67.411	58.472	58	59 135.7	36.745	31.873
14	108 036.6	67.131	58.229	59	57 478.1	35.715	30.979
15	107 553.1	66.830	57.969	60	55 802.8	34.674	30.076
16	107 037.0	66.510	57.690	61	54 110.2	33.622	29.164
17	106 488.5	66.169	57.395	62	52 400.9	32.560	28.243
18	105 907.7	65.808	57.082	63	50 675.4	31.488	27.313
19	105 294.7	65.427	56.751	64	48 934.3	30.406	26.374
20	104 649.8	65.026	56.404	65	47 178.0	29.315	25.428
21	103 973.2	64.606	56.039	66	45 407.1	28.215	24.473
22	103 265.0	64.166	55.657	67	43 622.2	27.106	23.511
23	102 525.4	63.706	55.259	68	41 823.8	25.988	22.542
24	101 754.6	63.227	54.843	69	40 012.4	24.862	21.566
25	100 953.0	62.729	54.411	70	38 188.6	23.729	20.583
26	100 120.6	62.212	53.963	71	36 353.0	22.589	19.593
27	99 257.8	61.676	53.498	72	34 506.2	21.441	18.598
28	98 364.8	61.121	53.016	73	32 648.6	20.287	17.597
29	97 441.9	60.548	52.519	74	30 780.9	19.126	16.590
30	96 489.3	59.956	52.006	75	28 903.6	17.960	15.578
31	05 507.3	59.345	51.476	76	27 017.4	16.788	14.562
32	94 496.2	58.717	50.931	77	25 122.8	15.611	13.541
33	93 456.3	58.071	50.371	78	23 220.4	14.428	12.515
34	92 387.9	57.407	49.795	79	21 310.8	13.242	11.486
35	91 291.3	56.726	49.204	80	19 394.6	12.051	10.453
36	90 166.8	56.027	48.598	81	17 472.4	10.857	9.417
37	89 014.8	55.311	47.977	82	15 544.7	9.659	8.378
38	87 835.6	54.578	47.341	83	13 612.2	8.458	7.337
39	86 629.6	53.829	46.691	84	11 675.5	7.255	6.293
40	85 397.0	53.063	46.027	85	9 735.1	6.049	5.247
41	84 138.4	52.281	45.349	86	7 791.7	4.841	4.200
42	82 854.0	51.483	44.656	87	5 845.9	3.632	3.151
43	81 544.2	50.669	43.950	88	3 898.3	2.422	2.101
44	80 209.4	49.840	43.231	89	1 949.4	1.211	1.051
45	78 850.0	48.995	42.498	90	0.0	0.000	0.000

SMITHSONIAN TABLES.

TABLE 61.

DURATION OF SUNSHINE AT DIFFERENT LATITUDES.

Declination of the Sun.	LATITUDE NORTH.								
	0°	5°	10°	15°	20°	25°	30°	35°	40°
	h. m.	h. m.	h. m.	h. m.	h. m.	h. m.	h. m.	h. m.	h. m.
−23° 27′	12 7	11 50	11 32	11 14	10 55	10 35	10 13	9 48	9 19
−23 20	12 7	11 50	11 32	11 14	10 56	10 36	10 14	9 49	9 20
−23 0	12 7	11 50	11 33	11 15	10 57	10 37	10 15	9 51	9 23
−22 40	12 7	11 50	11 33	11 16	10 58	10 38	10 17	9 53	9 26
−22 20	12 7	11 51	11 34	11 17	10 59	10 40	10 19	9 55	9 29
−22 0	12 7	11 51	11 34	11 18	11 0	10 41	10 20	9 58	9 31
−21 40	12 7	11 51	11 35	11 19	11 1	10 43	10 22	10 0	9 34
−21 20	12 7	11 52	11 35	11 19	11 2	10 44	10 24	10 2	9 37
−21 0	12 7	11 52	11 36	11 20	11 4	10 46	10 26	10 4	9 40
−20 40	12 7	11 52	11 37	11 21	11 5	10 47	10 28	10 6	9 42
−20 20	12 7	11 52	11 37	11 22	11 6	10 49	10 29	10 8	9 45
−20 0	12 7	11 53	11 38	11 23	11 7	10 50	10 31	10 11	9 47
−19 40	12 7	11 53	11 38	11 23	11 8	10 51	10 33	10 13	9 50
−19 20	12 7	11 53	11 39	11 24	11 9	10 53	10 35	10 15	9 53
−19 0	12 7	11 53	11 39	11 25	11 10	10 54	10 37	10 17	9 55
−18 40	12 7	11 54	11 40	11 26	11 11	10 55	10 38	10 19	9 58
−18 20	12 7	11 54	11 40	11 27	11 12	10 57	10 40	10 21	10 1
−18 0	12 7	11 54	11 41	11 28	11 13	10 58	10 42	10 23	10 3
−17 40	12 7	11 54	11 41	11 28	11 14	10 59	10 43	10 26	10 5
−17 20	12 7	11 55	11 42	11 29	11 15	11 1	10 45	10 28	10 8
−17 0	12 7	11 55	11 42	11 30	11 16	11 2	10 47	10 30	10 10
−16 40	12 7	11 55	11 43	11 31	11 17	11 4	10 49	10 32	10 13
−16 20	12 7	11 55	11 43	11 31	11 18	11 5	10 50	10 34	10 16
−16 0	12 7	11 56	11 44	11 32	11 19	11 6	10 52	10 36	10 18
−15 40	12 7	11 56	11 44	11 33	11 20	11 8	10 53	10 38	10 20
−15 20	12 7	11 56	11 45	11 34	11 21	11 9	10 55	10 40	10 23
−15 0	12 7	11 56	11 45	11 34	11 22	11 10	10 57	10 42	10 25
−14 40	12 7	11 57	11 46	11 35	11 23	11 11	10 59	10 44	10 28
−14 20	12 7	11 57	11 46	11 36	11 25	11 13	11 0	10 46	10 30
−14 0	12 7	11 57	11 47	11 37	11 26	11 14	11 2	10 48	10 32
−13 40	12 7	11 57	11 47	11 37	11 27	11 16	11 4	10 50	10 35
−13 20	12 7	11 58	11 48	11 38	11 28	11 17	11 5	10 52	10 37
−13 0	12 7	11 58	11 48	11 39	11 29	11 18	11 7	10 54	10 40
−12 40	12 7	11 58	11 49	11 40	11 30	11 19	11 8	10 56	10 42
−12 20	12 7	11 58	11 49	11 40	11 31	11 21	11 10	10 58	10 44
−12 0	12 7	11 58	11 50	11 41	11 32	11 22	11 11	11 0	10 47
−11 40	12 7	11 59	11 50	11 42	11 33	11 23	11 13	11 2	10 49
−11 20	12 7	11 59	11 51	11 43	11 34	11 25	11 15	11 4	10 52
−11 0	12 7	11 59	11 51	11 43	11 35	11 26	11 16	11 6	10 54
−10 40	12 7	11 59	11 52	11 44	11 36	11 27	11 18	11 8	10 56
−10 20	12 7	12 0	11 52	11 45	11 37	11 28	11 20	11 10	10 59
−10 0	12 7	12 0	11 53	11 46	11 38	11 30	11 21	11 12	11 1
− 9 40	12 7	12 0	11 53	11 46	11 39	11 31	11 23	11 14	11 3
− 9 20	12 7	12 0	11 54	11 47	11 40	11 32	11 24	11 16	11 5
− 9 0	12 7	12 1	11 54	11 47	11 41	11 34	11 26	11 17	11 8
− 8 40	12 7	12 1	11 55	11 48	11 42	11 35	11 28	11 19	11 10
− 8 20	12 7	12 1	11 55	11 49	11 43	11 36	11 29	11 21	11 12
− 8 0	12 7	12 1	11 56	11 50	11 44	11 37	11 31	11 23	11 14

TABLE 61.

DURATION OF SUNSHINE AT DIFFERENT LATITUDES.

Declination of the Sun.	LATITUDE NORTH.									
	42°	44°	46°	48°	50°	52°	54°	56°	58°	60°
	h. m.	h. m.	h. m.	h. m.	h. m.	h. m.	h. m.	h. m.	h. m.	h. m.
−23° 27′	9 7	8 53	8 38	8 22	8 4	7 44	7 22	6 56	6 27	5 52
−23 20	9 8	8 54	8 39	8 23	8 5	7 45	7 24	6 58	6 29	5 54
−23 0	9 11	8 58	8 43	8 28	8 10	7 50	7 29	7 4	6 36	6 2
−22 40	9 14	9 1	8 46	8 31	8 14	7 55	7 34	7 10	6 43	6 9
−22 20	9 17	9 4	8 50	8 35	8 18	8 0	7 39	7 16	6 49	6 17
−22 0	9 20	9 7	8 53	8 38	8 22	8 4	7 44	7 22	6 55	6 25
−21 40	9 23	9 10	8 57	8 42	8 26	8 9	7 49	7 27	7 1	6 32
−21 20	9 26	9 13	9 1	8 46	8 30	8 13	7 54	7 32	7 8	6 38
−21 0	9 28	9 17	9 4	8 50	8 34	8 18	7 59	7 38	7 14	6 46
−20 40	9 31	9 20	9 7	8 53	8 38	8 22	8 4	7 43	7 20	6 52
−20 20	9 34	9 23	9 11	8 57	8 42	8 26	8 8	7 49	7 25	6 59
−20 0	9 37	9 26	9 14	9 1	8 46	8 31	8 13	7 54	7 31	7 5
−19 40	9 40	9 29	9 17	9 4	8 50	8 35	8 18	7 59	7 37	7 12
−19 20	9 43	9 32	9 20	9 7	8 54	8 39	8 23	8 4	7 43	7 18
−19 0	9 46	9 35	9 24	9 11	8 58	8 43	8 27	8 9	7 48	7 25
−18 40	9 48	9 38	9 27	9 15	9 2	8 47	8 32	8 14	7 54	7 31
−18 20	9 51	9 41	9 30	9 19	9 6	8 52	8 36	8 19	7 59	7 37
−18 0	9 54	9 44	9 34	9 22	9 10	8 56	8 41	8 24	8 5	7 43
−17 40	9 56	9 47	9 37	9 25	9 13	9 0	8 45	8 29	8 10	7 49
−17 20	9 59	9 50	9 40	9 29	9 17	9 4	8 50	8 34	8 15	7 55
−17 0	10 2	9 53	9 43	9 32	9 21	9 8	8 54	8 38	8 20	8 1
−16 40	10 5	9 56	9 46	9 35	9 25	9 12	8 58	8 43	8 26	8 6
−16 20	10 7	9 59	9 49	9 39	9 28	9 16	9 2	8 47	8 31	8 12
−16 0	10 10	10 1	9 52	9 43	9 32	9 20	9 7	8 52	8 36	8 17
−15 40	10 12	10 4	9 55	9 46	9 35	9 24	9 11	8 57	8 41	8 23
−15 20	10 15	10 7	9 58	9 49	9 39	9 28	9 15	9 2	8 46	8 29
−15 0	10 18	10 10	10 1	9 52	9 43	9 31	9 19	9 6	8 51	8 34
−14 40	10 20	10 13	10 4	9 56	9 46	9 35	9 23	9 11	8 56	8 40
−14 20	10 23	10 16	10 7	9 59	9 49	9 39	9 28	9 15	9 1	8 45
−14 0	10 26	10 19	10 10	10 2	9 53	9 43	9 32	9 19	9 6	8 50
−13 40	10 28	10 21	10 13	10 5	9 56	9 47	9 36	9 24	9 11	8 56
−13 20	10 31	10 24	10 16	10 8	10 0	9 50	9 40	9 28	9 16	9 1
−13 0	10 33	10 26	10 19	10 11	10 3	9 54	9 44	9 33	9 20	9 6
−12 40	10 36	10 29	10 22	10 15	10 7	9 58	9 48	9 37	9 25	9 11
−12 20	10 38	10 32	10 25	10 18	10 10	10 1	9 52	9 41	9 30	9 17
−12 0	10 41	10 35	10 28	10 21	10 13	10 5	9 56	9 46	9 35	9 22
−11 40	10 44	10 38	10 31	10 25	10 17	10 9	10 0	9 50	9 39	9 27
−11 20	10 46	10 40	10 34	10 28	10 20	10 13	10 4	9 55	9 44	9 32
−11 0	10 49	10 43	10 37	10 31	10 23	10 16	10 8	9 59	9 49	9 37
−10 40	10 51	10 46	10 40	10 34	10 27	10 19	10 12	10 3	9 53	9 42
−10 20	10 53	10 49	10 43	10 37	10 31	10 23	10 16	10 7	9 58	9 47
−10 0	10 56	10 51	10 46	10 40	10 34	10 27	10 19	10 11	10 3	9 52
− 9 40	10 59	10 54	10 49	10 43	10 37	10 31	10 23	10 16	10 7	9 57
− 9 20	11 1	10 56	10 52	10 46	10 40	10 34	10 27	10 20	10 11	10 2
− 9 0	11 3	10 59	10 55	10 49	10 44	10 37	10 31	10 24	10 16	10 7
− 8 40	11 6	11 2	10 57	10 52	10 47	10 41	10 34	10 28	10 20	10 11
− 8 20	11 8	11 4	11 0	10 55	10 50	10 44	10 38	10 32	10 25	10 16
− 8 0	11 10	11 7	11 3	10 58	10 53	10 48	10 42	10 36	10 29	10 21

SMITHSONIAN TABLES.

TABLE 61.

DURATION OF SUNSHINE AT DIFFERENT LATITUDES.

Declination of the Sun.	LATITUDE NORTH.								
	0°	5°	10°	15°	20°	25°	30°	35°	40°
	h. m.	h. m.	h. m.	h. m.	h. m.	h. m.	h. m.	h. m.	h. m.
−8° 0′	12 7	12 1	11 55	11 50	11 44	11 37	11 31	11 23	11 14
−7 40	12 7	12 1	11 56	11 50	11 45	11 38	11 32	11 25	11 17
−7 20	12 7	12 1	11 56	11 51	11 46	11 40	11 34	11 27	11 19
−7 0	12 7	12 2	11 57	11 52	11 47	11 41	11 35	11 29	11 22
−6 40	12 7	12 2	11 57	11 53	11 48	11 42	11 37	11 31	11 24
−6 20	12 7	12 2	11 58	11 53	11 49	11 43	11 38	11 32	11 26
−6 0	12 7	12 2	11 58	11 54	11 50	11 45	11 40	11 34	11 28
−5 40	12 7	12 3	11 59	11 55	11 51	11 46	11 41	11 36	11 31
−5 20	12 7	12 3	11 59	11 55	11 52	11 47	11 43	11 38	11 33
−5 0	12 7	12 3	12 0	11 56	11 53	11 49	11 44	11 40	11 35
−4 40	12 7	12 3	12 0	11 57	11 54	11 50	11 46	11 42	11 37
−4 20	12 7	12 4	12 1	11 58	11 55	11 51	11 47	11 44	11 40
−4 0	12 7	12 4	12 1	11 58	11 56	11 52	11 49	11 46	11 42
−3 40	12 7	12 4	12 2	11 59	11 57	11 53	11 51	11 47	11 44
−3 20	12 7	12 4	12 2	12 0	11 58	11 55	11 52	11 49	11 46
−3 0	12 7	12 5	12 3	12 1	11 58	11 56	11 54	11 51	11 49
−2 40	12 7	12 5	12 3	12 1	11 59	11 58	11 55	11 53	11 51
−2 20	12 7	12 5	12 4	12 2	12 0	11 59	11 57	11 55	11 53
−2 0	12 7	12 5	12 4	12 3	12 1	12 0	11 58	11 57	11 55
−1 40	12 7	12 5	12 4	12 4	12 2	12 1	12 0	11 59	11 58
−1 20	12 7	12 6	12 5	12 4	12 3	12 2	12 2	12 1	12 0
−1 0	12 7	12 6	12 5	12 5	12 4	12 4	12 3	12 2	12 2
−0 40	12 7	12 6	12 6	12 5	12 5	12 5	12 5	12 4	12 4
−0 20	12 7	12 6	12 6	12 6	12 6	12 6	12 6	12 6	12 7
0 0	12 7	12 7	12 7	12 7	12 7	12 7	12 8	12 8	12 9
+0 20	12 7	12 7	12 7	12 8	12 8	12 8	12 9	12 10	12 11
0 40	12 7	12 7	12 8	12 8	12 9	12 10	12 11	12 12	12 13
1 0	12 7	12 7	12 8	12 9	12 10	12 11	12 13	12 14	12 15
1 20	12 7	12 8	12 9	12 10	12 11	12 13	12 14	12 16	12 17
1 40	12 7	12 8	12 9	12 10	12 12	12 14	12 16	12 17	12 20
2 0	12 7	12 8	12 10	12 11	12 13	12 15	12 17	12 19	12 22
2 20	12 7	12 8	12 10	12 12	12 14	12 16	12 19	12 21	12 25
2 40	12 7	12 9	12 11	12 13	12 15	12 17	12 20	12 23	12 27
3 0	12 7	12 9	12 11	12 13	12 16	12 19	12 22	12 25	12 29
3 20	12 7	12 9	12 12	12 14	12 17	12 20	12 23	12 27	12 31
3 40	12 7	12 9	12 12	12 15	12 18	12 21	12 25	12 29	12 33
4 0	12 7	12 10	12 13	12 16	12 19	12 22	12 26	12 31	12 35
4 20	12 7	12 10	12 13	12 16	12 20	12 23	12 28	12 32	12 38
4 40	12 7	12 10	12 14	12 17	12 21	12 25	12 29	12 34	12 40
5 0	12 7	12 10	12 14	12 18	12 22	12 26	12 31	12 36	12 43
5 20	12 7	12 10	12 15	12 19	12 23	12 28	12 32	12 38	12 45
5 40	12 7	12 11	12 15	12 19	12 24	12 29	12 34	12 40	12 47
6 0	12 7	12 11	12 16	12 20	12 25	12 30	12 35	12 42	12 49
6 20	12 7	12 11	12 16	12 21	12 26	12 31	12 37	12 44	12 52
6 40	12 7	12 11	12 16	12 22	12 27	12 32	12 39	12 46	12 54
7 0	12 7	12 12	12 17	12 22	12 28	12 34	12 40	12 48	12 56
7 20	12 7	12 12	12 17	12 23	12 29	12 35	12 42	12 50	12 58
7 40	12 7	12 12	12 18	12 23	12 30	12 36	12 43	12 52	13 1
8 0	12 7	12 13	12 18	12 24	12 31	12 38	12 45	12 53	13 3

TABLE 61

DURATION OF SUNSHINE AT DIFFERENT LATITUDES.

Declination of the Sun.	LATITUDE NORTH.									
	42°	44°	46°	48°	50°	52°	54°	56°	58°	60°
	h. m.	h. m.	h. m.	h. m.	h. m.	h. m.	h. m.	h. m.	h. m.	h. m.
−8° 0′	11 11	11 7	11 3	10 58	10 53	10 48	10 43	10 36	10 30	10 21
−7 40	11 13	11 10	11 5	11 1	10 57	10 52	10 46	10 40	10 34	10 26
−7 20	11 16	11 12	11 8	11 4	11 0	10 55	10 50	10 44	10 38	10 31
−7 0	11 19	11 15	11 11	11 7	11 3	10 59	10 54	10 48	10 42	10 35
−6 40	11 21	11 17	11 14	11 10	11 7	11 2	10 58	10 52	10 47	10 40
−6 20	11 23	11 20	11 17	11 13	11 10	11 5	11 1	10 56	10 51	10 45
−6 0	11 26	11 23	11 20	11 16	11 13	11 9	11 5	11 0	10 55	10 50
−5 40	11 28	11 25	11 23	11 19	11 16	11 13	11 8	11 4	10 59	10 55
−5 20	11 31	11 28	11 25	11 22	11 19	11 16	11 13	11 8	11 4	10 59
−5 0	11 33	11 31	11 28	11 25	11 23	11 19	11 16	11 12	11 8	11 4
−4 40	11 35	11 33	11 31	11 28	11 26	11 23	11 20	11 16	11 13	11 8
−4 20	11 38	11 36	11 34	11 31	11 29	11 26	11 23	11 20	11 17	11 13
−4 0	11 40	11 38	11 37	11 34	11 32	11 30	11 27	11 24	11 21	11 18
−3 40	11 43	11 41	11 39	11 37	11 35	11 33	11 31	11 28	11 26	11 22
−3 20	11 45	11 43	11 42	11 40	11 38	11 37	11 35	11 32	11 30	11 27
−3 0	11 47	11 46	11 45	11 43	11 42	11 40	11 38	11 36	11 34	11 32
−2 40	11 50	11 49	11 47	11 46	11 45	11 44	11 42	11 40	11 38	11 37
−2 20	11 52	11 51	11 50	11 49	11 48	11 47	11 46	11 44	11 43	11 41
−2 0	11 55	11 54	11 53	11 52	11 52	11 50	11 49	11 48	11 47	11 46
−1 40	11 57	11 56	11 55	11 55	11 55	11 54	11 53	11 52	11 51	11 50
−1 20	11 59	11 59	11 58	11 58	11 58	11 57	11 57	11 56	11 56	11 55
−1 0	12 2	12 2	12 1	12 1	12 1	12 1	12 1	12 0	12 0	11 59
−0 40	12 4	12 4	12 4	12 4	12 4	12 4	12 4	12 4	12 4	12 4
−0 20	12 7	12 7	12 7	12 7	12 7	12 7	12 8	12 8	12 8	12 9
+0 0	12 9	12 9	12 10	12 10	12 10	12 11	12 11	12 12	12 13	12 13
0 20	12 11	12 12	12 13	12 13	12 14	12 14	12 15	12 16	12 17	12 18
0 40	12 14	12 14	12 15	12 17	12 17	12 19	12 20	12 21	12 23
1 0	12 16	12 17	12 18	12 19	12 20	12 21	12 22	12 24	12 25	12 27
1 20	12 19	12 20	12 20	12 22	12 23	12 25	12 26	12 28	12 29	12 32
1 40	12 21	12 22	12 23	12 25	12 26	12 28	12 30	12 32	12 34	12 37
2 0	12 23	12 25	12 26	12 28	12 29	12 31	12 34	12 36	12 38	12 41
2 20	12 26	12 28	12 29	12 31	12 32	12 35	12 37	12 40	12 43	12 46
2 40	12 28	12 30	12 32	12 34	12 36	12 38	12 41	12 44	12 47	12 50
3 0	12 31	12 32	12 35	12 37	12 39	12 41	12 44	12 48	12 51	12 55
3 20	12 33	12 35	12 37	12 40	12 42	12 45	12 48	12 52	12 55	13 0
3 40	12 35	12 38	12 40	12 43	12 46	12 49	12 52	12 56	13 0	13 4
4 0	12 38	12 40	12 43	12 46	12 49	12 52	12 56	13 0	13 4	13 9
4 20	12 40	12 43	12 46	12 49	12 52	12 55	12 59	13 4	13 8	13 14
4 40	12 43	12 46	12 49	12 52	12 55	12 59	13 3	13 8	13 13	13 19
5 0	12 45	12 48	12 51	12 55	12 58	13 2	13 7	13 12	13 17	13 23
5 20	12 47	12 51	12 54	12 58	13 2	13 6	13 11	13 16	13 22	13 28
5 40	12 50	12 53	12 57	13 1	13 5	13 10	13 14	13 20	13 26	13 33
6 0	12 53	12 56	12 59	13 4	13 8	13 13	13 18	13 24	13 31	13 38
6 20	12 55	12 59	13 2	13 7	13 11	13 16	13 22	13 28	13 35	13 43
6 40	12 58	13 1	13 5	13 10	13 14	13 20	13 26	13 32	13 39	13 47
7 0	13 0	13 4	13 8	13 13	13 18	13 23	13 29	13 36	13 44	13 52
7 20	13 2	13 7	13 11	13 16	13 21	13 27	13 33	13 40	13 48	13 57
7 40	13 5	13 9	13 14	13 19	13 25	13 31	13 37	13 44	13 53	14 2
8 0	13 7	13 12	13 17	13 22	13 28	13 34	13 41	13 48	13 57	14 7

SMITHSONIAN TABLES.

TABLE 61.

DURATION OF SUNSHINE AT DIFFERENT LATITUDES.

Declination of the Sun.	LATITUDE NORTH.								
	0°	5°	10°	15°	20°	25°	30°	35°	40°
h. m.	h. m.	h. m.	h. m.	h. m.	h. m.	h. m.	h. m.	h. m.	h. m.
+8° 0′	12 7	12 13	12 18	12 24	12 31	12 38	12 45	12 53	13 3
8 20	12 7	12 13	12 19	12 25	12 32	12 39	12 47	12 55	13 5
8 40	12 7	12 13	12 19	12 26	12 33	12 40	12 48	12 57	13 8
9 0	12 7	12 13	12 20	12 26	12 34	12 41	12 50	12 59	13 10
9 20	12 7	12 13	12 20	12 27	12 35	12 43	12 52	13 1	13 13
9 40	12 7	12 14	12 21	12 28	12 36	12 44	12 53	13 3	13 14
10 0	12 7	12 14	12 21	12 29	12 37	12 45	12 55	13 5	13 17
10 20	12 7	12 14	12 22	12 29	12 38	12 47	12 56	13 7	13 19
10 40	12 7	12 14	12 22	12 30	12 39	12 48	12 58	13 9	13 22
11 0	12 7	12 15	12 23	12 31	12 40	12 49	12 59	13 11	13 24
11 20	12 7	12 15	12 23	12 32	12 41	12 50	13 1	13 13	13 26
11 40	12 7	12 15	12 24	12 32	12 42	12 52	13 2	13 15	13 29
12 0	12 7	12 15	12 24	12 33	12 43	12 53	13 4	13 17	13 31
12 20	12 7	12 16	12 25	12 34	12 44	12 55	13 6	13 19	13 34
12 40	12 7	12 16	12 25	12 35	12 45	12 56	13 8	13 21	13 36
13 0	12 7	12 16	12 26	12 35	12 46	12 57	13 9	13 23	13 38
13 20	12 7	12 16	12 26	12 36	12 47	12 58	13 11	13 25	13 41
13 40	12 7	12 17	12 27	12 37	12 48	13 0	13 13	13 27	13 43
14 0	12 7	12 17	12 27	12 38	12 49	13 1	13 14	13 29	13 46
14 20	12 7	12 17	12 28	12 39	12 50	13 2	13 16	13 31	13 48
14 40	12 7	12 17	12 28	12 40	12 51	13 4	13 17	13 33	13 51
15 0	12 7	12 18	12 29	12 40	12 52	13 5	13 19	13 35	13 53
15 20	12 7	12 18	12 29	12 41	12 53	13 7	13 21	13 37	13 56
15 40	12 7	12 18	12 30	12 41	12 54	13 8	13 23	13 39	13 58
16 0	12 7	12 19	12 30	12 42	12 55	13 9	13 25	13 41	14 1
16 20	12 7	12 19	12 31	12 43	12 56	13 11	13 26	13 43	14 3
16 40	12 7	12 19	12 31	12 44	12 58	13 12	13 28	13 45	14 6
17 0	12 7	12 19	12 32	12 45	12 59	13 13	13 29	13 47	14 8
17 20	12 7	12 20	12 32	12 46	13 0	13 15	13 31	13 50	14 11
17 40	12 7	12 20	12 33	12 46	13 1	13 16	13 33	13 52	14 14
18 0	12 7	12 20	12 33	12 47	13 2	13 17	13 35	13 54	14 16
18 20	12 7	12 20	12 34	12 48	13 3	13 19	13 37	13 56	14 19
18 40	12 7	12 21	12 34	12 49	13 4	13 20	13 38	13 58	14 22
19 0	12 7	12 21	12 35	12 50	13 5	13 22	13 40	14 0	14 24
19 20	12 7	12 21	12 35	12 51	13 6	13 23	13 42	14 2	14 26
19 40	12 7	12 22	12 36	12 52	13 7	13 25	13 44	14 5	14 29
20 0	12 7	12 22	12 36	12 52	13 8	13 26	13 46	14 7	14 32
20 20	12 7	12 22	12 37	12 53	13 10	13 28	13 47	14 10	14 35
20 40	12 7	12 22	12 37	12 54	13 11	13 29	13 49	14 12	14 37
21 0	12 7	12 23	12 38	12 55	13 12	13 31	13 51	14 14	14 40
21 20	12 7	12 23	12 39	12 56	13 13	13 32	13 53	14 16	14 43
21 40	12 7	12 23	12 39	12 56	13 14	13 34	13 55	14 19	14 46
22 0	12 7	12 24	12 40	12 57	13 16	13 35	13 56	14 21	14 49
22 20	12 7	12 24	12 41	12 58	13 17	13 37	13 58	14 23	14 52
22 40	12 7	12 24	12 41	12 59	13 18	13 38	14 0	14 25	14 54
23 0	12 7	12 25	12 42	13 0	13 19	13 40	14 2	14 28	14 57
23 20	12 7	12 25	12 42	13 1	13 20	13 41	14 4	14 30	15 0
23 27	12 7	12 25	12 43	13 1	13 20	13 41	14 5	14 31	15 1

SMITHSONIAN TABLES.

TABLE 61.

DURATION OF SUNSHINE AT DIFFERENT LATITUDES.

Declination of the Sun.	LATITUDE NORTH.									
	42°	44°	46°	48°	50°	52°	54°	56°	58°	60°
	h. m.	h. m.	h. m.	h. m.	h. m.	h. m.	h. m.	h. m.	h. m.	h. m.
+8° 0′	13 7	13 12	13 17	13 22	13 28	13 34	13 41	13 49	13 58	14 7
8 20	13 10	13 14	13 20	13 25	13 31	13 38	13 45	13 53	14 2	14 12
8 40	13 12	13 17	13 23	13 28	13 34	13 41	13 49	13 57	14 6	14 17
9 0	13 15	13 20	13 25	13 31	13 38	13 45	13 53	14 1	14 11	14 22
9 20	13 17	13 23	13 28	13 34	13 41	13 49	13 56	14 5	14 15	14 26
9 40	13 20	13 25	13 31	13 38	13 44	13 52	14 0	14 10	14 20	14 31
10 0	13 22	13 28	13 34	13 41	13 48	13 56	14 4	14 14	14 25	14 36
10 20	13 25	13 31	13 37	13 44	13 51	13 59	14 8	14 18	14 29	14 41
10 40	13 28	13 34	13 40	13 47	13 55	14 3	14 12	14 22	14 34	14 47
11 0	13 30	13 36	13 43	13 50	13 58	14 7	14 16	14 27	14 38	14 52
11 20	13 32	13 39	13 46	13 53	14 1	14 10	14 20	14 31	14 43	14 57
11 40	13 35	13 41	13 49	13 56	14 5	14 14	14 24	14 35	14 48	15 2
12 0	13 38	13 44	13 52	14 0	14 8	14 18	14 28	14 40	14 53	15 8
12 20	13 40	13 47	13 55	14 3	14 12	14 22	14 32	14 44	14 58	15 13
12 40	13 43	13 50	13 58	14 6	14 16	14 25	14 37	14 49	15 2	15 18
13 0	13 46	13 53	14 1	14 10	14 19	14 29	14 41	14 53	15 7	15 23
13 20	13 48	13 56	14 4	14 13	14 22	14 33	14 45	14 58	15 13	15 29
13 40	13 50	13 58	14 7	14 16	14 26	14 37	14 49	15 2	15 17	15 35
14 0	13 53	14 1	14 10	14 19	14 29	14 41	14 53	15 7	15 22	15 40
14 20	13 56	14 4	14 13	14 23	14 33	14 45	14 57	15 11	15 28	15 46
14 40	13 59	14 7	14 16	14 26	14 37	14 49	15 2	15 16	15 33	15 51
15 0	14 1	14 10	14 19	14 29	14 40	14 52	15 6	15 21	15 38	15 57
15 20	14 4	14 13	14 22	14 33	14 44	14 56	15 10	15 26	15 43	16 2
15 40	14 7	14 16	14 26	14 36	14 48	15 0	15 14	15 30	15 48	16 8
16 0	14 10	14 19	14 29	14 40	14 52	15 4	15 19	15 35	15 53	16 14
16 20	14 12	14 22	14 32	14 43	14 55	15 8	15 23	15 40	15 59	16 20
16 40	14 15	14 25	14 35	14 46	14 59	15 13	15 28	15 45	16 4	16 26
17 0	14 17	14 28	14 38	14 50	15 3	15 17	15 32	15 50	16 10	16 32
17 20	14 20	14 31	14 41	14 53	15 7	15 21	15 37	15 55	16 15	16 38
17 40	14 23	14 34	14 45	14 57	15 10	15 25	15 41	16 0	16 20	16 45
18 0	14 26	14 37	14 48	15 1	15 14	15 29	15 46	16 5	16 26	16 51
18 20	14 29	14 40	14 52	15 4	15 18	15 34	15 50	16 10	16 32	16 58
18 40	14 32	14 43	14 55	15 8	15 22	15 38	15 55	16 15	16 38	17 4
19 0	14 35	14 46	14 58	15 11	15 26	15 42	16 0	16 20	16 44	17 11
19 20	14 37	14 49	15 1	15 15	15 30	15 46	16 5	16 25	16 50	17 17
19 40	14 40	14 52	15 5	15 19	15 34	15 51	16 10	16 31	16 56	17 24
20 0	14 43	14 55	15 8	15 22	15 38	15 55	16 15	16 37	17 2	17 31
20 20	14 46	14 58	15 11	15 26	15 42	16 0	16 20	16 42	17 8	17 38
20 40	14 49	15 2	15 15	15 30	15 46	16 4	16 25	16 47	17 14	17 46
21 0	14 52	15 5	15 19	15 34	15 50	16 9	16 30	16 53	17 20	17 53
21 20	14 55	15 8	15 22	15 38	15 55	16 13	16 35	16 59	17 27	18 1
21 40	14 58	15 11	15 26	15 42	15 59	16 18	16 40	17 5	17 34	18 8
22 0	15 1	15 14	15 29	15 46	16 3	16 23	16 45	17 11	17 40	18 16
22 20	15 4	15 18	15 33	15 49	16 7	16 28	16 50	17 17	17 47	18 24
22 40	15 7	15 22	15 37	15 54	16 12	16 32	16 56	17 23	17 54	18 32
23 0	15 10	15 25	15 40	15 57	16 16	16 37	17 1	17 29	18 1	18 41
23 20	15 13	15 28	15 44	16 1	16 21	16 42	17 7	17 35	18 8	18 49
23 27	15 14	15 29	15 46	16 3	16 23	16 44	17 9	17 37	18 11	18 52

SMITHSONIAN TABLES.

TABLE 61.

DURATION OF SUNSHINE AT DIFFERENT LATITUDES.

Declination of the Sun.	LATITUDE NORTH.									
	60°	61°	62°	63°	64°	65°	66°	67°	68°	69°	70°
	h. m.	h. m.	h. m.	h. m.	h. m.	h. m.	h. m.	h. m.	h. m.	h. m.	h. m.
−23° 27′	5 52	5 31	5 8	4 42	4 11	3 34	2 46	1 29			
−23 20	5 55	5 34	5 12	4 46	4 16	3 40	2 53	1 41			
−23 0	6 2	5 43	5 21	4 56	4 28	3 53	3 11	2 11			
−22 40	6 10	5 51	5 30	5 6	4 39	4 7	3 27	2 35	0 59		
−22 20	6 17	5 59	5 39	5 16	4 50	4 20	3 43	2 56	1 43		
−22 0	6 25	6 7	5 47	5 25	5 1	4 32	3 58	3 14	2 13		
−21 40	6 32	6 14	5 56	5 34	5 11	4 43	4 11	3 31	2 38	1 1	
−21 20	6 39	6 22	6 4	5 43	5 20	4 55	4 24	3 47	2 59	1 45	
−21 0	6 46	6 29	6 12	5 52	5 30	5 5	4 36	4 1	3 18	2 16	
−20 40	6 52	6 37	6 20	6 1	5 40	5 16	4 48	4 16	3 35	2 41	1 2
−20 20	6 59	6 44	6 27	6 9	5 49	5 26	4 59	4 29	3 51	3 2	1 47
−20 0	7 5	6 51	6 34	6 17	5 58	5 35	5 10	4 41	4 6	3 22	2 19
−19 40	7 12	6 58	6 42	6 25	6 6	5 45	5 21	4 53	4 20	3 39	2 44
−19 20	7 18	7 4	6 49	6 33	6 14	5 54	5 31	5 5	4 34	3 55	3 6
−19 0	7 25	7 11	6 56	6 41	6 23	6 3	5 41	5 16	4 47	4 11	3 26
−18 40	7 31	7 17	7 4	6 48	6 31	6 12	5 51	5 26	4 59	4 25	3 44
−18 20	7 37	7 24	7 10	6 55	6 39	6 20	6 1	5 37	5 11	4 39	4 1
−18 0	7 43	7 31	7 17	7 3	6 47	6 29	6 10	5 47	5 22	4 52	4 16
−17 40	7 49	7 37	7 24	7 10	6 55	6 38	6 19	5 57	5 33	5 5	4 31
−17 20	7 55	7 43	7 31	7 17	7 2	6 46	6 28	6 7	5 43	5 17	4 45
−17 0	8 1	7 49	7 37	7 24	7 9	6 53	6 36	6 16	5 54	5 28	4 58
−16 40	8 6	7 55	7 44	7 31	7 17	7 1	6 44	6 26	6 4	5 40	5 11
−16 20	8 12	8 1	7 50	7 38	7 24	7 9	6 52	6 35	6 14	5 51	5 23
−16 0	8 17	8 7	7 56	7 44	7 31	7 17	7 1	6 44	6 24	6 2	5 35
−15 40	8 23	8 13	8 2	7 51	7 38	7 25	7 9	6 52	6 34	6 12	5 47
−15 20	8 29	8 19	8 8	7 58	7 45	7 32	7 17	7 1	6 43	6 22	5 59
−15 0	8 34	8 25	8 15	8 4	7 52	7 39	7 25	7 9	6 52	6 32	6 10
−14 40	8 40	8 31	8 21	8 10	7 59	7 46	7 32	7 17	7 1	6 42	6 20
−14 20	8 45	8 36	8 27	8 17	8 5	7 53	7 40	7 26	7 10	6 51	6 31
−14 0	8 50	8 42	8 33	8 23	8 12	8 1	7 47	7 34	7 18	7 1	6 41
−13 40	8 56	8 47	8 38	8 29	8 19	8 7	7 55	7 41	7 26	7 10	6 51
−13 20	9 1	8 53	8 44	8 35	8 25	8 14	8 2	7 49	7 35	7 19	7 1
−13 0	9 6	8 58	8 50	8 41	8 32	8 21	8 10	7 57	7 43	7 28	7 10
−12 40	9 11	9 4	8 56	8 47	8 38	8 28	8 17	8 5	7 51	7 37	7 20
−12 20	9 17	9 10	9 2	8 53	8 44	8 34	8 24	8 12	7 59	7 45	7 29
−12 0	9 22	9 15	9 7	8 59	8 50	8 41	8 31	8 20	8 7	7 53	7 38
−11 40	9 27	9 20	9 13	9 5	8 56	8 47	8 38	8 27	8 15	8 2	7 47
−11 20	9 32	9 25	9 19	9 11	9 3	8 54	8 44	8 34	8 23	8 10	7 56
−11 0	9 37	9 31	9 24	9 17	9 9	9 0	8 51	8 41	8 31	8 18	8 5
−10 40	9 42	9 36	9 29	9 22	9 15	9 7	8 58	8 49	8 38	8 26	8 14
−10 20	9 47	9 41	9 35	9 28	9 21	9 13	9 5	8 56	8 46	8 34	8 22
−10 0	9 52	9 46	9 40	9 34	9 27	9 19	9 11	9 3	8 53	8 42	8 31
− 9 40	9 57	9 51	9 46	9 40	9 33	9 26	9 18	9 10	9 0	8 50	8 39
− 9 20	10 2	9 56	9 51	9 45	9 39	9 32	9 25	9 16	9 8	8 58	8 47
− 9 0	10 7	10 2	9 56	9 50	9 44	9 38	9 31	9 23	9 15	9 5	8 55
− 8 40	10 11	10 7	10 2	9 56	9 50	9 44	9 37	9 30	9 22	9 13	9 3
− 8 20	10 16	10 12	10 7	10 2	9 56	9 50	9 44	9 37	9 29	9 21	9 11
− 8 0	10 21	10 17	10 12	10 7	10 2	9 56	9 50	9 43	9 36	9 28	9 19

SMITHSONIAN TABLES

TABLE 61.

DURATION OF SUNSHINE AT DIFFERENT LATITUDES.

Declination of the Sun.	LATITUDE NORTH.								
	71°	72°	73°	74°	75°	76°	77°	78°	79°	80°
	h. m.	h. m.	h. m.	h. m.	h. m.	h. m.	h. m.	h. m.	h. m.	h. m.
−23° 27′										
−23 20										
−23 0										
−22 40										
−22 20										
−22 0										
−21 40										
−21 20										
−21 0										
−20 40										
−20 20										
−20 0										
−19 40	1 3									
−19 20	1 50									
−19 0	2 22									
−18 40	2 47	1 5								
−18 20	3 10	1 52								
−18 0	3 30	2 25								
−17 40	3 49	2 52	1 6							
−17 20	4 6	3 14	1 55							
−17 0	4 22	3 35	2 29							
−16 40	4 37	3 54	2 56	1 8						
−16 20	4 52	4 12	3 20	1 58						
−16 0	5 6	4 28	3 41	2 32						
−15 40	5 19	4 44	4 1	3 1	1 10					
−15 20	5 32	4 59	4 19	3 25	2 2					
−15 0	5 44	5 13	4 36	3 47	2 37					
−14 40	5 56	5 27	4 52	4 7	3 6	1 13				
−14 20	6 8	5 40	5 7	4 26	3 31	2 5				
−14 0	6 19	5 52	5 21	4 43	3 54	2 42				
−13 40	6 29	6 5	5 35	5 0	4 14	3 12	1 15			
−13 20	6 40	6 17	5 49	5 16	4 34	3 38	2 10			
−13 0	6 51	6 29	6 2	5 31	4 52	4 2	2 48			
−12 40	6 1	6 40	6 15	5 45	5 9	4 23	3 19	1 18		
−12 20	7 11	6 50	6 27	5 59	5 25	4 43	3 46	2 15		
−12 0	7 21	7 1	6 39	6 13	5 41	5 2	4 10	2 55		
−11 40	7 31	7 12	6 51	6 26	5 56	5 19	4 32	3 27	1 21	
−11 20	7 40	7 23	7 3	6 38	6 11	5 38	4 53	3 55	2 20	
−11 0	7 50	7 33	7 14	6 51	6 25	5 54	5 13	4 20	3 2	
−10 40	7 59	7 43	7 25	7 3	6 34	6 9	5 31	4 43	3 35	1 25
−10 20	8 8	7 53	7 35	7 15	6 52	6 23	5 49	5 5	4 5	2 27
−10 0	8 17	8 3	7 46	7 27	7 4	6 38	6 6	5 25	4 31	3 10
− 9 40	8 26	8 13	7 56	7 38	7 17	6 52	6 22	5 44	4 56	3 46
− 9 20	8 35	8 22	8 7	7 50	7 29	7 6	6 38	6 3	5 19	4 17
− 9 0	8 44	8 31	8 17	8 1	7 41	7 20	6 53	6 21	5 40	4 44
− 8 40	8 53	8 41	8 27	8 11	7 53	7 33	7 8	6 38	6 0	5 10
− 8 20	9 1	8 50	8 37	8 22	8 5	7 46	7 22	6 55	6 19	5 34
− 8 0	9 10	8 59	8 47	8 33	8 17	7 59	7 36	7 11	6 38	5 56

SMITHSONIAN TABLES.

TABLE 61.
DURATION OF SUNSHINE AT DIFFERENT LATITUDES.

Declination of the Sun.		LATITUDE NORTH.									
		60°	61°	62°	63°	64°	65°	66°	67°	68°	69°	70°
		h. m.	h. m.	h. m.	h. m.	h. m.	h. m.	h. m.	h. m.	h. m.	h. m.	h. m.
−8°	0′	10 21	10 17	10 12	10 7	10 2	9 56	9 50	9 43	9 36	9 28	9 19
−7	40	10 26	10 22	10 17	10 13	10 8	10 2	9 56	9 50	9 43	9 35	9 27
−7	20	10 31	10 27	10 23	10 18	10 13	10 8	10 3	9 57	9 50	9 43	9 35
−7	0	10 35	10 32	10 28	10 23	10 19	10 14	10 9	10 4	9 57	9 50	9 43
−6	40	10 40	10 37	10 33	10 29	10 25	10 20	10 15	10 10	10 4	9 57	9 51
−6	20	10 45	10 42	10 38	10 34	10 31	10 26	10 22	10 16	10 11	10 5	9 58
−6	0	10 50	10 47	10 43	10 40	10 36	10 32	10 28	10 23	10 18	10 12	10 6
−5	40	10 55	10 52	10 49	10 45	10 41	10 38	10 34	10 29	10 25	10 19	10 14
−5	20	10 59	10 56	10 54	10 50	10 47	10 44	10 40	10 36	10 31	10 26	10 21
−5	0	11 4	11 1	10 59	10 56	10 53	10 50	10 46	10 42	10 38	10 34	10 29
−4	40	11 8	11 6	11 4	11 1	10 58	10 55	10 52	10 49	10 45	10 41	10 36
−4	20	11 13	11 11	11 9	11 7	11 4	11 1	10 58	10 55	10 52	10 48	10 44
−4	0	11 18	11 16	11 14	11 12	11 10	11 7	11 4	11 1	10 58	10 55	10 51
−3	40	11 22	11 21	11 19	11 17	11 15	11 13	11 10	11 8	11 5	11 2	10 59
−3	20	11 27	11 26	11 24	11 22	11 20	11 19	11 16	11 14	11 11	11 9	11 6
−3	0	11 32	11 31	11 29	11 28	11 26	11 24	11 22	11 20	11 18	11 16	11 13
−2	40	11 37	11 35	11 34	11 33	11 31	11 30	11 28	11 27	11 25	11 23	11 21
−2	20	11 41	11 40	11 39	11 38	11 37	11 36	11 34	11 33	11 32	11 30	11 28
−2	0	11 46	11 45	11 44	11 43	11 43	11 41	11 40	11 40	11 38	11 37	11 35
−1	40	11 50	11 50	11 49	11 49	11 48	11 47	11 46	11 46	11 45	11 44	11 43
−1	20	11 55	11 55	11 54	11 54	11 53	11 53	11 52	11 52	11 52	11 51	11 50
−1	0	11 59	11 59	11 59	11 59	11 59	11 59	11 58	11 58	11 58	11 58	11 58
−0	40	12 4	12 4	12 4	12 4	12 4	12 4	12 4	12 4	12 5	12 5	12 5
−0	20	12 9	12 9	12 9	12 10	12 10	12 10	12 10	12 11	12 11	12 12	12 12
0	0	12 13	12 14	12 14	12 15	12 15	12 16	12 16	12 17	12 18	12 19	12 19
+0	20	12 18	12 19	12 19	12 20	12 20	12 22	12 22	12 23	12 25	12 26	12 27
0	40	12 22	12 23	12 24	12 25	12 26	12 27	12 28	12 29	12 31	12 33	12 34
1	0	12 27	12 28	12 29	12 31	12 32	12 33	12 34	12 36	12 38	12 40	12 41
1	20	12 32	12 33	12 34	12 36	12 37	12 39	12 40	12 42	12 44	12 47	12 49
1	40	12 37	12 38	12 39	12 41	12 43	12 44	12 46	12 49	12 51	12 54	12 56
2	0	12 41	12 43	12 44	12 46	12 48	12 50	12 52	12 55	12 58	13 1	13 4
2	20	12 46	12 47	12 49	12 52	12 53	12 56	12 59	13 1	13 4	13 8	13 11
2	40	12 50	12 52	12 54	12 57	12 59	13 2	13 5	13 7	13 11	13 15	13 19
3	0	12 55	12 57	12 59	13 2	13 5	13 8	13 11	13 14	13 17	13 22	13 26
3	20	13 0	13 2	13 5	13 7	13 10	13 13	13 17	13 20	13 24	13 29	13 34
3	40	13 4	13 7	13 10	13 13	13 16	13 19	13 23	13 27	13 31	13 36	13 41
4	0	13 9	13 12	13 15	13 18	13 22	13 25	13 29	13 33	13 38	13 43	13 49
4	20	13 14	13 17	13 20	13 23	13 27	13 31	13 35	13 40	13 45	13 50	13 56
4	40	13 19	13 22	13 25	13 29	13 32	13 37	13 41	13 46	13 52	13 58	14 4
5	0	13 23	13 27	13 30	13 34	13 38	13 43	13 47	13 53	13 58	14 5	14 11
5	20	13 28	13 32	13 35	13 40	13 44	13 49	13 54	13 59	14 5	14 12	14 19
5	40	13 33	13 37	13 41	13 45	13 50	13 55	14 0	14 6	14 12	14 19	14 27
6	0	13 38	13 42	13 46	13 50	13 55	14 1	14 6	14 13	14 19	14 26	14 35
6	20	13 43	13 47	13 51	13 56	14 1	14 7	14 12	14 19	14 26	14 34	14 43
6	40	13 47	13 52	13 56	14 1	14 7	14 13	14 18	14 26	14 33	14 42	14 51
7	0	13 52	13 57	14 1	14 7	14 12	14 19	14 25	14 32	14 40	14 49	14 59
7	20	13 57	14 2	14 7	14 13	14 18	14 25	14 31	14 39	14 48	14 57	15 7
7	40	14 2	14 7	14 12	14 18	14 24	14 31	14 38	14 46	14 55	15 4	15 15
8	0	14 7	14 12	14 17	14 23	14 30	14 37	14 45	14 52	15 2	15 12	15 23

TABLE 61.

DURATION OF SUNSHINE AT DIFFERENT LATITUDES.

Declination of the Sun.	LATITUDE NORTH.									
	71°	72°	73°	74°	75°	76°	77°	78°	79°	80°
	h. m.	h. m.	h. m.	h. m.	h. m.	h. m.	h. m.	h. m.	h. m.	h. m.
−8° 0′	9 10	8 59	8 47	8 33	8 17	7 58	7 37	7 10	6 38	5 56
−7 40	9 18	9 08	8 56	8 43	8 28	8 11	7 50	7 26	6 56	6 18
−7 20	9 26	9 17	9 6	8 53	8 39	8 23	8 4	7 41	7 14	6 38
−7 0	9 35	9 26	9 16	9 3	8 50	8 35	8 17	7 56	7 31	6 58
−6 40	9 43	9 34	9 25	9 14	9 1	8 47	8 30	8 11	7 47	7 17
−6 20	9 51	9 43	9 34	9 24	9 12	8 59	8 43	8 25	8 3	7 36
−6 0	9 59	9 52	9 43	9 34	9 23	9 11	8 56	8 39	8 19	7 54
−5 40	10 7	10 1	9 53	9 44	9 34	9 22	9 9	8 53	8 34	8 11
−5 20	10 15	10 9	10 2	9 53	9 44	9 34	9 22	9 7	8 50	8 28
−5 0	10 23	10 17	10 11	10 3	9 55	9 45	9 34	9 20	9 5	8 46
−4 40	10 31	10 26	10 20	10 13	10 5	9 56	9 46	9 34	9 19	9 2
−4 20	10 39	10 34	10 29	10 22	10 15	10 7	9 58	9 47	9 34	9 18
−4 0	10 47	10 43	10 38	10 32	10 26	10 18	10 10	10 0	9 49	9 34
−3 40	10 55	10 51	10 46	10 41	10 36	10 29	10 22	10 13	10 3	9 50
−3 20	11 3	10 59	10 55	10 51	10 46	10 40	10 34	10 26	10 17	10 6
−3 0	11 11	11 8	11 4	11 0	10 56	10 51	10 45	10 39	10 31	10 22
−2 40	11 19	11 16	11 13	11 10	11 6	11 2	10 57	10 52	10 45	10 37
−2 20	11 26	11 24	11 22	11 19	11 16	11 13	11 8	11 4	10 59	10 52
−2 0	11 34	11 32	11 31	11 28	11 26	11 23	11 20	11 17	11 13	11 8
−1 40	11 42	11 41	11 39	11 38	11 36	11 34	11 32	11 29	11 26	11 23
−1 20	11 49	11 49	11 48	11 47	11 46	11 45	11 43	11 42	11 40	11 38
−1 0	11 57	11 57	11 56	11 56	11 56	11 55	11 55	11 55	11 54	11 53
−0 40	12 5	12 5	12 5	12 5	12 6	12 6	12 7	12 7	12 8	12 8
−0 20	12 13	12 13	12 14	12 15	12 16	12 17	12 18	12 20	12 21	12 23
0 0	12 20	12 22	12 22	12 24	12 26	12 28	12 29	12 32	12 35	12 38
+0 20	12 28	12 30	12 31	12 34	12 36	12 38	12 41	12 44	12 49	12 53
0 40	12 36	12 38	12 40	12 43	12 46	12 49	12 53	12 57	13 2	13 9
1 0	12 44	12 46	12 49	12 52	12 56	13 0	13 5	13 10	13 16	13 24
1 20	12 52	12 55	12 58	13 2	13 6	13 11	13 16	13 23	13 30	13 40
1 40	12 59	13 3	13 7	13 11	13 16	13 22	13 28	13 36	13 44	13 55
2 0	13 7	13 11	13 16	13 20	13 26	13 32	13 40	13 49	13 59	14 11
2 20	13 15	13 19	13 25	13 30	13 36	13 43	13 52	14 1	14 13	14 27
2 40	13 23	13 28	13 33	13 40	13 46	13 54	14 4	14 14	14 28	14 43
3 0	13 31	13 36	13 42	13 49	13 57	14 5	14 16	14 28	14 42	14 59
3 20	13 39	13 44	13 51	13 59	14 7	14 17	14 28	14 41	14 56	15 16
3 40	13 47	13 53	14 1	14 8	14 17	14 28	14 40	14 55	15 11	15 33
4 0	13 55	14 2	14 10	14 18	14 28	14 40	14 53	15 8	15 27	15 50
4 20	14 3	14 10	14 19	14 28	14 38	14 51	15 5	15 22	15 43	16 7
4 40	14 11	14 19	14 28	14 38	14 49	15 2	15 18	15 36	15 58	16 25
5 0	14 19	14 28	14 37	14 48	15 0	15 14	15 31	15 50	16 14	16 44
5 20	14 27	14 37	14 46	14 58	15 11	15 26	15 44	16 5	16 31	17 3
5 40	14 35	14 45	14 56	15 8	15 22	15 38	15 57	16 20	16 47	17 22
6 0	14 44	14 54	15 5	15 19	15 33	15 50	16 11	16 35	17 5	17 43
6 20	14 52	15 3	15 15	15 29	15 44	16 3	16 25	16 51	17 23	18 5
6 40	15 1	15 12	15 25	15 40	15 56	16 16	16 39	17 7	17 41	18 27
7 0	15 10	15 22	15 35	15 50	16 8	16 29	16 53	17 23	18 1	18 50
7 20	15 18	12 31	15 45	16 1	16 20	16 42	17 8	17 40	18 21	19 16
7 40	15 27	15 40	15 55	16 12	16 32	16 55	17 23	17 58	18 42	19 44
8 0	15 35	15 50	16 5	16 23	16 44	17 9	17 39	18 16	19 5	20 15

TABLE 61.
DURATION OF SUNSHINE AT DIFFERENT LATITUDES.

Declination of the Sun.	LATITUDE NORTH.										
	60°	61°	62°	63°	64°	65°	66°	67°	68°	69°	70°
	h. m.	h. m.	h. m.	h. m.	h. m.	h. m.	h. m.	h. m.	h. m.	h. m.	h. m.
+ 8° 0′	14 7	14 12	14 17	14 23	14 30	14 37	14 45	14 53	15 2	15 12	15 23
8 20	14 12	14 17	14 23	14 29	14 36	14 43	14 52	15 0	15 10	15 20	15 32
8 40	14 17	14 22	14 28	14 35	14 42	14 50	14 58	15 7	15 17	15 28	15 40
9 0	14 22	14 27	14 34	14 41	14 48	14 56	15 5	15 14	15 25	15 36	15 49
9 20	14 27	14 32	14 39	14 46	14 54	15 2	15 11	15 21	15 32	15 44	15 57
9 40	14 32	14 38	14 45	14 52	15 0	15 9	15 18	15 28	15 40	15 52	16 6
10 0	14 37	14 43	14 50	14 58	15 6	15 15	15 25	15 35	15 47	16 0	16 15
10 20	14 42	14 49	14 56	15 4	15 13	15 22	15 32	15 43	15 55	16 8	16 24
10 40	14 47	14 54	15 2	15 10	15 19	15 28	15 39	15 50	16 3	16 17	16 33
11 0	14 52	14 59	15 7	15 16	15 25	15 35	15 46	15 58	16 11	16 26	16 42
11 20	14 57	15 5	15 13	15 22	15 31	15 41	15 53	16 5	16 19	16 34	16 52
11 40	15 2	15 10	15 19	15 28	15 38	15 48	16 0	16 13	16 27	16 43	17 1
12 0	15 8	15 16	15 25	15 34	15 44	15 55	16 7	16 21	16 35	16 52	17 11
12 20	15 13	15 21	15 31	15 40	15 50	16 2	16 15	16 29	16 44	17 1	17 21
12 40	15 18	15 27	15 36	15 46	15 57	16 9	16 22	16 37	16 53	17 11	17 31
13 0	15 23	15 33	15 42	15 53	16 4	16 16	16 30	16 45	17 2	17 20	17 41
13 20	15 29	15 39	15 48	15 59	16 11	16 23	16 37	16 53	17 10	17 30	17 52
13 40	15 35	15 44	15 55	16 5	16 17	16 31	16 45	17 1	17 19	17 40	18 3
14 0	15 40	15 50	16 1	16 12	16 24	16 38	16 53	17 10	17 29	17 50	18 14
14 20	15 46	15 56	16 7	16 19	16 31	16 46	17 1	17 19	17 38	18 0	18 26
14 40	15 51	16 2	16 13	16 25	16 38	16 53	17 9	17 28	17 48	18 11	18 38
15 0	15 57	16 8	16 19	16 32	16 46	17 1	17 17	17 37	17 58	18 22	18 50
15 20	16 2	16 14	16 26	16 39	16 53	17 9	17 26	17 46	18 8	18 33	19 3
15 40	16 8	16 20	16 32	16 46	17 1	17 17	17 35	17 55	18 18	18 45	19 16
16 0	16 14	16 26	16 39	16 53	17 8	17 25	17 44	18 5	18 29	18 57	19 30
16 20	16 20	16 32	16 46	17 0	17 16	17 33	17 53	18 15	18 40	19 10	19 45
16 40	16 26	16 39	16 52	17 7	17 23	17 41	18 2	18 25	18 51	19 23	20 1
17 0	16 32	16 45	16 59	17 14	17 31	17 50	18 11	18 35	19 3	19 36	20 17
17 20	16 38	16 52	17 6	17 22	17 39	17 59	18 21	18 46	19 15	19 50	20 35
17 40	16 45	16 58	17 13	17 29	17 47	18 8	18 31	18 57	19 28	20 6	20 55
18 0	16 51	17 5	17 20	17 37	17 56	18 17	18 41	19 8	19 41	20 22	21 17
18 20	16 58	17 12	17 28	17 45	18 5	18 26	18 52	19 20	19 55	20 40	21 42
18 40	17 4	17 19	17 35	17 53	18 14	18 36	19 3	19 33	20 10	20 59	22 13
19 0	17 11	17 26	17 43	18 2	18 23	18 46	19 14	19 46	20 26	21 20	22 58
19 20	17 17	17 33	17 51	18 10	18 32	18 56	19 25	20 0	20 44	21 45	
19 40	17 24	17 41	17 59	18 19	18 41	19 7	19 37	20 14	21 3	22 16	
20 0	17 31	17 48	18 7	18 28	18 51	19 19	19 50	20 30	21 23	22 59	
20 20	17 38	17 56	18 15	18 37	19 1	19 30	20 4	20 47	24 47		
20 40	17 45	18 4	18 23	18 46	19 12	19 42	20 19	21 5	22 17		
21 0	17 52	14 11	18 32	18 56	19 23	19 25	20 34	21 26	23 1		
21 20	18 0	28 20	18 41	19 6	19 34	20 8	20 50	21 50			
21 40	18 .8	18 28	18 50	19 16	19 46	20 22	21 8	22 19			
22 0	18 16	18 37	19 0	19 27	19 58	20 37	21 29	23 2			
22 20	18 24	18 46	19 10	19 38	20 11	20 53	21 52				
22 40	18 32	18 55	19 20	19 50	20 25	21 11	22 21				
23 0	18 41	19 4	19 31	20 2	20 40	21 31	23 3				
23 20	18 49	19 13	19 41	20 14	20 56	21 54					
23 27	18 52	19 17	19 46	20 19	21 2	22 3					

SMITHSONIAN TABLES.

TABLE 61.
DURATION OF SUNSHINE AT DIFFERENT LATITUDES.

Declination of the Sun.	LATITUDE NORTH.				
	71°	72°	73°	74°	75°
	h. m.	h. m.	h. m.	h. m.	h. m.
+ 8° 0′	15 35	15 50	16 5	16 23	16 44
8 20	15 44	15 59	16 16	16 35	16 57
8 40	15 53	16 9	16 26	16 46	17 10
9 0	16 3	16 19	16 37	16 58	17 23
9 20	16 12	16 29	16 48	17 10	17 37
9 40	16 22	16 39	16 59	17 23	17 51
10 0	16 31	16 50	17 11	17 35	18 5
10 20	16 41	17 0	17 22	17 49	18 20
10 40	16 50	17 11	17 34	18 2	18 36
11 0	17 1	17 22	17 47	18 16	18 52
11 20	17 11	17 34	17 59	18 31	19 9
11 40	17 22	17 45	18 13	18 46	19 27
12 0	17 32	17 57	18 26	19 1	19 46
12 20	17 43	18 9	18 40	19 18	20 7
12 40	17 55	18 22	18 55	19 35	20 29
13 0	18 6	18 35	19 11	19 54	20 55
13 20	18 18	18 49	19 26	20 14	21 23
13 40	18 30	19 2	19 43	20 35	21 59
14 0	18 43	19 17	20 1	21 0	22 50
14 20	18 56	19 33	20 20	21 28	
14 40	19 10	19 49	20 41	22 2	
15 0	19 24	20 7	21 5	22 52	
15 20	19 40	20 26	21 32		
15 40	19 55	20 46	22 5		
16 0	20 13	21 10	22 54		
16 20	20 31	21 36			
16 40	20 51	22 8			
17 0	21 13	22 56			
17 20	21 39				
17 40	22 11				

	76°	77°	78°	79°	80°
+ 8° 0′	17 9	17 39	18 16	19 5	20 15
8 20	17 23	17 55	18 35	19 29	20 50
8 40	17 38	18 12	18 56	19 56	21 33
9 0	17 53	18 30	19 17	20 25	22 35
9 20	18 8	18 48	19 41	20 59	
9 40	18 25	19 8	20 6	21 40	
10 0	18 41	19 28	20 31	22 39	
10 20	18 59	19 50	21 6		
10 40	19 18	20 15	21 46		
11 0	19 38	20 41	22 43		
11 20	19 59	21 13			
11 40	20 23	21 50			
12 0	20 49	22 46			
12 20	21 19				
12 40	21 55				

TABLE 62.
DECLINATION OF THE SUN FOR THE YEAR 1894.

Day of Month.	Jan.	Feb.	Mar.
1	−22° 59′	−17° 1′	− 7° 29′
4	22 42	16 8	6 20
7	22 21	15 13	5 10
10	21 55	14 15	4 0
13	21 26	13 15	2 49
16	−20 53	−12 14	− 1 38
19	20 17	11 10	− 0 27
21	19 50	10 27	+ 0 21
24	19 8	9 21	1 32
27	18 23	8 14	2 42
30	17 35		3 52

	Apr.	May.	June.
1	+ 4° 39′	+15° 10′	+22° 6′
4	5 48	16 3	22 28
7	6 56	16 53	22 47
10	8 3	17 42	23 3
13	9 9	18 27	23 14
16	+10 13	+19 10	+23 22
19	11 16	19 50	23 27
21	11 57	20 15	23 27
24	12 57	20 49	23 25
27	13 55	21 21	23 20
30	14 51	21 49	23 11

	July.	Aug.	Sept.
1	+23° 7′	+17° 58′	+ 8° 12′
4	22 52	17 12	7 6
7	22 35	16 22	5 59
10	22 13	15 30	4 51
13	21 49	14 37	3 42
16	+21 21	+13 41	+ 2 33
19	20 49	12 43	1 23
21	20 27	12 3	+ 0 37
24	19 50	11 2	− 0 34
27	19 11	9 59	1 44
30	18 28	8 55	2 54

	Oct.	Nov.	Dec.
1	− 3° 17′	−14° 32′	−21° 52′
4	4 27	15 28	22 18
7	5 36	16 22	22 39
10	6 45	17 14	22 57
13	7 53	18 3	23 11
16	− 8 59	−18 49	−23 21
19	10 5	19 33	23 26
21	10 48	20 0	23 27
24	11 51	20 37	23 26
27	12 53	21 12	23 20
30	13 53	21 42	23 10

SMITHSONIAN TABLES.

TABLE 63.

RELATIVE INTENSITY OF SOLAR RADIATION.

Mean vertical intensity for 24 hours of solar radiation J and the solar constant A, in terms of the mean solar constant A_0.

Date.	Motion of the Sun in Longitude.	Relative Mean Vertical Intensity $\left(\frac{J}{A_0}\right)$. LATITUDE NORTH.										$\frac{A}{A_0}$.
		0°	10°	20°	30°	40°	50°	60°	70°	80°	90°	
Jan. 1	0°.99	0.303	0.265	0.220	0.169	0.117	0.066	0.018				1.0335
16	15.78	.307	.271	.229	.180	.129	.078	.028				1.0324
Feb. 1	31.54	.312	.282	.244	.200	.150	.100	.048	0.006			1.0288
15	45.34	.317	.293	.261	.223	.177	.118	.075	.027			1.0235
Mar. 1	59.14	.320	.303	.279	.245	.204	.158	.108	.056	0.013		1.0173
16	73.93	.321	.313	.296	.270	.236	.195	.148	.097	.057		1.0096
Apr. 1	89.70	.317	.319	.312	.295	.269	.235	.195	.148	.101	0.082	1.0009
16	104.49	.311	.321	.323	.315	.297	.271	.238	.201	.175	.177	0.9923
May 1	119.29	.303	.318	.330	.329	.320	.302	.278	.253	.255	.259	0.9841
16	134.05	.294	.318	.333	.339	.337	.327	.312	.298	.317	.322	0.9772
June 1	149.82	.287	.315	.334	.345	.349	.345	.337	.344	.360	.366	0.9714
16	164.60	.283	.313	.334	.348	.354	.353	.348	.361	.378	.384	0.9679
July 1	179.39	.283	.312	.333	.347	.352	.351	.345	.356	.373	.379	0.9666
16	194.13	.287	.314	.332	.342	.345	.340	.329	.331	.347	.352	0.9674
Aug. 1	209.94	.294	.316	.330	.334	.330	.318	.300	.282	.295	.300	0.9709
16	224.73	.303	.318	.325	.322	.310	.291	.264	.234	.227	.231	0.9760
Sept. 1	240.50	.310	.318	.316	.305	.285	.256	.220	.180	.139	.140	0.9828
16	255.29	.315	.315	.305	.284	.256	.220	.178	.130	.107	.043	0.9909
Oct. 1	270.07	.317	.308	.289	.261	.225	.183	.135	.084	.065		0.9995
16	284.86	.316	.298	.271	.236	.194	.147	.097	.047	.015		1.0080
Nov. 1	300.63	.312	.286	.251	.211	.164	.114	.063	.018			1.0164
16	315.42	.308	.276	.235	.190	.140	.089	.040				1.0235
Dec. 1	330.19	.304	.267	.224	.175	.124	.072	.024				1.0288
16	344.98	.302	.263	.218	.167	.115	.064	.016				1.0323
Year.....		0.305	0.301	0.289	0.268	0.241	0.209	0.173	0.144	0.133	0.126	

SMITHSONIAN TABLES.

CONVERSION OF LINEAR MEASURES.

Inches into millimetres . . . . TABLE 64

Millimetres into inches . . . . TABLE 65

Feet into metres . . . . TABLE 66

Metres into feet . . . . TABLE 67

Miles into kilometres . . . . TABLE 68

Kilometres into miles . . . . TABLE 69

Interconversion of nautical and statute miles . . . . . . TABLE 70

Continental measures of length with their metric and English
    equivalents . . . . TABLE 71

TABLE 64.
INCHES INTO MILLIMETRES.
1 inch = 25.40005 mm.

Inches.	.00	.01	.02	.03	.04	.05	.06	.07	.08	.09
	mm.	mm.	mm.	mm.	mm.	mm.	mm.	mm.	mm.	mm.
0.00	0.00	0.25	0.51	0.76	1.02	1.27	1.52	1.78	2.03	2.29
0.10	2.54	2.79	3.05	3.30	3.56	3.81	4.06	4.32	4.57	4.83
0.20	5.08	5.33	5.59	5.84	6.10	6.35	6.60	6.86	7.11	7.37
0.30	7.62	7.87	8.13	8.38	8.64	8.89	9.14	9.40	9.65	9.91
0.40	10.16	10.41	10.67	10.92	11.18	11.43	11.68	11.94	12.19	12.45
0.50	12.70	12.95	13.21	13.46	13.72	13.97	14.22	14.48	14.73	14.99
0.60	15.24	15.49	15.75	16.00	16.26	16.51	16.76	17.02	17.27	17.53
0.70	17.78	18.03	18.29	18.54	18.80	19.05	19.30	19.56	19.81	20.07
0.80	20.32	20.57	20.83	21.08	21.34	21.59	21.84	22.10	22.35	22.61
0.90	22.86	23.11	23.37	23.62	23.88	24.13	24.38	24.64	24.89	25.15
1.00	25.40	25.65	25.91	26.16	26.42	26.67	26.92	27.18	27.43	27.69
1.10	27.94	28.19	28.45	28.70	28.96	29.21	29.46	29.72	29.97	30.23
1.20	30.48	30.73	30.99	31.24	31.50	31.75	32.00	32.26	32.51	32.77
1.30	33.02	33.27	33.53	33.78	34.04	34.29	34.54	34.80	35.05	35.31
1.40	35.56	35.81	36.07	36.32	36.58	36.83	37.08	37.34	37.59	37.85
1.50	38.10	38.35	38.61	38.86	39.12	39.37	39.62	39.88	40.13	40.39
1.60	40.64	40.89	41.15	41.40	41.66	41.91	42.16	42.42	42.67	42.93
1.70	43.18	43.43	43.69	43.94	44.20	44.45	44.70	44.96	45.21	45.47
1.80	45.72	45.97	46.23	46.48	46.74	46.99	47.24	47.50	47.75	48.01
1.90	48.26	48.51	48.77	49.02	49.28	49.53	49.78	50.04	50.29	50.55
2.00	50.80	51.05	51.31	51.56	51.82	52.07	52.32	52.58	52.83	53.09
2.10	53.34	53.59	53.85	54.10	54.36	54.61	54.86	55.12	55.37	55.63
2.20	55.88	56.13	56.39	56.64	56.90	57.15	57.40	57.66	57.91	58.17
2.30	58.42	58.67	58.93	59.18	59.44	59.69	59.94	60.20	60.45	60.71
2.40	60.96	61.21	61.47	61.72	61.98	62.23	62.48	62.74	62.99	63.25
2.50	63.50	63.75	64.01	64.26	64.52	64.77	65.02	65.28	65.53	65.79
2.60	66.04	66.29	66.55	66.80	67.06	67.31	67.56	67.82	68.07	68.33
2.70	68.58	68.83	69.09	69.34	69.60	69.85	70.10	70.36	70.61	70.87
2.80	71.12	71.37	71.63	71.88	72.14	72.39	72.64	72.90	73.15	73.41
2.90	73.66	73.91	74.17	74.42	74.68	74.93	75.18	75.44	75.69	75.95
3.00	76.20	76.45	76.71	76.96	77.22	77.47	77.72	77.98	78.23	78.49
3.10	78.74	78.99	79.25	79.50	79.76	80.01	80.26	80.52	80.77	81.03
3.20	81.28	81.53	81.79	82.04	82.30	82.55	82.80	83.06	83.31	83.57
3.30	83.82	84.07	84.33	84.59	84.84	85.09	85.34	85.60	85.85	86.11
3.40	86.36	86.61	86.87	87.12	87.38	87.63	87.88	88.14	88.39	88.65
3.50	88.90	89.15	89.41	89.66	89.92	90.17	90.42	90.68	90.93	91.19
3.60	91.44	91.69	91.95	92.20	92.46	92.71	92.96	93.22	93.47	93.73
3.70	93.98	94.23	94.49	94.74	95.00	95.25	95.50	95.76	96.01	96.27
3.80	96.52	96.77	97.03	97.28	97.54	97.79	98.04	98.30	98.55	98.81
3.90	99.06	99.31	99.57	99.82	100.08	100.33	100.58	100.84	101.09	101.35
4.00	101.60	101.85	102.11	102.36	102.62	102.87	103.12	103.38	103.63	103.89
4.10	104.14	104.39	104.65	104.90	105.16	105.41	105.66	105.92	106.17	106.43
4.20	106.68	106.93	107.19	107.44	107.70	107.95	108.20	108.46	108.71	108.97
4.30	109.22	109.47	109.73	109.98	110.24	110.49	110.74	111.00	111.25	111.51
4.40	111.76	112.01	112.27	112.52	112.78	113.03	113.28	113.54	113.79	114.05
4.50	114.30	114.55	114.81	115.06	115.32	115.57	115.82	116.08	116.33	116.59
4.60	116.84	117.09	117.35	117.60	117.86	118.11	118.36	118.62	118.87	119.13
4.70	119.38	119.63	119.89	120.14	120.40	120.65	120.90	121.16	121.41	121.67
4.80	121.92	122.17	122.43	122.68	122.94	123.19	123.44	123.70	123.95	124.21
4.90	124.46	124.71	124.97	125.22	125.48	125.73	125.98	126.24	126.49	126.75
5.00	127.00	127.25	127.51	127.76	128.02	128.27	128.52	128.78	129.03	129.29

Proportional Parts.	Inch.	0.001	0.002	0.003	0.004	0.005	0.006	0.007	0.008	0.009
	mm.	0.025	0.051	0.076	0.102	0.127	0.152	0.178	0.203	0.229

SMITHSONIAN TABLES.

TABLE 64.

INCHES INTO MILLIMETRES.

1 inch = 25.40005 mm.

Inches.	.00	.01	.02	.03	.04	.05	.06	.07	.08	.09
	mm.	mm.	mm.	mm.	mm.	mm.	mm.	mm.	mm.	mm.
5.00	127.00	127.25	127.51	127.76	128.02	128.27	128.52	128.78	129.03	129.29
5.10	129.54	129.79	130.05	130.30	130.56	130.81	131.06	131.32	131.57	131.83
5.20	132.08	132.33	132.59	132.84	133.10	133.35	133.60	133.86	134.11	134.37
5.30	134.62	134.87	135.13	135.38	135.64	135.89	136.14	136.40	136.65	136.91
5.40	137.16	137.41	137.67	137.92	138.18	138.43	138.68	138.94	139.19	139.45
5.50	139.70	139.95	140.21	140.46	140.72	140.97	141.22	141.48	141.73	141.99
5.60	142.24	142.49	142.75	143.00	143.26	143.51	143.76	144.02	144.27	144.53
5.70	144.78	145.03	145.29	145.54	145.80	146.05	146.30	146.56	146.81	147.07
5.80	147.32	147.57	147.83	148.08	148.34	148.59	148.84	149.10	149.35	149.61
5.90	149.86	150.11	150.37	150.62	150.88	151.13	151.38	151.64	151.89	152.15
6.00	152.40	152.66	152.91	153.16	153.42	153.67	153.92	154.18	154.43	154.69
6.10	154.94	155.19	155.45	155.70	155.96	156.21	156.46	156.72	156.97	157.23
6.20	157.48	157.73	157.99	158.24	158.50	158.75	159.00	159.26	159.51	159.77
6.30	160.02	160.27	160.53	160.78	161.04	161.29	161.54	161.80	162.05	162.31
6.40	162.56	162.81	163.07	163.32	163.58	163.83	164.08	164.34	164.59	164.85
6.50	165.10	165.35	165.61	165.86	166.12	166.37	166.62	166.88	167.13	167.39
6.60	167.64	167.89	168.15	168.40	168.66	168.91	169.16	169.42	169.67	169.93
6.70	170.18	170.43	170.69	170.94	171.20	171.45	171.70	171.96	172.21	172.47
6.80	172.72	172.97	173.23	173.48	173.74	173.99	174.24	174.50	174.75	175.01
6.90	175.26	175.51	175.77	176.02	176.28	176.53	176.78	177.04	177.29	177.55
7.00	177.80	178.05	178.31	178.56	178.82	179.07	179.32	179.58	179.83	180.09
7.10	180.34	180.59	180.85	181.10	181.36	181.61	181.86	182.12	182.37	182.63
7.20	182.88	183.13	183.39	183.64	183.90	184.15	184.40	184.66	184.91	185.17
7.30	185.42	185.67	185.93	186.18	186.44	186.69	186.94	187.20	187.45	187.71
7.40	187.96	188.21	188.47	188.72	188.98	189.23	189.48	189.74	189.99	190.25
7.50	190.50	190.75	191.01	191.26	191.52	191.77	192.02	192.28	192.53	192.79
7.60	193.04	193.29	193.55	193.80	194.06	194.31	194.56	194.82	195.07	195.33
7.70	195.58	195.83	196.09	196.34	196.60	196.85	197.10	197.36	197.61	197.87
7.80	198.12	198.37	198.63	198.88	199.14	199.39	199.64	199.90	200.15	200.41
7.90	200.66	200.91	201.17	201.42	201.68	201.93	202.18	202.44	202.69	202.95
8.00	203.20	203.45	203.71	203.96	204.22	204.47	204.72	204.98	205.23	205.49
8.10	205.74	205.99	206.25	206.50	206.76	207.01	207.26	207.52	207.77	208.03
8.20	208.28	208.53	208.79	209.04	209.30	209.55	209.80	210.06	210.31	210.57
8.30	210.82	211.07	211.33	211.58	211.84	212.09	212.34	212.60	212.85	213.11
8.40	213.36	213.61	213.87	214.12	214.38	214.63	214.88	215.14	215.39	215.65
8.50	215.90	216.15	216.41	216.66	216.92	217.17	217.42	217.68	217.93	218.19
8.60	218.44	218.69	218.95	219.20	219.46	219.71	219.96	220.22	220.47	220.73
8.70	220.98	221.23	221.49	221.74	222.00	222.25	222.50	222.76	223.01	223.27
8.80	223.52	223.77	224.03	224.28	224.54	224.79	225.04	225.30	225.55	225.81
8.90	226.06	226.31	226.57	226.82	227.08	227.33	227.58	227.84	228.09	228.35
9.00	228.60	228.85	229.11	229.36	229.62	229.87	230.12	230.38	230.63	230.89
9.10	231.14	231.39	231.65	231.90	232.16	232.41	232.66	232.92	233.17	233.43
9.20	233.68	233.93	234.19	234.44	234.70	234.95	235.20	235.46	235.71	235.97
9.30	236.22	236.47	236.73	236.98	237.24	237.49	237.74	238.00	238.25	238.51
9.40	238.76	239.01	239.27	239.52	239.78	240.03	240.28	240.54	240.79	241.05
9.50	241.30	241.55	241.81	242.06	242.32	242.57	242.82	243.08	243.33	243.59
9.60	243.84	244.09	244.35	244.60	244.86	245.11	245.36	245.62	245.87	246.13
9.70	246.38	246.63	246.89	247.14	247.40	247.65	247.90	248.16	248.41	248.67
9.80	248.92	249.17	249.43	249.68	249.94	250.19	250.44	250.70	250.95	251.21
9.90	251.46	251.71	251.97	252.22	252.48	252.73	252.98	253.24	253.49	253.75
10.00	254.00	254.25	254.51	254.76	255.02	255.27	255.52	255.78	256.03	256.29

Proportional Parts.	Inch.	0.001	0.002	0.003	0.004	0.005	0.006	0.007	0.008	0.009
	mm.	0.025	0.051	0.076	0.102	0.127	0.152	0.178	0.203	0.229

TABLE 64.

INCHES INTO MILLIMETRES.

1 inch = 25.40005 mm.

Inches.	.00	.01	.02	.03	.04	.05	.06	.07	.08	.09
	mm.	mm.	mm.	mm.	mm.	mm.	mm.	mm.	mm.	mm.
10.00	254.00	254.25	254.51	254.76	255.02	255.27	255.52	255.78	256.03	256.29
10.10	256.54	256.79	257.05	257.30	257.56	257.81	258.06	258.32	258.57	258.83
10.20	259.08	259.33	259.59	259.84	260.10	260.35	260.60	260.86	261.11	261.37
10.30	261.62	261.87	262.13	262.38	262.64	262.89	263.14	263.40	263.65	263.91
10.40	264.16	264.41	264.67	264.92	265.18	265.43	265.68	265.94	266.19	266.45
10.50	266.70	266.95	267.21	267.46	267.72	267.97	268.22	268.48	268.73	268.99
10.60	269.24	269.49	269.75	270.00	270.26	270.51	270.76	271.02	271.27	271.53
10.70	271.78	272.03	272.29	272.54	272.80	273.05	273.30	273.56	273.81	274.07
10.80	274.32	274.57	274.93	275.08	275.34	275.59	275.84	276.10	276.35	276.61
10.90	276.86	277.11	277.37	277.62	277.88	278.13	278.38	278.64	278.89	279.15
11.00	279.40	279.65	279.91	280.16	280.42	280.67	280.92	281.18	281.43	281.69
11.10	281.94	282.19	282.45	282.70	282.96	283.21	283.46	283.72	283.97	284.23
11.20	284.48	284.73	284.99	285.24	285.50	285.75	286.00	286.26	286.51	286.77
11.30	287.02	287.27	287.53	287.78	288.04	288.29	288.54	288.80	289.05	289.31
11.40	289.56	289.81	290.07	290.32	290.58	290.83	291.08	291.34	291.59	291.85
11.50	292.10	292.35	292.61	292.86	293.12	293.37	293.62	293.88	294.13	294.39
11.60	294.64	294.89	295.15	295.40	295.66	295.91	296.16	296.42	296.67	296.93
11.70	297.18	297.43	297.69	297.94	298.20	298.45	298.70	298.96	299.21	299.47
11.80	299.72	299.97	300.23	300.48	300.74	300.99	301.24	301.50	301.75	302.01
11.90	302.26	302.51	302.77	303.02	303.28	303.53	303.78	304.04	304.29	304.55
12.00	304.80	305.05	305.31	305.56	305.82	306.07	306.32	306.58	306.83	307.09
12.10	307.34	307.59	307.85	308.10	308.36	308.61	308.86	309.12	309.37	309.63
12.20	309.88	310.13	310.39	310.64	310.90	311.15	311.40	311.66	311.91	312.17
12.30	312.42	312.67	312.93	313.18	313.44	313.69	313.94	314.20	314.45	314.71
12.40	314.96	315.21	315.47	315.72	315.98	316.23	316.48	316.74	316.99	317.25
12.50	317.50	317.75	318.01	318.26	318.52	318.77	319.02	319.28	319.53	319.79
12.60	320.04	320.29	320.55	320.80	321.06	321.31	321.56	321.82	322.07	322.33
12.70	322.58	322.83	323.09	323.34	323.60	323.85	324.10	324.36	324.61	324.87
12.80	325.12	325.37	325.63	325.88	326.14	326.39	326.64	326.90	327.15	327.41
12.90	327.66	327.91	328.17	328.42	328.68	328.93	329.18	329.44	329.69	329.95
13.00	330.20	330.45	330.71	330.96	331.22	331.47	331.72	331.98	332.23	332.49
13.10	332.74	332.99	333.25	333.50	333.76	334.01	334.26	334.52	334.77	335.03
13.20	335.28	335.53	335.79	336.04	336.30	336.55	336.80	337.06	337.31	337.57
13.30	337.82	338.07	338.33	338.58	338.84	339.09	339.34	339.60	339.85	340.11
13.40	340.36	340.61	340.87	341.12	341.38	341.63	341.88	342.14	342.39	342.65
13.50	342.90	343.15	343.41	343.66	343.92	344.17	344.42	344.68	344.93	345.19
13.60	345.44	345.69	345.95	346.20	346.46	346.71	346.96	347.22	347.47	347.73
13.70	347.98	348.23	348.49	348.74	349.00	349.25	349.50	349.76	350.01	350.27
13.80	350.52	350.77	351.03	351.28	351.54	351.79	352.04	352.30	352.55	352.81
13.90	353.06	353.31	353.57	353.82	354.08	354.33	354.58	354.84	355.09	355.35
14.00	355.60	355.85	356.11	356.36	356.62	356.87	357.12	357.38	357.63	357.89
14.10	358.14	358.39	358.65	358.90	359.16	359.41	359.66	359.92	360.17	360.43
14.20	360.68	360.93	361.19	361.44	361.70	361.95	362.20	362.46	362.71	362.97
14.30	363.22	363.47	363.73	363.98	364.24	364.49	364.74	365.00	365.25	365.51
14.40	365.76	366.01	366.27	366.52	366.78	367.03	367.28	367.54	367.79	368.05
14.50	368.30	368.55	368.81	369.06	369.32	369.57	369.82	370.08	370.33	370.59
14.60	370.84	371.09	371.35	371.60	371.86	372.11	372.36	372.62	372.87	373.13
14.70	373.38	373.63	373.89	374.14	374.40	374.65	374.90	375.16	375.41	375.67
14.80	375.92	376.17	376.43	376.68	376.94	377.19	377.44	377.70	377.95	378.21
14.90	378.46	378.71	378.97	379.22	379.48	379.73	379.98	380.24	380.49	380.75
15.00	381.00	381.25	381.51	381.76	382.02	382.27	382.52	382.78	383.03	383.29

Proportional Parts.	Inch.	0.001	0.002	0.003	0.004	0.005	0.006	0.007	0.008	0.009
	mm.	0.025	0.051	0.076	0.102	0.127	0.152	0.178	0.203	0.229

TABLE 64.

INCHES INTO MILLIMETRES.

1 inch = 25.40005 mm.

Inches.	.00	.01	.02	.03	.04	.05	.06	.07	.08	.09
	mm.	mm.	mm.	mm.	mm.	mm.	mm.	mm.	mm.	mm.
15.00	381.00	381.25	381.51	381.76	382.02	382.27	382.52	382.78	383.03	383.29
15.10	383.54	383.79	384.05	384.30	384.56	384.81	385.06	385.32	385.57	385.83
15.20	386.08	386.33	386.59	386.84	387.10	387.35	387.60	387.86	388.11	388.37
15.30	388.62	388.87	389.13	389.38	389.64	389.89	390.14	390.40	390.65	390.91
15.40	391.16	391.41	391.67	391.92	392.18	392.43	392.68	392.94	393.19	393.45
15.50	393.70	393.95	394.21	394.46	394.72	394.97	395.22	395.48	395.73	395.99
15.60	396.24	39.649	396.75	397.00	397.26	397.51	397.76	398.02	398.27	398.53
15.70	398.78	399.03	399.29	399.54	399.80	400.05	400.30	400.56	400.81	401.07
15.80	401.32	401.57	401.83	402.08	402.34	402.59	402.84	403.10	403.35	403.61
15.90	403.86	404.11	404.37	404.62	404.88	405.13	405.38	405.64	405.89	406.15
16.00	406.40	406.65	406.91	407.16	407.52	407.67	407.92	408.18	408.43	408.69
16.10	408.94	409.19	409.45	409.70	409.96	410.21	410.46	410.72	410.97	411.23
16.20	411.48	411.73	411.99	412.24	412.50	412.75	413.00	413.26	413.51	413.77
16.30	414.02	414.27	414.53	414.78	415.04	415.29	415.54	415.80	416.05	416.31
16.40	416.56	416.81	417.07	417.32	417.58	417.83	418.08	418.34	418.59	418.85
16.50	419.10	419.35	419.61	419.86	420.12	420.37	420.62	420.88	421.13	421.39
16.60	421.64	421.89	422.15	422.40	422.66	422.91	423.16	423.42	423.67	423.93
16.70	424.18	424.43	424.69	424.94	425.20	425.45	425.70	425.96	426.21	426.47
16.80	426.72	426.97	427.23	427.48	427.74	427.99	428.24	428.50	428.75	429.01
16.90	429.26	429.51	429.77	430.02	430.28	430.53	430.78	431.04	431.29	431.55
17.00	431.80	432.05	432.31	432.56	432.82	433.07	433.32	433.58	433.83	434.09
17.10	434.34	434.59	434.85	435.10	435.36	435.61	435.86	436.12	436.37	436.63
17.20	436.88	437.13	437.39	437.64	437.90	438.15	438.40	438.66	438.91	439.17
17.30	439.42	439.67	439.93	440.18	440.44	440.69	440.94	441.20	441.45	441.71
17.40	441.96	442.21	442.47	442.72	442.98	443.23	443.48	443.74	443.99	444.25
17.50	444.50	444.75	445.01	445.26	445.52	445.77	446.02	446.28	446.53	446.79
17.60	447.04	447.29	447.55	447.80	448.06	448.31	448.56	448.82	449.07	449.33
17.70	449.58	449.83	450.09	450.34	450.60	450.85	451.10	451.36	451.61	451.87
17.80	452.12	452.37	452.63	452.88	453.14	453.39	453.64	453.90	454.15	454.41
17.90	454.66	454.91	455.17	455.42	455.68	455.93	456.18	456.44	456.69	456.95
18.00	457.20	457.45	457.71	457.96	458.22	458.47	458.72	458.98	459.23	459.49
18.10	459.74	459.99	460.25	460.50	460.76	461.01	461.26	461.52	461.77	462.03
18.20	462.28	462.53	462.79	463.04	463.30	463.55	463.80	464.06	464.31	464.57
18.30	464.82	465.07	465.33	465.58	465.84	466.09	466.34	466.60	466.85	467.11
18.40	467.36	467.61	467.87	468.12	468.38	468.63	468.88	469.14	469.39	469.35
18.50	469.90	470.15	470.41	470.66	470.92	471.17	471.42	471.68	471.93	472.19
18.60	472.44	472.69	472.95	473.20	473.46	473.71	473.96	474.22	474.47	474.73
18.70	474.98	475.23	475.49	475.74	476.00	476.25	476.50	476.76	477.01	477.27
18.80	477.52	477.77	478.03	478.28	478.54	478.79	479.04	479.30	479.55	479.81
18.90	480.06	480.31	480.57	480.82	481.08	481.33	481.58	481.84	482.09	482.35
19.00	482.60	482.85	483.11	483.36	483.62	483.87	484.12	484.38	484.63	484.89
19.10	485.14	485.39	485.65	485.90	486.16	486.41	486.66	486.92	487.17	487.43
19.20	487.68	487.93	488.19	488.44	488.70	488.95	489.20	489.46	489.71	489.97
19.30	490.22	490.47	490.73	490.98	491.24	491.49	491.74	492.00	492.25	492.51
19.40	492.76	493.01	493.27	493.52	493.78	494.03	494.28	494.54	494.79	495.05
19.50	495.30	495.55	495.81	496.06	496.32	496.57	496.82	497.08	497.33	497.59
19.60	497.84	498.09	498.35	498.60	498.86	499.11	499.36	499.62	499.87	500.13
19.70	500.38	500.34	500.89	501.14	501.40	501.65	501.91	502.16	502.41	502.67
19.80	502.92	503.18	503.43	503.68	503.94	504.19	504.45	504.70	504.95	505.21
19.90	505.46	505.72	505.97	506.22	506.48	506.73	506.99	507.24	507.49	507.75
20.00	508.00	508.26	508.51	508.76	509.02	509.27	509.53	509.78	510.03	510.29

Proportional Parts.	Inch.	0.001	0.002	0.003	0.004	0.005	0.006	0.007	0.008	0.009
	mm.	0.025	0.051	0.076	0.102	0.127	0.152	0.178	0.203	0.229

TABLE 64.

INCHES INTO MILLIMETRES.

1 inch = 25.40005 mm.

Inches.	.00	.01	.02	.03	.04	.05	.06	.07	.08	.09
	mm.	mm.	mm.	mm.	mm.	mm.	mm.	mm.	mm.	mm.
20.00	508.00	508.26	508.51	508.76	509.02	509.27	509.53	509.78	510.03	510.29
20.10	510.54	510.80	511.05	511.30	511.56	511.81	512.07	512.32	512.57	512.83
20.20	513.08	513.34	513.59	513.84	514.10	514.35	514.61	514.86	515.11	515.37
20.30	515.62	515.88	516.13	516.38	516.64	516.89	517.15	517.40	517.65	517.91
20.40	518.16	518.42	518.67	518.92	519.18	519.43	519.69	519.94	520.19	520.45
20.50	520.70	520.96	521.21	521.46	521.72	521.97	522.23	522.48	522.73	522.99
20.60	523.24	523.50	523.75	524.00	524.26	524.51	524.77	525.02	525.27	525.53
20.70	525.78	526.04	526.29	526.54	526.80	527.05	527.31	527.56	527.81	528.07
20.80	528.32	528.58	528.83	529.08	529.34	529.59	529.85	530.10	530.35	530.61
20.90	530.86	531.12	531.37	531.62	531.88	532.13	532.39	532.64	532.89	533.15
21.00	533.40	533.66	533.91	534.16	534.42	534.67	534.93	535.18	535.43	535.69
21.10	535.94	536.20	536.45	536.70	536.96	537.21	537.47	537.72	537.98	538.23
21.20	538.48	538.74	538.99	539.24	539.50	539.75	540.01	540.26	540.51	540.77
21.30	541.02	541.28	541.53	541.78	542.04	542.29	542.55	542.80	543.05	543.31
21.40	543.56	543.82	544.07	544.32	544.58	544.83	545.09	545.34	545.59	545.85
21.50	546.10	546.36	546.61	546.86	547.12	547.37	547.63	547.88	548.13	548.39
21.60	548.64	548.90	549.15	549.40	549.66	549.91	550.17	550.42	550.67	550.93
21.70	551.18	551.44	551.69	551.94	552.20	552.45	552.71	552.96	553.21	553.47
21.80	553.72	553.98	554.23	554.48	554.74	554.99	555.25	555.50	555.75	556.01
21.90	556.26	556.52	556.77	557.02	557.28	557.53	557.79	558.04	558.29	558.55
22.00	558.80	559.06	559.31	559.56	559.82	560.07	560.03	560.58	560.83	561.09
22.10	561.34	561.60	561.85	562.10	562.36	562.61	562.87	563.12	563.37	563.63
22.20	563.88	564.14	564.39	564.64	564.90	565.15	565.41	565.66	565.91	566.17
22.30	566.42	566.68	566.93	567.18	567.44	567.69	567.95	568.20	568.45	568.71
22.40	568.96	569.22	569.47	569.72	569.98	570.23	570.49	570.74	570.99	571.25
22.50	571.50	571.76	572.01	572.26	572.52	572.77	573.03	573.28	573.53	573.79
22.60	574.04	574.30	574.55	574.80	575.06	575.31	575.57	575.82	576.07	576.33
22.70	576.58	576.84	577.09	577.34	577.60	577.85	578.11	578.36	578.61	578.87
22.80	579.12	579.38	579.63	579.88	580.14	580.39	580.65	580.90	581.15	581.41
22.90	581.66	581.92	582.17	582.42	582.68	582.93	583.19	583.44	583.69	583.95
23.00	584.20	584.46	584.71	584.96	585.22	585.47	585.73	585.98	586.23	586.49
23.10	586.74	587.00	587.25	587.50	587.76	588.01	588.27	588.52	588.77	589.03
23.20	589.28	589.54	589.79	590.04	590.30	590.55	590.81	591.06	591.31	591.57
23.30	591.82	592.08	592.33	592.58	592.84	593.09	593.35	593.60	593.85	594.11
23.40	594.36	594.62	594.87	595.12	595.38	595.63	595.89	596.14	596.39	596.65
23.50	596.90	597.16	597.41	597.66	597.92	598.17	598.43	598.68	598.93	599.19
23.60	599.44	599.70	599.95	600.20	600.46	600.71	600.97	601.22	601.47	601.73
23.70	601.98	602.24	602.49	602.74	603.00	603.25	603.51	603.76	604.01	604.27
23.80	604.52	604.78	605.03	605.28	605.54	605.79	606.05	606.30	606.55	606.81
23.90	607.06	607.32	607.57	607.82	608.08	608.33	608.59	608.84	609.09	609.35
24.00	609.60	609.86	610.11	610.36	610.62	610.87	611.13	611.38	611.63	611.89
24.10	612.14	612.40	612.65	612.90	613.16	613.41	613.67	613.92	614.17	614.43
24.20	614.68	614.94	615.19	615.44	615.70	615.95	616.21	616.46	616.71	616.97
24.30	617.22	617.48	617.73	617.98	618.24	618.49	618.75	619.00	619.25	619.51
24.40	619.76	620.02	620.27	620.52	620.78	621.03	621.29	621.54	621.79	622.05
24.50	622.30	622.56	622.81	623.06	623.32	623.57	623.83	624.08	624.33	624.59
24.60	624.84	625.10	625.35	625.60	625.86	626.11	626.37	626.62	626.87	627.13
24.70	627.38	627.64	627.89	628.14	628.40	628.65	628.91	629.16	629.41	629.67
24.80	629.92	630.18	630.43	630.68	630.94	631.19	631.45	631.70	631.95	632.21
24.90	632.46	632.72	632.97	633.22	633.48	633.73	633.99	634.24	634.49	634.75
25.00	635.00	635.26	635.51	635.76	636.02	636.27	636.53	636.78	637.03	637.29

Proportional Parts.	Inch.	0.001	0.002	0.003	0.004	0.005	0.006	0.007	0.008	0.009
	mm.	0.025	0.051	0.076	0.102	0.127	0.152	0.178	0.203	0.229

SMITHSONIAN TABLES.

TABLE 64
INCHES INTO MILLIMETRES.

1 inch = 25.40005 mm.

Inches.	.00	.01	.02	.03	.04	.05	.06	.07	.08	.09
	mm.	mm.	mm.	mm.	mm.	mm.	mm.	mm.	mm.	mm.
25.00	635.00	635.26	635.51	635.76	636.02	636.27	636.53	636.78	637.03	637.29
25.10	637.54	637.80	638.05	638.30	638.56	638.81	639.07	639.32	639.57	639.83
25.20	640.08	640.34	640.59	640.84	641.10	641.35	641.61	641.86	642.11	642.37
25.30	642.62	642.88	643.13	643.38	643.64	643.89	644.15	644.40	644.65	644.91
25.40	645.16	645.42	645.67	645.92	646.18	646.43	646.69	646.94	647.19	647.45
25.50	647.70	647.96	648.21	648.46	648.72	648.97	649.23	649.48	649.73	649.99
25.60	650.24	650.50	650.75	651.00	651.26	651.51	651.77	652.02	654.27	652.53
25.70	652.78	653.04	653.29	653.54	653.80	654.05	654.31	654.56	654.81	655.07
25.80	655.32	655.58	655.83	656.08	656.34	656.59	656.85	657.10	657.35	657.61
25.90	657.86	658.12	658.37	658.62	658.88	659.13	659.39	659.64	659.89	660.15
26.00	660.40	660.66	660.91	661.16	661.42	661.67	661.93	662.18	662.43	662.69
26.10	662.94	663.20	663.45	663.70	663.96	664.21	664.47	664.72	664.97	665.23
26.20	665.48	665.74	665.99	666.24	666.50	666.75	667.01	667.26	667.51	667.77
26.30	668.02	668.28	668.53	668.78	669.04	669.29	669.55	669.80	670.05	670.31
26.40	670.56	670.82	671.07	671.32	671.58	671.83	672.09	672.34	672.59	672.85
26.50	673.10	673.36	673.61	673.86	674.12	674.37	674.63	674.88	675.13	675.39
26.60	675.64	675.90	676.15	676.40	676.66	676.91	677.17	677.42	677.67	677.93
26.70	678.18	678.44	678.69	678.94	679.20	679.45	679.71	679.96	680.21	680.47
26.80	680.72	680.98	681.23	681.48	681.74	681.99	682.25	682.50	682.75	683.01
26.90	683.26	683.52	683.77	684.02	684.28	684.53	684.79	685.04	685.29	685.55
27.00	685.80	686.06	686.31	686.56	686.82	687.07	687.33	687.58	687.83	688.09
27.10	688.34	688.60	688.85	689.10	689.36	689.61	689.87	690.12	690.37	690.63
27.20	690.88	691.14	691.39	691.64	691.90	692.15	692.41	692.66	692.91	693.17
27.30	693.42	693.68	693.93	694.18	694.44	694.69	694.95	695.20	695.45	695.71
27.40	695.96	696.22	696.47	696.72	696.98	697.23	697.49	697.74	697.99	698.25
27.50	698.50	698.76	699.01	699.26	699.52	699.77	700.03	700.28	700.53	700.79
27.60	701.04	701.30	701.55	701.80	702.06	702.31	702.57	702.82	703.07	703.33
27.70	703.58	703.84	704.09	704.34	704.60	704.85	705.11	705.36	705.61	705.87
27.80	706.12	706.38	706.63	706.88	707.14	707.39	707.65	707.90	708.15	708.41
27.90	708.66	708.92	709.17	709.42	709.68	709.93	710.19	710.44	710.69	710.95
28.00	711.20	711.46	711.71	711.96	712.22	712.47	712.73	712.98	713.23	713.49
28.10	713.74	714.00	714.25	714.50	714.76	715.01	715.27	715.52	715.77	716.03
28.20	716.28	716.54	716.79	717.04	717.30	717.55	717.81	718.06	718.31	718.57
28.30	718.82	719.08	719.33	719.58	719.84	720.09	720.35	720.60	720.85	721.11
28.40	721.36	721.62	721.87	722.12	722.39	722.63	722.89	723.14	723.39	723.65
28.50	723.90	724.16	724.41	724.66	724.92	725.17	725.43	725.68	725.93	726.19
28.60	726.44	726.70	726.95	727.20	727.46	727.71	727.97	728.22	728.47	728.73
28.70	728.98	729.24	729.49	729.74	730.00	730.25	730.51	730.76	731.01	731.27
28.80	731.52	731.78	732.03	732.28	732.54	732.79	733.05	733.30	733.55	733.81
28.90	734.06	734.32	734.57	734.82	735.08	735.33	735.59	735.84	736.09	736.35
29.00	736.60	736.86	737.11	737.36	737.62	737.87	738.13	738.38	738.63	738.89
29.10	739.14	739.40	739.65	739.90	740.16	740.41	740.67	740.92	741.17	741.43
29.20	741.68	741.94	742.19	742.44	742.70	742.95	743.21	743.46	743.71	743.97
29.30	744.22	744.48	744.73	744.98	745.24	745.49	745.75	746.00	746.25	746.51
29.40	746.76	747.02	747.27	747.52	747.78	748.03	748.29	748.54	748.79	749.05
29.50	749.30	749.56	749.81	750.06	750.32	750.57	750.83	751.08	751.33	751.59
29.60	751.84	752.10	752.35	752.60	752.86	753.11	753.37	753.62	753.87	754.13
29.70	754.38	754.64	754.89	755.14	755.40	755.65	755.91	756.16	756.41	756.67
29.80	756.92	757.18	757.43	757.68	757.94	758.19	758.45	758.70	758.95	759.21
29.90	759.46	759.72	759.97	760.22	760.48	760.73	760.99	761.24	761.49	761.75
30.00	762.00	762.26	762.51	762.76	763.02	763.27	763.53	763.78	764.03	764.29

Proportional Parts.	Inch.	0.001	0.002	0.003	0.004	0.005	0.006	0.007	0.008	0.009
	mm.	0.025	0.051	0.076	0.102	0.127	0.152	0.178	0.203	0.229

SMITHSONIAN TABLES.

TABLE 64.

INCHES INTO MILLIMETRES.

1 inch = 25.40005 mm.

Inches.	.00	.01	.02	.03	.04	.05	.06	.07	.08	.09
	mm.	mm.	mm.	mm.	mm.	mm.	mm.	mm.	mm.	mm.
30.00	762.00	762.26	762.51	762.76	763.02	763.27	763.53	763.78	764.03	764.29
30.10	764.54	764.80	765.05	765.30	765.56	765.81	766.07	766.32	766.57	766.83
30.20	767.08	767.34	767.59	767.84	768.10	768.35	768.61	768.86	769.11	769.37
30.30	769.62	769.88	770.13	770.38	770.64	770.89	771.15	771.40	771.65	771.91
30.40	772.16	772.42	772.67	772.92	773.18	773.43	773.69	773.94	774.19	774.45
30.50	774.70	774.96	775.21	775.46	775.72	775.97	776.23	776.48	776.73	776.99
30.60	777.24	777.50	777.75	778.00	778.26	778.51	778.77	779.02	779.27	779.53
30.70	779.78	780.04	780.29	780.54	780.80	781.05	781.31	781.56	781.81	782.07
30.80	782.32	782.58	782.83	783.08	783.34	783.59	783.85	784.10	784.35	784.61
30.90	784.86	785.12	785.37	785.62	785.88	786.13	786.39	786.64	786.89	787.15
31.00	787.40	787.66	787.91	788.16	788.42	788.67	788.93	789.18	789.43	789.69
31.10	789.94	790.20	790.45	790.70	790.96	791.21	791.47	791.72	791.97	792.23
31.20	792.48	792.74	792.99	793.24	793.50	793.75	794.01	794.26	794.51	794.77
31.30	795.02	795.28	795.53	795.78	796.04	796.29	796.55	796.80	797.05	797.31
31.40	797.56	797.82	798.07	798.32	798.58	798.83	799.09	799.34	799.59	799.85
31.50	800.10	800.36	800.61	800.86	801.12	801.37	801.63	801.88	802.13	802.39
31.60	802.64	802.90	803.15	803.40	803.66	803.91	804.17	804.42	804.67	804.93
31.70	805.18	805.44	805.69	805.94	806.20	806.45	806.71	806.96	807.21	807.47
31.80	807.72	807.98	808.23	808.48	808.74	808.99	809.25	809.50	809.75	810.01
31.90	810.26	810.52	810.77	811.02	811.28	811.53	811.79	812.04	812.29	812.55
32.00	812.80									

Proportional Parts.	Inch.	0.001	0.002	0.003	0.004	0.005	0.006	0.007	0.008	0.009
	mm.	0.025	0.051	0.076	0.102	0.127	0.152	0.178	0.203	0.229

TABLE 65.
MILLIMETRES INTO INCHES.

1 mm. = 0.03937 inch.

Millimetres.	0	1	2	3	4	5	6	7	8	9
	Inches.	Inches.	Inches.	Inches.	Inches.	Inches.	Inches.	Inches.	Inches.	Inches.
0	0.0000	0.0394	0.0787	0.1181	0.1575	0.1968	0.2362	0.2756	0.3150	0.3543
10	0.3937	0.4331	0.4724	0.5118	0.5512	0.5906	0.6299	0.6693	0.7087	0.7480
20	0.7874	0.8268	0.8661	0.9055	0.9449	0.9842	1.0236	1.0630	1.1024	1.1417
30	1.1811	1.2205	1.2598	1.2992	1.3386	1.3780	1.4173	1.4567	1.4961	1.5354
40	1.5748	1.6142	1.6535	1.6929	1.7323	1.7716	1.8110	1.8504	1.8898	1.9291
50	1.9685	2.0079	2.0472	2.0866	2.1260	2.1654	2.2047	2.2441	2.2835	2.3228
60	2.3622	2.4016	2.4409	2.4803	2.5197	2.5590	2.5984	2.6378	2.6772	2.7165
70	2.7559	2.7953	2.8346	2.8740	2.9134	2.9528	2.9921	3.0315	3.0709	3.1102
80	3.1496	3.1890	3.2283	3.2677	3.3071	3.3464	3.3858	3.4252	3.4646	3.5039
90	3.5433	3.5828	3.6220	3.6614	3.7008	3.7402	3.7795	3.8189	3.8583	3.8976
100	3.9370	3.9764	4.0157	4.0551	4.0945	4.1338	4.1732	4.2126	4.2520	4.2913
110	4.3307	4.3701	4.4094	4.4488	4.4882	4.5276	4.5669	4.6063	4.6457	4.6850
120	4.7244	4.7638	4.8031	4.8425	4.8819	4.9212	4.9606	5.0000	5.0394	5.0787
130	5.1181	5.1575	5.1968	5.2362	5.2756	5.3150	5.3543	5.3937	5.4331	5.4724
140	5.5118	5.5512	5.5905	5.6299	5.6693	5.7086	5.7480	5.7874	5.8268	5.8661
150	5.9055	5.9449	5.9842	6.0236	6.0630	6.1024	6.1417	6.1811	6.2205	6.2598
160	6.2992	6.3386	6.3779	6.4173	6.4567	6.4960	6.5354	6.5748	6.6142	6.6535
170	6.6929	6.7323	6.7716	6.8110	6.8504	6.8898	6.9291	6.9685	7.0079	7.0472
180	7.0866	7.1260	7.1653	7.2047	7.2441	7.2834	7.3228	7.3622	7.4016	7.4409
190	7.4803	7.5197	7.5590	7.5984	7.6378	7.6772	7.7165	7.7559	7.7953	7.8346
200	7.8740	7.9134	7.9527	7.9921	8.0315	8.0708	8.1102	8.1496	8.1890	8.2283
210	8.2677	8.3071	8.3464	8.3858	8.4252	8.4646	8.5039	8.5433	8.5827	8.6220
220	8.6614	8.7008	8.7401	8.7795	8.8189	8.8582	8.8976	8.9370	8.9764	9.0157
230	9.0551	9.0945	9.1338	9.1732	9.2126	9.2520	9.2913	9.3307	9.3701	9.4094
240	9.4488	9.4882	9.5275	9.5669	9.6063	9.6456	9.6850	9.7244	9.7638	9.8031
250	9.8425	9.8819	9.9212	9.9606	10.0000	10.0394	10.0787	10.1181	10.1575	10.1968
260	10.2362	10.2756	10.3149	10.3543	10.3937	10.4330	10.4724	10.5118	10.5512	10.5905
270	10.6299	10.6693	10.7086	10.7480	10.7874	10.8268	10.8661	10.9055	10.9449	10.9842
280	11.0236	11.0630	11.1023	11.1417	11.1811	11.2204	11.2598	11.2992	11.3338	11.3779
290	11.4173	11.4568	11.4960	11.5354	11.5748	11.6142	11.6535	11.6929	11.7323	11.7716
300	11.8110	11.8504	11.8897	11.9291	11.9685	12.0078	12.0472	12.0866	12.1260	12.1653
310	12.2047	12.2441	12.2834	12.3228	12.3622	12.4016	12.4409	12.4803	12.5197	12.5590
320	12.5984	12.6378	12.6771	12.7165	12.7559	12.7952	12.8346	12.8740	12.9134	12.9527
330	12.9921	13.0315	13.0708	13.1102	13.1496	13.1890	13.2283	13.2677	13.3071	13.3464
340	13.3858	13.4252	13.4645	13.5039	13.5433	13.5826	13.6220	13.6614	13.7008	13.7401
350	13.7795	13.8189	13.8582	13.8976	13.9370	13.9764	14.0157	14.0551	14.0945	14.1338
360	14.1732	14.2126	14.2519	14.2913	14.3307	14.3700	14.4094	14.4488	14.4882	14.5275
370	14.5669	14.6063	14.6456	14.6850	14.7244	14.7638	14.8031	14.8425	14.8819	14.9212
380	14.9606	15.0000	15.0393	15.0787	15.1181	15.1574	15.1968	15.2362	15.2756	15.3149
390	15.3543	15.3937	15.4330	15.4724	15.5118	15.5512	15.5905	15.6299	15.6693	15.7086
400	15.7480	15.7874	15.8267	15.8661	15.9055	15.9448	15.9842	16.0236	16.0630	16.1023

Tenths of a millimetre.				Hundredths of a millimetre.			
mm.	Inch.	mm.	Inch.	mm.	Inch.	mm.	Inch.
0.1	0.0039	0.6	0.0236	0.01	0.0004	0.06	0.0024
.2	.0079	.7	.0276	.02	.0008	.07	.0028
.3	.0118	.8	.0315	.03	.0012	.08	.0031
.4	.0157	.9	.0354	.04	.0016	.09	.0035
.5	.0197	1.0	.0394	.05	.0020	.10	.0039

SMITHSONIAN TABLES.

TABLE 65.

MILLIMETRES INTO INCHES.

1 mm. = 0.03937 inch.

Milli-metres.	.0	.1	.2	.3	.4	.5	.6	.7	.8	.9
	Inches.	Inches.	Inches.	Inches.	Inches.	Inches.	Inches.	Inches.	Inches.	Inches.
400	15.748	15.752	15.756	15.760	15.764	15.768	15.772	15.776	15.779	15.783
401	15.787	15.791	15.795	15.799	15.803	15.807	15.811	15.815	15.819	15.823
402	15.827	15.831	15.835	15.839	15.842	15.846	15.850	15.854	15.858	15.862
403	15.866	15.870	15.874	15.878	15.882	15.886	15.890	15.894	15.898	15.902
404	15.905	15.909	15.913	15.917	15.921	15.925	15.929	15.933	15.937	15.941
405	15.945	15.949	15.953	15.957	15.961	15.965	15.968	15.972	15.976	15.980
406	15.984	15.988	15.992	15.996	16.000	16.004	16.008	16.012	16.016	16.020
407	16.024	16.028	16.031	16.035	16.039	16.043	16.047	16.051	16.055	16.059
408	16.063	16.067	16.071	16.075	16.079	16.083	16.087	16.091	16.094	16.098
409	16.102	16.106	16.110	16.114	16.118	16.122	16.126	16.130	16.134	16.138
410	16.142	16.146	16.150	16.154	16.157	16.161	16.165	16.169	16.173	16.177
411	16.181	16.185	16.189	16.193	16.197	16.201	16.205	16.209	16.213	16.217
412	16.220	16.224	16.228	16.232	16.236	16.240	16.244	16.248	16.252	16.256
413	16.260	16.264	16.268	16.272	16.276	16.279	16.283	16.287	16.291	16.295
414	16.299	16.303	16.307	16.311	16.315	16.319	16.323	16.327	16.331	16.335
415	16.339	16.342	16.346	16.350	16.354	16.358	16.362	16.366	16.370	16.374
416	16.378	16.382	16.386	16.390	16.394	16.398	16.402	16.405	16.409	16.413
417	16.417	16.421	16.425	16.429	16.433	16.437	16.441	16.445	16.449	16.453
418	16.457	16.461	16.465	16.468	16.472	16.476	16.480	16.484	16.488	16.492
419	16.496	16.500	16.504	16.508	16.512	16.516	16.520	16.524	16.528	16.531
420	16.535	16.539	16.543	16.547	16.551	16.555	16.559	16.563	16.567	16.571
421	16.575	16.579	16.583	16.587	16.591	16.594	16.598	16.602	16.606	16.610
422	16.614	16.618	16.622	16.626	16.630	16.634	16.638	16.642	16.646	16.650
423	16.654	16.657	16.661	16.665	16.669	16.673	16.677	16.681	16.685	16.689
424	16.693	16.697	16.701	16.705	16.709	16.713	16.717	16.720	16.724	16.728
425	16.732	16.736	16.740	16.744	16.748	16.752	16.756	16.760	16.764	16.768
426	16.772	16.776	16.779	16.783	16.787	16.791	16.795	16.799	16.803	16.807
427	16.811	16.815	16.819	16.823	16.827	16.831	16.835	16.839	16.842	16.846
428	16.850	16.854	16.858	16.862	16.866	16.870	16.874	16.878	16.882	16.886
429	16.890	16.894	16.898	16.902	16.905	16.909	16.913	16.917	16.921	16.925
430	16.929	16.933	16.937	16.941	16.945	16.949	16.953	16.957	16.961	16.965
431	16.968	16.972	16.976	16.980	16.984	16.988	16.992	16.996	17.000	17.004
432	17.008	17.012	17.016	17.020	17.024	17.028	17.031	17.035	17.039	17.043
433	17.047	17.051	17.055	17.059	17.063	17.067	17.071	17.075	17.079	17.083
434	17.087	17.091	17.094	17.098	17.102	17.106	17.110	17.114	17.118	17.122
435	17.126	17.130	17.134	17.138	17.142	17.146	17.150	17.154	17.157	17.161
436	17.165	17.169	17.173	17.177	17.181	17.185	17.189	17.193	17.197	17.201
437	17.205	17.209	17.213	17.217	17.220	17.224	17.228	17.232	17.236	17.240
438	17.244	17.248	17.252	17.256	17.260	17.264	17.268	17.272	17.276	17.279
439	17.283	17.287	17.291	17.295	17.299	17.303	17.307	17.311	17.315	17.319
440	17.323	17.327	17.331	17.335	17.339	17.342	17.346	17.350	17.354	17.358
441	17.362	17.366	17.370	17.374	17.378	17.382	17.386	17.390	17.394	17.398
442	17.402	17.405	17.409	17.413	17.417	17.421	17.425	17.429	17.433	17.437
443	17.441	17.445	17.449	17.453	17.457	17.461	17.465	17.468	17.472	17.476
444	17.480	17.484	17.488	17.492	17.496	17.500	17.504	17.508	17.512	17.516
445	17.520	17.524	17.528	17.531	17.535	17.539	17.543	17.547	17.551	17.555
446	17.559	17.563	17.567	17.571	17.575	17.579	17.583	17.587	17.591	17.594
447	17.598	17.602	17.606	17.610	17.614	17.618	17.622	17.626	17.630	17.634
448	17.638	17.642	17.646	17.650	17.654	17.657	17.661	17.665	17.669	17.673
449	17.677	17.681	17.685	17.689	17.693	17.697	17.701	17.705	17.709	17.713
450	17.717	17.720	17.724	17.728	17.732	17.736	17.740	17.744	17.748	17.752

SMITHSONIAN TABLES.

TABLE 65.

MILLIMETRES INTO INCHES.

1 mm. = 0.03937 inch.

Milli-metres.	.0	.1	.2	.3	.4	.5	.6	.7	.8	.9
	Inches.	Inches.	Inches.	Inches.	Inches.	Inches.	Inches.	Inches.	Inches.	Inches.
450	17.717	17.720	17.724	17.728	17.732	17.736	17.740	17.744	17.748	17.752
451	17.756	17.760	17.764	17.768	17.772	17.776	17.779	17.783	17.787	17.791
452	17.795	17.799	17.803	17.807	17.811	17.815	17.819	17.823	17.827	17.831
453	17.835	17.839	17.842	17.846	17.850	17.854	17.858	17.862	17.866	17.870
454	17.874	17.878	17.882	17.886	17.890	17.894	17.898	17.902	17.905	17.909
455	17.913	17.917	17.921	17.925	17.929	17.933	17.937	17.941	17.945	17.949
456	17.953	17.957	17.961	17.965	17.968	17.972	17.976	17.980	17.984	17.988
457	17.992	17.996	18.000	18.004	18.008	18.012	18.016	18.020	18.024	18.028
458	18.031	18.035	18.039	18.043	18.047	18.051	18.055	18.059	18.063	18.067
459	18.071	18.075	18.079	18.083	18.087	18.091	18.094	18.098	18.102	18.106
460	18.110	18.114	18.118	18.122	18.126	18.130	18.134	18.138	18.142	18.146
461	18.150	18.154	18.157	18.161	18.165	18.169	18.173	18.177	18.181	18.185
462	18.189	18.193	18.197	18.201	18.205	18.209	18.213	18.216	18.220	18.224
463	18.228	18.232	18.236	18.240	18.244	18.248	18.252	18.256	18.260	18.264
464	18.268	18.272	18.276	18.279	18.283	18.287	18.291	18.295	18.299	18.303
465	18.307	18.311	18.315	18.319	18.323	18.327	18.331	18.335	18.339	18.342
466	18.346	18.350	18.354	18.358	18.362	18.366	18.370	18.374	18.378	18.382
467	18.386	18.390	18.394	18.398	18.402	18.405	18.409	18.413	18.417	18.421
468	18.425	18.429	18.433	18.437	18.441	18.445	18.449	18.453	18.457	18.461
469	18.465	18.468	18.472	18.476	18.480	18.484	18.488	18.492	18.496	18.500
470	18.504	18.508	18.512	18.516	18.520	18.524	18.528	18.531	18.535	18.539
471	18.543	18.547	18.551	18.555	18.559	18.563	18.567	18.571	18.575	18.579
472	18.583	18.587	18.591	18.594	18.598	18.602	18.606	18.610	18.614	18.618
473	18.622	18.626	18.630	18.634	18.638	18.642	18.646	18.650	18.654	18.657
474	18.661	18.665	18.669	18.673	18.677	18.681	18.685	18.689	18.693	18.697
475	18.701	18.705	18.709	18.713	18.716	18.720	18.724	18.728	18.732	18.736
476	18.740	18.744	18.748	18.752	18.756	18.760	18.764	18.768	18.772	18.776
477	18.779	18.783	18.787	18.791	18.795	18.799	18.803	18.807	18.811	18.815
478	18.819	18.823	18.827	18.831	18.835	18.839	18.842	18.846	18.850	18.854
479	18.858	18.862	18.866	18.870	18.874	18.878	18.882	18.886	18.890	18.894
480	18.898	18.902	18.905	18.909	18.913	18.917	18.921	18.925	18.929	18.933
481	18.937	18.941	18.945	18.949	18.953	18.957	18.961	18.965	18.968	18.972
482	18.976	18.980	18.984	18.988	18.992	18.996	19.000	19.004	19.008	19.012
483	19.016	19.020	19.024	19.028	19.031	19.035	19.039	19.043	19.047	19.051
484	19.055	19.059	19.063	19.067	19.071	19.075	19.079	19.083	19.087	19.091
485	19.094	19.098	19.102	19.106	19.110	19.114	19.118	19.122	19.126	19.130
486	19.134	19.138	19.142	19.146	19.150	19.154	19.157	19.161	19.165	19.169
487	19.173	19.177	19.181	19.185	19.189	19.193	19.197	19.201	19.205	19.209
488	19.213	19.216	19.220	19.224	19.228	19.232	19.236	19.240	19.244	19.248
489	19.252	19.256	19.260	19.264	19.268	19.272	19.276	19.279	19.283	19.287
490	19.291	19.295	19.299	19.303	19.307	19.311	19.315	19.319	19.323	19.327
491	19.331	19.335	19.339	19.342	19.346	19.350	19.354	19.358	19.362	19.366
492	19.370	19.374	19.378	19.382	19.386	19.390	19.394	19.398	19.402	19.405
493	19.409	19.413	19.417	19.421	19.425	19.429	19.433	19.437	19.441	19.445
494	19.449	19.453	19.457	19.461	19.465	19.468	19.472	19.476	19.480	19.484
495	19.488	19.492	19.496	19.500	19.504	19.508	19.512	19.516	19.520	19.524
496	19.528	19.531	19.535	19.539	19.543	19.547	19.551	19.555	19.559	19.563
497	19.567	19.571	19.575	19.579	19.583	19.587	19.591	19.594	19.598	19.602
498	19.606	19.610	19.614	19.618	19.622	19.626	19.630	19.634	19.638	19.642
499	19.646	19.650	19.654	19.657	19.661	19.665	19.669	19.673	19.677	19.681
500	19.685	19.689	19.693	19.697	19.701	19.705	19.709	19.713	19.716	19.720

SMITHSONIAN TABLES.

TABLE 65

MILLIMETRES INTO INCHES.

1 mm. = 0.03937 inch.

Milli-metres.	.0	.1	.2	.3	.4	.5	.6	.7	.8	.9
	Inches.	Inches.	Inches.	Inches.	Inches.	Inches.	Inches.	Inches.	Inches.	Inches.
500	19.685	19.689	19.693	19.697	19.701	19.705	19.709	19.713	19.716	19.720
501	19.724	19.728	19.732	19.736	19.740	19.744	19.748	19.752	19.756	19.760
502	19.764	19.768	19.772	19.776	19.779	19.783	19.787	19.791	19.795	19.799
503	19.803	19.807	19.811	19.815	19.819	19.823	19.827	19.831	19.835	19.839
504	19.842	19.846	19.850	19.854	19.858	19.862	19.866	19.870	19.874	19.878
505	19.882	19.886	19.890	19.894	19.898	19.902	19.905	19.909	19.913	19.917
506	19.921	19.925	19.929	19.933	19.937	19.941	19.945	19.949	19.953	19.957
507	19.961	19.965	19.968	19.972	19.976	19.980	19.984	19.988	19.992	19.996
508	20.000	20.004	20.008	20.012	20.016	20.020	20.024	20.028	20.031	20.035
509	20.039	20.043	20.047	20.051	20.055	20.059	20.063	20.067	20.071	20.075
510	20.079	20.083	20.087	20.091	20.094	20.098	20.102	20.106	20.110	20.114
511	20.118	20.122	20.126	20.130	20.134	20.138	20.142	20.146	20.150	20.154
512	20.157	20.161	20.165	20.169	20.173	20.177	20.181	20.185	20.189	20.193
513	20.197	20.201	20.205	20.209	20.213	20.216	20.220	20.224	20.228	20.232
514	20.236	20.240	20.244	20.248	20.252	20.256	20.260	20.264	20.268	20.272
515	20.276	20.279	20.283	20.287	20.291	20.295	20.299	20.303	20.307	20.311
516	20.315	20.319	20.323	20.327	20.331	20.335	20.339	20.342	20.346	20.350
517	20.354	20.358	20.362	20.366	20.370	20.374	20.378	20.382	20.386	20.390
518	20.394	20.398	20.402	20.405	20.409	20.413	20.417	20.421	20.425	20.429
519	20.433	20.437	20.441	20.445	20.449	20.453	20.457	20.461	20.465	20.468
520	20.472	20.476	20.480	20.484	20.488	20.492	20.496	20.500	20.504	20.508
521	20.512	20.516	20.520	20.524	20.528	20.531	20.535	20.539	20.543	20.547
522	20.551	20.555	20.559	20.563	20.567	20.571	20.575	20.579	20.583	20.587
523	20.591	20.594	20.598	20.602	20.606	20.610	20.614	20.618	20.622	20.626
524	20.630	20.634	20.638	20.642	20.646	20.650	20.654	20.657	20.661	20.665
525	20.669	20.673	20.677	20.681	20.685	20.689	20.693	20.697	20.701	20.705
526	20.709	20.713	20.716	20.720	20.724	20.728	20.732	20.736	20.740	20.744
527	20.748	20.752	20.756	20.760	20.764	20.768	20.772	20.776	20.779	20.783
528	20.787	20.791	20.795	20.799	20.803	20.807	20.811	20.815	20.819	20.823
529	20.827	20.831	20.835	20.839	20.842	20.846	20.850	20.854	20.858	20.862
530	20.866	20.870	20.874	20.878	20.882	20.886	20.890	20.894	20.898	20.902
531	20.905	20.909	20.913	20.917	20.921	20.925	20.929	20.933	20.937	20.941
532	20.945	20.949	20.953	20.957	20.961	20.965	20.968	20.972	20.976	20.980
533	20.984	20.988	20.992	20.996	21.000	21.004	21.008	21.012	21.016	21.020
534	21.024	21.028	21.031	21.035	21.039	21.043	21.047	21.051	21.055	21.059
535	21.063	21.067	21.071	21.075	21.079	21.083	21.087	21.091	21.094	21.098
536	21.102	21.106	21.110	21.114	21.118	21.122	21.126	21.130	21.134	21.138
537	21.142	21.146	21.150	21.154	21.157	21.161	21.165	21.169	21.173	21.177
538	21.181	21.185	21.189	21.193	21.197	21.201	21.205	21.209	21.213	21.216
539	21.220	21.224	21.228	21.232	21.236	21.240	21.244	21.248	21.252	21.256
540	21.260	21.264	21.268	21.272	21.276	21.279	21.283	21.287	21.291	21.295
541	21.299	21.303	21.307	21.311	21.315	21.319	21.323	21.327	21.331	21.335
542	21.339	21.342	21.346	21.350	21.354	21.358	21.362	21.366	21.370	21.374
543	21.378	21.382	21.386	21.390	21.394	21.398	21.402	21.405	21.409	21.413
544	21.417	21.421	21.425	21.429	21.433	21.437	21.441	21.445	21.449	21.453
545	21.457	21.461	21.465	21.468	21.472	21.476	21.480	21.484	21.488	21.492
546	21.496	21.500	21.504	21.508	21.512	21.516	21.520	21.524	21.528	21.531
547	21.535	21.539	21.543	21.547	21.551	21.555	21.559	21.563	21.567	21.571
548	21.575	21.579	21.583	21.587	21.591	21.594	21.598	21.602	21.606	21.610
549	21.614	21.618	21.622	21.626	21.630	21.634	21.638	21.642	21.646	21.650
550	21.654	21.657	21.661	21.665	21.669	21.673	21.677	21.681	21.685	21.689

SMITHSONIAN TABLES.

TABLE 65.
MILLIMETRES INTO INCHES.
1 mm. = 0.03937 inch.

Milli-metres.	.0	.1	.2	.3	.4	.5	.6	.7	.8	.9
	Inches.	Inches.	Inches.	Inches.	Inches.	Inches.	Inches.	Inches.	Inches.	Inches.
550	21.654	21.657	21.661	21.665	21.669	21.673	21.677	21.681	21.685	21.689
551	21.693	21.697	21.701	21.705	21.709	21.713	21.716	21.720	21.724	21.728
552	21.732	21.736	21.740	21.744	21.748	21.752	21.756	21.760	21.764	21.768
553	21.772	21.776	21.779	21.783	21.787	21.791	21.795	21.799	21.803	21.807
554	21.811	21.815	21.819	21.823	21.827	21.831	21.835	21.839	21.842	21.846
555	21.850	21.854	21.858	21.862	21.866	21.870	21.874	21.878	21.882	21.886
556	21.890	21.894	21.898	21.902	21.905	21.909	21.913	21.917	21.921	21.925
557	21.929	21.933	21.937	21.941	21.945	21.949	21.953	21.957	21.961	21.965
558	21.968	21.972	21.976	21.980	21.984	21.988	21.992	21.996	22.000	22.004
559	22.008	22.012	22.016	22.020	22.024	22.028	22.031	22.035	22.039	22.043
560	22.047	22.051	22.055	22.059	22.063	22.067	22.071	22.075	22.079	22.083
561	22.087	22.091	22.094	22.098	22.102	22.106	22.110	22.114	22.118	22.122
562	22.126	22.130	22.134	22.138	22.142	22.146	22.150	22.153	22.157	22.161
563	22.165	22.169	22.173	22.177	22.181	22.185	22.189	22.193	22.197	22.201
564	22.205	22.209	22.213	22.216	22.220	22.224	22.228	22.232	22.236	22.240
565	22.244	22.248	22.252	22.256	22.260	22.264	22.268	22.272	22.276	22.279
566	22.283	22.287	22.291	22.295	22.299	22.303	22.307	22.311	22.315	22.319
567	22.323	22.327	22.331	22.335	22.339	22.342	22.346	22.350	22.354	22.358
568	22.362	22.366	22.370	22.374	22.378	22.382	22.386	22.390	22.394	22.398
569	22.402	22.405	22.409	22.413	22.417	22.421	22.425	22.429	22.433	22.437
570	22.441	22.445	22.449	22.453	22.457	22.461	22.465	22.468	22.472	22.476
571	22.480	22.484	22.488	22.492	22.496	22.500	22.504	22.508	22.512	22.516
572	22.520	22.524	22.528	22.531	22.535	22.539	22.543	22.547	22.551	22.555
573	22.559	22.563	22.567	22.571	22.575	22.579	22.583	22.587	22.591	22.594
574	22.598	22.602	22.606	22.610	22.614	22.618	22.622	22.626	22.630	22.634
575	22.638	22.642	22.646	22.650	22.653	22.657	22.661	22.665	22.669	22.673
576	22.677	22.681	22.685	22.689	22.693	22.697	22.701	22.705	22.709	22.713
577	22.716	22.720	22.724	22.728	22.732	22.736	22.740	22.744	22.748	22.752
578	22.756	22.760	22.764	22.768	22.772	22.776	22.779	22.783	22.787	22.791
579	22.795	22.799	22.803	22.807	22.811	22.815	22.819	22.823	22.827	22.831
580	22.835	22.839	22.842	22.846	22.850	22.854	22.858	22.862	22.866	22.870
581	22.874	22.878	22.882	22.886	22.890	22.894	22.898	22.902	22.905	22.909
582	22.913	22.917	22.921	22.925	22.929	22.933	22.937	22.941	22.945	22.949
583	22.953	22.957	22.961	22.965	22.968	22.972	22.976	22.980	22.984	22.988
584	22.992	22.996	23.000	23.004	23.008	23.012	23.016	23.020	23.024	23.028
585	23.031	23.035	23.039	23.043	23.047	23.051	23.055	23.059	23.063	23.067
586	23.071	23.075	23.079	23.083	23.087	23.091	23.094	23.098	23.102	23.106
587	23.110	23.114	23.118	23.122	23.126	23.130	23.134	23.138	23.142	23.146
588	23.150	23.153	23.157	23.161	23.165	23.169	23.173	23.177	23.181	23.185
589	23.189	23.193	23.197	23.201	23.205	23.209	23.213	23.216	23.220	23.224
590	23.228	23.232	23.236	23.240	23.244	23.248	23.252	23.256	23.260	23.264
591	23.268	23.272	23.276	23.279	23.283	23.287	23.291	23.295	23.299	23.303
592	23.307	23.311	23.315	23.319	23.323	23.327	23.331	23.335	23.339	23.342
593	23.346	23.350	23.354	23.358	23.362	23.366	23.370	23.374	23.378	23.382
594	23.386	23.390	23.394	23.398	23.402	23.405	23.409	23.413	23.417	23.421
595	23.425	23.429	23.433	23.437	23.441	23.445	23.449	23.453	23.457	23.461
596	23.465	23.468	23.472	23.476	23.480	23.484	23.488	23.492	23.496	23.500
597	23.504	23.508	23.512	23.516	23.520	23.524	23.528	23.531	23.535	23.539
598	23.543	23.547	23.551	23.555	23.559	23.563	23.567	23.571	23.575	23.579
599	23.583	23.587	23.591	23.594	23.598	23.602	23.606	23.610	23.614	23.618
600	23.622	23.626	23.630	23.634	23.638	23.642	23.646	23.650	23.653	23.657

TABLE 65.

MILLIMETRES INTO INCHES.

1 mm. = 0.03937 inch.

Milli-metres.	.0	.1	.2	.3	.4	.5	.6	.7	.8	.9
	Inches.	Inches.	Inches.	Inches.	Inches.	Inches.	Inches.	Inches.	Inches.	Inches.
600	23.622	23.626	23.630	23.634	23.638	23.642	23.646	23.650	23.653	23.657
601	23.661	23.665	23.669	23.673	23.677	23.681	23.685	23.689	23.693	23.697
602	23.701	23.705	23.709	23.713	23.716	23.720	23.724	23.728	23.732	23.736
603	23.740	23.744	23.748	23.752	23.756	23.760	23.764	23.768	23.772	23.776
604	23.779	23.783	23.787	23.791	23.795	23.799	23.803	23.807	23.811	23.815
605	23.819	23.823	23.827	23.831	23.835	23.839	23.842	23.846	23.850	23.854
606	23.858	23.862	23.866	23.870	23.874	23.878	23.882	23.886	23.890	23.894
607	23.898	23.902	23.905	23.909	23.913	23.917	23.921	23.925	23.929	23.933
608	23.937	23.941	23.945	23.949	23.953	23.957	23.961	23.965	23.968	23.972
609	23.976	23.980	23.984	23.988	23.992	23.996	24.000	24.004	24.008	24.012
610	24.016	24.020	24.024	24.028	24.031	24.035	24.039	24.043	24.047	24.051
611	24.055	24.059	24.063	24.067	24.071	24.075	24.079	24.083	24.087	24.091
612	24.094	24.098	24.102	24.106	24.110	24.114	24.118	24.122	24.126	24.130
613	24.134	24.138	24.142	24.146	24.150	24.153	24.157	24.161	24.165	24.169
614	24.173	24.177	24.181	24.185	24.189	24.193	24.197	24.201	24.205	24.209
615	24.213	24.216	24.220	24.224	24.228	24.232	24.236	24.240	24.244	24.248
616	24.252	24.256	24.260	24.264	24.268	24.272	24.276	24.279	24.283	24.287
617	24.291	24.295	24.299	24.303	24.307	24.311	24.315	24.319	24.323	24.327
618	24.331	24.335	24.339	24.342	24.346	24.350	24.354	24.358	24.362	24.366
619	24.370	24.374	24.378	24.382	24.386	24.390	24.394	24.398	24.402	24.405
620	24.409	24.413	24.417	24.421	24.425	24.429	24.433	24.437	24.441	24.445
621	24.449	24.453	24.457	24.461	24.465	24.468	24.472	24.476	24.480	24.484
622	24.488	24.492	24.496	24.500	24.504	24.508	24.512	24.516	24.520	24.524
623	24.528	24.531	24.535	24.539	24.543	24.547	24.551	24.555	24.559	24.563
624	24.567	24.571	24.575	24.579	24.583	24.587	24.591	24.594	24.598	24.602
625	24.606	24.610	24.614	24.618	24.622	24.626	24.630	24.634	24.638	24.642
626	24.646	24.650	24.653	24.657	24.661	24.665	24.669	24.673	24.677	24.681
627	24.685	24.689	24.693	24.697	24.701	24.705	24.709	24.713	24.716	24.720
628	24.724	24.728	24.732	24.736	24.740	24.744	24.748	24.752	24.756	24.760
629	24.764	24.768	24.772	24.776	24.779	24.783	24.787	24.791	24.795	24.799
630	24.803	24.807	24.811	24.815	24.819	24.823	24.827	24.831	24.835	24.839
631	24.842	24.846	24.850	24.854	24.858	24.862	24.866	24.870	24.874	24.878
632	24.882	24.886	24.890	24.894	24.898	24.902	24.905	24.909	24.913	24.917
633	24.921	24.925	24.929	24.933	24.937	24.941	24.945	24.949	24.953	24.957
634	24.961	24.965	24.968	24.972	24.976	24.980	24.984	24.988	24.992	24.996
635	25.000	25.004	25.008	25.012	25.016	25.020	25.024	25.028	25.031	25.035
636	25.039	25.043	25.047	25.051	25.055	25.059	25.063	25.067	25.071	25.075
637	25.079	25.083	25.087	25.091	25.094	25.098	25.102	25.106	25.110	25.114
638	25.118	25.122	25.126	25.130	25.134	25.138	25.142	25.146	25.150	25.153
639	25.157	25.161	25.165	25.169	25.173	25.177	25.181	25.185	25.189	25.193
640	25.197	25.201	25.205	25.209	25.213	25.216	25.220	25.224	25.228	25.232
641	25.236	25.240	25.244	25.248	25.252	25.256	25.260	25.264	25.268	25.272
642	25.276	25.279	25.283	25.287	25.291	25.295	25.299	25.303	25.307	25.311
643	25.315	25.319	25.323	25.327	25.331	25.335	25.339	25.342	25.346	25.350
644	25.354	25.358	25.362	25.366	25.370	25.374	25.378	25.382	25.386	25.390
645	25.394	25.398	25.402	25.405	25.409	25.413	25.417	25.421	25.425	25.429
646	25.433	25.437	25.441	25.445	25.449	25.453	25.457	25.461	25.465	25.468
647	25.472	25.476	25.480	25.484	25.488	25.492	25.496	25.500	25.504	25.508
648	25.512	25.516	25.520	25.524	25.528	25.531	25.535	25.539	25.543	25.547
649	25.551	25.555	25.559	25.563	25.567	25.571	25.575	25.579	25.583	25.587
650	25.591	25.594	25.598	25.602	25.606	25.610	25.614	25.618	25.622	25.626

TABLE 65.
MILLIMETRES INTO INCHES.
1 mm. = 0.03937 inch.

Milli-metres.	.0	.1	.2	.3	.4	.5	.6	.7	.8	.9
	Inches.	Inches.	Inches.	Inches.	Inches.	Inches.	Inches.	Inches.	Inches.	Inches.
650	25.591	25.594	25.598	25.602	25.606	25.610	25.614	25.618	25.622	25.626
651	25.630	25.634	25.638	25.642	25.646	25.650	25.653	25.657	25.661	25.665
652	25.669	25.673	25.677	25.681	25.685	25.689	25.693	25.697	25.701	25.705
653	25.709	25.713	25.716	25.720	25.724	25.728	25.732	25.736	25.740	25.744
654	25.748	25.752	25.756	25.760	25.764	25.768	25.772	25.776	25.779	25.783
655	25.787	25.791	25.795	25.799	25.803	25.807	25.811	25.815	25.819	25.823
656	25.827	25.831	25.835	25.839	25.842	25.846	25.850	25.854	25.858	25.862
657	25.866	25.870	25.874	25.878	25.882	25.886	25.890	25.894	25.898	25.902
658	25.905	25.909	25.913	25.917	25.921	25.925	25.929	25.933	25.937	25.941
659	25.945	25.949	25.953	25.957	25.961	25.965	25.968	25.972	25.976	25.980
660	25.984	25.988	25.992	25.996	26.000	26.004	26.008	26.012	26.016	26.020
661	26.024	26.028	26.031	26.035	26.039	26.043	26.047	26.051	26.055	26.059
662	26.063	26.067	26.071	26.075	26.079	26.083	26.087	26.090	26.094	26.098
663	26.102	26.106	26.110	26.114	26.118	26.122	26.126	26.130	26.134	26.138
664	26.142	26.146	26.150	26.153	26.157	26.161	26.165	26.169	26.173	26.177
665	26.181	26.185	26.189	26.193	26.197	26.201	26.205	26.209	26.213	26.216
666	26.220	26.224	26.228	26.232	26.236	26.240	26.244	26.248	26.252	26.256
667	26.260	26.264	26.268	26.272	26.276	26.279	26.283	26.287	26.291	26.295
668	26.299	26.303	26.307	26.311	26.315	26.319	26.323	26.327	26.331	26.335
669	26.339	26.342	26.346	26.350	26.354	26.358	26.362	26.366	26.370	26.374
670	26.378	26.382	26.386	26.390	26.394	26.398	26.402	26.405	26.409	26.413
671	26.417	26.421	26.425	26.429	26.433	26.437	26.441	26.445	26.449	26.453
672	26.457	26.461	26.465	26.468	26.472	26.476	26.480	26.484	26.488	26.492
673	26.496	26.500	26.504	26.508	26.512	26.516	26.520	26.524	26.528	26.531
674	26.535	26.539	26.543	26.547	26.551	26.555	26.559	26.563	26.567	26.571
675	26.575	26.579	26.583	26.587	26.590	26.594	26.598	26.602	26.606	26.610
676	26.614	26.618	26.622	26.626	26.630	26.634	26.638	26.642	26.646	26.650
677	26.653	26.657	26.661	26.665	26.669	26.673	26.677	26.681	26.685	26.689
678	26.693	26.697	26.701	26.705	26.709	26.713	26.716	26.720	26.724	26.728
679	26.732	26.736	26.740	26.744	26.748	26.752	26.756	26.760	26.764	26.768
680	26.772	26.776	26.779	26.783	26.787	26.791	26.795	26.799	26.803	26.807
681	26.811	26.815	26.819	26.823	26.827	26.831	26.835	26.838	26.842	26.846
682	26.850	26.854	26.858	26.862	26.866	26.870	26.874	26.878	26.882	26.886
683	26.890	26.894	26.898	26.902	26.905	26.909	26.913	26.917	26.921	26.925
684	26.929	26.933	26.937	26.941	26.945	26.949	26.953	26.957	26.961	26.965
685	26.968	26.972	26.976	26.980	26.984	26.988	26.992	26.996	27.000	27.004
686	27.008	27.012	27.016	27.020	27.024	27.028	27.031	27.035	27.039	27.043
687	27.047	27.051	27.055	27.059	27.063	27.067	27.071	27.075	27.079	27.083
688	27.087	27.090	27.094	27.098	27.102	27.106	27.110	27.114	27.118	27.122
689	27.126	27.130	27.134	27.138	27.142	27.146	27.150	27.153	27.157	27.161
690	27.165	27.169	27.173	27.177	27.181	27.185	27.189	27.193	27.197	27.201
691	27.205	27.209	27.213	27.216	27.220	27.224	27.228	27.232	27.236	27.240
692	27.244	27.248	27.252	27.256	27.260	27.264	27.268	27.272	27.276	27.279
693	27.283	27.287	27.291	27.295	27.299	27.303	27.307	27.311	27.315	27.319
694	27.323	27.327	27.331	27.335	27.339	27.342	27.346	27.350	27.354	27.358
695	27.362	27.366	27.370	27.374	27.378	27.382	27.386	27.390	27.394	27.398
696	27.402	27.405	27.409	27.413	27.417	27.421	27.425	27.429	27.433	27.437
697	27.441	27.445	27.449	27.453	27.457	27.461	27.465	27.468	27.472	27.476
698	27.480	27.484	27.488	27.492	27.496	27.500	27.504	27.508	27.512	27.516
699	27.520	27.524	27.528	27.531	27.535	27.539	27.543	27.547	27.551	27.555
700	27.559	27.563	27.567	27.571	27.575	27.579	27.583	27.587	27.590	27.594

SMITHSONIAN TABLES.

TABLE 65.

MILLIMETRES INTO INCHES.

1 mm. = 0.03937 inch.

Milli-metres.	.0	.1	.2	.3	.4	.5	.6	.7	.8	.9
	Inches.	Inches.	Inches.	Inches.	Inches.	Inches.	Inches.	Inches.	Inches.	Inches.
700	27.559	27.563	27.567	27.571	27.575	27.579	27.583	27.587	27.590	27.594
701	27.598	27.602	27.606	27.610	27.614	27.618	27.622	27.626	27.630	27.634
702	27.638	27.642	27.646	27.650	27.653	27.657	27.661	27.665	27.669	27.673
703	27.677	27.681	27.685	27.689	27.693	27.697	27.701	27.705	27.709	27.713
704	27.716	27.720	27.724	27.728	27.732	27.736	27.740	27.744	27.748	27.752
705	27.756	27.760	27.764	27.768	27.772	27.776	27.779	27.783	27.787	27.791
706	27.795	27.799	27.803	27.807	27.811	27.815	27.819	27.823	27.827	27.831
707	27.835	27.839	27.842	27.846	27.850	27.854	27.858	27.862	27.866	27.870
708	27.874	27.878	27.882	27.886	27.890	27.894	27.898	27.902	27.905	27.909
709	27.913	27.917	27.921	27.925	27.929	27.933	27.937	27.941	27.945	27.949
710	27.953	27.957	27.961	27.965	27.968	27.972	27.976	27.980	27.984	27.988
711	27.992	27.996	28.000	28.004	28.008	28.012	28.016	28.020	28.024	28.028
712	28.031	28.035	28.039	28.043	28.047	28.051	28.055	28.059	28.063	28.067
713	28.071	28.075	28.079	28.083	28.087	28.090	28.094	28.098	28.102	28.106
714	28.110	28.114	28.118	28.122	28.126	28.130	28.134	28.138	28.142	28.146
715	28.150	28.153	28.157	28.161	28.165	28.169	28.173	28.177	28.181	28.185
716	28.189	28.193	28.197	28.201	28.205	28.209	28.213	28.216	28.220	28.224
717	28.228	28.232	28.236	28.240	28.244	28.248	28.252	28.256	28.260	28.264
718	28.268	28.272	28.276	28.279	28.283	28.287	28.291	28.295	28.299	28.303
719	28.307	28.311	28.315	28.319	28.323	28.327	28.331	28.335	28.339	28.342
720	28.346	28.350	28.354	28.358	28.362	28.366	28.370	28.374	28.378	28.382
721	28.386	28.390	28.394	28.398	28.402	28.405	28.409	28.413	28.417	28.421
722	28.425	28.429	28.433	28.437	28.441	28.445	28.449	28.453	28.457	28.461
723	28.465	28.468	28.472	28.476	28.480	28.484	28.488	28.492	28.496	28.500
724	28.504	28.508	28.512	28.516	28.520	28.524	28.528	28.531	28.535	28.539
725	28.543	28.547	28.551	28.555	28.559	28.563	28.567	28.571	28.575	28.579
726	28.583	28.587	28.590	28.594	28.598	28.602	28.606	28.610	28.614	28.618
727	28.622	28.626	28.630	28.634	28.638	28.642	28.646	28.650	28.653	28.657
728	28.661	28.665	28.669	28.673	28.677	28.681	28.685	28.689	28.693	28.697
729	28.701	28.705	28.709	28.713	28.716	28.720	28.724	28.728	28.732	28.736
730	28.740	28.744	28.748	28.752	28.756	28.760	28.764	28.768	28.772	28.776
731	28.779	28.783	28.787	28.791	28.795	28.799	28.803	28.807	28.811	28.815
732	28.819	28.823	28.827	28.831	28.835	28.839	28.842	28.846	28.850	28.854
733	28.858	28.862	28.866	28.870	28.874	28.878	28.882	28.886	28.890	28.894
734	28.898	28.902	28.905	28.909	28.913	28.917	28.921	28.925	28.929	28.933
735	28.937	28.941	28.945	28.949	28.953	28.957	28.961	28.965	28.968	28.972
736	28.976	28.980	28.984	28.988	28.992	28.996	29.000	29.004	29.008	29.012
737	29.016	29.020	29.024	29.028	29.031	29.035	29.039	29.043	29.047	29.051
738	29.055	29.059	29.063	29.067	29.071	29.075	29.079	29.083	29.087	29.090
739	29.094	29.098	29.102	29.106	29.110	29.114	29.118	29.122	29.126	29.130
740	29.134	29.138	29.142	29.146	29.150	29.153	29.157	29.161	29.165	29.169
741	29.173	29.177	29.181	29.185	29.189	29.193	29.197	29.201	29.205	29.209
742	29.213	29.216	29.220	29.224	29.228	29.232	29.236	29.240	29.244	29.248
743	29.252	29.256	29.260	29.264	29.268	29.272	29.276	29.279	29.283	29.287
744	29.291	29.295	29.299	29.303	29.307	29.311	29.315	29.319	29.323	29.327
745	29.331	29.335	29.339	29.342	29.346	29.350	29.354	29.358	29.362	29.366
746	29.370	29.374	29.378	29.382	29.386	29.390	29.394	29.398	29.402	29.405
747	29.409	29.413	29.417	29.421	29.425	29.429	29.433	29.437	29.441	29.445
748	29.449	29.453	29.457	29.461	29.465	29.468	29.472	29.476	29.480	29.484
749	29.488	29.492	29.496	29.500	29.504	29.508	29.512	29.516	29.520	29.524
750	29.528	29.531	29.535	29.539	29.543	29.547	29.551	29.555	29.559	29.563

TABLE 65.
MILLIMETRES INTO INCHES.

1 mm. = 0.03937 inch.

Milli-metres.	.0	.1	.2	.3	.4	.5	.6	.7	.8	.9
	Inches.	Inches.	Inches.	Inches.	Inches.	Inches.	Inches.	Inches.	Inches.	Inches.
750	29.528	29.531	29.535	29.539	29.543	29.547	29.551	29.555	29.559	29.563
751	29.567	29.571	29.575	29.579	29.583	29.587	29.590	29.594	29.598	29.602
752	29.606	29.610	29.614	29.618	29.622	29.626	29.630	29.634	29.638	29.642
753	29.646	29.650	29.653	29.657	29.661	29.665	29.669	29.673	29.677	29.681
754	29.685	29.689	29.693	29.697	29.701	29.705	29.709	29.713	29.716	29.720
755	29.724	29.728	29.732	29.736	29.740	29.744	29.748	29.752	29.756	29.760
756	29.764	29.768	29.772	29.776	29.779	29.783	29.787	29.791	29.795	29.799
757	29.803	29.807	29.811	29.815	29.819	29.823	29.827	29.831	29.835	29.839
758	29.842	29.846	29.850	29.854	29.858	29.862	29.866	29.870	29.874	29.878
759	29.882	29.886	29.890	29.894	29.898	29.902	29.905	29.909	29.913	29.917
760	29.921	29.925	29.929	29.933	29.937	29.941	29.945	29.949	29.953	29.957
761	29.961	29.965	29.968	29.972	29.976	29.980	29.984	29.988	29.992	29.996
762	30.000	30.004	30.008	30.012	30.016	30.020	30.024	30.027	30.031	30.035
763	30.039	30.043	30.047	30.051	30.055	30.059	30.063	30.067	30.071	30.075
764	30.079	30.083	30.087	30.090	30.094	30.098	30.102	30.106	30.110	30.114
765	30.118	30.122	30.126	30.130	30.134	30.138	30.142	30.146	30.150	30.153
766	30.157	30.161	30.165	30.169	30.173	30.177	30.181	30.185	30.189	30.193
767	30.197	30.201	30.205	30.209	30.213	30.216	30.220	30.224	30.228	30.232
768	30.236	30.240	30.244	30.248	30.252	30.256	30.260	30.264	30.268	30.272
769	30.276	30.279	30.283	30.287	30.291	30.295	30.299	30.303	30.307	30.311
770	30.315	30.319	30.323	30.327	30.331	30.335	30.339	30.342	30.346	30.350
771	30.354	30.358	30.362	30.366	30.370	30.374	30.378	30.382	30.386	30.390
772	30.394	30.398	30.402	30.405	30.409	30.413	30.417	30.421	30.425	30.429
773	30.433	30.437	30.441	30.445	30.449	30.453	30.457	30.461	30.465	30.468
774	30.472	30.476	30.480	30.484	30.488	30.492	30.496	30.500	30.504	30.508
775	30.512	30.516	30.520	30.524	30.528	30.531	30.535	30.539	30.543	30.547
776	30.551	30.555	30.559	30.563	30.567	30.571	30.575	30.579	30.583	30.587
777	30.590	30.594	30.598	30.602	30.606	30.610	30.614	30.618	30.622	30.626
778	30.630	30.634	30.638	30.642	30.646	30.650	30.653	30.657	30.661	30.665
779	30.669	30.673	30.677	30.681	30.685	30.689	30.693	30.697	30.701	30.705
780	30.709	30.713	30.716	30.720	30.724	30.728	30.732	30.736	30.740	30.744
781	30.748	30.752	30.756	30.760	30.764	30.768	30.772	30.776	30.779	30.783
782	30.787	30.791	30.795	30.799	30.803	30.807	30.811	30.815	30.819	30.823
783	30.827	30.831	30.835	30.839	30.842	30.846	30.850	30.854	30.858	30.862
784	30.866	30.870	30.874	30.878	30.882	30.886	30.890	30.894	30.898	30.902
785	30.905	30.909	30.913	30.917	30.921	30.925	30.929	30.933	30.937	30.941
786	30.945	30.949	30.953	30.957	30.961	30.965	30.968	30.972	30.976	30.980
787	30.984	30.988	30.992	30.996	31.000	31.004	31.008	31.012	31.016	31.020
788	31.024	31.027	31.031	31.035	31.039	31.043	31.047	31.051	31.055	31.059
789	31.063	31.067	31.071	31.075	31.079	31.083	31.087	31.090	31.094	31.098
790	31.102	31.106	31.110	31.114	31.118	31.122	31.126	31.130	31.134	31.138
791	31.142	31.146	31.150	31.153	31.157	31.161	31.165	31.169	31.173	31.177
792	31.181	31.185	31.189	31.193	31.197	31.201	31.205	31.209	31.213	31.216
793	31.220	31.224	31.228	31.232	31.236	31.240	31.244	31.248	31.252	31.256
794	31.260	31.264	31.268	31.272	31.276	31.279	31.283	31.287	31.291	31.295
795	31.299	31.303	31.307	31.311	31.315	31.319	31.323	31.327	31.331	31.335
796	31.339	31.342	31.346	31.350	31.354	31.358	31.362	31.366	31.370	31.374
797	31.378	31.382	31.386	31.390	31.394	31.398	31.402	31.405	31.409	31.413
798	31.417	31.421	31.425	31.429	31.433	31.437	31.441	31.445	31.449	31.453
799	31.457	31.461	31.465	31.468	31.472	31.476	31.480	31.484	31.488	31.492
800	31.496	31.500	31.504	31.508	31.512	31.516	31.520	31.524	31.527	31.531

SMITHSONIAN TABLES.

TABLE 65.

MILLIMETRES INTO INCHES.
1 mm. = 0.03937 inch.

Milli-metres.	.0	.1	.2	.3	.4	.5	.6	.7	.8	.9
	Inches.	Inches.	Inches.	Inches.	Inches.	Inches.	Inches.	Inches.	Inches.	Inches.
800	31.496	31.500	31.504	31.508	31.512	31.516	31.520	31.524	31.527	31.531
801	31.535	31.539	31.543	31.547	31.551	31.555	31.559	31.563	31.567	31.571
802	31.575	31.579	31.583	31.587	31.590	31.594	31.598	31.602	31.606	31.610
803	31.614	31.618	31.622	31.626	31.630	31.634	31.638	31.642	31.646	31.650
804	31.653	31.657	31.661	31.665	31.669	31.673	31.677	31.681	31.685	31.689
805	31.693	31.697	31.701	31.705	31.709	31.713	31.716	31.720	31.724	31.728
806	31.732	31.736	31.740	31.744	31.748	31.752	31.756	31.760	31.764	31.768
807	31.772	31.776	31.779	31.783	31.787	31.791	31.795	31.799	31.803	31.807
808	31.811	31.815	31.819	31.823	31.827	31.831	31.835	31.839	31.842	31.846
809	31.850	31.854	31.858	31.862	31.866	31.870	31.874	31.878	31.882	31.886
810	31.890	31.894	31.898	31.902	31.905	31.909	31.913	31.917	31.921	31.925
811	31.929	31.933	31.937	31.941	31.945	31.949	31.953	31.957	31.961	31.965
812	31.968	31.972	31.976	31.980	31.984	31.988	31.992	31.996	32.000	32.004
813	32.008	32.012	32.016	32.020	32.024	32.027	32.031	32.035	32.039	32.043
814	32.047	32.051	32.055	32.059	32.063	32.067	32.071	32.075	32.079	32.083
815	32.087	32.090	32.094	32.098	32.102	32.106	32.110	32.114	32.118	32.122
816	32.126	32.130	32.134	32.138	32.142	32.146	32.150	32.153	32.157	32.161
817	32.165	32.169	32.173	32.177	32.181	32.185	32.189	32.193	32.197	32.201
818	32.205	32.209	32.213	32.216	32.220	32.224	32.228	32.232	32.236	32.240
819	32.244	32.248	32.252	32.256	32.260	32.264	32.268	32.272	32.276	32.279
820	32.283	32.287	32.291	32.295	32.299	32.303	32.307	32.311	32.315	32.319
821	32.323	32.327	32.331	32.335	32.339	32.342	32.346	32.350	32.354	32.358
822	32.362	32.366	32.370	32.374	32.378	32.382	32.386	32.390	32.394	32.398
823	32.402	32.405	32.409	32.413	32.417	32.421	32.425	32.429	32.433	32.437
824	32.441	32.445	32.449	32.453	32.457	32.461	32.465	32.468	32.472	32.476
825	32.480	32.484	32.488	32.492	32.496	32.500	32.504	32.508	32.512	32.516
826	32.520	32.524	32.527	32.531	32.535	32.539	32.543	32.547	32.551	32.555
827	32.559	32.563	32.567	32.571	32.575	32.579	32.583	32.587	32.590	32.594
828	32.598	32.602	32.606	32.610	32.614	32.618	32.622	32.626	32.630	32.634
829	32.638	32.642	32.646	32.650	32.653	32.657	32.661	32.665	32.669	32.673
830	32.677	32.681	32.685	32.689	32.693	32.697	32.701	32.705	32.709	32.713
831	32.716	32.720	32.724	32.728	32.732	32.736	32.740	32.744	32.748	32.752
832	32.756	32.760	32.764	32.768	32.772	32.776	32.779	32.783	32.787	32.791
833	32.795	32.799	32.803	32.807	32.811	32.815	32.819	32.823	32.827	32.831
834	32.835	32.839	32.842	32.846	32.850	32.854	32.858	32.862	32.866	32.870
835	32.874	32.878	32.882	32.886	32.890	32.894	32.898	32.902	32.905	32.909
836	32.913	32.917	32.921	32.925	32.929	32.933	32.937	32.941	32.945	32.949
837	32.953	32.957	32.961	32.965	32.968	32.972	32.976	32.980	32.984	32.988
838	32.992	32.996	33.000	33.004	33.008	33.012	33.016	33.020	33.024	33.027
839	33.031	33.035	33.039	33.043	33.047	33.051	33.055	33.059	33.063	33.067
840	33.071	33.075	33.079	33.083	33.087	33.090	33.094	33.098	33.102	33.106
841	33.110	33.114	33.118	33.122	33.126	33.130	33.134	33.138	33.142	33.146
842	33.150	33.153	33.157	33.161	33.165	33.169	33.173	33.177	33.181	33.185
843	33.189	33.193	33.197	33.201	33.205	33.209	33.213	33.216	33.220	33.224
844	33.228	33.232	33.236	33.240	33.244	33.248	33.252	33.256	33.260	33.264
845	33.268	33.272	33.276	33.279	33.283	33.287	33.291	33.295	33.299	33.303
846	33.307	33.311	33.315	33.319	33.323	33.327	33.331	33.335	33.339	33.342
847	33.346	33.350	33.354	33.358	33.362	33.366	33.370	33.374	33.378	33.382
848	33.386	33.390	33.394	33.398	33.402	33.405	33.409	33.413	33.417	33.421
849	33.425	33.429	33.433	33.437	33.441	33.445	33.449	33.453	33.457	33.461
850	33.464	33.468	33.472	33.476	33.480	33.484	33.488	33.492	33.496	33.500

TABLE 65.
MILLIMETRES INTO INCHES.
1 mm. = 0.03937 inch.

Milli-metres.	.0	.1	.2	.3	.4	.5	.6	.7	.8	.9
	Inches.	Inches.	Inches.	Inches.	Inches.	Inches.	Inches.	Inches.	Inches.	Inches.
850	33.464	33.468	33.472	33.476	33.480	33.484	33.488	33.492	33.496	33.500
851	33.504	33.508	33.512	33.516	33.520	33.524	33.527	33.531	33.535	33.539
852	33.543	33.547	33.551	33.555	33.559	33.563	33.567	33.571	33.575	33.579
853	33.583	33.587	33.590	33.594	33.598	33.602	33.606	33.610	33.614	33.618
854	33.622	33.626	33.630	33.634	33.638	33.642	33.646	33.650	33.653	33.657
855	33.661	33.665	33.669	33.673	33.677	33.681	33.685	33.689	33.693	33.697
856	33.701	33.705	33.709	33.713	33.716	33.720	33.724	33.728	33.732	33.736
857	33.740	33.744	33.748	33.752	33.756	33.760	33.764	33.768	33.772	33.776
858	33.779	33.783	33.787	33.791	33.795	33.799	33.803	33.807	33.811	33.815
859	33.819	33.823	33.827	33.831	33.835	33.839	33.842	33.846	33.850	33.854
860	33.858	33.862	33.866	33.870	33.874	33.878	33.882	33.886	33.890	33.894
861	33.898	33.902	33.905	33.909	33.913	33.917	33.921	33.925	33.929	33.933
862	33.937	33.941	33.945	33.949	33.953	33.957	33.961	33.964	33.968	33.972
863	33.976	33.980	33.984	33.988	33.992	33.996	34.000	34.004	34.008	34.012
864	34.016	34.020	34.024	34.027	34.031	34.035	34.039	34.043	34.047	34.051
865	34.055	34.059	34.063	34.067	34.071	34.075	34.079	34.083	34.087	34.090
866	34.094	34.098	34.102	34.106	34.110	34.114	34.118	34.122	34.126	34.130
867	34.134	34.138	34.142	34.146	34.150	34.153	34.157	34.161	34.165	34.169
868	34.173	34.177	34.181	34.185	34.189	34.193	34.197	34.201	34.205	34.209
869	34.213	34.216	34.220	34.224	34.228	34.232	34.236	34.240	34.244	34.248
870	34.252	34.256	34.260	34.264	34.268	34.272	34.276	34.279	34.283	34.287
871	34.291	34.295	34.299	34.303	34.307	34.311	34.315	34.319	34.323	34.327
872	34.331	34.335	34.339	34.342	34.346	34.350	34.354	34.358	34.362	34.366
873	34.370	34.374	34.378	34.382	34.386	34.390	34.394	34.398	34.402	34.405
874	34.409	34.413	34.417	34.421	34.425	34.429	34.433	34.437	34.441	34.445
875	34.449	34.453	34.457	34.461	34.464	24.468	34.472	34.476	34.480	34.484
876	34.488	34.492	34.496	34.500	34.504	34.508	34.512	34.516	34.520	34.524
877	34.527	34.531	34.535	34.539	34.543	34.547	34.551	34.555	34.559	34.563
878	34.567	34.571	34.575	34.579	34.583	34.587	34.590	34.594	34.598	34.602
879	34.606	34.610	34.614	34.618	34.622	34.626	34.630	34.634	34.638	34.642
880	34.646	34.650	34.653	34.657	34.661	34.665	34.669	34.673	34.677	34.681
881	34.685	34.689	34.693	34.697	34.701	34.705	34.709	34.713	34.716	34.720
882	34.724	34.728	34.732	34.736	34.740	34.744	34.748	34.752	34.756	34.760
883	34.764	34.768	34.772	34.776	34.779	34.783	34.787	34.791	34.795	34.799
884	34.803	34.807	34.811	34.815	34.819	34.823	34.827	34.831	34.835	34.839
885	34.842	34.846	34.850	34.854	34.858	34.862	34.866	34.870	34.874	34.878
886	34.882	34.886	34.890	34.894	34.898	34.902	34.905	34.909	34.913	34.917
887	34.921	34.925	34.929	34.933	34.937	34.941	34.945	34.949	34.953	34.957
888	34.961	34.964	34.968	34.972	34.976	34.980	34.984	34.988	34.992	34.996
889	35.000	35.004	35.008	35.012	35.016	35.020	35.024	35.027	35.031	35.035
890	35.039	35.043	35.047	35.051	35.055	35.059	35.063	35.067	35.071	35.075
891	35.079	35.083	35.087	35.090	35.094	35.098	35.102	35.106	35.110	35.114
892	35.118	35.122	35.126	35.130	35.134	35.138	35.142	35.146	35.150	35.153
893	35.157	35.161	35.165	35.169	35.173	35.177	35.181	35.185	35.189	35.193
894	35.197	35.201	35.205	35.209	35.213	35.216	35.220	35.224	35.228	35.232
895	35.236	35.240	35.244	35.248	35.252	35.256	35.260	35.264	35.268	35.272
896	35.276	35.279	35.283	35.287	35.291	35.295	35.299	35.303	35.307	35.311
897	35.315	35.319	35.323	35.327	35.331	35.335	35.339	35.342	35.346	35.350
898	35.354	35.358	35.362	35.366	35.370	35.374	35.378	35.382	35.386	35.390
899	35.394	35.398	35.402	35.405	35.409	35.413	35.417	35.421	35.425	35.429
900	35.433	35.437	35.441	35.445	35.449	35.453	35.457	35.461	35.464	35.468

SMITHSONIAN TABLES.

TABLE 65.

MILLIMETRES INTO INCHES.

1 mm. = 0.03937 inch.

Milli-metres.	.0	.1	.2	.3	.4	.5	.6	.7	.8	.9
	Inches.	Inches.	Inches.	Inches.	Inches.	Inches.	Inches.	Inches.	Inches.	Inches.
900	35.433	35.437	35.441	35.445	35.449	35.453	35.457	35.461	35.464	35.468
901	35.472	35.476	35.480	35.484	35.488	35.492	35.496	35.500	35.504	35.508
902	35.512	35.516	35.520	35.524	35.527	35.531	35.535	35.539	35.543	35.547
903	35.551	35.555	35.559	35.563	35.567	35.571	35.575	35.579	35.583	35.587
904	35.590	35.594	35.598	35.602	35.606	35.610	35.614	35.618	35.622	35.626
905	35.630	35.634	35.638	35.642	35.646	35.650	35.653	35.657	35.661	35.665
906	35.669	35.673	35.677	35.681	35.685	35.689	35.693	35.697	35.701	35.705
907	35.709	35.713	35.716	35.720	35.724	35.728	35.732	35.736	35.740	35.744
908	35.748	35.752	35.756	35.760	35.764	35.768	35.772	35.776	35.779	35.783
909	35.787	35.791	35.795	35.799	35.803	35.807	35.811	35.815	35.819	35.823
910	35.827	35.831	35.835	35.839	35.842	35.846	35.850	35.854	35.858	35.862
911	35.866	35.870	35.874	35.878	35.882	35.886	35.890	35.894	35.898	35.902
912	35.905	35.909	35.913	35.917	35.921	35.925	35.929	35.933	35.937	35.941
913	35.945	35.949	35.953	35.957	35.961	35.964	35.968	35.972	35.976	35.980
914	35.984	35.988	35.992	35.996	36.000	36.004	36.008	36.012	36.016	36.020
915	36.024	36.027	36.031	36.035	36.039	36.043	36.047	36.051	36.055	36.059
916	36.063	36.067	36.071	36.075	36.079	36.083	36.087	36.090	36.094	36.098
917	36.102	36.106	36.110	36.114	36.118	36.122	36.126	36.130	36.134	36.138
918	36.142	36.146	36.150	36.153	36.157	36.161	36.165	36.169	36.173	36.177
919	36.181	36.185	36.189	36.193	36.197	36.201	36.205	36.209	36.213	36.216
920	36.220	36.224	36.228	36.232	36.236	36.240	36.244	36.248	36.252	36.256
921	36.260	36.264	36.268	36.272	36.276	36.279	36.283	36.287	36.291	36.295
922	36.299	36.303	36.307	36.311	36.315	36.319	36.323	36.327	36.331	36.335
923	36.339	36.342	36.346	36.350	36.354	36.358	36.362	36.366	36.370	36.374
924	36.378	36.382	36.386	36.390	36.394	36.398	36.402	36.405	36.409	36.413
925	36.417	36.421	36.425	36.429	36.433	36.437	36.441	36.445	36.449	36.453
926	36.457	36.461	36.464	36.468	36.472	36.476	36.480	36.484	36.488	36.492
927	36.496	36.500	36.504	36.508	36.512	36.516	36.520	36.524	36.527	36.531
928	36.535	36.539	36.543	36.547	36.551	36.555	36.559	36.563	36.567	36.571
929	36.575	36.579	36.583	36.587	36.590	36.594	36.598	36.602	36.606	36.610
930	36.614	36.618	36.622	36.626	36.630	36.634	36.638	36.642	36.646	36.650
931	36.653	36.657	36.661	36.665	36.669	36.673	36.677	36.681	36.685	36.689
932	36.693	36.697	36.701	36.705	36.709	36.713	36.716	36.720	36.724	36.728
933	36.732	36.736	36.740	36.744	36.748	36.752	36.756	36.760	36.764	36.768
934	36.772	36.776	36.779	36.783	36.787	36.791	36.795	36.799	36.803	36.807
935	36.811	36.815	36.819	36.823	36.827	36.831	36.835	36.839	36.842	36.846
936	36.850	36.854	36.858	36.862	36.866	36.870	36.874	36.878	36.882	36.886
937	36.890	36.894	36.898	36.902	36.905	36.909	36.913	36.917	36.921	36.925
938	36.929	36.933	36.937	36.941	36.945	36.949	36.953	36.957	36.961	36.964
939	36.968	36.972	36.976	36.980	36.984	36.988	36.992	36.996	37.000	37.004
940	37.008	37.012	37.016	37.020	37.024	37.027	37.031	37.035	37.039	37.043
941	37.047	37.051	37.055	37.059	37.063	37.067	37.071	37.075	37.079	37.083
942	37.087	37.090	37.094	37.098	37.102	37.106	37.110	37.114	37.118	37.122
943	37.126	37.130	37.134	37.138	37.142	37.146	37.150	37.153	37.157	37.161
944	37.165	37.169	37.173	37.177	37.181	37.185	37.189	37.193	37.197	37.201
945	37.204	37.208	37.212	37.216	37.220	37.224	37.228	37.232	37.236	37.240
946	37.244	37.248	37.252	37.256	37.260	37.264	37.268	37.272	37.276	37.279
947	37.283	37.287	37.291	37.295	37.299	37.303	37.307	37.311	37.315	37.319
948	37.323	37.327	37.331	37.335	37.339	37.342	37.346	37.350	37.354	37.358
949	37.362	37.366	37.370	37.374	37.378	37.382	37.386	37.390	37.394	37.398
950	37.402	37.405	37.409	37.413	37.417	37.421	37.425	37.429	37.433	37.437

SMITHSONIAN TABLES.

TABLE 65.

MILLIMETRES INTO INCHES.

1 mm. = 0.03937 inch.

Milli-metres.	.0	.1	.2	.3	.4	.5	.6	.7	.8	.9
	Inches.	Inches.	Inches.	Inches.	Inches.	Inches.	Inches.	Inches.	Inches.	Inches.
950	37.402	37.405	37.409	37.413	37.417	37.421	37.425	37.429	37.433	37.437
951	37.441	37.445	37.449	37.453	37.457	37.461	37.464	37.468	37.472	37.476
952	37.480	37.484	37.488	37.492	37.496	37.500	37.504	37.508	37.512	37.516
953	37.520	37.524	37.527	37.531	37.535	37.539	37.543	37.547	37.551	37.555
954	37.559	37.563	37.567	37.571	37.575	37.579	37.583	37.587	37.590	37.594
955	37.598	37.602	37.606	37.610	37.614	37.618	37.622	37.626	37.630	37.634
956	37.638	37.642	37.646	37.650	37.653	37.657	37.661	37.665	37.669	37.673
957	37.677	37.681	37.685	37.689	37.693	37.697	37.701	37.705	37.709	37.713
958	37.716	37.720	37.724	37.728	37.732	37.736	37.740	37.744	37.748	37.752
959	37.756	37.760	37.764	37.768	37.772	37.776	37.779	37.783	37.787	37.791
960	37.795	37.799	37.803	37.807	37.811	37.815	37.819	37.823	37.827	37.831
961	37.835	37.839	37.842	37.846	37.850	37.854	37.858	37.862	37.866	37.870
962	37.874	37.878	37.882	37.886	37.890	37.894	37.898	37.901	37.905	37.909
963	37.913	37.917	37.921	37.925	37.929	37.933	37.937	37.941	37.945	37.949
964	37.953	37.957	37.961	37.964	37.968	37.972	37.976	37.980	37.984	37.988
965	37.992	37.996	38.000	38.004	38.008	38.012	38.016	38.020	38.024	38.027
966	38.031	38.035	38.039	38.043	38.047	38.051	38.055	38.059	38.063	38.067
967	38.071	38.075	38.079	38.083	38.087	38.090	38.094	38.098	38.102	38.106
968	38.110	38.114	38.118	38.122	38.126	38.130	38.134	38.138	38.142	38.146
969	38.150	38.153	38.157	38.161	38.165	38.169	38.173	38.177	38.181	38.185
970	38.189	38.193	38.197	38.201	38.205	38.209	38.213	38.216	38.220	38.224
971	38.228	38.232	38.236	38.240	38.244	38.248	38.252	38.256	38.260	38.264
972	38.268	38.272	38.276	38.279	38.283	38.287	38.291	38.295	38.299	38.303
973	38.307	38.311	38.315	38.319	38.323	38.327	38.331	38.335	38.339	38.342
974	38.346	38.350	38.354	38.358	38.362	38.366	38.370	38.374	38.378	38.382
975	38.386	38.390	38.394	38.398	38.401	38.405	38.409	38.413	38.417	38.421
976	38.425	38.429	38.433	38.437	38.441	38.445	38.449	38.453	38.457	38.461
977	38.464	38.468	38.472	38.476	38.480	38.484	38.488	38.492	38.496	38.500
978	38.504	38.508	38.512	38.516	38.520	38.524	38.527	38.531	38.535	38.539
979	38.543	38.547	38.551	38.555	38.559	38.563	38.567	38.571	38.575	38.579
980	38.583	38.587	38.590	38.594	38.598	38.602	38.606	38.610	38.614	38.618
981	38.622	38.626	38.630	38.634	38.638	38.642	38.646	38.650	38.653	38.657
982	38.661	38.665	38.669	38.673	38.677	38.681	38.685	38.689	38.693	38.697
983	38.701	38.705	38.709	38.713	38.716	38.720	38.724	38.728	38.732	38.736
984	38.740	38.744	38.748	38.752	38.756	38.760	38.764	38.768	38.772	38.776
985	38.780	38.783	38.787	38.791	38.795	38.799	38.803	38.807	38.811	38.815
986	38.819	38.823	38.827	38.831	38.835	38.839	38.842	38.846	38.850	38.854
987	38.858	38.862	38.866	38.870	38.874	38.878	38.882	38.886	38.890	38.894
988	38.898	38.901	38.905	38.909	38.913	38.917	38.921	38.925	38.929	38.933
989	38.937	38.941	38.945	38.949	38.953	38.957	38.961	38.964	38.968	38.972
990	38.976	38.980	38.984	38.988	38.992	38.996	39.000	39.004	39.008	39.012
991	39.016	39.020	39.024	39.027	39.031	39.035	39.039	39.043	39.047	39.051
992	39.055	39.059	39.063	39.067	39.071	39.075	39.079	39.083	39.087	39.090
993	39.094	39.098	39.102	39.106	39.110	39.114	39.118	39.122	39.126	39.130
994	39.134	39.138	39.142	39.146	39.150	39.153	39.157	39.161	39.165	39.169
995	39.173	39.177	39.181	39.185	39.189	39.193	39.197	39.201	39.205	39.209
996	39.213	39.216	39.220	39.224	39.228	39.232	39.236	39.240	39.244	39.248
997	39.252	39.256	39.260	39.264	39.268	39.272	39.276	39.279	39.283	39.287
998	39.291	39.295	39.299	39.303	39.307	39.311	39.315	39.319	39.323	39.327
999	39.331	39.335	39.339	39.342	39.346	39.350	39.354	39.358	39.362	39.366
1000	39.370	39.374	39.378	39.382	39.386	39.390	39.394	39.398	39.401	39.405

TABLE 66.

FEET INTO METRES.

1 foot = 0.3048006 metre.

Feet.	0	1	2	3	4	5	6	7	8	9
	m.	m.	m.	m.	m.	m.	m.	m.	m.	m.
0	0.000	0.305	0.610	0.914	1.219	1.524	1.829	2.134	2.438	2.743
10	3.048	3.353	3.658	3.962	4.267	4.572	4.877	5.182	5.486	5.791
20	6.096	6.401	6.706	7.010	7.315	7.620	7.925	8.230	8.534	8.839
30	9.144	9.449	9.754	10.058	10.363	10.668	10.973	11.278	11.582	11.887
40	12.192	12.497	12.802	13.106	13.411	13.716	14.021	14.326	14.630	14.935
50	15.240	15.545	15.850	16.154	16.459	16.764	17.069	17.374	17.678	17.983
60	18.288	18.593	18.898	19.202	19.507	19.812	20.117	20.422	20.726	21.031
70	21.336	21.641	21.946	22.250	22.555	22.860	23.165	23.470	23.774	24.079
80	24.384	24.689	24.994	25.298	25.603	25.908	26.213	26.518	26.822	27.127
90	27.432	27.737	28.042	28.346	28.651	28.956	29.261	29.566	29.870	30.175
	0	10	20	30	40	50	60	70	80	90
100	30.48	33.53	36.58	39.62	42.67	45.72	48.77	51.82	54.86	57.91
200	60.96	64.01	67.06	70.10	73.15	76.20	79.25	82.30	85.34	88.39
300	91.44	94.49	97.54	100.58	103.63	106.68	109.73	112.78	115.82	118.87
400	121.92	124.97	128.02	131.06	134.11	137.16	140.21	143.26	146.30	149.35
500	152.40	155.45	158.50	161.54	164.59	167.64	170.69	173.74	176.78	179.83
600	182.88	185.93	188.98	192.02	195.07	198.12	201.17	204.22	207.26	210.31
700	213.36	216.41	219.46	222.50	225.55	228.60	231.65	234.70	237.74	240.79
800	243.84	246.89	249.94	252.98	256.03	259.08	262.13	265.18	268.22	271.27
900	274.32	277.37	280.42	283.46	286.51	289.56	292.61	295.66	298.70	301.75
1000	304.80	307.85	310.90	313.94	316.99	320.04	323.09	326.14	329.18	332.23
1100	335.28	338.33	341.38	344.42	347.47	350.52	353.57	356.62	359.67	362.71
1200	365.76	368.81	371.86	374.90	377.95	381.00	384.05	387.10	390.14	393.19
1300	396.24	399.29	402.34	405.38	408.43	411.48	414.53	417.58	420.62	423.67
1400	426.72	429.77	432.82	435.86	438.91	441.96	445.01	448.06	451.10	454.15
1500	457.20	460.25	463.30	466.34	469.39	472.44	475.49	478.54	481.58	484.63
1600	487.68	490.73	493.78	496.82	499.87	502.92	505.97	509.02	512.07	515.11
1700	518.16	521.21	524.26	527.31	530.35	533.40	536.45	539.50	542.55	545.59
1800	548.64	551.69	554.74	557.79	560.83	563.88	566.93	569.98	573.03	576.07
1900	579.12	582.17	585.22	588.27	591.31	594.36	597.41	600.46	603.51	606.55
2000	609.60	612.65	615.70	618.75	621.79	624.84	627.89	630.94	633.99	637.03
2100	640.08	643.13	646.18	649.23	652.27	655.32	658.37	661.42	664.47	667.51
2200	670.56	673.61	676.66	679.71	682.75	685.80	688.85	691.90	694.95	697.99
2300	701.04	704.09	707.14	710.19	713.23	716.28	719.33	722.38	725.43	728.47
2400	731.52	734.57	737.62	740.67	743.71	746.76	749.81	752.86	755.91	758.95
2500	762.00	765.05	768.10	771.15	774.19	777.24	780.29	783.34	786.39	789.43
2600	792.48	795.53	798.58	801.63	804.67	807.72	810.77	813.82	816.87	819.91
2700	822.96	826.01	829.06	832.11	835.15	838.20	841.25	844.30	847.35	850.39
2800	853.44	856.49	859.54	862.59	865.63	868.68	871.73	874.78	877.83	880.87
2900	883.92	886.97	890.02	893.07	896.11	899.16	902.21	905.26	908.31	911.35
3000	914.40	917.45	920.50	923.55	926.59	929.64	932.69	935.74	938.79	941.83
3100	944.88	947.93	950.98	954.03	957.07	960.12	963.17	966.22	969.27	972.31
3200	975.36	978.41	981.46	984.51	987.55	990.60	993.65	996.70	999.75	1002.79
3300	1005.84	1008.89	1011.94	1014.99	1018.03	1021.08	1024.13	1027.18	1030.23	1033.27
3400	1036.32	1039.37	1042.42	1045.47	1048.51	1051.56	1054.61	1057.66	1060.71	1063.75
3500	1066.80	1069.85	1072.90	1075.95	1078.99	1082.04	1085.09	1088.14	1091.19	1094.23
3600	1097.28	1100.33	1103.38	1106.43	1109.47	1112.52	1115.57	1118.62	1121.67	1124.71
3700	1127.76	1130.81	1133.86	1136.91	1139.95	1143.00	1146.05	1149.10	1152.15	1155.19
3800	1158.24	1161.29	1164.34	1167.39	1170.43	1173.48	1176.53	1179.58	1182.63	1185.67
3900	1188.72	1191.77	1194.82	1197.87	1200.91	1203.96	1207.01	1210.06	1213.11	1216.15
4000	1219.20	1222.25	1225.30	1228.35	1231.39	1234.44	1237.49	1240.54	1243.59	1246.63

TABLE 66.
FEET INTO METRES.
1 foot = 0.3048006 metre.

Feet.	0	10	20	30	40	50	60	70	80	90
	m.	m.	m.	m.	m.	m.	m.	m.	m.	m.
4000	1219.2	1222.3	1225.3	1228.3	1231.4	1234.4	1237.5	1240.5	1243.6	1246.6
4100	1249.7	1252.7	1255.8	1258.8	1261.9	1264.9	1268.0	1271.0	1274.1	1277.1
4200	1280.2	1283.2	1286.3	1289.3	1292.4	1295.4	1298.5	1301.5	1304.5	1307.6
4300	1310.6	1313.7	1316.7	1319.8	1322.8	1325.9	1328.9	1332.0	1335.0	1338.1
4400	1341.1	1344.2	1347.2	1350.3	1353.3	1356.4	1359.4	1362.5	1365.5	1368.6
4500	1371.6	1374.7	1377.7	1380.7	1383.8	1386.8	1389.9	1392.9	1396.0	1399.0
4600	1402.1	1405.1	1408.2	1411.2	1414.3	1417.3	1420.4	1423.4	1426.5	1429.5
4700	1432.6	1435.6	1438.7	1441.7	1444.8	1447.8	1450.9	1453.9	1456.9	1460.0
4800	1463.0	1466.1	1469.1	1472.2	1475.2	1478.3	1481.3	1484.4	1487.4	1490.5
4900	1493.5	1496.6	1499.6	1502.7	1505.7	1508.8	1511.8	1514.9	1517.9	1521.0
5000	1524.0	1527.1	1530.1	1533.1	1536.2	1539.2	1542.3	1545.3	1548.4	1551.4
5100	1554.5	1557.5	1560.6	1563.6	1566.7	1569.7	1572.8	1575.8	1578.9	1581.9
5200	1585.0	1588.0	1591.1	1594.1	1597.2	1600.2	1603.3	1606.3	1609.3	1612.4
5300	1615.4	1618.5	1621.5	1624.6	1627.6	1630.7	1633.7	1636.8	1639.8	1642.9
5400	1645.9	1649.0	1652.0	1655.1	1658.1	1661.2	1664.2	1667.3	1670.3	1673.4
5500	1676.4	1679.5	1682.5	1685.5	1688.6	1691.6	1694.7	1697.7	1700.8	1703.8
5600	1706.9	1709.9	1713.0	1716.0	1719.1	1722.1	1725.2	1728.2	1731.3	1734.3
5700	1737.4	1740.4	1743.5	1746.5	1749.6	1752.6	1755.7	1758.7	1761.7	1764.8
5800	1767.8	1770.9	1773.9	1777.0	1780.0	1783.1	1786.1	1789.2	1792.2	1795.3
5900	1798.3	1801.4	1804.4	1807.5	1810.5	1813.6	1816.6	1819.7	1822.7	1825.8
6000	1828.8	1831.9	1834.9	1837.9	1841.0	1844.0	1847.1	1850.1	1853.2	1856.2
6100	1859.3	1862.3	1865.4	1868.4	1871.5	1874.5	1877.6	1880.6	1883.7	1886.7
6200	1889.8	1892.8	1895.9	1898.9	1902.0	1905.0	1908.1	1911.1	1914.1	1917.2
6300	1920.2	1923.3	1926.3	1929.4	1932.4	1935.5	1938.5	1941.6	1944.6	1947.7
6400	1950.7	1953.8	1956.8	1959.9	1962.9	1966.0	1969.0	1972.1	1975.1	1978.2
6500	1981.2	1984.3	1987.3	1990.3	1993.4	1996.4	1999.5	2002.5	2005.6	2008.6
6600	2011.7	2014.7	2017.8	2020.8	2023.9	2026.9	2030.0	2033.0	2036.1	2039.1
6700	2042.2	2045.2	2048.3	2051.3	2054.4	2057.4	2060.5	2063.5	2066.5	2069.6
6800	2072.6	2075.7	2078.7	2081.8	2084.8	2087.9	2090.9	2094.0	2097.0	2100.1
6900	2103.1	2106.2	2109.2	2112.3	2115.3	2118.4	2121.4	2124.5	2127.5	2130.6
7000	2133.6	2136.7	2139.7	2142.7	2145.8	2148.8	2151.9	2154.9	2158.0	2161.0
7100	2164.1	2167.1	2170.2	2173.2	2176.3	2179.3	2182.4	2185.4	2188.5	2191.5
7200	2194.6	2197.6	2200.7	2203.7	2206.8	2209.8	2212.9	2215.9	2218.9	2222.0
7300	2225.0	2228.1	2231.1	2234.2	2237.2	2240.3	2243.3	2246.4	2249.4	2252.5
7400	2255.5	2258.6	2261.6	2264.7	2267.7	2270.8	2273.8	2276.9	2279.9	2283.0
7500	2286.0	2289.1	2292.1	2295.1	2298.2	2301.2	2304.3	2307.3	2310.4	2313.4
7600	2316.5	2319.5	2322.6	2325.6	2328.7	2331.7	2334.8	2337.8	2340.9	2343.9
7700	2347.0	2350.0	2353.1	2356.1	2359.2	2362.2	2365.3	2368.3	2371.3	2374.4
7800	2377.4	2380.5	2383.5	2386.6	2389.6	2392.7	2395.7	2398.8	2401.8	2404.9
7900	2407.9	2411.0	2414.0	2417.1	2420.1	2423.2	2426.2	2429.3	2432.3	2435.4
8000	2438.4	2441.5	2444.5	2447.5	2450.6	2453.6	2456.7	2459.7	2462.8	2465.8
8100	2468.9	2471.9	2475.0	2478.0	2481.1	2484.1	2487.2	2490.2	2493.3	2496.3
8200	2499.4	2502.4	2505.5	2508.5	2511.6	2514.6	2517.7	2520.7	2523.7	2526.8
8300	2529.8	2532.9	2535.9	2539.0	2542.0	2545.1	2548.1	2551.2	2554.2	2557.3
8400	2560.3	2563.4	2566.4	2569.5	2572.5	2575.6	2578.6	2581.7	2584.7	2587.8
8500	2590.8	2593.9	2596.9	2599.9	2603.0	2606.0	2609.1	2612.1	2615.2	2618.2
8600	2621.3	2624.3	2627.4	2630.4	2633.5	2636.5	2639.6	2642.6	2645.7	2648.7
8700	2651.8	2654.8	2657.9	2660.9	2664.0	2667.0	2670.1	2673.1	2676.1	2679.2
8800	2682.2	2685.3	2688.3	2691.4	2694.4	2697.5	2700.5	2703.6	2706.6	2709.7
8900	2712.7	2715.8	2718.8	2721.9	2724.9	2728.0	2731.0	2734.1	2737.1	2740.2
9000	2743.2	2746.3	2749.3	2752.3	2755.4	2758.4	2761.5	2764.5	2767.6	2770.6

SMITHSONIAN TABLES.

TABLE 67.

METRES INTO FEET.

1 metre = 39.3700 inches = 3.280833 feet

Metres.	0	1	2	3	4	5	6	7	8	9
	Feet.	Feet.	Feet.	Feet.	Feet.	Feet.	Feet.	Feet.	Feet.	Feet.
0	0.00	3.28	6.56	9.84	13.12	16.40	19.68	22.97	26.25	29.53
10	32.81	36.09	39.37	42.65	45.93	49.21	52.49	55.77	59.05	62.34
20	65.62	68.90	72.18	75.46	78.74	82.02	85.30	88.58	91.86	95.14
30	98.42	101.71	104.99	108.27	111.55	114.83	118.11	121.39	124.67	127.95
40	131.23	134.51	137.79	141.08	144.36	147.64	150.92	154.20	157.48	160.76
50	164.04	167.32	170.60	173.88	177.16	180.45	183.73	187.01	190.29	193.57
60	196.85	200.13	203.41	206.69	209.97	213.25	216.53	219.82	223.10	226.38
70	229.66	232.94	236.22	239.50	242.78	246.06	249.34	252.62	255.90	259.19
80	262.47	265.75	269.03	272.31	275.59	278.87	282.15	285.43	288.71	291.99
90	295.27	298.56	301.84	305.12	308.40	311.68	314.96	318.24	321.52	324.80
100	328.08	331.36	334.64	337.93	341.21	344.49	347.77	351.05	354.33	357.61
110	360.89	364.17	367.45	370.73	374.01	377.30	380.58	383.86	387.14	390.42
120	393.70	396.98	400.26	403.54	406.82	410.10	413.38	416.67	419.95	423.23
130	426.51	429.79	433.07	436.35	439.63	442.91	446.19	449.47	452.75	456.04
140	459.32	462.60	465.88	469.16	472.44	475.72	479.00	482.28	485.56	488.84
150	492.12	495.41	498.69	501.97	505.25	508.53	511.81	515.09	518.37	521.65
160	524.93	528.21	531.49	534.78	538.06	541.34	544.62	547.90	551.18	554.46
170	557.74	561.02	564.30	567.58	570.86	574.15	577.43	580.71	583.99	587.27
180	590.55	593.83	597.11	600.39	603.67	606.95	610.23	613.52	616.80	620.08
190	623.36	626.64	629.92	633.20	636.48	639.76	643.04	646.32	649.60	652.89
200	656.17	659.45	662.73	666.01	669.29	672.57	675.85	679.13	682.41	685.69
210	688.97	692.26	695.54	698.82	702.10	705.38	708.66	711.94	715.22	718.50
220	721.78	725.06	728.34	731.63	734.91	738.19	741.47	744.75	748.03	751.31
230	754.59	757.87	761.15	764.43	767.71	771.00	774.28	777.56	780.84	784.12
240	787.40	790.68	793.96	797.24	800.52	803.80	807.08	810.37	813.65	816.93
250	820.21	823.49	826.77	830.05	833.33	836.61	839.89	843.17	846.45	849.74
260	853.02	856.30	859.58	862.86	866.14	869.42	872.70	875.98	879.26	882.54
270	885.82	889.11	892.39	895.67	898.95	902.23	905.51	908.79	912.07	915.35
280	918.63	921.91	925.19	928.48	931.76	935.04	938.32	941.60	944.88	948.16
290	951.44	954.72	958.00	961.28	964.56	967.85	971.13	974.41	977.69	980.97
300	984.25	987.53	990.81	994.09	997.37	1000.65	1003.93	1007.22	1010.50	1013.78
310	1017.06	1020.34	1023.62	1026.90	1030.18	1033.46	1036.74	1040.02	1043.30	1046.59
320	1049.87	1053.15	1056.43	1059.71	1062.99	1066.27	1069.55	1072.83	1076.11	1079.39
330	1082.67	1085.96	1089.24	1092.52	1095.80	1099.08	1102.36	1105.64	1109.92	1112.20
340	1115.48	1118.76	1122.04	1125.33	1128.61	1131.89	1135.17	1138.45	1141.73	1145.01
350	1148.29	1151.57	1154.85	1158.13	1161.41	1164.70	1167.98	1171.26	1174.54	1177.82
360	1181.10	1184.38	1187.66	1190.94	1194.22	1197.50	1200.78	1204.07	1207.35	1210.63
370	1213.91	1217.19	1220.47	1223.75	1227.03	1230.31	1233.59	1236.87	1240.15	1243.44
380	1246.72	1250.00	1253.28	1256.56	1259.84	1263.12	1266.40	1269.68	1272.96	1276.24
390	1279.52	1282.81	1286.09	1289.37	1292.65	1295.93	1299.21	1302.49	1305.77	1309.05
400	1312.33	1315.61	1318.89	1322.18	1325.46	1328.74	1332.02	1335.30	1338.58	1341.86
410	1345.14	1348.42	1351.70	1354.98	1358.26	1361.55	1364.83	1368.11	1371.39	1374.67
420	1377.95	1381.23	1384.51	1387.79	1391.07	1394.35	1397.63	1400.92	1404.20	1407.48
430	1410.76	1414.04	1417.32	1420.60	1423.88	1427.16	1430.44	1433.72	1437.00	1440.29
440	1443.57	1446.85	1450.13	1453.41	1456.69	1459.97	1463.25	1466.53	1469.81	1473.09
450	1476.37	1479.66	1482.94	1486.22	1489.50	1492.78	1496.06	1499.34	1502.62	1505.90
460	1509.18	1512.46	1515.74	1519.03	1522.31	1525.59	1528.87	1532.15	1535.43	1538.71
470	1541.99	1545.27	1548.55	1551.83	1555.11	1558.40	1561.68	1564.96	1568.24	1571.52
480	1574.80	1578.08	1581.36	1584.64	1587.92	1591.20	1594.48	1597.77	1601.05	1604.33
490	1607.61	1610.89	1614.17	1617.45	1620.73	1624.01	1627.29	1630.57	1633.85	1637.14
500	1640.42	1643.70	1646.98	1650.26	1653.54	1656.82	1660.10	1663.38	1666.66	1669.94

SMITHSONIAN TABLES.

TABLE 67.
METRES INTO FEET.
1 metre = 39.3700 inches = 3.280833 feet

Metres.	0	10	20	30	40	50	60	70	80	90
	Feet.	Feet.	Feet.	Feet.	Feet.	Feet.	Feet.	Feet.	Feet.	Feet.
500	1640.4	1673.2	1706.0	1738.8	1771.6	1804.5	1837.3	1870.1	1902.9	1935.7
600	1968.5	2001.3	2034.1	2066.9	2099.7	2132.5	2165.3	2198.2	2231.0	2263.8
700	2296.6	2329.4	2362.2	2395.0	2427.8	2460.6	2493.4	2526.2	2559.0	2591.9
800	2624.7	2657.5	2690.3	2723.1	2755.9	2788.7	2821.5	2854.3	2887.1	2919.9
900	2952.7	2985.6	3018.4	3051.2	3084.0	3116.8	3149.6	3182.4	3215.2	3248.0
1000	3280.8	3313.6	3346.4	3379.3	3412.1	3444.9	3477.7	3510.5	3543.3	3576.1
1100	3608.9	3641.7	3674.5	3707.3	3740.1	3773.0	3805.8	3838.6	3871.4	3904.2
1200	3937.0	3969.8	4002.6	4035.4	4068.2	4101.0	4133.8	4166.7	4199.5	4232.3
1300	4265.1	4297.9	4330.7	4363.5	4396.3	4429.1	4461.9	4494.7	4527.5	4560.4
1400	4593.2	4626.0	4658.8	4691.6	4724.4	4757.2	4790.0	4822.8	4855.6	4888.4
1500	4921.2	4954.1	4986.9	5019.7	5052.5	5085.3	5118.1	5150.9	5183.7	5216.5
1600	5249.3	5282.1	5314.9	5347.8	5380.6	5413.4	5446.2	5479.0	5511.8	5544.6
1700	5577.4	5610.2	5643.0	5675.8	5708.6	5741.5	5774.3	5807.1	5839.9	5872.7
1800	5905.5	5938.3	5971.1	6003.9	6036.7	6069.5	6102.3	6135.2	6168.0	6200.8
1900	6233.6	6266.4	6299.2	6332.0	6364.8	6397.6	6430.4	6463.2	6496.0	6528.9
2000	6561.7	6594.5	6627.3	6660.1	6692.9	6725.7	6758.5	6791.3	6824.1	6856.9
2100	6889.7	6922.6	6955.4	6988.2	7021.0	7053.8	7086.6	7119.4	7152.2	7185.0
2200	7217.8	7250.6	7283.4	7316.3	7349.1	7381.9	7414.7	7447.5	7480.3	7513.1
2300	7545.9	7578.7	7611.5	7644.3	7677.1	7710.0	7742.8	7775.6	7808.4	7841.2
2400	7874.0	7906.8	7939.6	7972.4	8005.2	8038.0	8070.8	8103.7	8136.5	8169.3
2500	8202.1	8234.9	8267.7	8300.5	8333.3	8366.1	8398.9	8431.7	8464.5	8497.4
2600	8530.2	8563.0	8595.8	8628.6	8661.4	8694.2	8727.0	8759.8	8792.6	8825.4
2700	8858.2	8891.1	8923.9	8956.7	8989.5	9022.3	9055.1	9087.9	9120.7	9153.5
2800	9186.3	9219.1	9251.9	9284.8	9317.6	9350.4	9383.2	9416.0	9448.8	9481.6
2900	9514.4	9547.2	9580.0	9612.8	9645.6	9678.5	9711.3	9744.1	9776.9	9809.7
3000	9842.5	9875.3	9908.1	9940.9	9973.7	10006.5	10039.3	10072.2	10105.0	10137.8
3100	10170.6	10203.4	10236.2	10269.0	10301.8	10334.6	10367.4	10400.2	10433.0	10465.9
3200	10498.7	10531.5	10564.3	10597.1	10629.9	10662.7	10695.5	10728.3	10761.1	10793.9
3300	10826.7	10859.6	10892.4	10925.2	10958.0	10990.8	11023.6	11056.4	11089.2	11122.0
3400	11154.8	11187.6	11220.4	11253.3	11286.1	11318.9	11351.7	11384.5	11417.3	11450.1
3500	11482.9	11515.7	11548.5	11581.3	11614.1	11647.0	11679.8	11712.6	11745.4	11778.2
3600	11811.0	11843.8	11876.6	11909.4	11942.2	11975.0	12007.8	12040.7	12073.5	12106.3
3700	12139.1	12171.9	12204.7	12237.5	12270.3	12303.1	12335.9	12368.7	12401.5	12434.4
3800	12467.2	12500.0	12532.8	12565.6	12598.4	12631.2	12664.0	12696.8	12729.6	12762.4
3900	12795.2	12828.1	12860.9	12893.7	12926.5	12959.3	12992.1	13024.9	13057.7	13090.5
4000	13123.3	13156.1	13188.9	13221.8	13254.6	13287.4	13320.2	13353.0	13385.8	13418.6
4100	13451.4	13484.2	13517.0	13549.8	13582.6	13615.5	13648.3	13681.1	13713.9	13746.7
4200	13779.5	13812.3	13845.1	13877.9	13910.7	13943.5	13976.3	14009.2	14042.0	14074.8
4300	14107.6	14140.4	14173.2	14206.0	14238.8	14271.6	14304.4	14337.2	14370.0	14402.9
4400	14435.7	14468.5	14501.3	14534.1	14566.9	14599.7	14632.5	14665.3	14698.1	14730.9
4500	14763.7	14796.6	14829.4	14862.2	14895.0	14927.8	14960.6	14993.4	15026.2	15059.0
4600	15091.8	15124.6	15157.4	15190.3	15223.1	15255.9	15288.7	15321.5	15354.3	15387.1
4700	15419.9	15452.7	15485.5	15518.3	15551.1	15584.0	15616.8	15649.6	15682.4	15715.2
4800	15748.0	15780.8	15813.6	15846.4	15879.2	15912.0	15944.8	15977.7	16010.5	16043.3
4900	16076.1	16108.9	16141.7	16174.5	16207.3	16240.1	16272.9	16305.7	16338.5	16371.4
5000	16404.2	16437.0	16469.8	16502.6	16535.4	16568.2	16601.0	16633.8	16666.6	16699.4

Tenths of a metre.	0.1	0.2	0.3	0.4	0.5	0.6	0.7	0.8	0.9
Feet.	0.328	0.656	0.984	1.312	1.640	1.968	2.297	2.625	2.953

SMITHSONIAN TABLES.

TABLE 68.

MILES INTO KILOMETRES.

1 mile = 1.609347 kilometres

Miles.	0	1	2	3	4	5	6	7	8	9
	km.	km.	km.	km.	km.	km.	km.	km.	km.	km.
0	0	2	3	5	6	8	10	11	13	14
10	16	18	19	21	23	24	26	27	29	31
20	32	34	35	37	39	40	42	43	45	47
30	48	50	51	53	55	56	58	60	61	63
40	64	66	68	69	71	72	74	76	77	79
50	80	82	84	85	87	89	90	92	93	95
60	97	98	100	101	103	105	106	108	109	111
70	113	114	116	117	119	121	122	124	126	127
80	129	130	132	134	135	137	138	140	142	143
90	145	146	148	150	151	153	154	156	158	159
100	161	163	164	166	167	169	171	172	174	175
110	177	179	180	182	183	185	187	188	190	192
120	193	195	196	198	200	201	203	204	206	208
130	209	211	212	214	216	217	219	220	222	224
140	225	227	229	230	232	233	235	237	238	240
150	241	243	245	246	248	249	251	253	254	256
160	257	259	261	262	264	266	267	269	270	272
170	274	275	277	278	280	282	283	285	286	288
180	290	291	293	295	296	298	299	301	303	304
190	306	307	309	311	312	314	315	317	319	320
200	322	323	325	327	328	330	332	333	335	336
210	338	340	341	343	344	346	348	349	351	352
220	354	356	357	359	360	362	364	365	367	369
230	370	372	373	375	377	378	380	381	383	385
240	386	388	389	391	393	394	396	398	399	401
250	402	404	406	407	409	410	412	414	415	417
260	418	420	422	423	425	426	428	430	431	433
270	435	436	438	439	441	443	444	446	447	449
280	451	452	454	455	457	459	460	462	463	465
290	467	468	470	472	473	475	476	478	480	481
300	483	484	486	488	489	491	492	494	496	497
310	499	501	502	504	505	507	509	510	512	513
320	515	517	518	520	521	523	525	526	528	529
330	531	533	534	536	538	539	541	542	544	546
340	547	549	550	552	554	555	557	558	560	562
350	563	565	566	568	570	571	573	575	576	578
360	579	581	583	584	586	587	589	591	592	594
370	595	597	599	600	602	604	605	607	608	610
380	612	613	615	616	618	620	621	623	624	626
390	628	629	631	632	634	636	637	639	641	642
400	644	645	647	649	650	652	653	655	657	658
410	660	661	663	665	666	668	669	671	673	674
420	676	678	679	681	682	684	686	687	689	690
430	692	694	695	697	698	700	702	703	705	706
440	708	710	711	713	715	716	718	719	721	723
450	724	726	727	729	731	732	734	735	737	739
460	740	742	744	745	747	748	750	752	753	755
470	756	758	760	761	763	764	766	768	769	771
480	772	774	776	778	779	781	782	784	785	787
490	789	790	792	793	795	797	798	800	801	803
500	805	806	808	809	811	813	814	816	818	819
510	821	822	824	826	827	829	830	832	834	835
520	837	838	840	842	843	845	847	848	850	851
530	853	855	856	858	859	861	863	864	866	867
540	869	871	872	874	875	877	879	880	882	884
550	885	887	888	890	892	893	895	896	898	900

SMITHSONIAN TABLES.

TABLE 68.

MILES INTO KILOMETRES.

Miles.	0	1	2	3	4	5	6	7	8	9
	km.	km.	km.	km.	km.	km.	km.	km.	km.	km.
550	885	887	888	890	892	893	895	896	898	900
560	901	903	904	906	908	909	911	912	914	916
570	917	919	921	922	924	925	927	929	930	932
580	933	935	937	938	940	941	943	945	946	948
590	950	951	953	954	956	958	959	961	962	964
600	966	967	969	970	972	974	975	977	978	980
610	982	983	985	987	988	990	991	993	995	996
620	998	999	1001	1003	1004	1006	1007	1009	1011	1012
630	1014	1015	1017	1019	1020	1022	1024	1025	1027	1028
640	1030	1032	1033	1035	1036	1038	1040	1041	1043	1044
650	1046	1048	1049	1051	1053	1054	1056	1057	1059	1061
660	1062	1064	1065	1067	1069	1070	1072	1073	1075	1077
670	1078	1080	1081	1083	1085	1086	1088	1090	1091	1093
680	1094	1096	1098	1099	1101	1102	1104	1106	1107	1109
690	1110	1112	1114	1115	1117	1118	1120	1122	1123	1125
700	1127	1128	1130	1131	1133	1135	1136	1138	1139	1141
710	1143	1144	1146	1147	1149	1151	1152	1154	1156	1157
720	1159	1160	1162	1164	1165	1167	1168	1170	1172	1173
730	1175	1176	1178	1180	1181	1183	1184	1186	1188	1189
740	1191	1193	1194	1196	1197	1199	1201	1202	1204	1205
750	1207	1209	1210	1212	1213	1215	1217	1218	1220	1221
760	1223	1225	1226	1228	1230	1231	1233	1234	1236	1238
770	1239	1241	1242	1244	1246	1247	1249	1250	1252	1254
780	1255	1257	1259	1260	1262	1263	1265	1267	1268	1270
790	1271	1273	1275	1276	1278	1279	1281	1283	1284	1286
800	1287	1289	1291	1292	1294	1296	1297	1299	1300	1302
810	1304	1305	1307	1308	1310	1312	1313	1315	1316	1318
820	1320	1321	1323	1324	1326	1328	1329	1331	1333	1334
830	1336	1337	1339	1341	1342	1344	1345	1347	1349	1350
840	1352	1353	1355	1357	1358	1360	1362	1363	1365	1366
850	1368	1370	1371	1373	1374	1376	1378	1379	1381	1382
860	1384	1386	1387	1389	1390	1392	1394	1395	1397	1399
870	1400	1402	1403	1405	1407	1408	1410	1411	1413	1415
880	1416	1418	1419	1421	1423	1424	1426	1427	1429	1431
890	1432	1434	1436	1437	1439	1440	1442	1444	1445	1447
900	1448	1450	1452	1453	1455	1456	1458	1460	1461	1463
910	1464	1466	1468	1469	1471	1473	1474	1476	1477	1479
920	1481	1482	1484	1485	1487	1489	1490	1492	1493	1495
930	1497	1498	1500	1502	1503	1505	1506	1508	1510	1511
940	1513	1514	1516	1518	1519	1521	1522	1524	1526	1527
950	1529	1530	1532	1534	1535	1537	1539	1540	1542	1543
960	1545	1547	1548	1550	1551	1553	1555	1556	1558	1559
970	1561	1563	1564	1566	1567	1569	1571	1572	1574	1576
980	1577	1579	1580	1582	1584	1585	1587	1588	1590	1592
990	1593	1595	1596	1598	1600	1601	1603	1605	1606	1608
1000	1609	1611	1613	1614	1616	1617	1619	1621	1622	1624

Miles.	km.	Miles.	km.	Miles.	km.	Miles.	km.
1000	1609	6000	9656	11000	17703	16000	25750
2000	3219	7000	11265	12000	19312	17000	27359
3000	4828	8000	12875	13000	20922	18000	28968
4000	6437	9000	14484	14000	22531	19000	30578
5000	8047	10000	16093	15000	24140	20000	32187

TABLE 69.

KILOMETRES INTO MILES.

1 kilometre = 0.621370 mile.

Kilo-metres.	0	1	2	3	4	5	6	7	8	9
	Miles.	Miles.	Miles.	Miles.	Miles.	Miles.	Miles.	Miles.	Miles.	Miles.
0	0.0	0.6	1.2	1.9	2.5	3.1	3.7	4.3	5.0	5.6
10	6.2	6.8	7.5	8.1	8.7	9.3	9.9	10.6	11.2	11.8
20	12.4	13.0	13.7	14.3	14.9	15.5	16.2	16.8	17.4	18.0
30	18.6	19.3	19.9	20.5	21.1	21.7	22.4	23.0	23.6	24.2
40	24.9	25.5	26.1	26.7	27.3	28.0	28.6	29.2	29.8	30.4
50	31.1	31.7	32.3	32.9	33.6	34.2	34.8	35.4	36.0	36.7
60	37.3	37.9	38.5	39.1	39.8	40.4	41.0	41.6	42.3	42.9
70	43.5	44.1	44.7	45.4	46.0	46.6	47.2	47.8	48.5	49.1
80	49.7	50.3	51.0	51.6	52.2	52.8	53.4	54.1	54.7	55.3
90	55.9	56.5	57.2	57.8	58.4	59.0	59.7	60.3	60.9	61.5
100	62.1	62.8	63.4	64.0	64.6	65.2	65.9	66.5	67.1	67.7
110	68.4	69.0	69.6	70.2	70.8	71.5	72.1	72.7	73.3	73.9
120	74.6	75.2	75.8	76.4	77.0	77.7	78.3	78.9	79.5	80.2
130	80.8	81.4	82.0	82.6	83.3	83.9	84.5	85.1	85.7	86.4
140	87.0	87.6	88.2	88.9	89.5	90.1	90.7	91.3	92.0	92.6
150	93.2	93.8	94.4	95.1	95.7	96.3	96.9	97.6	98.2	98.8
160	99.4	100.0	100.7	101.3	101.9	102.5	103.1	103.8	104.4	105.0
170	105.6	106.3	106.9	107.5	108.1	108.7	109.4	110.0	110.6	111.2
180	111.8	112.5	113.1	113.7	114.3	115.0	115.6	116.2	116.8	117.4
190	118.1	118.7	119.3	119.9	120.5	121.2	121.8	122.4	123.0	123.7
200	124.3	124.9	125.5	126.1	126.8	127.4	128.0	128.6	129.2	129.9
210	130.5	131.1	131.7	132.4	133.0	133.6	134.2	134.8	135.5	136.1
220	136.7	137.3	137.9	138.6	139.2	139.8	140.4	141.1	141.7	142.3
230	142.9	143.5	144.2	144.8	145.4	146.0	146.6	147.3	147.9	148.5
240	149.1	149.8	150.4	151.0	151.6	152.2	152.9	153.5	154.1	154.7
250	155.3	156.0	156.6	157.2	157.8	158.4	159.1	159.7	160.3	160.9
260	161.6	162.2	162.8	163.4	164.0	164.7	165.3	165.9	166.5	167.1
270	167.8	168.4	169.0	169.6	170.3	170.9	171.5	172.1	172.7	173.4
280	174.0	174.6	175.2	175.8	176.5	177.1	177.7	178.3	179.0	179.6
290	180.2	180.8	181.4	182.1	182.7	183.3	183.9	184.5	185.2	185.8
300	186.4	187.0	187.7	188.3	188.9	189.5	190.1	190.8	191.4	192.0
310	192.6	193.3	193.9	194.5	195.1	195.7	196.4	197.0	197.6	198.2
320	198.8	199.5	200.1	200.7	201.3	201.9	202.6	203.2	203.8	204.4
330	205.1	205.7	206.3	206.9	207.5	208.2	208.8	209.4	210.0	210.6
340	211.3	211.9	212.5	213.1	213.8	214.4	215.0	215.6	216.2	216.9
350	217.5	218.1	218.7	219.3	220.0	220.6	221.2	221.8	222.5	223.1
360	223.7	224.3	224.9	225.6	226.2	226.8	227.4	228.0	228.7	229.3
370	229.9	230.5	231.1	231.8	232.4	233.0	233.6	234.3	234.9	235.5
380	236.1	236.7	237.4	238.0	238.6	239.2	239.8	240.5	241.1	241.7
390	242.3	243.0	243.6	244.2	244.8	245.4	246.1	246.7	247.3	247.9
400	248.5	249.2	249.8	250.4	251.0	251.7	252.3	252.9	253.5	254.1
410	254.8	255.4	256.0	256.6	257.2	257.9	258.5	259.1	259.7	260.4
420	261.0	261.6	262.2	262.8	263.5	264.1	264.7	265.3	265.9	266.6
430	267.2	267.8	268.4	269.1	269.7	270.3	270.9	271.5	272.2	272.8
440	273.4	274.0	274.6	275.3	275.9	276.5	277.1	277.8	278.4	279.0
450	279.6	280.2	280.9	281.5	282.1	282.7	283.3	284.0	284.6	285.2
460	285.8	286.5	287.1	287.7	288.3	288.9	289.6	290.2	290.8	291.4
470	292.0	292.7	293.3	293.9	294.5	295.2	295.8	296.4	297.0	297.6
480	298.3	298.9	299.5	300.1	300.7	301.4	302.0	302.6	303.2	303.8
490	304.5	305.1	305.7	306.3	307.0	307.6	308.2	308.8	309.4	310.1
500	310.7	311.3	311.9	312.5	313.2	313.8	314.4	315.0	315.7	316.3
510	316.9	317.5	318.1	318.8	319.4	320.0	320.6	321.2	321.9	322.5
520	323.1	323.7	324.4	325.0	325.6	326.2	326.8	327.5	328.1	328.7
530	329.3	329.9	330.6	331.2	331.8	332.4	333.1	333.7	334.3	334.9
540	335.5	336.2	336.8	337.4	338.0	338.6	339.3	339.9	340.5	341.1

TABLE 69

KILOMETRES INTO MILES.

Kilo-metres.	0	1	2	3	4	5	6	7	8	9
	Miles.	Miles.	Miles.	Miles.	Miles.	Miles.	Miles.	Miles.	Miles.	Miles.
550	341.8	342.4	343.0	343.6	344.2	344.9	345.5	346.1	346.7	347.3
560	348.0	348.6	349.2	349.8	350.5	351.1	351.7	352.3	352.9	353.6
570	354.2	354.8	355.4	356.0	356.7	357.3	357.9	358.5	359.2	359.8
580	360.4	361.0	361.6	362.3	362.9	363.5	364.1	364.7	365.4	366.0
590	366.6	367.2	367.9	368.5	369.1	369.7	370.3	371.0	371.6	372.2
600	372.8	373.4	374.1	374.7	375.3	375.9	376.6	377.2	377.8	378.4
610	379.0	379.7	380.3	380.9	381.5	382.1	382.8	383.4	384.0	384.6
620	385.2	385.9	386.5	387.1	387.7	388.4	389.0	389.6	390.2	390.8
630	391.5	392.1	392.7	393.3	393.9	394.6	395.2	395.8	396.4	397.1
640	397.7	398.3	398.9	399.5	400.2	400.8	401.4	402.0	402.6	403.3
650	403.9	404.5	405.1	405.8	406.4	407.0	407.6	408.2	408.9	409.5
660	410.1	410.7	411.3	412.0	412.6	413.2	413.8	414.5	415.1	415.7
670	416.3	416.9	417.6	418.2	418.8	419.4	420.0	420.7	421.3	421.9
680	422.5	423.2	423.8	424.4	425.0	425.6	426.3	426.9	427.5	428.1
690	428.7	429.4	430.0	430.6	431.2	431.9	432.5	433.1	433.7	434.3
700	435.0	435.6	436.2	436.8	437.4	438.1	438.7	439.3	439.9	440.6
710	441.2	441.8	442.4	443.0	443.7	444.3	444.9	445.5	446.1	446.8
720	447.4	448.0	448.6	449.3	449.9	450.5	451.1	451.7	452.4	453.0
730	453.6	454.2	454.8	455.5	456.1	456.7	457.3	457.9	458.6	459.2
740	459.8	460.4	461.1	461.7	462.3	462.9	463.5	464.2	464.8	465.4
750	466.0	466.6	467.3	467.9	468.5	469.1	469.8	470.4	471.0	471.6
760	472.2	472.9	473.5	474.1	474.7	475.3	476.0	476.6	477.2	477.8
770	478.5	479.1	479.7	480.3	480.9	481.6	482.2	482.8	483.4	484.0
780	484.7	485.3	485.9	486.5	487.2	487.8	488.4	489.0	489.6	490.3
790	490.9	491.5	492.1	492.7	493.4	494.0	494.6	495.2	495.9	496.5
800	497.1	497.7	498.3	499.0	499.6	500.2	500.8	501.4	502.1	502.7
810	503.3	503.9	504.6	505.2	505.8	506.4	507.0	507.7	508.3	508.9
820	509.5	510.1	510.8	511.4	512.0	512.6	513.3	513.9	514.5	515.1
830	515.7	516.4	517.0	517.6	518.2	518.8	519.5	520.1	520.7	521.3
840	522.0	522.6	523.2	523.8	524.4	525.1	525.7	526.3	526.9	527.5
850	528.2	528.8	529.4	530.0	530.6	531.3	531.9	532.5	533.1	533.8
860	534.4	535.0	535.6	536.2	536.9	537.5	538.1	538.7	539.3	540.0
870	540.6	541.2	541.8	542.5	543.1	543.7	544.3	544.9	545.6	546.2
880	546.8	547.4	548.0	548.7	549.3	549.9	550.5	551.2	551.8	552.4
890	553.0	553.6	554.3	554.9	555.5	556.1	556.7	557.4	558.0	558.6
900	559.2	559.9	560.5	561.1	561.7	562.3	563.0	563.6	564.2	564.8
910	565.4	566.1	566.7	567.3	567.9	568.6	569.2	569.8	570.4	571.0
920	571.7	572.3	572.9	573.5	574.1	574.8	575.4	576.0	576.6	577.3
930	577.9	578.5	579.1	579.7	580.4	581.0	581.6	582.2	582.8	583.5
940	584.1	584.7	585.3	586.0	586.6	587.2	587.8	588.4	589.1	589.7
950	590.3	590.9	591.5	592.2	592.8	593.4	594.0	594.7	595.3	595.9
960	596.5	597.1	597.8	598.4	599.0	599.6	600.2	600.9	601.5	602.1
970	602.7	603.4	604.0	604.6	605.2	605.8	606.5	607.1	607.7	608.3
980	608.9	609.6	610.2	610.8	611.4	612.0	612.7	613.3	613.9	614.5
990	615.2	615.8	616.4	617.0	617.6	618.3	618.9	619.5	620.1	620.7
1000	621.4	622.0	622.6	623.2	623.9	624.5	625.1	625.7	626.3	627.0

km.	Miles.	km.	Miles.	km.	Miles.	km.	Miles.
1000	621.4	6000	3728.2	11000	6835.1	16000	9941.9
2000	1242.7	7000	4349.6	12000	7456.4	17000	10563.3
3000	1864.1	8000	4971.0	13000	8077.8	18000	11184.7
4000	2485.5	9000	5592.3	14000	8699.2	19000	11806.0
5000	3106.8	10000	6213.7	15000	9320.5	20000	12427.4

SMITHSONIAN TABLES.

TABLE 70.

INTERCONVERSION OF NAUTICAL AND STATUTE MILES.

1 nautical mile* = 6080.27 feet.

Nautical Miles.	Statute Miles.	Statute Miles.	Nautical Miles.
1	1.1516	1	0.8684
2	2.3031	2	1.7368
3	3.4547	3	2.6052
4	4.6062	4	3.4736
5	5.7578	5	4.3420
6	6.9093	6	5.2104
7	8.0609	7	6.0788
8	9.2124	8	6.9472
9	10.3640	9	7.8155

* As defined by the United States Coast Survey.

TABLE 71.

CONTINENTAL MEASURES OF LENGTH WITH THEIR METRIC AND ENGLISH EQUIVALENTS.

The asterisk (*) indicates that the measure is obsolete or seldom used.

Measure.	Metric Equivalent.	English Equivalent.
El (Netherlands)	1 metre.	3.2808 feet.
Fathom, Swedish = 6 feet	1.7814 "	5.8445 "
Foot, Austrian*	0.31608 "	1.0370 "
old French*	0.32484 "	1.0657 "
Russian	0.30480 "	1 "
Rheinlandisch or Rhenish (Prussia*, Denmark, Norway*).	0.31385 "	1.0297 "
Swedish*	0.2969 "	0.9741 "
Spanish* = ⅓ vara	0.2786 "	0.9140 "
*Klafter, Wiener (Vienna)	1.89648 "	6.2221 "
*Line, old French = $\frac{1}{144}$ foot	0.22558 cm.	0.0888 inch.
Mile, Austrian post* = 24000 feet	7.58594 km.	4.714 statute miles.
German sea	1.852 "	1.1508 " "
Swedish = 36000 feet	10.69 "	6.642 " "
Norwegian = 36000 feet	11.2986 "	7.02 " "
Netherlands (mijl)	1 "	0.6214 " "
Prussian (law of 1868)	7.500 "	4.660 " "
Danish	7.5324 "	4.6804 " "
Palm, Netherlands	0.1 metre.	0.3281 feet.
*Rode, Danish	3.7662 "	12.356 "
*Ruthe, Prussian, Norwegian	3.7662 "	12.356 "
Sagene (Russian)	2.1336 "	7 "
*Toise, old French = 6 feet	1.9490 "	6.3943 "
*Vara, Spanish	0.8359 "	2.7424 "
Mexican	0.8380 "	2.7293 "
Werst, or versta (Russian) = 500 sagene	1.0668 km.	3.500 "

SMITHSONIAN TABLES.

CONVERSION OF MEASURES OF TIME AND ANGLE.

Arc into time . . . . TABLE 72

Time into arc . . . . TABLE 73

Days into decimals of a year and angle . . . . TABLE 74

Hours, minutes and seconds into decimals of a day . . . . TABLE 75

Decimals of a day into hours, minutes and seconds . . . . TABLE 76

Minutes and seconds into decimals of an hour . . . . . TABLE 77

Mean time at apparent noon . . . . TABLE 78

Sidereal time into mean solar time . . . . TABLE 79

Mean solar time into sidereal time . . . . TABLE 80

TABLE 72.
ARC INTO TIME.

°	h. m.	°	h. m.	°	h. m.	°	h. m.	°	h. m.	°	h. m.	°	h. m.	′	m. s.	″	s.
0	0 0	60	4 0	120	8 0	180	12 0	240	16 0	300	20 0			0	0 0	0	0.000
1	0 4	61	4 4	121	8 4	181	12 4	241	16 4	301	20 4			1	0 4	1	0.067
2	0 8	62	4 8	122	8 8	182	12 8	242	16 8	302	20 8			2	0 8	2	0.133
3	0 12	63	4 12	123	8 12	183	12 12	243	16 12	303	20 12			3	0 12	3	0.200
4	0 16	64	4 16	124	8 16	184	12 16	244	16 16	304	20 16			4	0 16	4	0.267
5	0 20	65	4 20	125	8 20	185	12 20	245	16 20	305	20 20			5	0 20	5	0.333
6	0 24	66	4 24	126	8 24	186	12 24	246	16 24	306	20 24			6	0 24	6	0.400
7	0 28	67	4 28	127	8 28	187	12 28	247	16 28	307	20 28			7	0 28	7	0.467
8	0 32	68	4 32	128	8 32	188	12 32	248	16 32	308	20 32			8	0 32	8	0.533
9	0 36	69	4 36	129	8 36	189	12 36	249	16 36	309	20 36			9	0 36	9	0.600
10	0 40	70	4 40	130	8 40	190	12 40	250	16 40	310	20 40			10	0 40	10	0.667
11	0 44	71	4 44	131	8 44	191	12 44	251	16 44	311	20 44			11	0 44	11	0.733
12	0 48	72	4 48	132	8 48	192	12 48	252	16 48	312	20 48			12	0 48	12	0.800
13	0 52	73	4 52	133	8 52	193	12 52	253	16 52	313	20 52			13	0 52	13	0.867
14	0 56	74	4 56	134	8 56	194	12 56	254	16 56	314	20 56			14	0 56	14	0.933
15	1 0	75	5 0	135	9 0	195	13 0	255	17 0	315	21 0			15	1 0	15	1.000
16	1 4	76	5 4	136	9 4	196	13 4	256	17 4	316	21 4			16	1 4	16	1.067
17	1 8	77	5 8	137	9 8	197	13 8	257	17 8	317	21 8			17	1 8	17	1.133
18	1 12	78	5 12	138	9 12	198	13 12	258	17 12	318	21 12			18	1 12	18	1.200
19	1 16	79	5 16	139	9 16	199	13 16	259	17 16	319	21 16			19	1 16	19	1.267
20	1 20	80	5 20	140	9 20	200	13 20	260	17 20	320	21 20			20	1 20	20	1.333
21	1 24	81	5 24	141	9 24	201	13 24	261	17 24	321	21 24			21	1 24	21	1.400
22	1 28	82	5 28	142	9 28	202	13 28	262	17 28	322	21 28			22	1 28	22	1.467
23	1 32	83	5 32	143	9 32	203	13 32	263	17 32	323	21 32			23	1 32	23	1.533
24	1 36	84	5 36	144	9 36	204	13 36	264	17 36	324	21 36			24	1 36	24	1.600
25	1 40	85	5 40	145	9 40	205	13 40	265	17 40	325	21 40			25	1 40	25	1.667
26	1 44	86	5 44	146	9 44	206	13 44	266	17 44	326	21 44			26	1 44	26	1.733
27	1 48	87	5 48	147	9 48	207	13 48	267	17 48	327	21 48			27	1 48	27	1.800
28	1 52	88	5 52	148	9 52	208	13 52	268	17 52	328	21 52			28	1 52	28	1.867
29	1 56	89	5 56	149	9 56	209	13 56	269	17 56	329	21 56			29	1 56	29	1.933
30	2 0	90	6 0	150	10 0	210	14 0	270	18 0	330	22 0			30	2 0	30	2.000
31	2 4	91	6 4	151	10 4	211	14 4	271	18 4	331	22 4			31	2 4	31	2.067
32	2 8	92	6 8	152	10 8	212	14 8	272	18 8	332	22 8			32	2 8	32	2.133
33	2 12	93	6 12	153	10 12	213	14 12	273	18 12	333	22 12			33	2 12	33	2.200
34	2 16	94	6 16	154	10 16	214	14 16	274	18 16	334	22 16			34	2 16	34	2.267
35	2 20	95	6 20	155	10 20	215	14 20	275	18 20	335	22 20			35	2 20	35	2.333
36	2 24	96	6 24	156	10 24	216	14 24	276	18 24	336	22 24			36	2 24	36	2.400
37	2 28	97	6 28	157	10 28	217	14 28	277	18 28	337	22 28			37	2 28	37	2.467
38	2 32	98	6 32	158	10 32	218	14 32	278	18 32	338	22 32			38	2 32	38	2.533
39	2 36	99	6 36	159	10 36	219	14 36	279	18 36	339	22 36			39	2 36	39	2.600
40	2 40	100	6 40	160	10 40	220	14 40	280	18 40	340	22 40			40	2 40	40	2.667
41	2 44	101	6 44	161	10 44	221	14 44	281	18 44	341	22 44			41	2 44	41	2.733
42	2 48	102	6 48	162	10 48	222	14 48	282	18 48	342	22 48			42	2 48	42	2.800
43	2 52	103	6 52	163	10 52	223	14 52	283	18 52	343	22 52			43	2 52	43	2.867
44	2 56	104	6 56	164	10 56	224	14 56	284	18 56	344	22 56			44	2 56	44	2.933
45	3 0	105	7 0	165	11 0	225	15 0	285	19 0	345	23 0			45	3 0	45	3.000
46	3 4	106	7 4	166	11 4	226	15 4	286	19 4	346	23 4			46	3 4	46	3.067
47	3 8	107	7 8	167	11 8	227	15 8	287	19 8	347	23 8			47	3 8	47	3.133
48	3 12	108	7 12	168	11 12	228	15 12	288	19 12	348	23 12			48	3 12	48	3.200
49	3 16	109	7 16	169	11 16	229	15 16	289	19 16	349	23 16			49	3 16	49	3.267
50	3 20	110	7 20	170	11 20	230	15 20	290	19 20	350	23 20			50	3 20	50	3.333
51	3 24	111	7 24	171	11 24	231	15 24	291	19 24	351	23 24			51	3 24	51	3.400
52	3 28	112	7 28	172	11 28	232	15 28	292	19 28	352	23 28			52	3 28	52	3.467
53	3 32	113	7 32	173	11 32	233	15 32	293	19 32	353	23 32			53	3 32	53	3.533
54	3 36	114	7 36	174	11 36	234	15 36	294	19 36	354	23 36			54	3 36	54	3.600
55	3 40	115	7 40	175	11 40	235	15 40	295	19 40	355	23 40			55	3 40	55	3.667
56	3 44	116	7 44	176	11 44	236	15 44	296	19 44	356	23 44			56	3 44	56	3.733
57	3 48	117	7 48	177	11 48	237	15 48	297	19 48	357	23 48			57	3 48	57	3.800
58	3 52	118	7 52	178	11 52	238	15 52	298	19 52	358	23 52			58	3 52	58	3.867
59	3 56	119	7 56	179	11 56	239	15 56	299	19 56	359	23 56			59	3 56	59	3.933
60	4 0	120	8 0	180	12 0	240	16 0	300	20 0	360	24 0			60	4 0	60	4.000

SMITHSONIAN TABLES.

TABLE 73.

TIME INTO ARC.

Hours into Arc.

Time.	Arc.	Time.	Arc.	Time.	Arc.	Time.	Arc.	Time.	Arc.
hrs.	°	hrs.	°	hrs.	°	hrs.	°	hrs.	°
1	15	5	75	9	135	13	195	17	255
2	30	6	90	10	150	14	210	18	270
3	45	7	105	11	165	15	225	19	285
4	60	8	120	12	180	16	240	20	300

Time.	Arc.
21	315
22	330
23	345
24	360

Minutes of Time into Arc.

m.	°	′	m.	°	′	m.	°	′
1	0	15	21	5	15	41	10	15
2	0	30	22	5	30	42	10	30
3	0	45	23	5	45	43	10	45
4	1	0	24	6	0	44	11	0
5	1	15	25	6	15	45	11	15
6	1	30	26	6	30	46	11	30
7	1	45	27	6	45	47	11	45
8	2	0	28	7	0	48	12	0
9	2	15	29	7	15	49	12	15
10	2	30	30	7	30	50	12	30
11	2	45	31	7	45	51	12	45
12	3	0	32	8	0	52	13	0
13	3	15	33	8	15	53	13	15
14	3	30	34	8	30	54	13	30
15	3	45	35	8	45	55	13	45
16	4	0	36	9	0	56	14	0
17	4	15	37	9	15	57	14	15
18	4	30	38	9	30	58	14	30
19	4	45	39	9	45	59	14	45
20	5	0	40	10	0	60	15	0

Seconds of Time into Arc.

s.	′	″	s.	′	″	s.	′	″
1	0	15	21	5	15	41	10	15
2	0	30	22	5	30	42	10	30
3	0	45	23	5	45	43	10	45
4	1	0	24	6	0	44	11	0
5	1	15	25	6	15	45	11	15
6	1	30	26	6	30	46	11	30
7	1	45	27	6	45	47	11	45
8	2	0	28	7	0	48	12	0
9	2	15	29	7	15	49	12	15
10	2	30	30	7	30	50	12	30
11	2	45	31	7	45	51	12	45
12	3	0	32	8	0	52	13	0
13	3	15	33	8	15	53	13	15
14	3	30	34	8	30	54	13	30
15	3	45	35	8	45	55	13	45
16	4	0	36	9	0	56	14	0
17	4	15	37	9	15	57	14	15
18	4	30	38	9	30	58	14	30
19	4	45	39	9	45	59	14	45
20	5	0	40	10	0	60	15	0

Hundredths of a Second of Time into Arc.

Hundredths of a Second of Time.	.00	.01	.02	.03	.04	.05	.06	.07	.08	.09
s.	″	″	″	″	″	″	″	″	″	″
0.00	0.00	0.15	0.30	0.45	0.60	0.75	0.90	1.05	1.20	1.35
.10	1.50	1.65	1.80	1.95	2.10	2.25	2.40	2.55	2.70	2.85
.20	3.00	3.15	3.30	3.45	3.60	3.75	3.90	4.05	4.20	4.35
.30	4.50	4.65	4.80	4.95	5.10	5.25	5.40	5.55	5.70	5.85
.40	6.00	6.15	6.30	6.45	6.60	6.75	6.90	7.05	7.20	7.35
0.50	7.50	7.65	7.80	7.95	8.10	8.25	8.40	8.55	8.70	8.85
.60	9.00	9.15	9.30	9.45	9.60	9.75	9.90	10.05	10.20	10.35
.70	10.50	10.65	10.80	10.95	11.10	11.25	11.40	11.55	11.70	11.85
.80	12.00	12.15	12.30	12.45	12.60	12.75	12.90	13.05	13.20	13.35
.90	13.50	13.65	13.80	13.95	14.10	14.25	14.40	14.55	14.70	14.85

SMITHSONIAN TABLES.

TABLE 74.

DAYS INTO DECIMALS OF A YEAR AND ANGLE.

Day of Year.	Decimal of a Year.	Angle.	Day of Month. Common Year.	Day of Month. Bissextile Year.	Day of Year.	Decimal of a Year.	Angle.	Day of Month. Common Year.	Day of Month. Bissextile Year.
1	0.00000	0° 0'	*Jan.* 1	*Jan.* 1	51	0.13689	49° 17'	*Feb.* 20	*Feb.* 20
2	.00274	0 59	2	2	52	.13963	50 16	21	21
3	.00548	1 58	3	3	53	.14237	51 15	22	22
4	.00821	2 57	4	4	54	.14511	52 14	23	23
5	0.01095	3 57	5	5	55	0.14784	53 13	24	24
6	.01369	4 56	6	6	56	.15058	54 13	25	25
7	.01643	5 55	7	7	57	.15332	55 12	26	26
8	.01916	6 54	8	8	58	.15606	56 11	27	27
9	.02190	7 53	9	9	59	.15880	57 10	28	28
10	0.02464	8 52	10	10	60	0.16153	58 9	*Mar.* 1	29
11	.02738	9 51	11	11	61	.16427	59 8	2	*Mar.* 1
12	.03011	10 51	12	12	62	.16701	60 7	3	2
13	.03285	11 50	13	13	63	.16975	61 7	4	3
14	.03559	12 49	14	14	64	.17248	62 6	5	4
15	0.03833	13 48	15	15	65	0.17522	63 5	6	5
16	.04107	14 47	16	16	66	.17796	64 4	7	6
17	.04381	15 46	17	17	67	.18070	65 3	8	7
18	.04654	16 45	18	18	68	.18344	66 2	9	8
19	.04928	17 44	19	19	69	.18617	67 1	10	9
20	0.05202	18 44	20	20	70	0.18891	68 0	11	10
21	.05476	19 43	21	21	71	.19165	69 0	12	11
22	.05749	20 42	22	22	72	.19439	69 59	13	12
23	.06023	21 41	23	23	73	.19713	70 58	14	13
24	.06297	22 40	24	24	74	.19986	71 57	15	14
25	0.06571	23 39	25	25	75	0.20260	72 56	16	15
26	.06845	24 38	26	26	76	.20534	73 55	17	16
27	.07118	25 38	27	27	77	.20808	74 54	18	17
28	.07392	26 37	28	28	78	.21081	75 54	19	18
29	.07666	27 36	29	29	79	.21355	76 53	20	19
30	0.07940	28 35	30	30	80	0.21629	77 52	21	20
31	.08214	29 34	31	31	81	.21903	78 51	22	21
32	.08487	30 33	*Feb.* 1	*Feb.* 1	82	.22177	79 50	23	22
33	.08761	31 32	2	2	83	.22450	80 49	24	23
34	.09035	32 32	3	3	84	.22724	81 48	25	24
35	0.09309	33 31	4	4	85	0.22998	82 48	26	25
36	.09582	34 30	5	5	86	.23272	83 47	27	26
37	.09856	35 29	6	6	87	.23546	84 46	28	27
38	.10130	36 28	7	7	88	.23819	85 45	29	28
39	.10404	37 27	8	8	89	.24093	86 44	30	29
40	0.10678	38 26	9	9	90	0.24367	87 43	31	30
41	.10951	39 26	10	10	91	.24641	88 42	*Apr.* 1	31
42	.11225	40 25	11	11	92	.24914	89 42	2	*Apr.* 1
43	.11499	41 24	12	12	93	.25188	90 41	3	2
44	.11773	42 23	13	13	94	.25462	91 40	4	3
45	0.12047	43 22	14	14	95	0.25736	92 39	5	4
46	.12320	44 21	15	15	96	.26010	93 38	6	5
47	.12594	45 20	16	16	97	.26283	94 37	7	6
48	.12868	46 19	17	17	98	.26557	95 36	8	7
49	.13142	47 19	18	18	99	.26831	96 35	9	8
50	0.13415	48 18	19	19	100	0.27105	97 35	10	9

TABLE 74.

DAYS INTO DECIMALS OF A YEAR AND ANGLE.

Day of Year.	Decimal of a Year.	Angle.	Day of Month. Common Year.	Day of Month. Bissextile Year.	Day of Year.	Decimal of a Year.	Angle.	Day of Month. Common Year.	Day of Month. Bissextile Year.
101	0.27379	98° 34'	Apr. 11	Apr. 10	151	0.41068	147° 51'	May 31	May 30
102	.27652	99 33	12	11	152	.41342	148 50	June 1	31
103	.27926	100 32	13	12	153	.41615	149 49	2	June 1
104	.28200	101 31	14	13	154	.41889	150 48	3	2
105	0.28474	102 30	15	14	155	0.42163	151 47	4	3
106	.28747	103 29	16	15	156	.42437	152 46	5	4
107	.29021	104 29	17	16	157	.42710	153 45	6	5
108	.29295	105 28	18	17	158	.42984	154 45	7	6
109	.29569	106 27	19	18	159	.43258	155 44	8	7
110	0.29843	107 26	20	19	160	0.43532	156 43	9	8
111	.30116	108 25	21	20	161	.43806	157 42	10	9
112	.30390	109 24	22	21	162	.44079	158 41	11	10
113	.30664	110 23	23	22	163	.44353	159 40	12	11
114	.30938	111 23	24	23	164	.44627	160 39	13	12
115	0.31211	112 22	25	24	165	0.44901	161 39	14	13
116	.31485	113 21	26	25	166	.45175	162 38	15	14
117	.31759	114 20	27	26	167	.45448	163 37	16	15
118	.32033	115 19	28	27	168	.45722	164 36	17	16
119	.32307	116 18	29	28	169	.45996	165 35	18	17
120	0.32580	117 17	30	29	170	0.46270	166 34	19	18
121	.32854	118 17	May 1	30	171	.46543	167 33	20	19
122	.33128	119 16	2	May 1	172	.46817	168 33	21	20
123	.33402	120 15	3	2	173	.47091	169 32	22	21
124	.33676	121 14	4	3	174	.47365	170 31	23	22
125	0.33949	122 13	5	4	175	0.47639	171 30	24	23
126	.34223	123 12	6	5	176	.47912	172 29	25	24
127	.34497	124 11	7	6	177	.48186	173 28	26	25
128	.34771	125 10	8	7	178	.48460	174 27	27	26
129	.35044	126 10	9	8	179	.48734	175 26	28	27
130	0.35318	127 9	10	9	180	0.49008	176 26	29	28
131	.35592	128 8	11	10	181	.49281	177 25	30	29
132	.35866	129 7	12	11	182	.49555	178 24	July 1	30
133	.36140	130 6	13	12	183	.49829	179 23	2	July 1
134	.36413	131 5	14	13	184	.50103	180 22	3	2
135	0.36687	132 4	15	14	185	0.50376	181 21	4	3
136	.36961	133 4	16	15	186	.50650	182 20	5	4
137	.37235	134 3	17	16	187	.50924	183 20	6	5
138	.37509	135 2	18	17	188	.51198	184 19	7	6
139	.37782	136 1	19	18	189	.51472	185 18	8	7
140	0.38056	137 0	20	19	190	0.51745	186 17	9	8
141	.38330	137 59	21	20	191	.52019	187 16	10	9
142	.38604	138 58	22	21	192	.52293	188 15	11	10
143	.38877	139 58	23	22	193	.52567	189 14	12	11
144	.39151	140 57	24	23	194	.52841	190 14	13	12
145	0.39425	141 56	25	24	195	0.53114	191 13	14	13
146	.39699	142 55	26	25	196	.53388	192 12	15	14
147	.39973	143 54	27	26	197	.53662	193 11	16	15
148	.40246	144 53	28	27	198	.53936	194 10	17	16
149	.40520	145 52	29	28	199	.54209	195 9	18	17
150	0.40794	146 51	30	29	200	0.54483	196 8	19	18

TABLE 74.

DAYS INTO DECIMALS OF A YEAR AND ANGLE.

Day of Year.	Decimal of a Year.	Angle.	Day of Month. Common Year.	Day of Month. Bissextile Year.	Day of Year.	Decimal of a Year.	Angle.	Day of Month. Common Year.	Day of Month. Bissextile Year.
201	0.54757	197° 8'	*July* 20	*July* 19	251	0.68446	246° 24'	*Sept.* 8	*Sept.* 7
202	.55031	198 7	21	20	252	.68720	247 24	9	8
203	.55305	199 6	22	21	253	.68994	248 23	10	9
204	.55578	200 5	23	22	254	.69268	249 22	11	10
205	0.55852	201 4	24	23	255	0.69541	250 21	12	11
206	.56126	202 3	25	24	256	.69815	251 20	13	12
207	.56400	203 2	26	25	257	.70089	252 19	14	13
208	.56674	204 1	27	26	258	.70363	253 18	15	14
209	.56947	205 1	28	27	259	.70637	254 17	16	15
210	0.57221	206 0	29	28	260	0.70910	255 17	17	16
211	.57495	206 59	30	29	261	.71184	256 16	18	17
212	.57769	207 58	31	30	262	.71458	257 15	19	18
213	.58042	208 57	*Aug.* 1	31	263	.71732	258 14	20	19
214	.58316	209 56	2	*Aug.* 1	264	.72005	259 13	21	20
215	0.58590	210 55	3	2	265	0.72279	260 12	22	21
216	.58864	211 55	4	3	266	.72553	261 11	23	22
217	.59138	212 54	5	4	267	.72827	262 11	24	23
218	.59411	213 53	6	5	268	.73101	263 10	25	24
219	.59685	214 52	7	6	269	.73374	264 9	26	25
220	0.59959	215 51	8	7	270	0.73648	265 8	27	26
221	.60233	216 50	9	8	271	.73922	266 7	28	27
222	.60507	217 49	10	9	272	.74196	267 6	29	28
223	.60780	218 49	11	10	273	.74470	268 5	30	29
224	.61054	219 48	12	11	274	.74743	269 5	*Oct.* 1	30
225	0.61328	220 47	13	12	275	0.75017	270 4	2	*Oct.* 1
226	.61602	221 46	14	13	276	.75291	271 3	3	2
227	.61875	222 45	15	14	277	.75565	272 2	4	3
228	.62149	223 44	16	15	278	.75838	273 1	5	4
229	.62423	224 43	17	16	279	.76112	274 0	6	5
230	0.62697	225 43	18	17	280	0.76386	274 59	7	6
231	.62971	226 42	19	18	281	.76660	275 59	8	7
232	.63244	227 41	20	19	282	.76934	276 58	9	8
233	.63518	228 40	21	20	283	.77207	277 57	10	9
234	.63792	229 39	22	21	284	.77481	278 56	11	10
235	0.64066	230 38	23	22	285	0.77755	279 55	12	11
236	.64339	231 37	24	23	286	.78029	280 54	13	12
237	.64613	232 36	25	24	287	.78303	281 53	14	13
238	.64887	233 36	26	25	288	.78576	282 52	15	14
239	.65161	234 35	27	26	289	.78850	283 52	16	15
240	0.65435	235 34	28	27	290	0.79124	284 51	17	16
241	.65708	236 33	29	28	291	.79398	285 50	18	17
242	.65982	237 32	30	29	292	.79671	286 49	19	18
243	.66256	238 31	31	30	293	.79945	287 48	20	19
244	.66530	239 30	*Sept.* 1	31	294	.80219	288 47	21	20
245	0.66804	240 30	2	*Sept.* 1	295	0.80493	289 46	22	21
246	.67077	241 29	3	2	296	.80767	290 46	23	22
247	.67351	242 28	4	3	297	.81040	291 45	24	23
248	.67625	243 27	5	4	298	.81314	292 44	25	24
249	.67899	244 26	6	5	299	.81588	293 43	26	25
250	0.68172	245 25	7	6	300	0.81862	294 42	27	26

SMITHSONIAN TABLES.

TABLE 74.

DAYS INTO DECIMALS OF A YEAR AND ANGLE.

Day of Year.	Decimal of a Year.	Angle.	Day of Month. Common Year	Day of Month. Bissextile Year.	Day of Year.	Decimal of a Year.	Angle.	Day of Month. Common Year	Day of Month. Bissextile Year.
301	0.82136	295° 41'	Oct. 28	Oct. 27	351	0.95825	344° 58'	Dec. 17	Dec. 16
302	.82409	296 40	29	28	352	.96099	345 57	18	17
303	.82683	297 40	30	29	353	.96372	346 56	19	18
304	.82957	298 39	31	30	354	.96646	347 56	20	19
305	0.83231	299 38	Nov. 1	31	355	0.96920	348 55	21	20
306	.83504	300 37	2	Nov. 1	356	.97194	349 54	22	21
307	.83778	301 36	3	2	357	.97467	350 53	23	22
308	.84052	302 35	4	3	358	.97741	351 52	24	23
309	.84326	303 34	5	4	359	.98015	352 51	25	24
310	0.84600	304 34	6	5	360	0.98289	353 50	26	25
311	.84873	305 33	7	6	361	.98563	354 50	27	26
312	.85147	306 32	8	7	362	.98836	355 49	28	27
313	.85421	307 31	9	8	363	.99110	356 48	29	28
314	.85695	308 30	10	9	364	.99384	357 47	30	29
315	0.85969	309 29	11	10	365	0.99658	358 46	31	30
316	.86242	310 28	12	11	366	.99932	359 45		31
317	.86516	311 27	13	12					
318	.86790	312 27	14	13	Conversion for Hours.			Conversion for Minutes.	
319	.87064	313 26	15	14					
320	0.87337	314 25	16	15	Hrs.	Dec. of Year.	Angle.	Min.	Dec. of Year.	Angle.
321	.87611	315 24	17	16						
322	.87885	316 23	18	17						
323	.88159	317 22	19	18	1	0.00011	2.5	1	0.00000	0.04
324	.88433	318 21	20	19	2	23	4.9	2	0	.08
					3	34	7.4	3	1	.12
325	0.88706	319 21	21	20	4	46	9.9	4	1	.16
326	.88980	320 20	22	21						
327	.89254	321 19	23	22	5	0.00057	12.3	5	0.00001	0.21
328	.89528	322 18	24	23	6	68	14.8	6	1	.25
329	.89802	323 17	25	24	7	80	17.2	7	1	.29
330	0.90075	324 16	26	25	8	91	19.7	8	2	.33
331	.90349	325 15	27	26	9	103	22.2	9	2	.37
332	.90623	326 15	28	27						
333	.90897	327 14	29	28	10	0.00114	24.6	10	0.00002	0.41
334	.91170	328 13	30	29	11	126	27.1	20	4	.82
335	0.91444	329 12	Dec. 1	30	12	137	29.6	30	6	1.23
336	.91718	330 11	2	Dec. 1	13	148	32.0	40	8	1.64
337	.91992	331 10	3	2	14	160	34.5	50	10	2.05
338	.92266	332 9	4	3						
339	.92539	333 9	5	4	15	0.00171	37.0	60	0.00011	2.46
					16	183	39.4			
340	0.92813	334 8	6	5	17	194	41.9			
341	.93087	335 7	7	6	18	205	44.4			
342	.93361	336 6	8	7	19	217	46.8			
343	.93634	337 5	9	8						
344	.93908	338 4	10	9	20	0.00228	49.3			
345	0.94182	339 3	11	10	21	240	51.7			
346	.94456	340 2	12	11	22	251	54.2			
347	.94730	341 2	13	12	23	262	56.7			
348	.95003	342 1	14	13	24	274	59.1			
349	.95277	343 0	15	14						
350	0.95551	343 59	16	15						

SMITHSONIAN TABLES

TABLE 75.
HOURS, MINUTES AND SECONDS INTO DECIMALS OF A DAY.

Hours.	Day.	Min.	Day.	Min.	Day.	Sec.	Day.	Sec.	Day.
1	0.041 667	1	0.000 694	31	0.021 528	1	0.000 012	31	0.000 359
2	.083 333	2	.001 389	32	.022 222	2	.000 023	32	.000 370
3	.125 000	3	.002 083	33	.022 917	3	.000 035	33	.000 382
4	.166 667	4	.002 778	34	.023 611	4	.000 046	34	.000 394
5	0.208 333	5	0.003 472	35	0.024 305	5	0.000 058	35	0.000 405
6	.250 000	6	.004 167	36	.025 000	6	.000 069	36	.000 417
7	.291 667	7	.004 861	37	.025 694	7	.000 081	37	.000 428
8	.333 333	8	.005 556	38	.026 389	8	.000 093	38	.000 440
9	.375 000	9	.006 250	39	.027 083	9	.000 104	39	.000 451
10	0.416 667	10	0.006 944	40	0.027 778	10	0.000 116	40	0.000 463
11	.458 333	11	.007 639	41	.028 472	11	.000 127	41	.000 475
12	.500 000	12	.008 333	42	.029 167	12	.000 139	42	.000 486
13	.541 667	13	.009 028	43	.029 861	13	.000 150	43	.000 498
14	.583 333	14	.009 722	44	.030 556	14	.000 162	44	.000 509
15	0.625 000	15	0.010 417	45	0.031 250	15	0.000 174	45	0.000 521
16	.666 667	16	.011 111	46	.031 944	16	.000 185	46	.000 532
17	.708 333	17	.011 806	47	.032 639	17	.000 197	47	.000 544
18	.750 000	18	.012 500	48	.033 333	18	.000 208	48	.000 556
19	.791 667	19	.013 194	49	.034 028	19	.000 220	49	.000 567
20	0.833 333	20	0.013 889	50	0.034 722	20	0.000 231	50	0.000 579
21	.875 000	21	.014 583	51	.035 417	21	.000 243	51	.000 590
22	.916 667	22	.015 278	52	.036 111	22	.000 255	52	.000 602
23	.958 333	23	.015 972	53	.036 806	23	.000 266	53	.000 613
24	1.000 000	24	.016 667	54	.037 500	24	.000 278	54	.000 625
		25	0.017 361	55	0.038 194	25	0.000 289	55	0.000 637
		26	.018 056	56	.038 889	26	.000 301	56	.000 648
		27	.018 750	57	.039 583	27	.000 313	57	.000 660
		28	.019 444	58	.040 278	28	.000 324	58	.000 671
		29	.020 139	59	.040 972	29	.000 336	59	.000 683
		30	0.020 833	60	0.041 667	30	0.000 347	60	0.000 694

TABLE 76.
DECIMALS OF A DAY INTO HOURS, MINUTES AND SECONDS.

Hundredths of a Day.				Ten Thousandths of a Day.			Millionths of a Day.	
d.	h.	m.	s.	d.	min.	sec.	d.	sec.
0.01		14	24	0.0001		8.64	0.000001	0.09
.02		28	48	2		17.28	2	0.17
.03		43	12	3		25.92	3	0.26
.04		57	36	4		34.56	4	0.35
0.05	1	12	0	0.0005		43.20	0.000005	0.43
.06	1	26	24	6		51.84	6	0.52
.07	1	40	48	7	1	0.48	7	0.60
.08	1	55	12	8	1	9.12	8	0.69
.09	2	9	36	9	1	17.76	9	0.78
0.10	2	24	0	0.0010	1	26.40	0.000010	0.86
.20	4	48	0	20	2	52.80	20	1.73
.30	7	12	0	30	4	19.20	30	2.59
.40	9	36	0	40	5	45.60	40	3.46
0.50	12	0	0	0.0050	7	12.00	0.000050	4.32
.60	14	24	0	60	8	38.40	60	5.18
.70	16	48	0	70	10	4.80	70	6.05
.80	19	12	0	80	11	31.20	80	6.91
.90	21	36	0	90	12	57.60	90	7.78

SMITHSONIAN TABLES.

TABLE 77.

MINUTES AND SECONDS INTO DECIMALS OF AN HOUR.

Min.	Decimals of an hour.	Min.	Decimals of an hour.	Sec.	Decimals of an hour.	Sec.	Decimals of an hour.
1	0.016 667	31	0.516 667	1	0.000 278	31	0.008 611
2	.033 333	32	.533 333	2	.000 556	32	.008 889
3	.050 000	33	.550 000	3	.000 833	33	.009 167
4	.066 667	34	.566 667	4	.001 111	34	.009 444
5	0.083 333	35	0.583 333	5	0.001 389	35	0.009 722
6	.100 000	36	.600 000	6	.001 667	36	.010 000
7	.116 667	37	.616 667	7	.001 944	37	.010 278
8	.133 333	38	.633 333	8	.002 222	38	.010 556
9	.150 000	39	.650 000	9	.002 500	39	.010 833
10	0.166 667	40	0.666 667	10	0.002 778	40	0.011 111
11	.183 333	41	.683 333	11	.003 056	41	.011 389
12	.200 000	42	.700 000	12	.003 333	42	.011 667
13	.216 667	43	.716 667	13	.003 611	43	.011 944
14	.233 333	44	.733 333	14	.003 889	44	.012 222
15	0.250 000	45	0.750 000	15	0.004 167	45	0.012 500
16	.266 667	46	.766 667	16	.004 444	46	.012 778
17	.283 333	47	.783 333	17	.004 722	47	.013 056
18	.300 000	48	.800 000	18	.005 000	48	.013 333
19	.316 667	49	.816 667	19	.005 278	49	.013 611
20	0.333 333	50	0.833 333	20	0.005 556	50	0.013 889
21	.350 000	51	.850 000	21	.005 833	51	.014 167
22	.366 667	52	.866 667	22	.006 111	52	.014 444
23	.383 333	53	.883 333	23	.006 389	53	.014 722
24	.400 000	54	.900 000	24	.006 667	54	.015 000
25	0.416 667	55	0.916 667	25	0.006 944	55	0.015 278
26	.433 333	56	.933 333	26	.007 222	56	.015 556
27	.450 000	57	.950 000	27	.007 500	57	.015 833
28	.466 667	58	.966 667	28	.007 778	58	.016 111
29	.483 333	59	.983 333	29	.008 056	59	.016 389
30	0.500 000	60	1.000 000	30	0.008 333	60	0.016 667

TABLE 78.

MEAN TIME AT APPARENT NOON.

Day of Month.	JAN. h. m.	FEB. h. m.	MAR. h. m.	APR. h. m.	MAY. h. m.	JUNE. h. m.
1	12 4	12 14	12 12	12 4	11 57	11 58
8	12 7	12 14	12 11	12 2	11 56	11 59
16	12 10	12 14	12 9	12 0	11 56	12 0
24	12 12	12 13	12 6	11 58	11 57	12 2

	JULY. h. m.	AUG. h. m.	SEPT. h. m.	OCT. h. m.	NOV. h. m.	DEC. h. m.
1	12 3	12 6	12 0	11 50	11 44	11 50
8	12 5	12 5	11 58	11 48	11 44	11 53
16	12 6	12 4	11 55	11 46	11 45	11 56
24	12 6	12 2	11 52	11 45	11 47	12 0

TABLE 79.
SIDEREAL TIME INTO MEAN SOLAR TIME.

The tabular values are to be *subtracted* from a sidereal time interval.

Hrs.	Reduction to Mean Time.	Min.	Reduction to Mean Time.	Min.	Reduction to Mean Time.
h.	m. s.	m.	s.	m.	s.
1	0 9.83	1	0.16	31	5.08
2	0 19.66	2	0.33	32	5.24
3	0 29.49	3	0.49	33	5.41
4	0 39.32	4	0.66	34	5.57
5	0 49.15	5	0.82	35	5.73
6	0 58.98	6	0.98	36	5.90
7	1 8.81	7	1.15	37	6.06
8	1 18.64	8	1.31	38	6.23
9	1 28.47	9	1.47	39	6.39
10	1 38.30	10	1.64	40	6.55
11	1 48.13	11	1.80	41	6.72
12	1 57.96	12	1.97	42	6.88
13	2 7.78	13	2.13	43	7.05
14	2 17.61	14	2.29	44	7.21
15	2 27.44	15	2.46	45	7.37
16	2 37.27	16	2.62	46	7.54
17	2 47.10	17	2.79	47	7.70
18	2 56.93	18	2.95	48	7.86
19	3 6.76	19	3.11	49	8.03
20	3 16.59	20	3.28	50	8.19
21	3 26.42	21	3.44	51	8.36
22	3 36.25	22	3.60	52	8.52
23	3 46.08	23	3.77	53	8.68
24	3 55.91	24	3.93	54	8.85
		25	4.10	55	9.01
		26	4.26	56	9.17
		27	4.42	57	9.34
		28	4.59	58	9.50
		29	4.75	59	9.67
		30	4.92	60	9.83

TABLE 80.
MEAN SOLAR TIME INTO SIDEREAL TIME.

The tabular values are to be *added* to a mean solar time interval.

Hrs.	Reduction to Sidereal Time.	Min.	Reduction to Sidereal Time.	Min.	Reduction to Sidereal Time.
h.	m. s.	m.	s.	m.	s.
1	0 9.86	1	0.16	31	5.09
2	0 19.71	2	0.33	32	5.26
3	0 29.57	3	0.49	33	5.42
4	0 39.43	4	0.66	34	5.59
5	0 49.28	5	0.82	35	5.75
6	0 59.14	6	0.99	36	5.91
7	1 9.00	7	1.15	37	6.08
8	1 18.85	8	1.31	38	6.24
9	1 28.71	9	1.48	39	6.41
10	1 38.57	10	1.64	40	6.57
11	1 48.42	11	1.81	41	6.74
12	1 58.28	12	1.97	42	6.90
13	2 8.13	13	2.14	43	7.06
14	2 17.99	14	2.30	44	7.23
15	2 27.85	15	2.46	45	7.39
16	2 37.70	16	2.63	46	7.56
17	2 47.56	17	2.79	47	7.72
18	2 57.42	18	2.96	48	7.89
19	3 7.27	19	3.12	49	8.05
20	3 17.13	20	3.29	50	8.21
21	3 26.99	21	3.45	51	8.38
22	3 36.84	22	3.61	52	8.54
23	3 46.70	23	3.78	53	8.71
24	3 56.56	24	3.94	54	8.87
		25	4.11	55	9.04
		26	4.27	56	9.20
		27	4.43	57	9.36
		28	4.60	58	9.53
		29	4.76	59	9.69
		30	4.93	60	9.86

Reduction for Seconds—sidereal or mean solar.

The tabular values are to be { *subtracted* from a sidereal / *added* to a mean solar } time interval.

Sidereal or Mean Time.	0	1	2	3	4	5	6	7	8	9
s.	s.	s.	s.	s.	s.	s.	s.	s.	s.	s.
0	0.00	0.00	0.01	0.01	0.01	0.01	0.02	0.02	0.02	0.03
10	.03	.03	.03	.04	.04	.04	.04	.05	.05	.05
20	.06	.06	.06	.06	.07	.07	.07	.07	.08	.08
30	.08	.09	.09	.09	.09	.10	.10	.10	.10	.11
40	.11	.11	.12	.12	.12	.12	.13	.13	.13	.13
50	0.14	0.14	0.14	0.15	0.15	0.15	0.15	0.16	0.16	0.16

SMITHSONIAN TABLES.

MISCELLANEOUS TABLES.

Density of air at different temperatures Fahrenheit . . . . TABLE 81

Density of air at different humidities and pressures — English measures.
    Term for humidity: auxiliary to Table 83 . . . . . . TABLE 82
    Values of $\dfrac{h}{29.921} = \dfrac{b - 0.378\,e}{29.921}$ 83

Density of air at different temperatures Centigrade . . . . TABLE 84

Density of air at different humidities and pressures — Metric measures.
    Term for humidity: auxiliary to Table 86 . . . . . TABLE 85
    Values of $\dfrac{h}{760} = \dfrac{b - 0.378\,e}{760}$ 86

Conversion of avoirdupois pounds and ounces into kilogrammes   TABLE 87
Conversion of kilogrammes into avoirdupois pounds and ounces   88

Conversion of grains into grammes . . . . TABLE 89
Conversion of grammes into grains   90

Conversion of units of magnetic intensity . . . . TABLE 91

Quantity of water corresponding to given depths of rainfall . TABLE 92

Dates of Dove's pentades . . . . TABLE 93

Division by 28 of numbers from 28 to 867 972 . . . . . . TABLE 94
Division by 29 of numbers from 29 to 898 971   95
Division by 31 of numbers from 31 to 960 969   96

Natural sines and cosines . . . . TABLE 97

Natural tangents and cotangents . . . . TABLE 98

Logarithms of numbers . . . . TABLE 99

LIST OF METEOROLOGICAL STATIONS . . . . TABLE 100

TABLE 81.

DENSITY OF AIR AT DIFFERENT TEMPERATURES FAHRENHEIT.

$$\delta_t = \frac{0.00129305}{1 + 0.0020389\,(t - 32°)}.$$

1 cubic centimetre of dry air at the temperature 32° F. and pressure 760 mm., and under the standard value of gravity at latitude 45° and sea-level, weighs 0.00129305 gramme.

Temperature.	δ_t	Log δ_t	Temperature.	δ_t	Log δ_t	Temperature.	δ_t	Log δ_t
F.	0.00	− 10	F.	0.00	− 10	F.	0.00	− 10
−45°	15339	7.18579	30°	12983	7.11339	75°	11888	7.07512
−40	15155	.18056	31	12957	.11250	76	11866	.07430
−35	14977	.17541	32	12931	.11162	77	11844	.07349
−30	14802	.17031	33	12904	.11073	78	11822	.07268
−25	14631	.16527	34	12878	.10985	79	11800	.07187
				0.00			0.00	
−20	14464	7.16029	35	12852	7.10897	80	11778	7.07107
−18	14398	.15831	36	12826	.10809	81	11756	.07026
−16	14333	.15634	37	12800	.10721	82	11734	.06946
−14	14269	.15439	38	12774	.10633	83	11713	.06865
−12	14205	.15244	39	12749	.10546	84	11691	.06785
	0.00			0.00			0.00	
−10	14142	7.15050	40	12723	7.10459	85	11670	7.06705
−8	14079	.14856	41	12698	.10372	86	11648	.06625
−6	14017	.14664	42	12672	.10285	87	11627	.06546
−4	13955	.14472	43	12647	.10198	88	11605	.06466
−2	13894	.14282	44	12622	.10112	89	11584	.06387
	0.00			0.00			0.00	
+0	13833	7.14092	45	12597	7.10025	90	11563	7.06307
1	13803	.13997	46	12572	.09939	91	11542	.06228
2	13773	.13903	47	12547	.09853	92	11521	.06149
3	13743	.13808	48	12522	.09767	93	11500	.06070
4	13713	.13714	49	12497	.09682	94	11479	.05992
	0.00			0.00			0.00	
5	13684	7.13621	50	12473	7.09596	95	11458	7.05913
6	13654	.13527	51	12448	.09511	96	11438	.05835
7	13625	.13434	52	12424	.09426	97	11418	.05757
8	13596	.13340	53	12400	.09341	98	11397	.05678
9	13567	.13247	54	12375	.09256	99	11376	.05600
	0.00			0.00			0.00	
10	13538	7.13155	55	12351	7.09171	100	11356	7.05523
11	13509	.13062	56	12327	.09087	101	11336	.05445
12	13480	.12970	57	12303	.09002	102	11315	.05367
13	13452	.12877	58	12280	.08918	103	11295	.05290
14	13423	.12785	59	12256	.08834	104	11275	.05213
	0.00			0.00			0.00	
15	13395	7.12694	60	12232	7.08750	105	11255	7.05136
16	13367	.12602	61	12209	.08667	106	11235	.05058
17	13338	.12510	62	12185	.08583	107	11215	.04982
18	13310	.12419	63	12162	.08500	108	11196	.04905
19	13282	.12328	64	12138	.08416	109	11176	.04828
	0.00			0.00			0.00	
20	13255	7.12237	65	12115	7.08334	110	11156	7.04752
21	13227	.12147	66	12092	.08251	112	11117	.04599
22	13200	.12056	67	12069	.08168	114	11078	.04447
23	13172	.11966	68	12046	.08085	116	11040	.04296
24	13145	.11876	69	12023	.08003	118	11001	.04145
	0.00			0.00			0.00	
25	13118	7.11786	70	12001	7.07921	120	10963	7.03994
26	13091	.11696	71	11978	.07839	125	10870	.03621
27	13064	.11606	72	11956	.07757	130	10776	.03248
28	13037	.11517	73	11933	.07675	135	10686	.02883
29	13010	.11428	74	11910	.07593	140	10597	.02518

SMITHSONIAN TABLES.

TABLE 82.

DENSITY OF AIR AT DIFFERENT HUMIDITIES AND PRESSURES.
ENGLISH MEASURES.

Term for Humidity: Values of $0.378e$. Auxiliary to Table 83.

$e =$ Vapor pressure in inches.

Dew-Point.	Vapor Pressure. e.	$0.378e$.	Dew-Point.	Vapor Pressure. e.	$0.378e$.	Dew-Point.	Vapor Pressure. e.	$0.378e$.	Dew-Point.	Vapor Pressure. e.	$0.378e$.
F.	Inch.	Inch.	F.	Inch.	Inch.	F.	Inch.	Inch.	F.	Inches.	Inches.
−40°	0.0054	0.002	5°	0.057	0.021	50°	0.360	0.136	95°	1.645	0.622
−39	.0058	.002	6	.059	.022	51	.373	.141	96	1.696	.641
−38	.0061	.002	7	.062	.023	52	.387	.146	97	1.749	.661
−37	.0065	.002	8	.065	.025	53	.402	.152	98	1.803	.682
−36	.0069	.003	9	.068	.026	54	.417	.158	99	1.859	.703
−35	0.0073	0.003	10	0.071	0.027	55	0.432	0.163	100	1.916	0.724
−34	.0077	.003	11	.074	.028	56	.448	.169	101	1.975	.747
−33	.0082	.003	12	.078	.029	57	.465	.176	102	2.035	.769
−32	.0087	.003	13	.081	.031	58	.481	.182	103	2.097	.793
−31	.0092	.003	14	.085	.032	59	.499	.189	104	2.160	.816
−30	0.0097	0.004	15	0.088	0.033	60	0.517	0.195	105	2.225	0.841
−29	.0103	.004	16	.092	.035	61	.536	.203	106	2.292	.866
−28	.0109	.004	17	.096	.036	62	.555	.210	107	2.360	.892
−27	.0115	.004	18	.101	.038	63	.575	.217	108	2.431	.919
−26	.0121	.005	19	.105	.040	64	.595	.225	109	2.503	.946
−25	0.0128	0.005	20	0.110	0.042	65	0.616	0.233	110	2.576	0.974
−24	.0135	.005	21	.114	.043	66	.638	.241	111	2.652	1.002
−23	.0142	.005	22	.119	.045	67	.661	.250	112	2.730	1.031
−22	.0150	.006	23	.124	.047	68	.684	.259	113	2.810	1.062
−21	.0158	.006	24	.130	.049	69	.707	.267	114	2.891	1.093
−20	0.0167	0.006	25	0.135	0.051	70	0.732	0.277	115	2.975	1.125
−19	.0176	.007	26	.141	.053	71	.757	.286	116	3.061	1.157
−18	.0185	.007	27	.147	.056	72	.783	.296	117	3.148	1.190
−17	.0195	.007	28	.153	.058	73	.810	.306	118	3.239	1.224
−16	.0205	.008	29	.159	.060	74	.838	.317	119	3.331	1.259
−15	0.0216	0.008	30	0.166	0.063	75	0.866	0.327	120	3.425	1.295
−14	.0227	.009	31	.173	.065	76	.896	.339	121	3.522	1.331
−13	.0239	.009	32	.180	.068	77	.926	.350	122	3.621	1.369
−12	.0251	.009	33	.187	.071	78	.957	.362	123	3.723	1.407
−11	.0264	.010	34	.195	.074	79	.989	.374	124	3.827	1.447
−10	0.0277	0.010	35	0.203	0.077	80	1.022	0.386	125	3.933	1.487
−9	.0292	.011	36	.211	.080	81	1.056	.399	126	4.042	1.528
−8	.0306	.012	37	.219	.083	82	1.091	.412	127	4.154	1.570
−7	.0322	.012	38	.228	.086	83	1.127	.426	128	4.268	1.613
−6	.0338	.013	39	.237	.090	84	1.163	.440	129	4.385	1.658
−5	0.0354	0.013	40	0.246	0.093	85	1.201	0.454	130	4.504	1.703
−4	.0372	.014	41	.256	.097	86	1.241	.469	131	4.627	1.749
−3	.0390	.015	42	.266	.101	87	1.281	.484	132	4.752	1.796
−2	.0409	.015	43	.276	.105	88	1.322	.500	133	4.880	1.844
−1	.0429	.016	44	.287	.109	89	1.364	.516	134	5.011	1.894
0	0.0449	0.017	45	0.298	0.113	90	1.408	0.532	135	5.145	1.945
+1	.0471	.018	46	.310	.117	91	1.453	.549	136	5.282	1.997
2	.0493	.019	47	.322	.122	92	1.499	.567	137	5.422	2.050
3	.0517	.020	48	.334	.126	93	1.546	.584	138	5.565	2.104
4	.0541	.020	49	.347	.131	94	1.595	.603	139	5.712	2.159
5	0.0567	0.021	50	0.360	0.136	95	1.645	0.622	140	5.862	2.216

SMITHSONIAN TABLES.

TABLE 83.
DENSITY OF AIR AT DIFFERENT HUMIDITIES AND PRESSURES.
ENGLISH MEASURES.

Values of $\dfrac{h}{29.921}$. $\qquad \dfrac{\delta}{\delta_o} = \dfrac{h}{29.921} = \dfrac{b - 0.378\,e}{29.921}$.

$b =$ Barometric pressure in inches; $e =$ Vapor pressure in inches.

h.	$\dfrac{h}{29.921}$	Log $\dfrac{h}{29.921}$	h.	$\dfrac{h}{29.291}$	Log $\dfrac{h}{29.921}$	h.	$\dfrac{h}{29.921}$	Log $\dfrac{h}{29.921}$
Inch's.		−10	Inches.		−10	Inches.		−10
10.0	0.3342	9.52402	15.0	0.5013	9.70012	20.0	0.6684	9.82505
10.1	.3376	.52835	15.1	.5047	.70300	20.1	.6718	.82722
10.2	.3409	.53262	15.2	.5080	.70587	20.2	.6751	.82938
10.3	.3442	.53686	15.3	.5113	.70871	20.3	.6784	.83152
10.4	.3476	.54106	15.4	.5147	.71154	20.4	.6818	.83365
10.5	0.3509	9.54521	15.5	0.5180	9.71435	20.5	0.6851	9.83578
10.6	.3543	.54933	15.6	.5214	.71715	20.6	.6885	.83789
10.7	.3576	.55341	15.7	.5247	.71992	20.7	.6918	.83999
10.8	.3609	.55745	15.8	.5281	.72268	20.8	.6952	.84209
10.9	.3643	.56145	15.9	.5314	.72542	20.9	.6985	.84417
11.0	0.3676	9.56542	16.0	0.5347	9.72814	21.0	0.7018	9.84624
11.1	.3710	.56935	16.1	.5381	.73085	21.1	.7052	.84831
11.2	.3743	.57324	16.2	.5414	.73354	21.2	.7085	.85036
11.3	.3777	.57710	16.3	.5448	.73621	21.3	.7119	.85240
11.4	.3810	.58093	16.4	.5481	.73887	21.4	.7152	.85444
11.5	0.3843	9.58472	16.5	0.5515	9.74151	21.5	0.7186	9.85646
11.6	.3877	.58848	16.6	.5548	.74413	21.6	.7219	.85848
11.7	.3910	.59221	16.7	.5581	.74674	21.7	.7252	.86048
11.8	.3944	.59591	16.8	.5615	.74933	21.8	.7286	.86248
11.9	.3977	.59957	16.9	.5648	.75191	21.9	.7319	.86447
12.0	0.4011	9.60321	17.0	0.5682	9.75447	22.0	0.7353	9.86645
12.1	.4044	.60681	17.1	.5715	.75702	22.1	.7386	.86842
12.2	.4077	.61038	17.2	.5748	.75955	22.2	.7420	.87038
12.3	.4111	.61393	17.3	.5782	.76207	22.3	.7453	.87233
12.4	.4144	.61745	17.4	.5815	.76457	22.4	.7486	.87427
12.5	0.4178	9.62093	17.5	0.5849	9.76706	22.5	0.7520	9.87621
12.6	.4211	.62439	17.6	.5882	.76954	22.6	.7553	.87813
12.7	.4244	.62782	17.7	.5916	.77200	22.7	.7587	.88005
12.8	.4278	.63123	17.8	.5949	.77444	22.8	.7620	.88196
12.9	.4311	.63461	17.9	.5982	.77687	22.9	.7653	.88386
13.0	0.4345	9.63797	18.0	0.6016	9.77930	23.0	0.7687	9.88575
13.1	.4378	.64130	18.1	.6049	.78170	23.1	.7720	.88764
13.2	.4412	.64460	18.2	.6083	.78410	23.2	.7754	.88951
13.3	.4445	.64788	18.3	.6116	.78648	23.3	.7787	.89138
13.4	.4478	.65113	18.4	.6149	.78884	23.4	.7821	.89324
13.5	0.4512	9.65436	18.5	0.6183	9.79120	23.5	0.7854	9.89509
13.6	.4545	.65756	18.6	.6216	.79354	23.6	.7887	.89693
13.7	.4579	.66074	18.7	.6250	.79587	23.7	.7921	.89877
13.8	.4612	.66390	18.8	.6283	.79818	23.8	.7954	.90060
13.9	.4646	.66704	18.9	.6317	.80049	23.9	.7988	.90242
14.0	0.4679	9.67015	19.0	0.6350	9.80278	24.0	0.8021	9.90424
14.1	.4712	.67324	19.1	.6383	.80506	24.1	.8054	.90604
14.2	.4746	.67631	19.2	.6417	.80733	24.2	.8088	.90784
14.3	.4779	.67936	19.3	.6450	.80958	24.3	.8121	.90963
14.4	.4813	.68239	19.4	.6484	.81183	24.4	.8155	.91141
14.5	0.4846	9.68539	19.5	0.6517	9.81406	24.5	0.8188	9.91319
14.6	.4879	.68837	19.6	.6551	.81628	24.6	.8222	.91496
14.7	.4913	.69134	19.7	.6584	.81849	24.7	.8255	.91672
14.8	.4946	.69429	19.8	.6617	.82069	24.8	.8289	.91848
14.9	.4980	.69721	19.9	.6651	.82288	24.9	.8322	.92022

SMITHSONIAN TABLES.

TABLE 83.

DENSITY OF AIR AT DIFFERENT HUMIDITIES AND PRESSURES.
ENGLISH MEASURES.

Values of $\dfrac{h}{29.921}$. $\qquad \dfrac{\delta}{\delta_0} = \dfrac{h}{29.921} = \dfrac{b - 0.378\,e}{29.921}$.

$b =$ Barometric pressure in inches; $\quad e =$ Vapor pressure in inches.

h.	$\dfrac{h}{29.921}$	Log $\dfrac{h}{29.921}$	h.	$\dfrac{h}{29.921}$	Log $\dfrac{h}{29.921}$	h.	$\dfrac{h}{29.921}$	Log $\dfrac{h}{29.921}$
Inches.		-10	Inches.		-10	Inches.		-10
25.00	0.8355	9.92196	27.25	0.9107	9.95939	29.50	0.9859	9.99385
25.05	.8372	.92283	27.30	.9124	.96019	29.55	.9876	.99458
25.10	.8389	.92370	27.35	.9141	.96098	29.60	.9893	.99532
25.15	.8405	.92456	27.40	.9157	.96177	29.65	.9909	.99605
25.20	.8422	.92542	27.45	.9174	.96256	29.70	.9926	.99678
25.25	0.8439	9.92628	27.50	0.9191	9.96336	29.75	0.9943	9.99751
25.30	.8456	.92714	27.55	.9208	.96414	29.80	.9960	.99824
25.35	.8472	.92800	27.60	.9224	.96493	29.85	.9976	.99897
25.40	.8489	.92886	27.65	.9241	.96572	29.90	.9993	.99970
25.45	.8506	.92971	27.70	.9258	.96650	29.95	1.0010	0.00042
25.50	0.8522	9.93056	27.75	0.9274	9.96728	30.00	1.0026	0.00115
25.55	.8539	.93141	27.80	.9291	.96807	30.05	1.0043	.00187
25.60	.8556	.93226	27.85	.9308	.96885	30.10	1.0060	.00259
25.65	.8573	.93311	27.90	.9325	.96963	30.15	1.0076	.00331
25.70	.8589	.93396	27.95	.9341	.97040	30.20	1.0093	.00403
25.75	0.8606	9.93480	28.00	0.9358	9.97118	30.25	1.0110	0.00475
25.80	.8623	.93564	28.05	.9375	.97195	30.30	1.0127	.00547
25.85	.8639	.93648	28.10	.9391	.97273	30.35	1.0143	.00618
25.90	.8656	.93732	28.15	.9408	.97350	30.40	1.0160	.00690
25.95	.8673	.93816	28.20	.9425	.97427	30.45	1.0177	.00761
26.00	0.8690	9.93900	28.25	0.9441	9.97504	30.50	1.0193	0.00832
26.05	.8706	.93983	28.30	.9458	.97581	30.55	1.0210	.00903
26.10	.8723	.94066	28.35	.9475	.97657	30.60	1.0227	.00975
26.15	.8740	.94149	28.40	.9492	.97734	30.65	1.0244	.01045
26.20	.8756	.94233	28.45	.9508	.97810	30.70	1.0260	.01116
26.25	0.8773	9.94315	28.50	0.9525	9.97887	30.75	1.0277	0.01187
26.30	.8790	.94398	28.55	.9542	.97963	30.80	1.0294	.01257
26.35	.8806	.94480	28.60	.9558	.98039	30.85	1.0310	.01328
26.40	.8823	.94563	28.65	.9575	.98115	30.90	1.0327	.01398
26.45	.8840	.94645	28.70	.9592	.98191	30.95	1.0344	.01468
26.50	0.8857	9.94727	28.75	0.9609	9.98266	31.00	1.0361	0.01539
26.55	.8873	.94809	28.80	.9625	.98342	31.05	1.0377	.01608
26.60	.8890	.94891	28.85	.9642	.98417	31.10	1.0394	.01678
26.65	.8907	.94972	28.90	.9659	.98492	31.15	1.0411	.01748
26.70	.8924	.95054	28.95	.9675	.98567	31.20	1.0427	.01818
26.75	0.8940	9.95135	29.00	0.9692	9.98642	31.25	1.0444	0.01887
26.80	.8957	.95216	29.05	.9709	.98717	31.30	1.0461	.01957
26.85	.8974	.95297	29.10	.9726	.98792	31.35	1.0478	.02026
26.90	.8990	.95378	29.15	.9742	.98866	31.40	1.0494	.02095
26.95	.9007	.95458	29.20	.9759	.98941	31.45	1.0511	.02164
27.00	0.9024	9.95539	29.25	0.9776	9.99015	31.50	1.0528	0.02233
27.05	.9040	.95619	29.30	.9792	.99089	31.55	1.0544	.02302
27.10	.9057	.95699	29.35	.9809	.99163	31.60	1.0561	.02371
27.15	.9074	.95779	29.40	.9826	.99237	31.65	1.0578	.02439
27.20	.9091	.95859	29.45	.9843	.99311	31.70	1.0594	.02508

SMITHSONIAN TABLES.

TABLE 84.

DENSITY OF AIR AT DIFFERENT TEMPERATURES CENTIGRADE.

$$\delta_{t,\,760} = \frac{0.00129305}{1 + 0.003670\,t}.$$

1 cubic metre of dry air at the temperature 0° C. and pressure 760 mm., and under the standard value of gravity at latitude 45° and sea level, weighs 1.29305 kilogramme.

t.	$\delta_{t,\,760}$	Log $\delta_{t,\,760}$	t.	$\delta_{t,\,760}$	Log $\delta_{t,\,760}$	t.	$\delta_{t,\,760}$	Log $\delta_{t,\,760}$
C.	0.00	− 10	C.	0.00	− 10	C.	0.00	− 10
−34°	14774	7.16950	− 4.5	13148	7.11885	18.0	12129	7.08383
−33	14712	.16768	− 4.0	13123	.11804	18.5	12108	8309
−32	14651	.16587	− 3.5	13099	.11723	19.0	12088	8234
−31	14590	.16407	− 3.0	13074	.11642	19.5	12067	8160
	0.00			0.00			0.00	
−30	14530	7.16227	− 2.5	13050	7.11562	20.0	12046	7.08085
−29	14471	.16049	− 2.0	13026	.11481	20.5	12026	8011
−28	14412	.15871	− 1.5	13002	.11401	21.0	12005	7937
−27	14353	.15693	− 1.0	12978	.11321	21.5	11985	7863
−26	14295	.15517	− 0.5	12954	.11241	22.0	11965	7789
	0.00			0.00			0.00	
−25	14237	7.15341	0.0	12931	7.11162	22.5	11944	7.07716
−24	14179	.15166	+ 0.5	12907	.11082	23.0	11924	7642
−23	14123	.14991	1.0	12884	.11006	23.5	11904	7569
−22	14066	.14818	1.5	12860	.10923	24.0	11884	7496
−21	14010	.14645	2.0	12836	.10844	24.5	11864	7422
	0.00			0.00			0.00	
−20.0	13955	7.14472	2.5	12813	7.10765	25.0	11844	7.07349
−19.5	13927	.14386	3.0	12790	.10686	25.5	11824	7276
−19.0	13900	.14301	3.5	12766	.10607	26.0	11804	7204
−18.5	13872	.14215	4.0	12744	.10529	26.5	11784	7131
−18.0	13845	.14130	4.5	12720	.10450	27.0	11765	7058
	0.00			0.00			0.00	
−17.5	13818	7.14044	5.0	12698	7.10372	27.5	11745	7.06986
−17.0	13791	.13959	5.5	12675	.10294	28.0	11726	6913
−16.5	13764	.13874	6.0	12652	.10216	28.5	11706	6841
−16.0	13737	.13790	6.5	12629	.10138	29.0	11687	6769
−15.5	13710	.13705	7.0	12607	.10069	29.5	11667	6697
	0.00			0.00			0.00	
−15.0	13684	7.13621	7.5	12584	7.09982	30.0	11648	7.06625
−14.5	13657	.13536	8.0	12562	9905	30.5	11629	6554
−14.0	13631	.13452	8.5	12539	9828	31.0	11610	6482
−13.5	13604	.13368	9.0	12517	9750	31.5	11591	6411
−13.0	13578	.13285	9.5	12495	9673	32.0	11572	6340
	0.00			0.00			0.00	
−12.5	13552	7.13201	10.0	12473	7.09596	32.5	11553	7.06268
−12.0	13526	.13117	10.5	12451	9519	33.0	11534	6197
−11.5	13500	.13034	11.0	12429	9443	33.5	11515	6126
−11.0	13473	.12951	11.5	12407	9366	34.0	11496	6055
−10.5	13449	.12868	12.0	12385	9290	34.5	11477	5984
	0.00			0.00			0.00	
−10.0	13423	7.12785	12.5	12363	7.09214	35.0	11459	7.05913
− 9.5	13398	.12703	13.0	12342	9137	35.5	11440	5843
− 9.0	13372	.12620	13.5	12320	9061	36.0	11421	5772
− 8.5	13347	.12538	14.0	12299	8986	36.5	11403	5702
− 8.0	13322	.12456	14.5	12277	8910	37.0	11385	5632
	0.00			0.00			0.00	
− 7.5	13297	7.12374	15.0	12256	7.08834	37.5	11366	7.05562
− 7.0	13271	.12292	15.5	12235	8759	38.0	11348	5492
− 6.5	13246	.12210	16.0	12213	8683	38.5	11330	5422
− 6.0	13222	.12128	16.5	12192	8608	39.0	11311	5352
− 5.5	13197	.12047	17.0	12171	8533	39.5	11293	5282
	0.00			0.00			0.00	
− 5.0	13172	7.11966	17.5	12150	7.08458	40.0	11275	7.05213

Smithsonian Tables.

TABLE 84.
DENSITY OF AIR AT DIFFERENT TEMPERATURES CENTIGRADE.
(*Continued.*)

t.	$\delta_{t,760}$	Log $\delta_{t,760}$	t.	$\delta_{t,760}$	Log $\delta_{t,760}$	t.	$\delta_{t,760}$	Log $\delta_{t,760}$
C.	0.00	−10	C.	0.00	−10	C.	0.00	−10
40°	11275	7.05213	50°	10926	7.03845	60°	10597	7.02518
41	11239	.05074	51	10892	.03710	61	10565	.02388
42	11204	.04936	52	10858	.03576	62	10534	.02258
43	11168	.04798	53	10825	.03443	63	10502	.02128
44	11133	.04660	54	10792	.03309	64	10471	.01999
	0.00			0.00			0.00	
45	11098	7.04523	55	10759	7.03177	65	10440	7.01870
46	11063	.04387	56	10726	.03044	66	10409	.01742
47	11028	.04251	57	10694	.02912	67	10379	.01614
48	10994	.04115	58	10661	.02780	68	10348	.01486
49	10960	.03980	59	10629	.02649	69	10318	.01358

TABLE 85.
DENSITY OF AIR AT DIFFERENT HUMIDITIES AND PRESSURES.
METRIC MEASURES.
Term for humidity: values of $0.378\,e$. Auxiliary to Table 86.
$e =$ vapor pressure in mm.

Dew-point.	Vapor Pressure. e	$0.378\,e$	Dew-point.	Vapor Pressure. e	$0.378\,e$	Dew-point.	Vapor Pressure. e	$0.378\,e$
C.	mm.	mm.	C.	mm.	mm.	C.	mm.	mm.
−30°	0.38	0.14	0°	4.57	1.73	30°	31.51	11.91
29	.42	.16	1	4.91	1.86	31	33.37	12.61
28	.46	.17	2	5.27	1.99	32	35.32	13.35
27	.50	.19	3	5.66	2.14	33	37.37	14.13
26	.55	.21	4	6.07	2.29	34	39.52	14.94
−25	0.61	0.23	5	6.51	2.46	35	41.78	15.79
24	.66	.25	6	6.97	2.63	36	44.16	16.69
23	.73	.28	7	7.47	2.82	37	46.65	17.63
22	.79	.30	8	7.99	3.02	38	49.26	18.62
21	.87	.33	9	8.55	3.23	39	52.00	19.66
−20	0.94	0.36	10	9.14	3.45	40	54.87	20.74
19	1.03	.39	11	9.77	3.69	41	57.87	21.86
18	1.12	.42	12	10.43	3.94	42	61.02	23.06
17	1.22	.46	13	11.14	4.21	43	64.31	24.31
16	1.32	.50	14	11.88	4.49	44	67.76	25.61
−15	1.44	0.54	15	12.67	4.79	45	71.36	26.97
14	1.56	.59	16	13.51	5.11	46	75.13	28.40
13	1.69	.64	17	14.40	5.44	47	79.07	29.89
12	1.84	.70	18	15.33	5.79	48	83.19	31.45
11	1.99	.75	19	16.32	6.17	49	87.49	33.07
−10	2.15	0.81	20	17.36	6.56	50	91.98	34.77
9	2.33	.88	21	18.47	6.98	51	96.66	36.54
8	2.51	.95	22	19.63	7.42	52	101.55	38.39
7	2.72	1.03	23	20.86	7.89	53	106.65	40.31
6	2.93	1.11	24	22.15	8.37	54	111.97	42.32
−5	3.16	1.19	25	23.52	8.89	55	117.52	44.42
4	3.41	1.29	26	24.96	9.43	56	123.29	46.60
3	3.67	1.39	27	26.47	10.01	57	129.31	48.88
2	3.95	1.49	28	28.07	10.61	58	135.58	51.25
1	4.25	1.61	29	29.74	11.24	59	142.10	53.71

SMITHSONIAN TABLES.

TABLE 88.

DENSITY OF AIR AT DIFFERENT HUMIDITIES AND PRESSURES.
METRIC MEASURES.

Values of $\dfrac{h}{760}$. $\dfrac{\delta}{\delta_0} = \dfrac{h}{760} = \dfrac{b - 0.378e}{760}$.

b = Barometric pressure in mm.; e = Vapor pressure in mm.

h.	$\dfrac{h}{760}$.	Log $\dfrac{h}{760}$.	h.	$\dfrac{h}{760}$.	Log $\dfrac{h}{760}$.	h.	$\dfrac{h}{760}$.	Log $\dfrac{h}{760}$.
mm.		-10	mm.		-10	mm.		-10
300	0.3947	9.59631	400	0.5263	9.72125	450	0.5921	9.77240
302	.3974	.59919	401	.5276	.72233	451	.5934	.77336
304	.4000	.60206	402	.5289	.72341	452	.5947	.77432
306	.4026	.60491	403	.5303	.72449	453	.5961	.77528
308	.4053	.60774	404	.5316	.72557	454	.5974	.77624
310	0.4079	9.61055	405	0.5329	9.72664	455	0.5987	9.77720
312	.4105	.61334	406	.5342	.72771	456	.6000	.77815
314	.4132	.61612	407	.5355	.72878	457	.6013	.77910
316	.4158	.61887	408	.5369	.72985	458	.6026	.78005
318	.4184	.62161	409	.5382	.73091	459	.6040	.78100
320	0.4211	9.62434	410	0.5395	9.73197	460	0.6053	9.78194
322	.4237	.62704	411	.5408	.73303	461	.6066	.78289
324	.4263	.62973	412	.5421	.73408	462	.6079	.78383
326	.4289	.63240	413	.5434	.73514	463	.6092	.78477
328	.4316	.63506	414	.5447	.73619	464	.6105	.78570
330	0.4342	9.63770	415	0.5461	9.73723	465	0.6118	9.78664
332	.4368	.64032	416	.5474	.73828	466	.6132	.78757
334	.4395	.64293	417	.5487	.73932	467	.6145	.78850
336	.4421	.64552	418	.5500	.74036	468	.6158	.78943
338	.4447	.64810	419	.5513	.74140	469	.6171	.79036
340	0.4474	9.65066	420	0.5526	9.74244	470	0.6184	9.79128
342	.4500	.65321	421	.5540	.74347	471	.6197	.79221
344	.4526	.65574	422	.5553	.74450	472	.6210	.79313
346	.4553	.65826	423	.5566	.74553	473	.6224	.79405
348	.4579	.66076	424	.5579	.74655	474	.6237	.79496
350	0.4605	9.66325	425	0.5592	9.74758	475	0.6250	9.79588
352	.4632	.66573	426	.5605	.74860	476	.6263	.79679
354	.4658	.66819	427	.5618	.74961	477	.6276	.79770
356	.4684	.67064	428	.5632	.75063	478	.6289	.79861
358	.4711	.67307	429	.5645	.75164	479	.6303	.79952
360	0.4737	9.67549	430	0.5658	9.75265	480	0.6316	9.80043
362	.4763	.67790	431	.5671	.75366	481	.6329	.80133
364	.4789	.68029	432	.5684	.75467	482	.6342	.80223
366	.4816	.68267	433	.5697	.75567	483	.6355	.80313
368	.4842	.68503	434	.5711	.75668	484	.6368	.80403
370	0.4868	9.68739	435	0.5724	9.75768	485	0.6382	9.80493
372	.4895	.68973	436	.5737	.75867	486	.6395	.80582
374	.4921	.69206	437	.5750	.75967	487	.6408	.80672
376	.4947	.69437	438	.5763	.76066	488	.6421	.80761
378	.4974	.69668	439	.5776	.76165	489	.6434	.80850
380	0.5000	9.69897	440	0.5790	9.76264	490	0.6447	9.80938
382	.5026	.70125	441	.5803	.76362	491	.6461	.81027
384	.5053	.70352	442	.5816	.76461	492	.6474	.81115
386	.5079	.70577	443	.5829	.76559	493	.6487	.81203
388	.5105	.70802	444	.5842	.76657	494	.6500	.81291
390	0.5132	9.71025	445	0.5855	9.76755	495	0.6513	9.81379
392	.5158	.71247	446	.5868	.76852	496	.6526	.81467
394	.5184	.71468	447	.5882	.76949	497	.6540	.81556
396	.5211	.71688	448	.5895	.77046	498	.6553	.81642
398	.5237	.71907	449	.5908	.77143	499	.6566	.81729

SMITHSONIAN TABLES.

TABLE 86.

DENSITY OF AIR AT DIFFERENT HUMIDITIES AND PRESSURES.
METRIC MEASURES.

Values of $\dfrac{h}{760}$. $\dfrac{\delta}{\delta_0} = \dfrac{h}{760} = \dfrac{b - 0.378e}{760}$.

$b =$ Barometric pressure in mm.; $e =$ Vapor pressure in mm.

h.	$\dfrac{h}{760}$.	Log $\dfrac{h}{760}$.	h.	$\dfrac{h}{760}$.	Log $\dfrac{h}{760}$.	h.	$\dfrac{h}{760}$.	Log $\dfrac{h}{760}$.
mm.		−10	mm.		−10	mm.		−10
500	0.6579	9.81816	550	0.7237	9.85955	600	0.7895	9.89734
501	.6592	.81902	551	.7250	.86034	601	.7908	.89806
502	.6605	.81989	552	.7263	.86112	602	.7921	.89878
503	.6618	.82075	553	.7276	.86191	603	.7934	.89950
504	.6632	.82162	554	.7290	.86270	604	.7947	.90022
505	0.6645	9.82248	555	0.7303	9.86348	605	0.7961	9.90094
506	.6658	.82334	556	.7316	.86426	606	.7974	.90166
507	.6671	.82419	557	.7329	.86504	607	.7987	.90238
508	.6684	.82505	558	.7342	.86582	608	.8000	.90309
509	.6697	.82590	559	.7355	.86660	609	.8013	.90380
510	0.6711	9.82676	560	0.7368	9.86737	610	0.8026	9.90452
511	.6724	.82761	561	.7382	.86815	611	.8040	.90523
512	.6737	.82846	562	.7395	.86892	612	.8053	.90594
513	.6750	.82930	563	.7408	.86969	613	.8066	.90665
514	.6763	.83015	564	.7421	.87046	614	.8079	.90735
515	0.6776	9.83099	565	0.7434	9.87123	615	0.8092	9.90806
516	.6789	.83184	566	.7447	.87200	616	.8105	.90877
517	.6803	.83268	567	.7461	.87277	617	.8118	.90947
518	.6816	.83352	568	.7474	.87353	618	.8132	.91017
519	.6829	.83435	569	.7487	.87430	619	.8145	.91088
520	0.6842	9.83519	570	0.7500	9.87506	620	0.8158	9.91158
521	.6855	.83602	571	.7513	.87582	621	.8171	.91228
522	.6869	.83686	572	.7526	.87658	622	.8184	.91298
523	.6882	.83769	573	.7540	.87734	623	.8197	.91367
524	.6895	.83852	574	.7553	.87810	624	.8211	.91437
525	0.6908	9.83934	575	0.7566	9.87885	625	0.8224	9.91507
526	.6921	.84017	576	.7579	.87961	626	.8237	.91576
527	.6934	.84100	577	.7592	.88036	627	.8250	.91645
528	.6947	.84182	578	.7605	.88111	628	.8263	.91715
529	.6961	.84264	579	.7618	.88186	629	.8276	.91784
530	0.6974	9.84346	580	0.7632	9.88261	630	0.8289	9.91853
531	.6987	.84428	581	.7645	.88336	631	.8303	.91922
532	.7000	.84510	582	.7658	.88411	632	.8316	.91990
533	.7013	.84591	583	.7671	.88486	633	.8329	.92059
534	.7026	.84673	584	.7684	.88560	634	.8342	.92128
535	0.7040	9.84754	585	0.7697	9.88634	635	0.8355	9.92196
536	.7053	.84835	586	.7711	.88708	636	.8368	.92264
537	.7066	.84916	587	.7724	.88782	637	.8382	.92332
538	.7079	.84997	588	.7737	.88856	638	.8395	.92401
539	.7092	.85078	589	.7750	.88930	639	.8408	.92469
540	0.7105	9.85158	590	0.7763	9.89004	640	0.8421	9.92537
541	.7118	.85238	591	.7776	.89077	641	.8434	.92604
542	.7132	.85318	592	.7789	.89151	642	.8447	.92672
543	.7145	.85399	593	.7803	.89224	643	.8461	.92740
544	.7158	.85478	594	.7816	.89297	644	.8474	.92807
545	0.7171	9.85558	595	0.7829	9.89370	645	0.8487	9.92875
546	.7184	.85638	596	.7842	.89443	646	.8500	.92942
547	.7197	.85717	597	.7855	.89516	647	.8513	.93009
548	.7211	.85797	598	.7868	.89589	648	.8526	.93076
549	.7224	.85876	599	.7882	.89662	649	.8539	.93143

SMITHSONIAN TABLES.

TABLE 86.
DENSITY OF AIR AT DIFFERENT HUMIDITIES AND PRESSURES.
METRIC MEASURES.

Values of $\frac{h}{760}$. $\qquad \frac{\delta}{\delta_0} = \frac{h}{760} = \frac{b - 0.378e}{760}$.

$b =$ Barometric pressure in mm.; $e =$ Vapor pressure in mm.

h.	$\frac{h}{760}$.	Log $\frac{h}{760}$.	h.	$\frac{h}{760}$.	Log $\frac{h}{760}$.	h.	$\frac{h}{760}$.	Log $\frac{h}{760}$.
mm.		− 10	mm.		− 10	mm.		− 10
650	0.8553	9.93210	700	0.9211	9.96428	750	0.9868	9.99425
651	.8566	.93277	701	.9224	.96490	751	.9882	.99483
652	.8579	.93341	702	.9237	.96552	752	.9895	.99540
653	.8592	.93410	703	.9250	.96614	753	.9908	.99598
654	.8605	.93476	704	.9263	.96676	754	.9921	.99656
655	0.8618	9.93543	705	0.9276	9.96738	755	0.9934	9.99713
656	.8632	.93609	706	.9289	.96799	756	.9947	.99771
657	.8645	.93675	707	.9303	.96860	757	.9961	.99828
658	.8658	.93741	708	.9316	.96922	758	.9974	.99886
659	.8671	.93807	709	.9329	.96983	759	.9987	.99943
660	0.8684	9.93873	710	0.9342	9.97044	760	1.0000	0.00000
661	.8697	.93939	711	.9355	.97106	761	.0013	.00057
662	.8711	.94004	712	.9368	.97167	762	.0026	.00114
663	.8724	.94070	713	.9382	.97228	763	.0039	.00171
664	.8737	.94135	714	.9395	.97288	764	.0053	.00228
665	0.8750	9.94201	715	0.9408	9.97349	765	1.0066	0.00285
666	.8763	.94266	716	.9421	.97410	766	.0079	.00342
667	.8776	.94331	717	.9434	.97470	767	.0092	.00398
668	.8790	.94396	718	.9447	.97531	768	.0105	.00455
669	.8803	.94461	719	.9461	.97592	769	.0118	.00511
670	0.8816	9.94526	720	0.9474	9.97652	770	1.0132	0.00568
671	.8829	.94591	721	.9487	.97712	771	.0145	.00624
672	.8842	.94656	722	.9500	.97772	772	.0158	.00680
673	.8855	.94720	723	.9513	.97832	773	.0171	.00736
674	.8869	.94785	724	.9526	.97892	774	.0184	.00793
675	0.8882	9.94849	725	0.9539	9.97952	775	1.0197	0.00849
676	.8895	.94913	726	.9553	.98012	776	.0211	.00905
677	.8908	.94978	727	.9566	.98072	777	.0224	.00961
678	.8921	.95042	728	.9579	.98132	778	.0237	.01017
679	.8934	.95106	729	.9592	.98191	779	.0250	.01072
680	0.8947	9.95170	730	0.9605	9.98250	780	1.0263	0.01128
681	.8960	.95233	731	.9618	.98310	781	.0276	.01184
682	.8974	.95297	732	.9632	.98370	782	.0289	.01239
683	.8987	.95361	733	.9645	.98429	783	.0303	.01295
684	.9000	.95424	734	.9658	.98488	784	.0316	.01350
685	0.9013	9.95488	735	0.9671	9.98547	785	1.0329	0.01406
686	.9026	.95551	736	.9684	.98606	786	.0342	.01461
687	.9039	.95614	737	.9697	.98665	787	.0355	.01516
688	.9053	.95677	738	.9711	.98724	788	.0368	.01571
689	.9066	.95740	739	.9724	.98783	789	.0382	.01626
690	0.9079	9.95804	740	0.9737	9.98842	790	1.0395	0.01681
691	.9092	.95866	741	.9750	.98900	791	.0408	.01736
692	.9105	.95929	742	.9763	.98959	792	.0421	.01791
693	.9118	.95992	743	.9776	.99018	793	.0434	.01846
694	.9132	.96054	744	.9789	.99076	794	.0447	.01901
695	0.9145	9.96117	745	0.9803	9.99134	795	1.0461	0.01955
696	.9158	.96180	746	.9816	.99192	796	.0474	.02010
697	.9171	.96242	747	.9829	.99251	797	.0487	.02064
698	.9184	.96304	748	.9842	.99309	798	.0500	.02119
699	.9197	.96366	749	.9855	.99367	799	.0513	.02173

SMITHSONIAN TABLES.

TABLE 87.

AVOIRDUPOIS POUNDS AND OUNCES INTO KILOGRAMMES.

1 avoirdupois pound = 0.4535924 kilogramme.
1 avoirdupois ounce = 0.0283495 kilogramme.

Pounds.	.0	.1	.2	.3	.4	.5	.6	.7	.8	.9
	kg.	kg.	kg.	kg.	kg.	kg.	kg.	kg.	kg.	kg.
0	0.0000	0.0454	0.0907	0.1361	0.1814	0.2268	0.2722	0.3175	0.3629	0.4082
1	0.4536	0.4990	0.5443	0.5897	0.6350	0.6804	0.7257	0.7711	0.8165	0.8618
2	0.9072	0.9525	0.9979	1.0433	1.0886	1.1340	1.1793	1.2247	1.2701	1.3154
3	1.3608	1.4061	1.4515	1.4969	1.5422	1.5876	1.6329	1.6783	1.7237	1.7690
4	1.8144	1.8597	1.9051	1.9504	1.9958	2.0412	2.0865	2.1319	2.1772	2.2226
5	2.2680	2.3133	2.3587	2.4040	2.4494	2.4948	2.5401	2.5855	2.6308	2.6762
6	2.7216	2.7669	2.8123	2.8576	2.9030	2.9484	2.9937	3.0391	3.0844	3.1298
7	3.1751	3.2205	3.2659	3.3112	3.3566	3.4019	3.4473	3.4927	3.5380	3.5834
8	3.6287	3.6741	3.7195	3.7648	3.8102	3.8555	3.9009	3.9463	3.9916	4.0370
9	4.0823	4.1277	4.1731	4.2184	4.2638	4.3091	4.3545	4.3998	4.4452	4.4906

Ounces.	.0	.1	.2	.3	.4	.5	.6	.7	.8	.9
	kg.	kg.	kg.	kg.	kg.	kg.	kg.	kg.	kg.	kg.
0	0.0000	0.0028	0.0057	0.0085	0.0113	0.0142	0.0170	0.0198	0.0227	0.0255
1	.0283	.0312	.0340	.0369	.0397	.0425	.0454	.0482	.0510	.0539
2	.0567	.0595	.0624	.0652	.0680	.0709	.0737	.0765	.0794	.0822
3	.0850	.0879	.0907	.0936	.0964	.0992	.1021	.1049	.1077	.1106
4	.1134	.1162	.1191	.1219	.1247	.1276	.1304	.1332	.1361	.1389
5	0.1417	0.1446	0.1474	0.1503	0.1531	0.1559	0.1588	0.1616	0.1644	0.1673
6	.1701	.1729	.1758	.1786	.1814	.1843	.1871	.1899	.1928	.1956
7	.1984	.2013	.2041	.2070	.2098	.2126	.2155	.2183	.2211	.2240
8	.2268	.2296	.2325	.2353	.2381	.2410	.2438	.2466	.2495	.2523
9	.2551	.2580	.2608	.2637	.2665	.2693	.2722	.2750	.2778	.2807
10	0.2835	0.2863	0.2892	0.2920	0.2948	0.2977	0.3005	0.3033	0.3062	0.3090
11	.3118	.3147	.3175	.3203	.3232	.3260	.3289	.3317	.3345	.3374
12	.3402	.3430	.3459	.3487	.3515	.3544	.3572	.3600	.3629	.3657
13	.3685	.3714	.3742	.3770	.3799	.3827	.3856	.3884	.3912	.3941
14	.3969	.3997	.4026	.4054	.4082	.4111	.4139	.4167	.4196	.4224
15	.4252	.4281	.4309	.4337	.4366	.4394	.4423	.4451	.4479	.4508

TABLE 88.
KILOGRAMMES INTO AVOIRDUPOIS POUNDS AND OUNCES.
1 kilogramme = 2.204622 avoirdupois pounds.

Kilo-grammes.	0.0	0.1	0.2	0.3	0.4	0.5	0.6	0.7	0.8	0.9
	Av. lbs.	Av. lbs.	Av. lbs.	Av. lbs.	Av. lbs.	Av. lbs.	Av. lbs.	Av. lbs.	Av. lbs.	Av. lbs.
0	0.000	0.220	0.441	0.661	0.882	1.102	1.323	1.543	1.764	1.984
1	2.205	2.425	2.646	2.866	3.086	3.307	3.527	3.748	3.968	4.189
2	4.409	4.630	4.850	5.071	5.291	5.512	5.732	5.952	6.173	6.393
3	6.614	6.834	7.055	7.275	7.496	7.716	7.937	8.157	8.378	8.598
4	8.818	9.039	9.259	9.480	9.700	9.921	10.141	10.362	10.582	10.803
5	11.023	11.244	11.464	11.684	11.905	12.125	12.346	12.566	12.787	13.007
6	13.228	13.448	13.669	13.889	14.110	14.330	14.551	14.771	14.991	15.212
7	15.432	15.653	15.873	16.094	16.314	16.535	16.755	16.976	17.196	17.417
8	17.637	17.857	18.078	18.298	18.519	18.739	18.960	19.180	19.401	19.621
9	19.842	20.062	20.283	20.503	20.723	20.944	21.164	21.385	21.605	21.826

Tenths of a Kilogramme into Ounces.				Hundredths of a Kilogramme into Decimals of a Pound and Ounces.			
kg.	Oz.	kg.	Oz.	kg.	Av. lbs. Oz.	kg.	Av. lbs. Oz.
0.1	3.5274	0.6	21.1644	0.01	0.022 = 0.35	0.06	0.132 = 2.12
.2	7.0548	.7	24.6918	.02	.044 = 0.71	.07	.154 = 2.47
.3	10.5822	.8	28.2192	.03	.066 = 1.06	.08	.176 = 2.82
.4	14.1096	.9	31.7466	.04	.088 = 1.41	.09	.198 = 3.17
.5	17.6370	1.0	35.2740	.05	.110 = 1.76	.10	.220 = 3.53

TABLE 89.
GRAINS INTO GRAMMES.
1 grain = 0.06479892 gramme.

Grains.	0	1	2	3	4	5	6	7	8	9
	gram's.	gram's.	gram's.	gram's.	gram's.	gram's.	gram's.	gram's.	gram's.	gram's.
0	0.0000	0.0648	0.1296	0.1944	0.2592	0.3240	0.3888	0.4536	0.5184	0.5832
10	0.6480	0.7128	0.7776	0.8424	0.9072	0.9720	1.0368	1.1016	1.1664	1.2312
20	1.2960	1.3608	1.4256	1.4904	1.5552	1.6200	1.6848	1.7496	1.8144	1.8792
30	1.9440	2.0088	2.0736	2.1384	2.2032	2.2680	2.3328	2.3976	2.4624	2.5272
40	2.5920	2.6568	2.7216	2.7864	2.8512	2.9160	2.9808	3.0456	3.1103	3.1751
50	3.2399	3.3047	3.3695	3.4343	3.4991	3.5639	3.6287	3.6935	3.7583	3.8231
60	3.8879	3.9527	4.0175	4.0823	4.1471	4.2119	4.2767	4.3415	4.4063	4.4711
70	4.5359	4.6007	4.6655	4.7303	4.7951	4.8599	4.9247	4.9895	5.0543	5.1191
80	5.1839	5.2487	5.3135	5.3783	5.4431	5.5079	5.5727	5.6375	5.7023	5.7671
90	5.8319	5.8967	5.9615	6.0263	6.0911	6.1559	6.2207	6.2855	6.3503	6.4151

Tenths of a Grain.				Hundredths of a Grain.			
Grain.	gramme.	Grain.	gramme.	Grain.	gramme.	Grain.	gramme.
0.1	0.0065	0.6	0.0389	0.01	0.0006	0.06	0.0039
.2	.0130	.7	.0454	.02	.0013	.07	.0045
.3	.0194	.8	.0518	.03	.0019	.08	.0052
.4	.0259	.9	.0583	.04	.0026	.09	.0058
.5	.0324	1.0	.0648	.05	.0032	.10	.0065

SMITHSONIAN TABLES.

TABLE 90.

GRAMMES INTO GRAINS.

1 gramme = 15.432351 grains.

Grammes.	.0	.1	.2	.3	.4	.5	.6	.7	.8	.9
	Grains.	Grains.	Grains.	Grains.	Grains.	Grains.	Grains.	Grains.	Grains.	Grains.
0	0.00	1.54	3.09	4.63	6.17	7.72	9.26	10.80	12.35	13.89
1	15.43	16.98	18.52	20.06	21.61	23.15	24.69	26.24	27.78	29.32
2	30.86	32.41	33.95	35.49	37.04	38.58	40.12	41.67	43.21	44.75
3	46.30	47.84	49.38	50.93	52.47	54.01	55.56	57.10	58.64	60.19
4	61.73	63.27	64.82	66.36	67.90	69.45	70.99	72.53	74.08	75.62
5	77.16	78.71	80.25	81.79	83.33	84.88	86.42	87.96	89.51	91.05
6	92.59	94.14	95.68	97.22	98.77	100.31	101.85	103.40	104.94	106.48
7	108.03	109.57	111.11	112.66	114.20	115.74	117.29	118.83	120.37	121.92
8	123.46	125.00	126.55	128.09	129.63	131.18	132.72	134.26	135.80	137.35
9	138.89	140.43	141.98	143.52	145.06	146.61	148.15	149.69	151.24	152.78

	0	1	2	3	4	5	6	7	8	9
	Grains.	Grains.	Grains.	Grains.	Grains.	Grains.	Grains.	Grains.	Grains.	Grains.
0	0.00	15.43	30.86	46.30	61.73	77.16	92.59	108.03	123.46	138.89
10	154.32	169.76	185.19	200.62	216.05	231.49	246.92	262.35	277.78	293.21
20	308.65	324.08	339.51	354.94	370.38	385.81	401.24	416.67	432.11	447.54
30	462.97	478.40	493.84	509.27	524.70	540.13	555.56	571.00	586.43	601.86
40	617.29	632.73	648.16	663.59	679.02	694.46	709.89	725.32	740.75	756.19
50	771.62	787.05	802.48	817.91	833.35	848.78	864.21	879.64	895.08	910.51
60	925.94	941.37	956.81	972.24	987.67	1003.10	1018.54	1033.97	1049.40	1064.83
70	1080.26	1095.70	1111.13	1126.56	1141.99	1157.43	1172.86	1188.29	1203.72	1219.16
80	1234.59	1250.02	1265.45	1280.89	1296.32	1311.75	1327.18	1342.62	1358.05	1373.48
90	1388.91	1404.34	1419.78	1435.21	1450.64	1466.07	1481.51	1496.94	1512.37	1527.80

gramme.	Grain.	gramme.	Grain.	gramme.	Grain.	gramme.	Grain.
0.01	0.154	0.06	0.926	0.001	0.015	0.006	0.093
.02	.309	.07	1.080	.002	.031	.007	.108
.03	.463	.08	1.235	.003	.046	.008	.123
.04	.617	.09	1.389	.004	.062	.009	.139
.05	.772	.10	1.543	.005	.077	.010	.154

TABLE 91.

CONVERSION OF UNITS OF MAGNETIC INTENSITY.

English Units.	Dynes.	Dynes.	English Units.
1	0.046 108	0.1	2.168 82
2	.092 216	.2	4.337 64
3	.138 324	.3	6.506 46
4	.184 432	.4	8.675 28
5	0.230 540	0.5	10.844 10
6	.276 648	.6	13.012 92
7	.322 756	.7	15.181 74
8	.368 864	.8	17.350 56
9	.414 972	.9	19.519 38

The *English unit* of magnetic intensity is the force which acting for 1 second on a unit of magnetism, associated with a mass of 1 grain, produces a velocity of 1 foot per second.

The *C. G. S.* unit of magnetic intensity is the *dyne*—the force which, acting on one gramme for one second, generates a velocity of 1 centimetre per second.

The dimensions of magnetic intensity are $[M^{\frac{1}{2}}/L^{\frac{1}{2}}T]$.

TABLE 92.

QUANTITY OF RAINFALL CORRESPONDING TO GIVEN DEPTHS.

1 inch of rainfall = 22624.0417 imperial gallons per acre.   1 inch of rainfall = 113.3068 tons per acre.
= 226613.713 lbs. per acre.   = 72516.3878 tons per sq. mile.

Depth of Rainfall.	Imperial Gallons per acre.	Tons per square Mile.	Depth of Rainfall.	Imperial Gallons per acre.	Tons per square Mile.	Depth of Rainfall.	Imperial Gallons per acre.	Tons per square Mile.
Inches.			Inches.			Inches.		
0.00	—	—	0.20	4524.80	14503.27	0.40	9049.61	29006.55
.01	226.24	725.16	.21	4751.04	15228.44	.41	9275.85	29731.71
.02	452.48	1450.32	.22	4977.28	15953.60	.42	9502.09	30456.88
.03	678.72	2175.49	.23	5203.52	16678.76	.43	9728.33	31182.04
.04	904.96	2900.65	.24	5429.77	17403.93	.44	9954.57	31907.21
0.05	1131.20	3625.81	0.25	5656.01	18129.09	0.45	10180.81	32632.37
.06	1357.44	4350.98	.26	5882.25	18854.26	.46	10407.05	33357.53
.07	1583.68	5076.14	.27	6108.49	19579.42	.47	10633.29	34082.70
.08	1809.92	5801.31	.28	6334.73	20304.58	.48	10859.53	34807.86
.09	2036.16	6526.47	.29	6560.97	21029.75	.49	11085.77	35533.03
0.10	2262.40	7251.63	0.30	6787.21	21754.91	0.50	11312.02	36258.19
.11	2488.64	7976.80	.31	7013.45	22480.08	.60	13574.42	43509.83
.12	2714.88	8701.96	.32	7239.69	23205.24	.70	15836.82	50761.47
.13	2941.12	9427.13	.33	7465.93	23930.40	.80	18099.23	58013.11
.14	3167.36	10152.29	.34	7692.17	24655.57	.90	20361.63	65264.74
0.15	3393.60	10877.45	0.35	7918.41	25380.73	1.00	22624.04	72516.38
.16	3619.84	11602.62	.36	8144.65	26105.89	2.00	45248.08	145032.77
.17	3846.08	12327.78	.37	8370.89	26831.06	3.00	67872.12	217549.16
.18	4072.32	13052.94	.38	8597.13	27556.22	4.00	90496.16	290065.55
.19	4298.56	13778.11	.39	8823.37	28281.39	5.00	113120.20	362581.93
0.20	4524.80	14503.27	0.40	9049.61	29006.55	6.00	135744.24	435098.32

TABLE 93.

DATES OF DOVE'S PENTADES.

No. of Pentade.	Epoch of the Year.	No. of Pentade.	Epoch of the Year.	No. of Pentade.	Epoch of the Year.	No. of Pentade.	Epoch of the Year.
1	Jan. 1 to 5	19	Apr. 1 to 5	37	June 30 to July 4	55	Sept. 28 to Oct. 2
2	6 10	20	6 10	38	July 5 9	56	Oct. 3 7
3	11 15	21	11 15	39	10 14	57	8 12
4	16 20	22	16 20	40	15 19	58	13 17
5	Jan. 21 to 25	23	Apr. 21 to 25	41	July 20 to 24	59	Oct. 18 to 22
6	26 30	24	26 30	42	25 29	60	23 27
7	31 Feb. 4	25	May 1 5	43	30 Aug. 3	61	28 Nov. 1
8	Feb. 5 9	26	6 10	44	Aug. 4 8	62	Nov. 2 6
9	10 14	27	11 15	45	9 13	63	7 11
10	Feb. 15 to 19	28	May 16 to 20	46	Aug. 14 to 18	64	Nov. 12 to 16
11	20 24	29	21 25	47	19 23	65	17 21
12*	25 Mar. 1	30	26 30	48	24 28	66	22 26
13	Mar. 2 6	31	31 June 4	49	29 Sept. 2	67	27 Dec. 1
14	7 11	32	June 5 9	50	Sept. 3 7	68	Dec. 2 6
						69	7 11
15	Mar. 12 to 16	33	June 10 to 14	51	Sept. 8 to 12	70	12 16
16	17 21	34	15 19	52	13 17	71	17 21
17	22 26	35	20 24	53	18 22	72	22 26
18	27 31	36	25 29	54	23 27	73	27 31

*In the bissextile year the 12th pentade contains six days.

TABLE 94
DIVISION BY 28 OF NUMBERS FROM 28 TO 867972.

	0	100	200	300	400	500	600	700	800	900						
0		28	56	84	112	140	168	196	224	252	D. Q.	00 00	28 01	56 02	84 03	840 30
1		29	57	85	113	141	169	197	225	253	D. Q.	12 04	40 05	68 06	96 07	812 29
2		30	58	86	114	142	170	198	226	254	D. Q.	24 08	52 09	80 10		784 28
3		31	59	87	115	143	171	199	227	255	D. Q.	08 11	36 12	64 13	92 14	756 27
4		32	60	88	116	144	172	200	228	256	D. Q.	20 15	48 16	76 17		728 26
5		33	61	89	117	145	173	201	229	257	D. Q.	04 18	32 19	60 20	88 21	700 25
6		34	62	90	118	146	174	202	230	258	D. Q.	16 22	44 23	72 24		672 24
7		35	63	91	119	147	175	203	231	259	D. Q.	00 25	28 26	56 27	84 28	644 23
8		36	64	92	120	148	176	204	232	260	D. Q.	12 29	40 30	68 31	96 32	616 22
9		37	65	93	121	149	177	205	233	261	D. Q.	24 33	52 34	80 35		588 21
10		38	66	94	122	150	178	206	234	262	D. Q.	08 36	36 37	64 38	92 39	560 20
11		39	67	95	123	151	179	207	235	263	D. Q.	20 40	48 41	76 42		532 19
0		100	200	300	400	500	600	700	800	900						
12		40	68	96	124	152	180	208	236	264	D. Q.	04 43	32 44	60 45	88 46	504 18
13		41	69	97	125	153	181	209	237	265	D. Q.	16 47	44 48	72 49		476 17
14		42	70	98	126	154	182	210	238	266	D. Q.	00 50	28 51	56 52	84 53	448 16
15		43	71	99	127	155	183	211	239	267	D. Q.	12 54	40 55	68 56	96 57	420 15
16		44	72	100	128	156	184	212	240	268	D. Q.	24 58	52 59	80 60		392 14
17		45	73	101	129	157	185	213	241	269	D. Q.	08 61	36 62	64 63	92 64	364 13
18		46	74	102	130	158	186	214	242	270	D. Q.	20 65	48 66	76 67		336 12
19		47	75	103	131	159	187	215	243	271	D. Q.	04 68	32 69	60 70	88 71	308 11
20		48	76	104	132	160	188	216	244	272	D. Q.	16 72	44 73	72 74		280 10
21		49	77	105	133	161	189	217	245	273	D. Q.	00 75	28 76	56 77	84 78	252 9
22		50	78	106	134	162	190	218	246	274	D. Q.	12 79	40 80	68 81	96 82	224 8
23		51	79	107	135	163	191	219	247	275	D. Q.	24 83	52 84	80 85		196 7
24		52	80	108	136	164	192	220	248	276	D. Q.	08 86	36 87	64 88	92 89	168 6
25		53	81	109	137	165	193	221	249	277	D. Q.	20 90	48 91	76 92		140 5
26		54	82	110	138	166	194	222	250	278	D. Q.	04 93	32 94	60 95	88 96	112 4
27		55	83	111	139	167	195	223	251	279	D. Q.	16 97	44 98	72 99		84 3
0		100	200	300	400	500	600	700	800	900						

SMITHSONIAN TABLES.

TABLE 95.
DIVISION BY 29 OF NUMBERS FROM 29 TO 6989/1.

	0	100	200	300	400	500	600	700	800	900						
0		29	58	87	116	145	174	203	232	261	D.	00	29	58	87	870
											Q.	00	01	02	03	30
1		30	59	88	117	146	175	204	233	262	D.	16	45	74		841
											Q.	04	05	06		29
2		31	60	89	118	147	176	205	234	263	D.	03	32	61	90	812
											Q.	07	08	09	10	28
3		32	61	90	119	148	177	206	235	264	D.	19	48	77		783
											Q.	11	12	13		27
4		33	62	91	120	149	178	207	236	265	D.	06	35	64	93	754
											Q.	14	15	16	17	26
5		34	63	92	121	150	179	208	237	266	D.	22	51	80		725
											Q.	18	19	20		25
6		35	64	93	122	151	180	209	238	267	D.	09	38	67	96	696
											Q.	21	22	23	24	24
7		36	65	94	123	152	181	210	239	268	D.	25	54	83		667
											Q.	25	26	27		23
8		37	66	95	124	153	182	211	240	269	D.	12	41	70	99	638
											Q.	28	29	30	31	22
9		38	67	96	125	154	183	212	241	270	D.	28	57	86		609
											Q.	32	33	34		21
10		39	68	97	126	155	184	213	242	271	D.	15	44	73		580
											Q.	35	36	37		20
11		40	69	98	127	156	185	214	243	272	D.	02	31	60	89	551
											Q.	38	39	40	41	19
12		41	70	99	128	157	186	215	244	273	D.	18	47	76		522
											Q.	42	43	44		18
13		42	71	100	129	158	187	216	245	274	D.	05	34	63	92	493
											Q.	45	46	47	48	17
14		43	72	101	130	159	188	217	246	275	D.	21	50	79		464
0		100	200	300	400	500	600	700	800	900	Q.	49	50	51		16
15		44	73	102	131	160	189	218	247	276	D.	08	37	66	95	435
											Q.	52	53	54	55	15
16		45	74	103	132	161	190	219	248	277	D.	24	53	82		406
											Q.	56	57	58		14
17		46	75	104	133	162	191	220	249	278	D.	11	40	69	98	377
											Q.	59	60	61	62	13
18		47	76	105	134	163	192	221	250	279	D.	27	56	85		348
											Q.	63	64	65		12
19		48	77	106	135	164	193	222	251	280	D.	14	43	72		319
											Q.	66	67	68		11
20		49	78	107	136	165	194	223	252	281	D.	01	30	59	88	290
											Q.	69	70	71	72	10
21		50	79	108	137	166	195	224	253	282	D.	17	46	75		261
											Q.	73	74	75		9
22		51	80	109	138	167	196	225	254	283	D.	04	33	62	91	232
											Q.	76	77	78	79	8
23		52	81	110	139	168	197	226	255	284	D.	20	49	78		203
											Q.	80	81	82		7
24		53	82	111	140	169	198	227	256	285	D.	07	36	65	94	174
											Q.	83	84	85	86	6
25		54	83	112	141	170	199	228	257	286	D.	23	52	81		145
											Q.	87	88	89		5
26		55	84	113	142	171	200	229	258	287	D.	10	39	68	97	116
											Q.	90	91	92	93	4
27		56	85	114	143	172	201	230	259	288	D.	26	55	84		87
											Q.	94	95	96		3
28		57	86	115	144	173	202	231	260	289	D.	13	42	71		58
0		100	200	300	400	500	600	700	800	900	Q.	97	98	99		2

SMITHSONIAN TABLES.

DIVISION BY 31 OF NUMBERS FROM 31 TO 960969.

	0	100	200	300	400	500	600	700	800	900						
0		31	62	93	124	155	186	217	248	279	D.	00	31	62	93	930
											Q.	00	01	02	03	30
1		32	63	94	125	156	187	218	249	280	D.	24	55	86		899
											Q.	04	05	06		29
2		33	64	95	126	157	188	219	250	281	D.	17	48	79		868
											Q.	07	08	09		28
3		34	65	96	127	158	189	220	251	282	D.	10	41	72		837
											Q.	10	11	12		27
4		35	66	97	128	159	190	221	252	283	D.	03	34	65	96	806
											Q.	13	14	15	16	26
5		36	67	98	129	160	191	222	253	284	D.	27	58	89		775
											Q.	17	18	19		25
6		37	68	99	130	161	192	223	254	285	D.	20	51	82		744
											Q.	20	21	22		24
7		38	69	100	131	162	193	224	255	286	D.	13	44	75		713
											Q.	23	24	25		23
8		39	70	101	132	163	194	225	256	287	D.	06	37	68	99	682
											Q.	26	27	28	29	22
9		40	71	102	133	164	195	226	257	288	D.	30	61	92		651
											Q.	30	31	32		21
10		41	72	103	134	165	196	227	258	289	D.	23	54	85		620
											Q.	33	34	35		20
11		42	73	104	135	166	197	228	259	290	D.	16	47	78		589
											Q.	36	37	38		19
12		43	74	105	136	167	198	229	260	291	D.	09	40	71		558
											Q.	39	40	41		18
13		44	75	106	137	168	199	230	261	292	D.	02	33	64	95	527
											Q.	42	43	44	45	17
14		45	76	107	138	169	200	231	262	293	D.	26	57	88		496
											Q.	46	47	48		16
15		46	77	108	139	170	201	232	263	294	D.	19	50	81		465
0		100	200	300	400	500	600	700	800	900	Q.	49	50	51		15
16		47	78	109	140	171	202	233	264	295	D.	12	43	74		434
											Q.	52	53	54		14
17		48	79	110	141	172	203	234	265	296	D.	05	36	67	98	403
											Q.	55	56	57	58	13
18		49	80	111	142	173	204	235	266	297	D.	29	60	91		372
											Q.	59	60	61		12
19		50	81	112	143	174	205	236	267	298	D.	22	53	84		341
											Q.	62	63	64		11
20		51	82	113	144	175	206	237	268	299	D.	15	46	77		310
											Q.	65	66	67		10
21		52	83	114	145	176	207	238	269	300	D.	08	39	70		279
											Q.	68	69	70		9
22		53	84	115	146	177	208	239	270	301	D.	01	32	63	94	248
											Q.	71	72	73	74	8
23		54	85	116	147	178	209	240	271	302	D.	25	56	87		217
											Q.	75	76	77		7
24		55	86	117	148	179	210	241	272	303	D.	18	49	80		186
											Q.	78	79	80		6
25		56	87	118	149	180	211	242	273	304	D.	11	42	73		155
											Q.	81	82	83		5
26		57	88	119	150	181	212	243	274	305	D.	04	35	66	97	124
											Q.	84	85	86	87	4
27		58	89	120	151	182	213	244	275	306	D.	28	59	90		93
											Q.	88	89	90		3
28		59	90	121	152	183	214	245	276	307	D.	21	52	83		62
											Q.	91	92	93		2
29		60	91	122	153	184	215	246	277	308	D.	14	45	76		31
											Q.	94	95	96		1
30		61	92	123	154	185	216	247	278	309	D.	07	38	69		
0		100	200	300	400	500	600	700	800	900	Q.	97	98	99		

SMITHSONIAN TABLES.

TABLE 97.
NATURAL SINES AND COSINES.
Natural Sines.

Angle.	0′	10′	20′	30′	40′	50′	60′	Angle.	Prop. Parts for 1′.
0°	.0000 00	.0029 09	.0058 18	.0087 27	.0116 35	.0145 44	.0174 52	89°	
1	.0174 52	.0203 6	.0232 7	.0261 8	.0290 8	.0319 9	.0349 0	88	
2	.0349 0	.0378 1	.0407 1	.0436 2	.0465 3	.0494 3	.0523 4	87	
3	.0523 4	.0552 4	.0581 4	.0610 5	.0639 5	.0668 5	.0697 6	86	
4	.0697 6	.0726 6	.0755 6	.0784 6	.0813 6	.0842 6	.0871 6	85	
5	.0871 6	.0900 5	.0929 5	.0958 5	.0987 4	.1016 4	.1045 3	84	
6	.1045 3	.1074 2	.1103 1	.1132 0	.1160 9	.1189 8	.1218 7	83	
7	.1218 7	.1247 6	.1276 4	.1305 3	.1334	.1363	.1392	82	
8	.1392	.1421	.1449	.1478	.1507	.1536	.1564	81	2.9
9	.1564	.1593	.1622	.1650	.1679	.1708	.1736	80	2.9
10	.1736	.1765	.1794	.1822	.1851	.1880	.1908	79	2.9
11	.1908	.1937	.1965	.1994	.2022	.2051	.2079	78	2.9
12	.2079	.2108	.2136	.2164	.2193	.2221	.2250	77	2.8
13	.2250	.2278	.2306	.2334	.2363	.2391	.2419	76	2.8
14	.2419	.2447	.2476	.2504	.2532	.2560	.2588	75	2.8
15	.2588	.2616	.2644	.2672	.2700	.2728	.2756	74	2.8
16	.2756	.2784	.2812	.2840	.2868	.2896	.2924	73	2.8
17	.2924	.2952	.2979	.3007	.3035	.3062	.3090	72	2.8
18	.3090	.3118	.3145	.3173	.3201	.3228	.3256	71	2.8
19	.3256	.3283	.3311	.3338	.3365	.3393	.3420	70	2.7
20	.3420	.3448	.3475	.3502	.3529	.3557	.3584	69	2.7
21	.3584	.3611	.3638	.3665	.3692	.3719	.3746	68	2.7
22	.3746	.3773	.3800	.3827	.3854	.3881	.3907	67	2.7
23	.3907	.3934	.3961	.3987	.4014	.4041	.4067	66	2.7
24	.4067	.4094	.4120	.4147	.4173	.4200	.4226	65	2.7
25	.4226	.4253	.4279	.4305	.4331	.4358	.4384	64	2.6
26	.4384	.4410	.4436	.4462	.4488	.4514	.4540	63	2.6
27	.4540	.4566	.4592	.4617	.4643	.4669	.4695	62	2.6
28	.4695	.4720	.4746	.4772	.4797	.4823	.4848	61	2.6
29	.4848	.4874	.4899	.4924	.4950	.4975	.5000	60	2.5
30	.5000	.5025	.5050	.5075	.5100	.5125	.5150	59	2.5
31	.5150	.5175	.5200	.5225	.5250	.5275	.5299	58	2.5
32	.5299	.5324	.5348	.5373	.5398	.5422	.5446	57	2.5
33	.5446	.5471	.5495	.5519	.5544	.5568	.5592	56	2.4
34	.5592	.5616	.5640	.5664	.5688	.5712	.5736	55	2.4
35	.5736	.5760	.5783	.5807	.5831	.5854	.5878	54	2.4
36	.5878	.5901	.5925	.5948	.5972	.5995	.6018	53	2.3
37	.6018	.6041	.6065	.6088	.6111	.6134	.6157	52	2.3
38	.6157	.6180	.6202	.6225	.6248	.6271	.6293	51	2.3
39	.6293	.6316	.6338	.6361	.6383	.6406	.6428	50	2.3
40	.6428	.6450	.6472	.6494	.6517	.6539	.6561	49	2.2
41	.6561	.6583	.6604	.6626	.6648	.6670	.6691	48	2.2
42	.6691	.6713	.6734	.6756	.6777	.6799	.6820	47	2.2
43	.6820	.6841	.6862	.6884	.6905	.6926	.6947	46	2.1
44	.6947	.6967	.6988	.7009	.7030	.7050	.7071	45	2.1
	60′	50′	40′	30′	20′	10′	0′	Angle.	

Natural Cosines.

NATURAL SINES AND COSINES.
Natural Sines.

TABLE 97.

Angle.	0′	10′	20′	30′	40′	50′	60′	Angle.	Prop. Parts for 1′.
45°	.7071	.7092	.7112	.7133	.7153	.7173	.7193	44°	2.0
46	.7193	.7214	.7234	.7254	.7274	.7294	.7314	43	2.0
47	.7314	.7333	.7353	.7373	.7392	.7412	.7431	42	2.0
48	.7431	.7451	.7470	.7490	.7509	.7528	.7547	41	1.9
49	.7547	.7566	.7585	.7604	.7623	.7642	.7660	40	1.9
50	.7660	.7679	.7698	.7716	.7735	.7753	.7771	39	1.9
51	.7771	.7790	.7808	.7826	.7844	.7862	.7880	38	1.8
52	.7880	.7898	.7916	.7934	.7951	.7969	.7986	37	1.8
53	.7986	.8004	.8021	.8039	.8056	.8073	.8090	36	1.7
54	.8090	.8107	.8124	.8141	.8158	.8175	.8192	35	1.7
55	.8192	.8208	.8225	.8241	.8258	.8274	.8290	34	1.6
56	.8290	.8307	.8323	.8339	.8355	.8371	.8387	33	1.6
57	.8387	.8403	.8418	.8434	.8450	.8465	.8480	32	1.6
58	.8480	.8496	.8511	.8526	.8542	.8557	.8572	31	1.5
59	.8572	.8587	.8601	.8616	.8631	.8646	.8660	30	1.5
60	.8660	.8675	.8689	.8704	.8718	.8732	.8746	29	1.4
61	.8746	.8760	.8774	.8788	.8802	.8816	.8829	28	1.4
62	.8829	.8843	.8857	.8870	.8884	.8897	.8910	27	1.4
63	.8910	.8923	.8936	.8949	.8962	.8975	.8988	26	1.3
64	.8988	.9001	.9013	.9026	.9038	.9051	.9063	25	1.3
65	.9063	.9075	.9088	.9100	.9112	.9124	.9135	24	1.2
66	.9135	.9147	.9159	.9171	.9182	.9194	.9205	23	1.2
67	.9205	.9216	.9228	.9239	.9250	.9261	.9272	22	1.1
68	.9272	.9283	.9293	.9304	.9315	.9325	.9336	21	1.1
69	.9336	.9346	.9356	.9367	.9377	.9387	.9397	20	1.0
70	.9397	.9407	.9417	.9426	.9436	.9446	.9455	19	1.0
71	.9455	.9465	.9474	.9483	.9492	.9502	.9511	18	0.9
72	.9511	.9520	.9528	.9537	.9546	.9555	.9563	17	0.9
73	.9563	.9572	.9580	.9588	.9596	.9605	.9613	16	0.8
74	.9613	.9621	.9628	.9636	.9644	.9652	.9659	15	0.8
75	.9659	.9667	.9674	.9681	.9689	.9696	.9703	14	0.7
76	.9703	.9710	.9717	.9724	.9730	.9737	.9744	13	0.7
77	.9744	.9750	.9757	.9763	.9769	.9775	.9781	12	0.6
78	.9781	.9787	.9793	.9799	.9805	.9811	.9816	11	0.6
79	.9816	.9822	.9827	.9833	.9838	.9843	.9848	10	0.5
80	.9848	.9853	.9858	.9863	.9868	.9872	.9877	9	0.5
81	.9877	.9881	.9886	.9890	.9894	.9899	.9903	8	0.4
82	.9903	.9907	.9911	.9914	.9918	.9922	.9925	7	0.4
83	.9925	.9929	.9932	.9936	.9939	.9942	.9945	6	0.3
84	.9945	.9948	.9951	.9954	.9957	.9959	.9962	5	0.3
85	.9962	.9964	.9967	.9969	.9971	.9974	.9976	4	0.2
86	.9976	.9978	.9980	.9981	.9983	.9985	.9986	3	0.2
87	.9986	.9988	.9989	.9990	.9992	.9993	.9994	2	0.1
88	.9994	.9995	.9996	.9997	.9997	.9998	.9998	1	0.1
89	.9998	.9999	.9999	1.0000	1.0000	1.0000	1.0000	0	0.0
	60′	50′	40′	30′	20′	10′	0′	Angle.	

Natural Cosines.

TABLE 98.

NATURAL TANGENTS AND COTANGENTS.

Natural Tangents.

Angle.	0'	10'	20'	30'	40'	50'	60'	Angle.	Prop. Parts for 1'.
0°	.00000	.00291	.00582	.00873	.01164	.01455	.01746	89°	2.9
1	.01746	.02036	.02328	.02619	.02910	.03201	.03492	88	2.9
2	.03492	.03783	.04075	.04366	.04658	.04949	.05241	87	2.9
3	.05241	.05533	.05824	.06116	.06408	.06700	.06993	86	2.9
4	.06993	.07285	.07578	.07870	.08163	.08456	.08749	85	2.9
5	.08749	.09042	.09335	.09629	.09923	.10216	.10510	84	2.9
6	.10510	.10805	.11099	.11394	.11688	.11983	.12278	83	2.9
7	.12278	.12574	.12869	.13165	.1346	.1376	.1405	82	3.0
8	.1405	.1435	.1465	.1495	.1524	.1554	.1584	81	3.0
9	.1584	.1614	.1644	.1673	.1703	.1733	.1763	80	3.0
10	.1763	.1793	.1823	.1853	.1883	.1914	.1944	79	3.0
11	.1944	.1974	.2004	.2035	.2065	.2095	.2126	78	3.0
12	.2126	.2156	.2186	.2217	.2247	.2278	.2309	77	3.1
13	.2309	.2339	.2370	.2401	.2432	.2462	.2493	76	3.1
14	.2493	.2524	.2555	.2586	.2617	.2648	.2679	75	3.1
15	.2679	.2711	.2742	.2773	.2805	.2836	.2867	74	3.1
16	.2867	.2899	.2931	.2962	.2994	.3026	.3057	73	3.2
17	.3057	.3089	.3121	.3153	.3185	.3217	.3249	72	3.2
18	.3249	.3281	.3314	.3346	.3378	.3411	.3443	71	3.2
19	.3443	.3476	.3508	.3541	.3574	.3607	.3640	70	3.3
20	.3640	.3673	.3706	.3739	.3772	.3805	.3839	69	3.3
21	.3839	.3872	.3906	.3939	.3973	.4006	.4040	68	3.4
22	.4040	.4074	.4108	.4142	.4176	.4210	.4245	67	3.4
23	.4245	.4279	.4314	.4348	.4383	.4417	.4452	66	3.5
24	.4452	.4487	.4522	.4557	.4592	.4628	.4663	65	3.5
25	.4663	.4699	.4734	.4770	.4806	.4841	.4877	64	3.6
26	.4877	.4913	.4950	.4986	.5022	.5059	.5095	63	3.6
27	.5095	.5132	.5169	.5206	.5243	.5280	.5317	62	3.7
28	.5317	.5354	.5392	.5430	.5467	.5505	.5543	61	3.8
29	.5543	.5581	.5619	.5658	.5696	.5735	.5774	60	3.8
30	.5774	.5812	.5851	.5890	.5930	.5969	.6009	59	3.9
31	.6009	.6048	.6088	.6128	.6168	.6208	.6249	58	4.0
32	.6249	.6289	.6330	.6371	.6412	.6453	.6494	57	4.1
33	.6494	.6536	.6577	.6619	.6661	.6703	.6745	56	4.2
34	.6745	.6787	.6830	.6873	.6916	.6959	.7002	55	4.3
35	.7002	.7046	.7089	.7133	.7177	.7221	.7265	54	4.4
36	.7265	.7310	.7355	.7400	.7445	.7490	.7536	53	4.5
37	.7536	.7581	.7627	.7673	.7720	.7766	.7813	52	4.6
38	.7813	.7860	.7907	.7954	.8002	.8050	.8098	51	4.7
39	.8098	.8146	.8195	.8243	.8292	.8342	.8391	50	4.9
40	.8391	.8441	.8491	.8541	.8591	.8642	.8693	49	5.0
41	.8693	.8744	.8796	.8847	.8899	.8952	.9004	48	5.2
42	.9004	.9057	.9110	.9163	.9217	.9271	.9325	47	5.4
43	.9325	.9380	.9435	.9490	.9545	.9601	.9657	46	5.5
44	.9657	.9713	.9770	.9827	.9884	.9942	1.0000	45	5.7
	60'	50'	40'	30'	20'	10'	0'	Angle.	

Natural Cotangents

TABLE 98.

NATURAL TANGENTS AND COTANGENTS.

Natural Tangents.

Angle.	0′	10′	20′	30′	40′	50′	60′	Angle.	Prop. Parts for 1′.
45°	1.0000	1.0058	1.0117	1.0176	1.0235	1.0295	1.0355	44°	5.9
46	1.0355	1.0416	1.0477	1.0538	1.0599	1.0661	1.0724	43	6.1
47	1.0724	1.0786	1.0850	1.0913	1.0977	1.1041	1.1106	42	6.4
48	1.1106	1.1171	1.1237	1.1303	1.1369	1.1436	1.1504	41	6.6
49	1.1504	1.1571	1.1640	1.1708	1.1778	1.1847	1.1918	40	6.9
50	1.1918	1.1988	1.2059	1.2131	1.2203	1.2276	1.2349	39	7.2
51	1.2349	1.2423	1.2497	1.2572	1.2647	1.2723	1.2799	38	7.5
52	1.2799	1.2876	1.2954	1.3032	1.3111	1.3190	1.3270	37	7.9
53	1.3270	1.3351	1.3432	1.3514	1.3597	1.3680	1.3764	36	8.2
54	1.3764	1.3848	1.3934	1.4019	1.4106	1.4193	1.4281	35	8 6
55	1.4281	1.4370	1.4460	1.4550	1.4641	1.4733	1.4826	34	9.1
56	1.4826	1.4919	1.5013	1.5108	1.5204	1.5301	1.5399	33	9.6
57	1.5399	1.5497	1.5597	1.5697	1.5798	1.5900	1.6003	32	10.1
58	1.6003	1.6107	1.6212	1.6319	1.6426	1.6534	1.6643	31	10.7
59	1.6643	1.6753	1.6864	1.6977	1.7090	1.7205	1.7321	30	11.3
60	1.7321	1.7437	1.7556	1.7675	1.7796	1.7917	1.8040	29	12.0
61	1.8040	1.8165	1.8291	1.8418	1.8546	1.8676	1.8807	28	12.8
62	1.8807	1.8940	1.9074	1.9210	1.9347	1.9486	1.9626	27	13.6
63	1.9626	1.9768	1.9912	2.0057	2.0204	2.0353	2.0503	26	14.6
64	2.0503	2.0655	2.0809	2.0965	2.1123	2.1283	2.1445	25	15.7
65	2.1445	2.1609	2.1775	2.1943	2.2113	2.2286	2.2460	24	16.9
66	2.2460	2.2637	2.2817	2.2998	2.3183	2.3369	2.3559	23	18.3
67	2.3559	2.3750	2.3945	2.4142	2.4342	2.4545	2.4751	22	19.9
68	2.4751	2.4960	2.5172	2.5386	2.5605	2.5826	2.6051	21	21.7
69	2.6051	2.6279	2.6511	2.6746	2.6985	2.7228	2.7475	20	23.7
70	2.7475	2.7725	2.7980	2.8239	2.8502	2.8770	2.9042	19	
71	2.9042	2.9319	2.9600	2.9887	3.0178	3.0475	3.0777	18	
72	3.0777	3.1084	3.1397	3.1716	3.2041	3.2371	3.2709	17	
73	3.2709	3.3052	3.3402	3.3759	3.4124	3.4495	3.4874	16	
74	3.4874	3.5261	3.5656	3.6059	3.6470	3.6891	3.7321	15	
75	3.7321	3.7760	3.8208	3.8667	3.9136	3.9617	4.0108	14	
76	4.0108	4.0611	4.1126	4.1653	4.2193	4.2747	4.3315	13	
77	4.3315	4.3897	4.4494	4.5107	4.5736	4.6382	4.7046	12	
78	4.7046	4.7729	4.8430	4.9152	4.9894	5.0658	5.1446	11	
79	5.1446	5.2257	5.3093	5.3955	5.4845	5.5764	5.6713	10	
80	5.6713	5.7694	5.8708	5.9758	6.0844	6.1970	6.3138	9	
81	6.3138	6.4348	6.5606	6.6912	6.8269	6.9682	7.1154	8	
82	7.1154	7.2687	7.4287	7.5958	7.7704	7.9530	8.1443	7	
83	8.1443	8.3450	8.5555	8.7769	9.0098	9.2553	9.5144	6	
84	9.5144	9.7882	10.0780	10.3854	10.7119	11.0594	11.4301	5	
85	11.4301	11.8262	12.2505	12.7062	13.1969	13.7267	14.3007	4	
86	14.3007	14.9244	15.6048	16.3499	17.1693	18.0750	19.0811	3	
87	19.0811	20.2056	21.4704	22.9038	24.5418	26.4316	28.6363	2	
88	28.6363	31.2416	34.3678	38.1885	42.9641	49.1039	57.2900	1	
89	57.2900	68.7501	85.9398	114.5887	171.8854	343.7737	∞	0	
	60′	50′	40′	30′	20′	10′	0′	Angle.	

Natural Cotangents.

TABLE 99.
LOGARITHMS OF NUMBERS.

N.	0	1	2	3	4	5	6	7	8	9	d.	Prop. Parts.
0	......	0000	3010	4771	6021	6990	7782	8451	9031	9542		
1	0000	0414	0792	1139	1461	1761	2041	2304	2553	2788		
2	3010	3222	3424	3617	3802	3979	4150	4314	4472	4624		
3	4771	4914	5051	5185	5315	5441	5563	5682	5798	5911		**43** | **42** | **41** | **40**
4	6021	6128	6232	6335	6435	6532	6628	6721	6812	6902		1 4,3 4,2 4,1 4,0
5	6990	7076	7160	7243	7324	7404	7482	7559	7634	7709		2 8,6 8,4 8,2 8,0
6	7782	7853	7924	7993	8062	8129	8195	8261	8325	8388		3 12,9 12,6 12,3 12,0
7	8451	8513	8573	8633	8692	8751	8808	8865	8921	8976		4 17,2 16,8 16,4 16,0
8	9031	9085	9138	9191	9243	9294	9345	9395	9445	9494		5 21,5 21,0 20,5 20,0
9	9542	9590	9638	9685	9731	9777	9823	9868	9912	9956		6 25,8 25,2 24,6 24,0
												7 30,1 29,4 28,7 28,0
												8 34,4 33,6 32,8 32,0
												9 38,7 37,8 36,9 36,0
10	0000	0043	0086	0128	0170	0212	0253	0294	0334	0374	41	
11	0414	0453	0492	0531	0569	0607	0645	0682	0719	0755	38	
12	0792	0828	0864	0899	0934	0969	1004	1038	1072	1106	35	**39** | **38** | **37** | **36**
13	1139	1173	1206	1239	1271	1303	1335	1367	1399	1430	32	1 3,9 3,8 3,7 3,6
14	1461	1492	1523	1553	1584	1614	1644	1673	1703	1732	30	2 7,8 7,6 7,4 7,2
15	1761	1790	1818	1847	1875	1903	1931	1959	1987	2014	28	3 11,7 11,4 11,1 10,8
16	2041	2068	2095	2122	2148	2175	2201	2227	2253	2279	26	4 15,6 15,2 14,8 14,4
17	2304	2330	2355	2380	2405	2430	2455	2480	2504	2529	25	5 19,5 19,0 18,5 18,0
18	2553	2577	2601	2625	2648	2672	2695	2718	2742	2765	24	6 23,4 22,8 22,2 21,6
19	2788	2810	2833	2856	2878	2900	2923	2945	2967	2989	22	7 27,3 26,6 25,9 25,2
												8 31,2 30,4 29,6 28,8
												9 35,1 34,2 33,3 32,4
20	3010	3032	3054	3075	3096	3118	3139	3160	3181	3201	21	
21	3222	3243	3263	3284	3304	3324	3345	3365	3385	3404	20	
22	3424	3444	3464	3483	3502	3522	3541	3560	3579	3598	19	**35** | **34** | **33** | **32**
23	3617	3636	3655	3674	3692	3711	3729	3747	3766	3784	18	1 3,5 3,4 3,3 3,2
24	3802	3820	3838	3856	3874	3892	3909	3927	3945	3962	18	2 7,0 6,8 6,6 6,4
25	3979	3997	4014	4031	4048	4065	4082	4099	4116	4133	17	3 10,5 10,2 9,9 9,6
26	4150	4166	4183	4200	4216	4232	4249	4265	4281	4298	16	4 14,0 13,6 13,2 12,8
27	4314	4330	4346	4362	4378	4393	4409	4425	4440	4456	16	5 17,5 17,0 16,5 16,0
28	4472	4487	4502	4518	4533	4548	4564	4579	4594	4609	15	6 21,0 20,4 19,8 19,2
29	4624	4639	4654	4669	4683	4698	4713	4728	4742	4757	15	7 24,5 23,8 23,1 22,4
												8 28,0 27,2 26,4 25,6
												9 31,5 30,6 29,7 28,8
30	4771	4786	4800	4814	4829	4843	4857	4871	4886	4900	14	
31	4914	4928	4942	4955	4969	4983	4997	5011	5024	5038	14	
32	5051	5065	5079	5092	5105	5119	5132	5145	5159	5172	13	**31** | **30** | **29** | **28**
33	5185	5198	5211	5224	5237	5250	5263	5276	5289	5302	13	1 3,1 3,0 2,9 2,8
34	5315	5328	5340	5353	5366	5378	5391	5403	5416	5428	13	2 6,2 6,0 5,8 5,6
35	5441	5453	5465	5478	5490	5502	5514	5527	5539	5551	12	3 9,3 9,0 8,7 8,4
36	5563	5575	5587	5599	5611	5623	5635	5647	5658	5670	12	4 12,4 12,0 11,6 11,2
37	5682	5694	5705	5717	5729	5740	5752	5763	5775	5786	12	5 15,5 15,0 14,5 14,0
38	5798	5809	5821	5832	5843	5855	5866	5877	5888	5899	11	6 18,6 18,0 17,4 16,8
39	5911	5922	5933	5944	5955	5966	5977	5988	5999	6010	11	7 21,7 21,0 20,3 19,6
												8 24,8 24,0 23,2 22,4
												9 27,9 27,0 26,1 25,2
40	6021	6031	6042	6053	6064	6075	6085	6096	6107	6117	11	
41	6128	6138	6149	6160	6170	6180	6191	6201	6212	6222	10	
42	6232	6243	6253	6263	6274	6284	6294	6304	6314	6325	10	**27** | **26** | **25** | **24**
43	6335	6345	6355	6365	6375	6385	6395	6405	6415	6425	10	1 2,7 2,6 2,5 2,4
44	6435	6444	6454	6464	6474	6484	6493	6503	6513	6522	10	2 5,4 5,2 5,0 4,8
45	6532	6542	6551	6561	6571	6580	6590	6599	6609	6618	10	3 8,1 7,8 7,5 7,2
46	6628	6637	6646	6656	6665	6675	6684	6693	6702	6712	9	4 10,8 10,4 10,0 9,6
47	6721	6730	6739	6749	6758	6767	6776	6785	6794	6803	9	5 13,5 13,0 12,5 12,0
48	6812	6821	6830	6839	6848	6857	6866	6875	6884	6893	9	6 16,2 15,6 15,0 14,4
49	6902	6911	6920	6928	6937	6946	6955	6964	6972	6981	9	7 18,9 18,2 17,5 16,8
												8 21,6 20,8 20,0 19,2
												9 24,3 23,4 22,5 21,6
50	6990	6998	7007	7016	7024	7033	7042	7050	7059	7067	9	
N.	0	1	2	3	4	5	6	7	8	9	d.	Prop. Parts.

SMITHSONIAN TABLES.

TABLE 99.
LOGARITHMS OF NUMBERS.

N.	0	1	2	3	4	5	6	7	8	9	d.
50	6990	6998	7007	7016	7024	7033	7042	7050	7059	7067	9
51	7076	7084	7093	7101	7110	7118	7126	7135	7143	7152	8
52	7160	7168	7177	7185	7193	7202	7210	7218	7226	7235	8
53	7243	7251	7259	7267	7275	7284	7292	7300	7308	7316	8
54	7324	7332	7340	7348	7356	7364	7372	7380	7388	7396	8
55	7404	7412	7419	7427	7435	7443	7451	7459	7466	7474	8
56	7482	7490	7497	7505	7513	7520	7528	7536	7543	7551	8
57	7559	7566	7574	7582	7589	7597	7604	7612	7619	7627	8
58	7634	7642	7649	7657	7664	7672	7679	7686	7694	7701	7
59	7709	7716	7723	7731	7738	7745	7752	7760	7767	7774	7
60	7782	7789	7796	7803	7810	7818	7825	7832	7839	7846	7
61	7853	7860	7868	7875	7882	7889	7896	7903	7910	7917	7
62	7924	7931	7938	7945	7952	7959	7966	7973	7980	7987	7
63	7993	8000	8007	8014	8021	8028	8035	8041	8048	8055	7
64	8062	8069	8075	8082	8089	8096	8102	8109	8116	8122	7
65	8129	8136	8142	8149	8156	8162	8169	8176	8182	8189	7
66	8195	8202	8209	8215	8222	8228	8235	8241	8248	8254	7
67	8261	8267	8274	8280	8287	8293	8299	8306	8312	8319	6
68	8325	8331	8338	8344	8351	8357	8363	8370	8376	8382	6
69	8388	8395	8401	8407	8414	8420	8426	8432	8439	8445	6
70	8451	8457	8463	8470	8476	8482	8488	8494	8500	8506	6
71	8513	8519	8525	8531	8537	8543	8549	8555	8561	8567	6
72	8573	8579	8585	8591	8597	8603	8609	8615	8621	8627	6
73	8633	8639	8645	8651	8657	8663	8669	8675	8681	8686	6
74	8692	8698	8704	8710	8716	8722	8727	8733	8739	8745	6
75	8751	8756	8762	8768	8774	8779	8785	8791	8797	8802	6
76	8808	8814	8820	8825	8831	8837	8842	8848	8854	8859	6
77	8865	8871	8876	8882	8887	8893	8899	8904	8910	8915	6
78	8921	8927	8932	8938	8943	8949	8954	8960	8965	8971	6
79	8976	8982	8987	8993	8998	9004	9009	9015	9020	9025	5
80	9031	9036	9042	9047	9053	9058	9063	9069	9074	9079	5
81	9085	9090	9096	9101	9106	9112	9117	9122	9128	9133	5
82	9138	9143	9149	9154	9159	9165	9170	9175	9180	9186	5
83	9191	9196	9201	9206	9212	9217	9222	9227	9232	9238	5
84	9243	9248	9253	9258	9263	9269	9274	9279	9284	9289	5
85	9294	9299	9304	9309	9315	9320	9325	9330	9335	9340	5
86	9345	9350	9355	9360	9365	9370	9375	9380	9385	9390	5
87	9395	9400	9405	9410	9415	9420	9425	9430	9435	9440	5
88	9445	9450	9455	9460	9465	9469	9474	9479	9484	9489	5
89	9494	9499	9504	9509	9513	9518	9523	9528	9533	9538	5
90	9542	9547	9552	9557	9562	9566	9571	9576	9581	9586	5
91	9590	9595	9600	9605	9609	9614	9619	9624	9628	9633	5
92	9638	9643	9647	9652	9657	9661	9666	9671	9675	9680	5
93	9685	9689	9694	9699	9703	9708	9713	9717	9722	9727	5
94	9731	9736	9741	9745	9750	9754	9759	9763	9768	9773	5
95	9777	9782	9786	9791	9795	9800	9805	9809	9814	9818	5
96	9823	9827	9832	9836	9841	9845	9850	9854	9859	9863	4
97	9868	9872	9877	9881	9886	9890	9894	9899	9903	9908	4
98	9912	9917	9921	9926	9930	9934	9939	9943	9948	9952	4
99	9956	9961	9965	9969	9974	9978	9983	9987	9991	9996	4
100	0000	0004	0009	0013	0017	0022	0026	0030	0035	0039	4
N.	0	1	2	3	4	5	6	7	8	9	d.

Prop. Parts.

	23	22	21	20	19
1	2,3	2,2	2,1	2,0	1,9
2	4,6	4,4	4,2	4,0	3,8
3	6,9	6,6	6,3	6,0	5,7
4	9,2	8,8	8,4	8,0	7,6
5	11,5	11,0	10,5	10,0	9,5
6	13,8	13,2	12,6	12,0	11,4
7	16,1	15,4	14,7	14,0	13,3
8	18,4	17,6	16,8	16,0	15,2
9	20,7	19,8	18,9	18,0	17,1

	18	17	16	15	14
1	1,8	1,7	1,6	1,5	1,4
2	3,6	3,4	3,2	3,0	2,8
3	5,4	5,1	4,8	4,5	4,2
4	7,2	6,8	6,4	6,0	5,6
5	9,0	8,5	8,0	7,5	7,0
6	10,8	10,2	9,6	9,0	8,4
7	12,6	11,9	11,2	10,5	9,8
8	14,4	13,6	12,8	12,0	11,2
9	16,2	15,3	14,4	13,5	12,6

	13	12	11	10	9
1	1,3	1,2	1,1	1,0	0,9
2	2,6	2,4	2,2	2,0	1,8
3	3,9	3,6	3,3	3,0	2,7
4	5,2	4,8	4,4	4,0	3,6
5	6,5	6,0	5,5	5,0	4,5
6	7,8	7,2	6,6	6,0	5,4
7	9,1	8,4	7,7	7,0	6,3
8	10,4	9,6	8,8	8,0	7,2
9	11,7	10,8	9,9	9,0	8,1

	8	7	6	5	4
1	0,8	0,7	0,6	0,5	0,4
2	1,6	1,4	1,2	1,0	0,8
3	2,4	2,1	1,8	1,5	1,2
4	3,2	2,8	2,4	2,0	1,6
5	4,0	3,5	3,0	2,5	2,0
6	4,8	4,2	3,6	3,0	2,4
7	5,6	4,9	4,2	3,5	2,8
8	6,4	5,6	4,8	4,0	3,2
9	7,2	6,3	5,4	4,5	3,6

SMITHSONIAN TABLES.

LIST OF METEOROLOGICAL STATIONS.

NORTH AMERICA—
   Canada . . . . PAGE 244
   Central America 244
   Greenland 244
   Mexico 244
   United States 245
   West Indies 244

SOUTH AMERICA . . . . PAGE 246

EUROPE—
   Austro-Hungary . . . . PAGE 247
   Belgium 248
   British Isles 248
   Denmark 249
   France 249
   Germany 250
   Greece 248
   Holland 248
   Italy 251
   Norway 249
   Portugal 253
   Roumania 248
   Russia 251
   Spain 253
   Sweden 249
   Switzerland 253
   Turkey 248

ASIA . . . . PAGE 254

AUSTRALASIA . . . . PAGE 256

AFRICA AND NEIGHBORING ISLANDS . . . . PAGE 256

INTERNATIONAL POLAR STATIONS . . . . PAGE 257

MISCELLANEOUS ISLANDS . . . . PAGE 257

TABLE 100.
LIST OF METEOROLOGICAL STATIONS.
(The asterisk * designates stations of the first order.)

	Latitude.	Longitude from Greenwich.	Height above Sea-level.	
NORTH AMERICA.			Feet.	m.
CANADA.				
Father Point	48° 31′ N.	68° 28′ W.	20	6
*Frederickton	45 57	66 38	164	50
*Halifax..................	44 39	63 36	122	37
*Kingston	44 14	76 29	307	94
*Montreal.................	45 30	73 35	187	57
Parry Sound...............	45 19	80 0	641	195
Qu'Appelle	50 44	103 42		
*Quebec...................	46 48	71 13	293	89
*Saint John...............	45 17	66 3	116	35
*Sydney...................	46 8	60 10	37	11
*Toronto	43 29	79 23	350	107
*Westminster..............	49 12	122 53	33	10
*Winnepeg	49 51	97 7	758	231
*Woodstock................	43 8	80 47	980	299
CENTRAL AMERICA. (*See* MEXICO.)				
GREENLAND.				
Godthaab	64 11 N.	51 46 W.	36	11
Iviktut...................	61 12	48 11	16	5
Upernivik	72 47	55 53	39	12
MEXICO, CENTRAL AMERICA, WEST INDIES, *ETC.*				
Bermuda, West Indies	32 18 N.	64 47 W.	151	46
Guanajuato, Mexico	21 0	101 15	6759	2060
*Habana, Cuba.............	23 8	76 35	62	19
Kingston, West Indies	17 58	76 48	10	3
Leon, Mexico	21 7	101 41	5899	1798
Mazatlan, Mexico...........	23 11	106 25	249	76
Mexico (City of)	19 26	99 8	7487	2282
*Nassau, Bahamas	25 5	77 21	44	13
New Castle, Jamaica	18 6	76 42	3800	1158
Pabellon, Mexico	22 4	102 12	6312	1924
Port au Prince, Haiti......	18 34	72 21	118	36
Puebla, Mexico	19 2	98 11	7119	2170
St. Thomas, West Indies	18 20	64 56	131	40
Saltillo, Mexico	25 25	100 38	5358	1633
San Luis Potosi, Mexico.....	22 9	100 58	6201	1890
San Salvador, Central America ..	13 44	89 9	2156	657
Santiago, Cuba	19 55	75 50	21	6
Tacubaya, Mexico...........	19 24	99 12	7621	2323
Vera Cruz, Mexico	19 12	96 8	23	7
Zacatecas, Mexico	22 47	100 15	8189	2496

TABLE 100.

LIST OF METEOROLOGICAL STATIONS.

(The asterisk * designates stations of the first order.)

	Latitude.	Longitude from Greenwich.	Height above Sea-level.	
			Feet.	m.
WEST INDIES. (*See* Mexico.)				
UNITED STATES.				
*Abilene, Texas	32° 23' N.	99° 40' W.	1748	533
*Albany, New York	42 39	73 45	85	26
*Alpena, Michigan	45 5	83 30	609	186
*Atlanta, Georgia	33 45	84 23	1131	345
*Augusta, Georgia	33 28	81 54	209	64
*Bismarck, North Dakota	46 47	100 38	1698	518
*Blue Hill, Massachusetts	42 13	71 7	640	195
*Boston, Massachusetts	42 21	71 4	125	38
*Buffalo, New York	42 53	78 53	690	210
*Chicago, Illinois............	41 52	87 38	824	251
*Cincinnati, Ohio	39 6	84 30	628	191
*Cleveland, Ohio............	41 30	81 42	751	229
*Columbus, Ohio............	39 58	83 0	837	255
*Davenport, Iowa	41 30	90 38	613	187
*Denver, Colorado	39 45	105 0	5287	1612
*Des Moines, Iowa	41 35	93 37	869	265
*Detroit, Michigan	42 20	83 3	724	221
*Dodge City, Kansas..........	37 45	100 0	2523	769
*Duluth, Minnesota	46 48	92 6	656	200
*Eastport, Maine.............	44 54	66 59	53	16
*El Paso, Texas	31 47	106 30	3796	1157
*Fort Assiniboine, Montana.....	48 32	109 42	2690	820
*Galveston, Texas	29 18	94 50	42	13
*Hamilton, Mount, California ...	37 20	121 39	4300	1311
*Helena, Montana	46 34	112 4	4118	1255
*Huron, South Dakota........	44 21	98 14	1310	399
*Indianapolis, Indiana.........	39 46	86 10	766	234
*Jacksonville, Florida.........	30 20	81 39	43	13
*Kansas City, Missouri	39 5	94 37	963	294
*Keeler, California...........	36 35	117 50	3622	1104
*Key West, Florida	24 34	81 49	22	7
*Knoxville, Tennessee.........	35 56	83 58	980	299
*Lynchburg, Virginia	37 25	79 9	685	209
*Manistee, Michigan..........	44 13	86 16	615	187
*Marquette, Michigan	46 34	87 24	734	224
*Memphis, Tennessee	35 9	90 3	330	101
*Milwaukee, Wisconsin	43 2	87 54	673	205
*Moorhead, Minnesota	46 52	96 44	935	285
*Nantucket, Massachusetts	41 17	70 6	14	4
*Nashville, Tennessee	36 10	86 47	553	169
*New Orleans, Louisiana	29 58	90 4	54	16
*New York City, (*Weather Bureau*).	40 43	74 0	185	56
*New York, (*Central Park*)	40 46	73 58	97	30
*Norfolk, Virginia	36 51	76 17	43	13

Smithsonian Tables.

TABLE 100.
LIST OF METEOROLOGICAL STATIONS.
(The asterisk * designates stations of the first order.)

	Latitude.		Longitude from Greenwich.		Height above Sea-level.	
UNITED STATES. *(Continued.)*					Feet.	m.
* Olympia, Washington	47°	3′ N.	122°	53′ W.	44	13
* Omaha, Nebraska	41	16	95	56	1113	339
* Philadelphia, (*Girard College*)	39	58	75	11	112	34
* Philadelphia, (*Weather Bureau*)	39	57	75	9	117	36
* Pike's Peak, Colorado	38	50	105	2	14134	4308
* Pittsburg, Pennsylvania	40	32	80	2	847	258
* Portland, Oregon	45	32	122	43	80	24
* Rochester, New York	43	8	77	42	523	159
* Roseburg, Oregon	43	13	123	20	523	159
* St. Louis, Missouri	38	38	90	12	571	174
* St. Paul, Minnesota	44	58	93	3	851	259
* Salt Lake City, Utah	40	46	111	54	4345	1324
* San Diego, California	32	43	117	10	93	28
* San Francisco, California	37	48	122	26	109	33
* Santa Fé, New Mexico	35	41	105	57	7026	2142
* Sault de Ste. Marie, Michigan	46	28	84	22	642	196
* Savannah, Georgia	32	5	81	5	87	26
Sitka, Alaska	57	3	135	19	63	19
* Spokane, Washington	47	40	117	25	1938	591
* Tampa, Florida	27	57	82	27	36	11
* Toledo, Ohio	41	40	83	34	674	205
Unalaska, Alaska	53	53	166	32	13	4
* Vicksburg, Mississippi	32	22	90	53	254	77
* Washington City, (*Weather Bureau*)	38	54	77	3	112	34
* Washington City, (*Naval Obs'v'y*)	38	54	77	3	110	33
Washington, Mount, N. H.	44	16	71	18	6279	1914
* Wilmington, North Carolina	34	14	77	57	78	24
* Yuma, Arizona	32	45	114	36	141	43
SOUTH AMERICA.						
Arequipa	16	22 S.	71	22 W.	8050	2454
Bahia-Blanca, Argentine Republic	38	44 S.	62	11	49	15
Bogota, United States of Colombia	4	36 N.	73	15		
Buenos Aires, Argentine Republic	34	36 S.	58	22	72	22
Caldera, Chile	27	3 S.	70	53	85	26
Caracas, Venezuela	10	31 N.	66	55		
Catamarca, Argentine Republic	28	28 S.	65	56	1788	545
Cayenne, French Guiana	4	56 N.	52	21		
Charchani, Peru, (Arequipa)					16650	5075
Concordia, Argentine Republic	31	23 S.	58	4	200	61
Coquimbo, Chile	29	56 S.	71	21	72	22
Cordoba, Argentine Republic	31	25 S.	64	12	1434	437
Corrientes, Argentine Republic	27	28 S.	58	50	253	77
El Misti, Peru, (Arequipa)					19300	5883
Georgetown, British Guiana	6	47 N.	58	9		
Iquique, Chile	20	12 S.	70	11	26	8
La Plata, Argentine Republic	34	55 S.	57	54		

TABLE 100.

LIST OF METEOROLOGICAL STATIONS.
(The asterisk * designates stations of the first order.)

	Latitude.		Longitude from Greenwich.		Height above Sea-level.	
					Feet.	m.
SOUTH AMERICA. *(Continued.)*						
Lima, Peru (Unanue)	12°	4' S.	77°	3' W.	520	158
Matanzas, Argentine Republic	34	49	58	37		
Montevidio, Uruguay	34	53	56	15		
Natal, Brazil	5	50	35	11		
Paramaribo, Dutch Guiana	5	49 N.	55	22		
Paraná, Argentine Republic	31	43 S.	60	16	256	78
Potosi, Bolivia	19	35	65	35	13324	4061
Punta Arenas, Chile	53	10	70	54		
Quito, Equador	0	14	78	45	9541	2908
Rio de Janeiro, Brazil	22	54	43	10		
Rio de Janeiro, Brazil, (new)					3500	1067
Rioja, Argentine Republic	29	19	67	10	1772	540
Santa Cruz de la Sierra, Bolivia	17	50	63	0		
Santiago, Chile	33	27	70	41	1703	519
Sao Paulo, Brazil	23	37	46	40		
Valdivia, Chile	39	49	73	16	39	12
Valparaiso, Chile	33	2	71	39	131	40
Villa Colon, Uruguay	34	50	59	19	134	41
Villa Formoza, Argentine Republic	26	12	58	6	269	82
EUROPE. **AUSTRO-HUNGARY.**						
* Agram (*Zágráb*)	45	49 N.	15	59 E.	535	163
* Barzdorf	50	23	17	5	827	252
Bregenz	47	31	9	45	1352	412
Brünn	49	11	16	36	758	231
* Budapest	47	30	19	2	502	153
* Czernowitz	48	18	25	56	774	236
* Eger	50	5	12	22	1519	463
* Fiume	45	19	14	27	16	5
Gleichenberg	46	53'	15	55	974	297
Görz	45	57	13 · 37		299	91
Gries	46	30	11	20	958	292
Krakow	50	4	19	57	722	220
Kremsmünster	48	3	14	8	1280	390
Lemberg	49	50	24	2	978	298
Lesina	43	10	16	27	30	9
Lienz	46	50	12	46	2231	680
* Obir	46	30	14	29	6716	2047
O-Gyalla	47	53	18	12	364	111
* Pisek	49	19	14	9	1280	390
Pola	44	52	13	50	105	32
Prag	50	5	14	25	663	202
* Prerau	49	27	17	27	705	215
* Riva	45	53	10	50	295	90
* Salzburg	47	48	13	2	1434	437

TABLE 100.
LIST OF METEOROLOGICAL STATIONS.
(The asterisk * designates stations of the first order.)

	Latitude.	Longitude from Greenwich.	Height above Sea-level.	
			Feet.	m.
AUSTRO-HUNGARY. *(Continued.)*				
Schafberg	47° 47′ N.	13° 26′ E.	5827	1776
Sonnblick	47 3	12 57	10154	3095
*Triest	45 39	13 46	85	26
*Wien	48 15	16 21	663	202
Zágráb *(see Agram)*				
GREECE, ROUMANIA, TURKEY.				
Athens, Greece	37 58 N.	23 45 E.		
Bagdad, Asiatic Turkey	33 19	44 26		
Beirut, Turkey	33 54	35 28	112	34
*Bucarest, Roumania	44 25	26 6	285	87
Constantinople, Turkey	41 2	28 59		
Samsoun, Asiatic Turkey	41 18	36 19	26	8
Sinaia, Roumania	45 21	25 34	2822	860
Sinope, Turkey	42 1	35 19	49	15
Sulina, Roumania	45 9	29 40	7	2
Trebizond, Asiatic Turkey	41 1	39 45	92	28
BELGIUM AND HOLLAND.				
Arlon, Belgium	49 40 N.	5 48 E.	1286	392
Bruxelles, Belgium	50 51	4 22	177	54
Furnes, Belgium	51 4	2 40	10	3
*Groningen, Holland	53 13	6 34	49	15
*Helder, Holland	52 57	4 45	0	0
*Liège, Belgium	50 37	5 34	200	61
Maeseyck, Belgium	51 6	5 48	115	35
Maestricht, Holland	50 51	5 41	164	50
*Ostende, Belgium	51 14	2 55	16	5
*Utrecht, Holland	52 5	5 7	43	13
BRITISH ISLES.				
*Aberdeen	57 10 N.	2 6 W.	88	27
*Armagh	54 21	6 39	196	60
*Ben Nevis	56 48	5 8	4406	1343
Dublin	53 22	6 21	155	47
Dundee	56 28	2 56	160	49
Edinburgh	55 56	3 11		
*Falmouth	50 9	5 4	183	56
*Glasgow	55 53	4 18	180	55
*Kew	51 28	0 19	34	10
Londonderry	55 0	7 19	220	67
*Markree Castle	54 11	8 27	122	37
*Oxford	51 46	1 20	212	65
Greenwich Observatory	51 29	0 0	159	48

SMITHSONIAN TABLES.

TABLE 100.
LIST OF METEOROLOGICAL STATIONS.
(The asterisk * designates stations of the first order.)

	Latitude.	Longitude from Greenwich.	Height above Sea-level.	
			Feet.	m.
BRITISH ISLES. *(Continued.)*				
Southampton	50° 55' N.	1° 24' W.	78	24
Southbourne	50 44	1 48	295	90
* Stonyhurst	53 51	2 28	375	114
* Valencia	51 55	10 18	23	7
York	53 57	1 5	167	51
DENMARK, NORWAY, SWEDEN.				
Bodö, Norway	67 17 N.	14 24 E.	23	7
Carlshamn, Sweden	56 10	14 52	30	9
* Christiania, Norway	59 55	10 43	82	25
Christiansund, Norway	63 7	7 45	52	16
Dovre, Norway	62 5	9 7	2110	643
Fanö, Denmark	55 27	8 24	20	6
Florö, Norway	61 36	5 2	26	8
Haparanda, Sweden	65 50	24 9	30	9
Hernösand, Sweden	62 38	17 57	49	15
Kjöbenhavn, Denmark	55 41	12 36	43	13
Skagen, Denmark	57 44	10 38	10	3
Skudesnes, Norway	59 9	5 16	13	4
Stockholm, Sweden	59 21	18 4	144	44
* Upsal, Sweden	59 52	17 38		
* Vandrup, Denmark	55 25	9 18	131	40
FRANCE.				
Bagnères-de-Bigorre	43 4 N.	0 9 E.	1795	547
Besançon	47 14	5 59 E.	896	273
Bordeaux	44 50	0 31 W.		
Brest	48 24	4 30 W.	210	64
Cherbourg	49 39	1 30 W.		
Chamonix	45 55	7 2 E.	3406	1038
Dunkerque	51 3	2 22 E.	23	7
Langres	47 52	5 20 E.	1529	466
* Lyon	45 41	4 47 E.	981	299
* Marseille	43 17	5 23 E.	246	75
Mont Blanc (*Haute Savoie*)	45 50	7 2 E.	15780	4810
* Mont Ventoux	44 17	5 16 E.	6234	1900
Nantes	47 13	1 33 W.	135	41
Nice	43 43	7 18 E.	1115	340
* Paris, (*Parc de Saint-Maur*)	48 49	2 30 E.	161	49
* Paris, (*Tour Eiffel*)	48 52	2 18 E.	1027	313
Paris, (*Montsouris*)	48 49	2 20 E.		
* Perpignan	42 42	2 53 E.	105	32
* Pic-du-Midi	42 57	0 8 E.	9380	2859
Puy-de-Dome, (*Plaine*)	45 46	3 5 E.	1273	388
* Puy-de-Dome, (*Sommet*)	45 47	2 57 E.	4813	1467
* Saint-Martin-de-Hinx	43 35	1 16 W.	131	40
* Toulouse	43 37	1 26 E.	636	194

TABLE 100.
LIST OF METEOROLOGICAL STATIONS.
(The asterisk * designates stations of the first order.)

	Latitude.	Longitude from Greenwich.	Height above Sea-level.	
GERMANY.			Feet.	m.
Bamberg, Bavaria	49° 54' N.	10° 53' E.	817	249
Berlin, Prussia	52 30	13 23	161	49
Borkum, Prussia	53 35	6 40	33	10
Bremen	53 51	8 48	13	4
Breslau, Prussia	51 7	17 2	482	147
Bromberg, Prussia	53 8	18 0	138	42
Chemnitz, Saxony	50 50	12 55	1037	316
Danzig, Prussia	54 21	18 40	72	22
Dresden, Saxony	51 2	13 44	390	119
Eichberg, Prussia	50 55	15 48	1145	349
Freiberg, Saxony	50 55	13 21	1335	407
Friedrichshafen, Württemberg	47 39	9 28	1335	407
Göttingen, Prussia	51 32	9 56	492	150
Halle, Prussia	51 29	11 38	364	111
*Hamburg	53 33	9 58	85	26
Heidelberg, Baden	49 25	8 42	394	120
Hirschberg, Bavaria	47 40	11 42	4954	1510
Hohenpeissenberg, Bavaria	47 48	11 1	3261	994
Jena, Saxony	50 56	11 35	525	160
*Kaiserslautern, Bavaria	49 27	7 46	794	242
Karlsruhe, Baden	49 1	8 25	407	124
Kassel, Prussia	51 19	9 30	669	204
*Keitum, Prussia	54 54	8 22	30	9
Kiel, Prussia	54 20	10 9	154	47
Leipzig, Saxony	51 20	12 23	390	119
*Magdeburg, Prussia	52 8	11 38	177	54
Mannheim, Baden	49 29	8 28	367	112
*Memel, Prussia	55 43	21 7	13	4
Metz, Lorraine	49 7	6 10	600	183
Mülhausen, Alsace	47 45	7 20	787	240
*München, Bavaria	48 9	11 36	1736	529
*Neufahrwasser, Prussia	54 24	18 40	13	4
Nürnberg, Bavaria	49 27	11 4	1033	315
Regensburg, Bavaria	49 1	12 6	1175	358
Rostock, Mecklenburg	54 5	12 7	72	22
Rügenwaldermüude, Prussia	54 26	16 23	13	4
Schneekoppe, Prussia	50 44	15 44	5259	1603
Strassburg, Alsace	48 35	7 45	472	144
Stuttgart, Württemberg	48 47	9 10	879	268
*Swinemünde, Prussia	53 56	14 16	33	10
Wendelstein, Bavaria	47 42	12 1	5666	1727
Wilhelmshaven, Oldenburg	53 32	8 9	26	8
Würzburg, Bavaria	49 48	9 56	587	179
*Wustrow, Mecklenburg	54 21	12 24	23	7
HOLLAND. (*See* BELGIUM.)				

TABLE 100.
LIST OF METEOROLOGICAL STATIONS.
(The asterisk * designates stations of the first order.)

	Latitude.	Longitude from Greenwich.	Height above Sea-level.	
ITALY.			Feet.	m.
Agnone	41° 48′ N.	14° 22′ E.	2644	806
Allessandria	44 54	8 37	322	98
Bologna	44 30	11 20	279	85
Catania, Sicily	37 5	14 55	102	31
Cosenza	39 19	16 17	840	256
Firenze	43 46	11 15	240	73
Genova	44 24	8 55	177	54
Milano	45 28	9 11	482	147
Modena	44 39	10 56	210	64
Moncalieri	45 0	7 41	846	258
Napoli	40 50	14 10	187	57
Palermo	38 7	13 21	233	71
Parma	44 48	10 19	295	90
Riposto	37 41	15 14	46	14
*Roma	41 54	12 29	164	50
Siracusa	37 3	15 15	72	22
*Torino	45 4	7 41	902	275
Venezia	45 26	12 20	69	21
Verona	45 26	11 1	217	66
NORWAY. (*See* DENMARK.)				
PORTUGAL. (*See* SPAIN.)				
ROUMANIA. (*See* GREECE.)				
RUSSIA.				
Alexandrowka, Siberia	50° 50′ N.	142° 7′ E.	52	16
Astrachan	46 21	48 2	46	14
Baranowo	56 25	38 36	597	182
Barnaul, Siberia	53 20	83 47	459	140
Beresow, Siberia	63 56	65 4	105	32
Bogoslowsk	59 45	60 1	617	188
Brest-Litowsk	52 5	23 40	443	135
Brjansk	53 15	34 22	656	200
Dorpat	58 23	26 43	210	64
Elissawetgrad	48 31	32 17	407	124
Enisseisk, Siberia	58 27	92 6	279	85
Eriwan	40 10	44 30	3261	994
Gudaur	42 28	44 28	7251	2210
*Helsingfors, Finland	60 10	24 57	66	20
*Irkutsk, Siberia	52 16	104 19	1611	491
Kaluga	54 31	36 16	643	196
Kargopol	61 30	38 57	440	134

TABLE 100.
LIST OF METEOROLOGICAL STATIONS.
(The asterisk * designates stations of the first order.)

RUSSIA. (Continued.)	Latitude.		Longitude from Greenwich.		Height above Sea-level.	
					Feet.	m.
Kars	40° 37′	N.	43°	5′ E.	5722	1744
Kasan	55 47		49	8	226	69
*Katherinenburg	56 50		60	38	928	283
Kiew	50 27		30	30	600	183
Krassnyj-Koljadin	50 56		33	3	538	164
Libau	56 31		21	1	20	6
Lugan	48 35		39	20	164	50
Malyj-Usen	50 31		47	37	95	29
Marchinskae, Siberia	62 10		129	43	518	158
Melitopol	46 51		35	23	56	17
Mesen	65 50		44	16	52	16
Moskau	55 46		37	40	469	143
Nertschinsk, (Hüttenw.) Siberia	51 19		119	37	2156	657
Nikolaewsk a. A., Siberia	53 8		140	45	85	26
Nikolaewskij-Sawod, Siberia	55 55		101	28	1198	365
Nikolsk	59 32		45	27	486	148
Noshowka	57 5		54	45	387	118
Noworossijsk	44 43		37	46	66	20
Obdorsk	66 31		66	35	121	37
Odessa	46 29		30	44	213	65
Omsk, Siberia	54 58		73	20	292	89
Orenburg	51 45		55	6	354	108
*Pawlowsk	59 41		30	29	131	40
Pensa	53 11		45	1	722	220
Perm	58 1		56	16	384	117
Pernau	58 23		24	30	33	10
Petrosawodsk	61 47		34	23	220	67
Petrowsk	42 59		47	31	33	10
Pinsk	52 7		26	6	459	140
Pjatigorsk	44 3		43	5	1657	505
Pleskau	57 49		28	20	148	45
Polibino	53 44		52	56	322	98
Rostow, a. D.	47 13		39	43	292	89
Rykowskoe, Siberia	50 47		142	55	449	137
*St. Petersburg	59 56		30	16	20	6
Schenkursk	62 6		42	54	138	42
Skopin	53 49		39	33	512	156
Slatoust	55 10		59	41	1476	450
Ssimferopol	44 57		34	6	882	269
Ssmolensk	54 47		32	4	692	211
Ssofijskij Priisk, Siberia	52 27		134	7		
Ssolowezkij-Kloster	65 1		35	45	39	12
Staro-Ssidorowa, Siberia	55 26		65	10	344	105
Tambow	52 44		41	28	433	132
Tara, Siberia	56 54		74	17	259	79
*Tiflis	41 43		44	48	1342	409

SMITHSONIAN TABLES.

TABLE 100.

LIST OF METEOROLOGICAL STATIONS.

(The asterisk * designates stations of the first order.)

	Latitude.	Longitude from Greenwich.	Height above Sea-level.	
			Feet.	m.
RUSSIA. *(Continued.)*				
Tjumeu, Siberia	57° 10′ N.	65° 32′ E.	272	83
Tobolsk, Siberia	58 12	68 14	171	52
Tomsk, Siberia	56 30	84 58	305	93
Tunka, Siberia	51 45	102 33	2434	742
Uman	48 45	30 13	735	224
Uralsk	51 12	51 22	98	30
Urjupinskaja	50 48	42 0	302	92
Ust-Ssyssolsk	61 40	50 51	413	126
Walaam, Finland	61 23	30 57	141	43
Warschau	52 13	21 2	390	119
Wernyj, Siberia	43 16	76 53	2402	732
Wilna	54 41	25 18	348	106
Wjatka	58 36	49 41	587	179
Wladikawkas	43 2	44 41	2244	684
Wologda	59 14	39 53	387	118
Wyschnij-Wolotschek	57 35	34 34	545	166
SPAIN AND PORTUGAL.				
Barcelona, Spain	41 22 N.	2 10 E.	69	21
Cádiz, Spain	36 31	6 18 W.		
*Coimbra, Portugal	40 12	8 25 W.	459	140
Gibralter	36 6	5 21 W.	53	16
*Lisboa, Portugal	38 43	9 9 W.	312	95
Madrid, Spain	40 24	3 41 W.	2149	655
Oporto, Portugal	41 9	8 27 W.	279	85
Oviedo, Spain	43 23	5 48 W.	801	244
San Fernando, Spain	36 28	0 25 W.	92	28
*Sierra da Estrella, Portugal	40 25	7 35 W.	4728	1441
Valencia, Spain	39 28	0 22 W.	59	18
SWEDEN. *(See* DENMARK.)				
SWITZERLAND.				
Altstätten	47 23 N.	9 33 E.	1542	470
Altdorf	46 53	8 39	1588	484
Basel	47 33	7 35	912	278
*Bern	46 57	7 26	1880	573
Castasegna	46 20	9 31	2297	700
Chaumont	47 1	6 59	3701	1128
Gäbris	47 23	9 28	4111	1253
Genf	46 12	6 9	1339	408
Lugano	46 0	8 57	902	275
Neuenburg	47 0	6 57	1601	488
Rigi-Kulm	47 3	8 30	5873	1790
*St. Bernhard	45 52	7 11	8130	2478

TABLE 100.
LIST OF METEOROLOGICAL STATIONS.
(The asterisk * designates stations of the first order.)

	Latitude.	Longitude from Greenwich.	Height above Sea-level.	
			Feet.	m.
SWITZERLAND. *(Continued.)*				
*Säntis	47° 15′ N.	9° 20′ E.	8202	2500
Sils-Maria	46 26	9 46	5938	1810
Zürich	47 23	8 33	1542	470
TURKEY. (See GREECE.)				
ASIA.				
[*The Stations are in India unless otherwise indicated. For Siberian Stations, see* RUSSIA.]				
Aden, Arabia	12 45 N.	45 3 E.	94	29
Ajmere	26 28	74 37	1611	491
Akyab	20 28	92 57	20	6
*Allahabad	25 26	81 52	309	94
Amini Divi	11 6	72 48	15	5
Bangalore	12 59	77 38	2981	909
Belgaum	15 52	74 42	2524	769
Bellary	15 9	76 57	1475	450
Benares	25 20	83 2	267	81
Berhampore	24 6	88 17	66	20
Bhamo	24 12	96 58		
*Bombay	18 54	72 49	37	11
Bushirè, Persia	28 59	50 49	25	8
*Calcutta	22 32	88 20	21	6
Chamba	32 34	76 10	3005	916
Chemulpo, Corea	37 29	126 33	30	9
Chittagong	22 21	91 50	87	26
Colombo	6 56	79 52	40	12
Cuttack	20 29	85 54	80	24
Dacca	23 43	90 27	22	7
Deesa	24 16	72 14	466	142
Delhi	28 40	77 16	718	219
Dhubri	26 7	89 50	115	35
Diamond Island	15 52	94 19	41	12
Fusan, Corea	35 6	129 30		
Hakodate, Japan	41 46	140 44	10	3
Hiroshima, Japan	34 23	132 27	14	4
*Hong-Kong, China	22 18	114 11	110	34
Hyderabad	25 25	68 27	117	36
Indore	22 44	75 53	1823	556
Jeypore	26 55	75 50	1431	436
Jhansi	25 27	78 37	840	256
Jubbulpore	23 9	79 59	1341	409
Kagoshima, Japan	31 35	130 33	13	4
Kanazawa, Japan	36 33	136 40	95	29
Kandy	7 18	80 40	1696	517
Kaschgar, China	39 25	76 7	3999	1219
Katmandu	27 42	85 12	4388	1338
Kelung, China	25 20	121 46	33	10

LIST OF METEOROLOGICAL STATIONS.

(The asterisk * designates stations of the first order.)

	Latitude.	Longitude from Greenwich.	Height above Sea-level.	
ASIA. *(Continued.)*			Feet.	m.
Kioto, Japan	35° 1' N.	135° 46' E.	161	49
Kurrachee	24 47	67 4	49	15
*Lahore	31 34	74 20	702	214
Leh	34 10	77 42	11503	3506
Lucknow	26 50	81 0	369	112
Madras	13 4	80 14	22	7
Mandalay	21 59	96 8		
Mangalore	12 52	74 54	26	8
Matsuyama, Japan	33 50	132 45	105	32
Mergui	12 11	98 38	96	29
Moulmein	16 29	97 40	94	29
Mussooree	30 28	78 7	6881	2097
Nagasaki, Japan	32 44	129 52	190	58
Nagoya, Japan	35 10	136 55	49	15
Nagpur	21 9	79 11	1025	312
Nemuro, Japan	43 20	145 35	89	27
Niigata, Japan	37 55	139 3	85	26
Oita, Japan	33 13	131 36	26	8
Osaka, Japan	34 42	135 31	23	7
Patna	25 37	85 14	183	56
Peking, China	39 57	116 28	125	38
Peshawar	34 2	71 37	1110	338
Poona	18 28	74 10	1840	561
Quetta, Beluchistan	30 11	67 3	5502	1677
Raipur	21 15	81 41	960	293
Rajkot	22 17	70 52	429	131
Rangoon	16 46	96 12	41	12
Sakai, Japan	35 33	133 14	7	2
Sapporo, Japan	43 4	141 22	56	17
Saugor Island	21 39	88 5	25	8
Silchar	24 49	92 50	104	32
Simla	31 6	77 12	7048	2148
Si-wan-tse, China	40 59	115 18	3904	1190
Söul, Corea	37 35	127 7	118	36
Soya, Japan	45 31	141 55	79	24
Surat	21 13	72 46	36	11
Taku, China	38 59	117 40	33	10
Tezpur	26 36	92 50	251	76
Tokio, Japan	35 41	139 45	69	21
Trichinopoly	10 50	78 44	255	78
Udan, China	44 35	111 10		
Urga, China	47 55	106 50	3773	1150
Vizagapatam	17 42	83 22	31	9
Wakayama, Japan	34 14	135 9	49	15
Yuensan, Corea	39 10	127 25		
*Zi-Ka-Wei, China	31 12	119 6	23	7

TABLE 100.
LIST OF METEOROLOGICAL STATIONS.
(The asterisk * designates stations of the first order.)

	Latitude.	Longitude from Greenwich.	Height above Sea-level.	
			Feet.	m.
AUSTRALASIA.				
Adelaide, South Australia	34° 57' S.	138° 35' E.		
Albany, West Australia	35 2	117 54	88	27
Alice Springs, South Australia	23 38	133 37	2100	640
Auckland, New Zealand	36 50	174 51	258	79
*Batavia, Java	6 11	106 50	26	8
*Boulia, Queensland	22 55	139 38		
Bourke, New South Wales	30 3	145 58	347	106
*Brisbane, Queensland	27 28	153 6	137	42
*Burketown, Queensland	17 48	139 34		
*Cooktown, Queensland	15 28	145 17		
Derby, West Australia	17 18	123 39	17	5
Eucla, South Australia	31 45	128 58	7	2
Hobart, Tasmania	42 53	147 20	190	58
*Mackay, Queensland	21 9	149 13		
Malacca, Straits Settlements	2 10 N.	102 14	12	4
*Manila, Philippine Islands	14 35 N.	120 58	46	14
Melbourne, Victoria	37 50 S.	145 0	91	28
Penang, Straits Settlement	5 2 N.	100 20	20	6
Perth, West Australia	31 57 S.	115 52	47	14
Port Darwin, South Australia	12 28 S.	130 51	70	21
Province Wellesley, Straits Settlement.	5 22 N.	100 30	43	13
Singapore, Straits Settlement	1 17 N.	103 51	10	3
*Sydney, New South Wales	33 52 S.	151 12	155	47
*Thargomindah, Queensland	27 58 S.	143 43		
*Thursday Island, Queensland	10 34 S.	142 12		
Wellington, New Zealand	41 16 S.	174 47	140	43
AFRICA AND NEIGHBORING ISLANDS.				
Alexandria, Egypt	31 12 N.	29 53 E.	62	19
Assab, Abyssinia	12 59	42 45	36	11
Alger, Algeria	36 47	3 4	125	38
Biskra, Algeria	34 51	5 40	400	122
Bizerte, Tunis	37 17	9 50	20	6
Cairo, Egypt	30 5	31 17		
Cape Town, Cape Colony	33 56 S.	18 29	40	12
Ceres, Cape Colony	33 22 S.	19 20	1493	455
Constantine, Algeria	36 22 N.	6 37	2165	660
Cradock, Cape Colony	32 11 S.	25 38	2856	870
Fort Napier, Natal	29 36 S.	30 23	2200	671
Fort National, Algeria	36 38 N.	4 12	3005	916
Gabès, Tunis	33 53 N.	10 7	33	10
Ghardaia, Algeria	32 35 N.	3 40	1706	520
Grahamstown, Cape Colony	33 20 S.	26 33	1800	549
Ismailia, Egypt	30 36 N.	32 16	30	9
Kimberley, Cape Colony	28 43 S.	26 46	4050	1234

TABLE 100.

LIST OF METEOROLOGICAL STATIONS.
(The asterisk * designates stations of the first order.)

	Latitude.	Longitude from Greenwich.	Height above Sea-level.	
AFRICA AND NEIGHBORING ISLANDS. (*Continued.*)			Feet.	m.
Laghouat, Algeria	33° 48′ N.	2° 51′ E.	2454	748
Memours, Algeria	35 6 N.	1 51 W.	13	4
Oran, Algeria	35 42 N.	0 39 W.	197	60
Port Elizabeth, Cape Colony . . .	33 57 S.	25 37 E.	181	55
Port-Saïd, Egypt	31 16 N.	32 18 E.	20	6
Queenstown, Cape Colony	31 51 S.	26 51 E.	3500	1067
* St. Paul de Loando, Angola . . .	8 49 S.	13 7 E.	194	59
Sierra Leone, Senegambia	8 30 N.	13 9 W.	224	68
Sidi-Bel-Abbés, Algeria	35 2 N.	0 39 W.	1562	476
Suez, Egypt	29 59 N.	32 30 E.	10	3
Tamatave, Madagascar	18 10 S.	49 25 E.	10	3
Tananarive, Madagascar	18 55 S.	47 36 E.	4593	1400
Tripoli	32 53 N.	13 11 E.	66	20
Vivi, Congo	5 40 S.	13 49 E.	364	111
INTERNATIONAL POLAR STATIONS.				
Bossekop, (*Norway*)	69 57 N.	23 15 E.		
Dicksonhavn, (*Holland*)	73 30	81 0 E.		
Fort Rae, (*Great Britain*)	62 39	115 44 W.		
Godthaab, (*Denmark*)	64 11	51 44 W.		
Jan Mayen, (*Austria*)	70 59	8 28 W.		
Kingua-Fjord, Cumberland Sound, (*Germany*).	66 36	67 9 W.		
Lady Franklin Bay, (*United States*)	81 44	64 45 W.		
Nowaja Semlja, (*Russia*)	72 30	52 45 E.		
Orange Baie, Cape Horn, (*France*)	55 31 S.	70 25 W.		
Point Barrow, (*United States*) . .	71 23 N.	156 40 W.		
Sagastyr, Lena River, (*Russia*) . .	73 23 N.	124 5 E.		
Sodankylä, (*Finland*)	67 27 N.	26 36 E.		
Spitzbergen,(*Sweden*),Cap Thordsen	78 28 N.	15 42 E.		
Süd-Georgien, (*Germany*)	54 31 S.	36 0 W.		
MISCELLANEOUS ISLANDS.				
Barbados	13 8 N.	59 40 W.	31	9
Honolulu, Hawaiian Islands . . .	21 18	157 50 W.	50	15
La Canée, Crête	35 30	24 0 E.	141	43
Las Palmas, Canaries	27 28	15 27 W.	30	9
Malta, Mediterranean	35 54	14 31 E.	70	21
Massaua, Red Sea	15 36	39 27 E.	10	3
* Port Louis, Mauritius	20 6 S.	57 33 E.	180	55
* St. Helena	15 55 S.	5 43 W.	40	12
Sainte-Croix, Teneriffe	28 29 N.	16 21 W.	118	36
Stanley, Falkland Islands	51 41 S.	57 51 W.		
Stykkisholm, Iceland	65 5 N.	22 46 W.	36	11
Thorshavn, Färoë Islands	62 2 N.	6 44 W.	30	9

SMITHSONIAN TABLES

APPENDIX.

CONSTANTS.

Numerical Constants.

	Number.	Logarithm.
Base of natural (Naperian) logarithms,	$e = 2.7182818$	0.4342945
Log e, modulus of common logarithms,	$M = 0.4342945$	$9.6377843 - 10$
Circumference of circle in degrees,	$= 360$	2.5563025
" " " in minutes,	$= 21\,600$	4.3344538
" " " in seconds,	$= 1\,296\,000$	6.1126050
Circumference of circle, diameter unity,	$\pi = 3.14159265$	0.4971499

Number.	Logarithm.		Number.	Logarithm.
$2\pi = 6.2831853$	0.7981799	$1/\pi^2 =$	0.1013212	$9.0057003 - 10$
$\dfrac{\pi}{3} = 1.0471976$	0.0200286	$\sqrt{\pi} =$	1.7724539	0.2485749
		$\dfrac{1}{\sqrt{\pi}} =$	0.5641896	$9.7514251 - 10$
$\dfrac{1}{\pi} = 0.3183099$	$9.5028501 - 10$	$\sqrt{2} =$	1.4142136	0.1505150
$\pi^2 = 9.8696044$	0.9942997	$\sqrt{3} =$	1.7320508	0.2385607

The arc of a circle equal to its radius is
in degrees, $\rho° = 180/\pi$	$= 57°.29578$	1.7581226
in minutes, $\rho' = 60\,\rho°$	$= 3\,437.7468'$	3.5362739
in seconds, $\rho'' = 60\,\rho'$	$= 206\,264.8''$	5.3144251

For a circle of unit radius, the
arc of $1° = 1/\rho°$	$= 0.017\,4533$	$8.2418774 - 10$
arc of $1' = 1/\rho'$	$= 0.000\,2909$	$6.4637261 - 10$
arc (or sine) of $1'' = 1/\rho''$	$= 0.000\,00485$	$4.6855749 - 10$

Geodetical Constants.

Dimensions of the earth (Clarke's spheroid, 1866) and derived quantities:

Equatorial semi-axis in feet,	$a = 20926062.$	7.3206875
in miles,	$a = 3963.3$	3.5980536
Polar semi-axis in feet,	$b = 20855121.$	7.3192127
in miles,	$b = 3949.8$	3.5965788
(Eccentricity)$^2 = \dfrac{a^2 - b^2}{a^2}$	$e^2 = 0.00676866$	$7.8305030 - 10$
Flattening $= \dfrac{a - b}{a}$	$\varepsilon = 1/294.9784$	
Perimeter of meridian ellipse,	$= 24\,859.76$	miles.
Circumference of equator,	$= 24\,901.96$	"
Area of earth's surface,	$= 196\,940\,400$	square miles.
Mean density of the earth (HARKNESS)	$= 5.576 \pm 0.016.$	
Surface density " " "	$= 2.56 \pm 0.16.$	

Acceleration of gravity (HARKNESS):

g_ϕ (cm. per second) $= 980.60\,(1 - 0.002662 \cos 2\phi)$, for latitude ϕ and sea level.

g, at equator $= 977.99$; g, at Washington $= 980.07$; g, at Paris $= 980.94$.
g, at poles $= 983.21$; g, at Greenwich $= 981.17$;

Length of the seconds pendulum (HARKNESS):

$l = 39.012540 + 0.208268 \sin^2 \phi$ inches $= 0.990910 + 0.005290 \sin^2 \phi$ metres.

APPENDIX.

CONSTANTS.— Continued.

Astronomical Constants (HARKNESS).

Sidereal year = 365.256 357 8 mean solar days. Tropical year = 365.2422 d.
Sidereal day = 23^h 56^m 4.100^s mean solar time.
Mean solar day = 24^h 3^m 56.546^s sidereal time.
Mean distance of the earth from the sun = 92 800 000 miles.

Physical Constants.

Velocity of light (HARKNESS) = 186 337 miles per second = 299 878 km. per second.
Velocity of sound through dry air = 1090 $\sqrt{1 + 0.00367\, t^\circ\, C.}$ feet per second.
Weight of distilled water, free from air, barometer 30 inches:

Volume.	Weight in grains.		Weight in grammes.	
	62° F.	4° C.	62° F.	4° C.
1 cubic inch (determination of 1890)	252.286	252.568	16.3479	16.3662
1 cubic centimetre (1890)	15.3953	15.4125	0.9976	0.9987
1 cubic foot (1890) at 62° F.	62.2786 lbs.			

A standard atmosphere is the pressure of a vertical column of pure mercury whose height is 760 mm. and temperature 0° C., under standard gravity at latitude 45° and at sea level.

1 standard atmosphere = 1033 grammes per sq. cm. = 14.7 pounds per sq. inch.
Pressure of mercurial column 1 inch high = 34.5 grammes per sq. cm. = 0.491 pounds per sq. inch.

Weight of dry air (containing 0.0004 of its weight of carbonic acid):
  1 cubic centimetre at temperature 32° F. and pressure 760 mm. and under the standard value of gravity weighs 0.00129305 gramme.

Density of mercury at 0° C. (compared with water of maximum density under atmospheric pressure) = 13.5956.

Freezing point of mercury = $-38^\circ.5$ C. (REGNAULT, 1862.)

Coefficient of expansion of air (at const. pressure of 760^{mm}) for 1° C. (DO.): 0.003670.

Coefficient of expansion of mercury for Centigrade temperatures (BROCH):
  $\Delta = \Delta_0 (1 - 0.000\,181\,792\,t - 0.000\,000\,000\,175\,t^2 - .000\,000\,000\,035\,116\,t^3)$.

Coefficient of linear expansion of brass for 1° C., $\beta = 0.000\,0174$ to $0.000\,0190$.

Coefficient of cubical expansion of glass for 1° C., $\gamma = 0.000\,021$ to $0.000\,028$.
  Ordinary glass (RECKNAGEL): at 10° C., $\gamma = 0.000\,0255$; at 100°, $\gamma = 0.000\,0276$.

Specific heat of dry air compared with an equal weight of water:
  at constant pressure, $Kp = 0.2374$ (from 0° to 100° C., REGNAULT).
  at constant volume, $Kv = 0.1689$.

Ratio of the two specific heats of air (RÖNTGEN): $Kp/Kv = 1.4053$.

Thermal conductivity of air (GRAETZ): $k = 0.000\,0484\, (1 + 0.001\,85\, t^\circ\, C.)\, \frac{\text{gramme}}{\text{cm. sec.}}$.
[The quantity of heat that passes in unit time through unit area of a plate of unit thickness, when its opposite faces differ in temperature by one degree.]

Latent heat of liquefaction of ice (BUNSEN) = 80.025 mass-degrees, C.
Latent heat of vaporization of water = $606.5 - 0.695\, t^\circ\, C$.

Absolute zero of temperature (THOMSON, Heat, *Encyc. Brit.*): $-273^\circ.0$ C. = $-459^\circ.4$ F.

Mechanical equivalent of heat*:
  1 pound-degree, F. (the British thermal unit) = about 778 foot-pounds.
  1 pound-degree, C. = 1400 foot-pounds.
  1 calorie or kilogramme-degree, C. = 3087 foot-pounds = 426.8 kilogram-metres = 4187 joules (for $g = 981$ cm.).

* Based on Prof. Rowland's determinations. (*Proc. Am. Acad. Arts and Sci.*, 1880.)

APPENDIX.

SYNOPTIC CONVERSION OF ENGLISH AND METRIC UNITS.
English to Metric.

	Metric equivalents.	Logarithms.
Units of length.		
1 inch.	2.54000 centimetres.	0.404 835
1 foot.	0.304801 metre.	9.484 016 − 10
1 yard.	0.914402 "	9.961 137 − 10
1 mile.	1.60935 kilometres.	0.206 650
Units of area.		
1 square inch.	6.4516 square centimetres.	0.809 669
1 square foot.	929.034 " "	2.968 032
1 square yard.	0.83613 square metre.	9.922 274 − 10
1 acre.	0.404687 hectares.	9.607 120 − 10
1 square mile.	2.5900 square kilometres.	0.413 300
" "	259 hectares.	2.413 300
Units of volume.		
1 cubic inch.	16.3872 cubic centimetres.	1.214 504
1 cubic foot.	0.028317 cubic metres or steres.	8.452 047 − 10
1 cubic yard.	0.76456 cubic metres or steres.	9.883 411 − 10
Units of capacity.		
1 gallon (U. S.) = 231 cubic inches.	3.78544 litres.	0.578 116
1 quart (U. S.)	0.94636 litres.	9.976 056 − 10
1 Imperial gallon (British).	4.5468 litres.	0.657 709
277.463 cubic inches (1890).		
1 bushel (U. S.) = 2150.42 cubic inches.	35.2393 litres.	1.547 027
1 bushel (British).	36.3477 litres.	1.560 477
Units of mass.		
1 grain.	64.7989 milligrammes.	1.811 568
1 pound avoirdupois.	0.4535924 kilogrammes.	9.656 666 − 10
1 ounce avoirdupois.	28.3495 grammes.	1.452 546
1 ounce troy.	31.1035 grammes.	1.492 809
1 ton (2240 lbs.).	1.01605 tonnes.	0.006 914

Units of velocity.

1 foot per sec. (0.6818 miles per hr.) = 0.30480 metres per sec. = 1.0973 km. per hr.
1 mile per hr. (1.46667 feet per sec.) = 0.44704 metres per sec. = 1.6093 km. per hr.

Units of force.

1 poundal.	13825.5 dynes.	4.140 682
Weight of 1 grain (for $g = 981$ cm.).	63.57 dynes.	1.803 237
Weight of 1 pound av. (for $g = 981$ cm.).	4.45×10^5 dynes.	5.648 335

Units of stress—in gravitation measure.

1 pound per square inch =	70.307 grammes per sq. centimetre.	1.846 997
1 pound per square foot =	4.8824 kilogrammes per sq. metre.	0.688 634

Units of work—in absolute measure.

1 foot-poundal.	421 403 ergs.	5.624 697

—in gravitation measure.

1 foot-pound (for $g = 981$ cm.) = 1356.3×10^4 ergs = 0.138255 kilogram-metres.

Units of activity (rate of doing work).

1 foot-pound per minute (for $g = 981$ cm.) = 0.022605 watts.
1 horse-power (33 000 foot-pounds per min.) = 746 watts = 1.01387 force de cheval.

Units of heat.

1 pound-degree, F.	= 252 small calories or gramme-degrees, C.
1 pound-degree, C.	= 1.8 pound-degrees, F.

APPENDIX.

SYNOPTIC CONVERSION OF ENGLISH AND METRIC UNITS.
Metric to English.

	English equivalents.		Logarithms.
Units of length.			
1 metre (10^6 microns).	39.3700	inches.	1.595 165
"	3.28083	feet.	0.515 984
"	1.09361	yards.	0.038 863
1 kilometre.	0.62137	miles.	9.793 350 − 10
Units of area.			
1 square centimetre.	0.15500	square inches.	9.190 331 − 10
1 square metre.	10.7639	square feet.	1.031 968
" "	1.19599	square yards.	0.077 726
1 are (= 100 square metres).	119.599	square yards.	2.077 726
1 hectare.	2.47104	acres.	0.392 880
1 square kilometre.	0.38610	square miles.	9.586 700 − 10
Units of volume.			
1 cubic centimetre.	0.0610234	cubic inches.	8.785 496 − 10
1 cubic metre or stere.	35.3145	cubic feet.	1.547 953
" " "	1.30794	cubic yards.	0.116 589
Units of capacity.			
1 litre (61.023 cubic inches).	0.26417	gallons (U. S.).	9.421 884 − 10
"	1.05668	quarts (U. S.).	0.023 944
"	0.21993	Imp. gallons (British).	9.342 291 − 10
1 hectolitre.	2.83774	bushels (U. S.).	0.452 973
"	2.7512	bushels (British).	0.439 523
Units of mass.			
1 gramme.	15.4324	grains.	1.188 432
1 kilogramme.	2.20462	pounds avoirdupois.	0.343 334
"	35.274	ounces avoirdupois.	1.547 454
"	32.1507	ounces troy.	1.507 191
1 tonne.	0.98421	tons (2240 lbs.).	9.993 086 − 10
Units of velocity.			
1 metre per second.	3.2808	feet per second.	0.515 984
" " "	2.2369	miles per hour.	0.349 653
1 km. per hr. (0.2778 m. per sec.)	0.62137	miles per hour.	9.793 350 − 10

Units of force.

1 dyne (weight of $(981)^{-1}$ grammes, for $g = 981$ cm.) $= 7.2330 \times 10^{-5}$ poundals.

Units of stress—in gravitation measure.

1 gramme per square centimetre. 0.014223 pounds per sq. inch.
1 kilogramme per square metre. 0.20482 pounds per sq. foot.
1 standard atmosphere. 14.7 pounds per sq. inch. (*See* def. p. 259.)

Units of work—in absolute measure.

1 erg. 2.3730×10^{-8} foot-poundals.
 1 megalerg $= 10^6$ ergs; 1 joule $= 10^7$ ergs.

—In gravitation measure.

1 kilogram-metre (for $g = 981$ cm.) $= 981 \times 10^5$ ergs $= 7.2330$ foot-pounds.

Units of activity (rate of doing work).

1 watt. 44.2385 foot-pounds per minute, for $g = 981$ cm.
 1 watt $=$ 1 joule per sec. $= 0.10194$ kilogram-metre per sec., for $g = 981$ cm.
1 force de cheval $= 75$ kilogram-metres per sec. $= 735\frac{3}{4}$ watts $= 0.98632$ horse-power.

Units of heat.

1 calorie or kilogramme-degree $= 3.968$ pound-degrees, $F. = 2.2046$ pound-degrees, $C.$
1 small calorie or therm, or gramme-degree $= 0.001$ calorie or kilogramme-degree.

APPENDIX.

DIMENSIONS OF PHYSICAL QUANTITIES.

L = length; M = mass; T = time.

Quantity.	Dimensions.	Quantity.	Dimensions.
Area.	$[L^2]$	Momentum.	$[L\,M\,T^{-1}]$
Volume.	$[L^3]$	Moment of Inertia.	$[M\,L^2]$
Mass.	$[M]$	Force.	$[L\,M\,T^{-2}]$
Density.	$[M\,L^{-3}]$	Stress (per unit area).	$[L^{-1}\,M\,T^{-2}]$
Velocity.	$[L\,T^{-1}]$	Work or Energy.	$[L^2\,M\,T^{-2}]$
Acceleration.	$[L\,T^{-2}]$	Rate of Working (power)	$[L^2\,M\,T^{-3}]$
Angle.	$[0]$	Heat.	$[L^2\,M\,T^{-2}]$
Angular Velocity.	$[T^{-1}]$	Thermal Conductivity.	$[L^{-1}\,M\,T^{-1}]$

In Electrostatics.

Quantity	Symbol.	Dimensions in electrostatic system.
Quantity of Electricity.	e	$[L^{3/2}\,M^{1/2}\,T^{-1}]$
Surface Density: quantity per unit area.	σ	$[L^{-1/2}\,M^{1/2}\,T^{-1}]$
Difference of Potential: quantity of work required to move a quantity of electricity; (work done) ÷ (quantity moved).	E	$[L^{1/2}\,M^{1/2}\,T^{-1}]$
Electric Force, or Electro-motive Intensity: (quantity) ÷ (distance^2).	F	$[L^{-1/2}\,M^{1/2}\,T^{-1}]$
Capacity of an accumulator: $e \div E$.	C or q	$[L]$
Specific Inductive Capacity.	k	$[0]$

In Magnetics.

Quantity	Symbol.	Dimensions in electro-magnetic system.
Quantity of Magnetism, or Strength of Pole.	m	$[L^{3/2}\,M^{1/2}\,T^{-1}]$
Strength or Intensity of Field: (quantity) ÷ (distance^2).	S	$[L^{-1/2}\,M^{1/2}\,T^{-1}]$
Magnetic Force.	$\mathfrak{H}$	$[L^{-1/2}\,M^{1/2}\,T^{-1}]$
Magnetic Moment: (quantity) × (length).	ml	$[L^{5/2}\,M^{1/2}\,T^{-1}]$
Intensity of Magnetization: magnetic moment per unit volume.	I	$[L^{-1/2}\,M^{1/2}\,T^{-1}]$
Magnetic Potential: work done in moving a quantity of magnetism; (work done) ÷ (quantity moved).	V or Ω	$[L^{1/2}\,M^{1/2}\,T^{-1}]$
Magnetic Inductive Capacity.	μ	$[0]$

In Electro-magnetics.

Quantity	Symbol.	Dimensions in electro-magnetic system.	Name of practical unit.
Intensity of Current.	i	$[L^{1/2}\,M^{1/2}\,T^{-1}]$	Ampere.
Quantity of Electricity conveyed by current: (intensity) × (time).	e	$[L^{1/2}\,M^{1/2}]$	Coulomb.
Potential, or difference of potential: (work done) ÷ quantity of electricity upon which work is done.	E	$[L^{1/2}\,M^{1/2}\,T^{-2}]$	Volt.
Electric Force: the mechanical force acting on electro-magnetic unit of quantity; (mechanical force) ÷ (quantity).	$\mathfrak{E}$	$[L^{1/2}\,M^{1/2}\,T^{-2}]$	
Resistance of a conductor: $E \div i$.	R	$[L\,T^{-1}]$	Ohm.
Capacity: quantity of electricity stored up per unit potential-difference produced by it.	q	$[L^{-1}\,T^{2}]$	Farad.
Specific Conductivity: the intensity of current passing across unit area under the action of unit electric force.		$[L^{-2}\,T]$	
Specific Resistance: the reciprocal of specific conductivity.	r	$[L^2\,T^{-1}]$	

INTERNATIONAL METEOROLOGICAL SYMBOLS.

The International Meteorological Congress, held at Vienna, in September, 1873, decided that it was desirable to introduce for various meteorological conditions, symbols which should be independent of any national language and therefore universally intelligible. From the symbols and abbreviations then in use among different nations, the Permanent Committee of the Congress selected a number for international use. The symbols were modified by the Congress at Munich, in 1891, and the abbreviations for clouds by the Conference at Upsala, in 1894.

References :
"Summary of Resolutions of the Vienna Congress, Appendix K." p. 64. Prepared by Mr. Robert H. Scott, Secretary.
"Bericht über die Int. Meteor. Conferenz in München, 1891," p. 19.
"Report of the Int. Met. Conference at Munich," p. 20.
"Circular of the U. S. Weather Bureau, January 1, 1894."

The intensity of the condition is indicated by the small figures ^{0} and ^{2} which are used as exponents of the symbols. Zero (^{0}) denotes very slight intensity; two (^{2}) strong or marked. Absence of an exponent indicates moderate intensity. A dash (—) indicates continuance.

Example.   *Translation.*

⬤^{0} Light rain.

⬤   Moderate rain.

⬤^{2} Heavy rain.

1 ⚡ 9 p. — 10 p. in E.   1st. Silent lightning from 9–10 p. m. in the E.

4 ⬤^{2} — 10 a ; ⚡ 3 p. — 5 p.   4th. Heavy rain ended 10 a. m.; thunderstorm from 3 to 5 p. m.

16 ∞^{2} a ; ✳^{2} 11.30 a — 2.50 p.   16th. Dense haze in the morning; heavy snow from 11.30 a. m. to 2.50 p. m.

The time of occurrence is expressed in hours; morning and afternoon by a. and p. respectively. The hours are counted from 0 to 12 commencing with midnight.

Where tables are printed, **maximum** and **minimum** values will be in heavy-faced type.

Absence of precipitation is denoted by a dot (.), and amounts less than .01 inch (formerly marked T) are recorded .00.

°	means	Degree.	Mi.	means	Miles.
F	"	Fahrenheit.	Kil.	"	Kilometers.
C	"	Centigrade.	N.	"	Nimbus.
Ci.	"	Cirrus.	AS.	"	Alto-stratus.
Ci. Cu.	"	Cirro-cumulus.	CuN.	"	Cumulo-nimbus.
Ci. S.	"	Cirro-stratus.	Fr. Cu.	"	Fracto-cumulus.
A. Cu.	"	Alto-cumulus.	Fr. N.	"	Fracto-nimbus.
Cu.	"	Cumulus.	Fr. S.	"	Fracto-stratus.
S. Cu.	"	Strato-cumulus.	Scf.	"	Stratus cumuliformis.
S	"	Stratus.	Ncf.	"	Nimbus cumuliformis
Max.	"	Maximum.	MCu.	"	Mammato cumulus.
Min.	"	Minimum.			

1. ◉ RAINFALL—Indicates that an appreciable quantity of rain (one hundredth of an inch or more) has fallen during the day or since the last observation; also, that the day is a rainy day as distinguished from snowy or clear days.

2. ✲ SNOWFALL—Indicates that an appreciable quantity of snow has fallen during the day. ✲° may be used to denote flurries of snow.

3. ▲ HAILSTONES—Hard semi-transparent ice, whether small or large, crystalline or rounded. ▲° small quantity of hailstones; ▲² large quantity of hailstones.

4. △ SLEET—Or pellets of snow or soft hail without any crystalline structure. This symbol is used by the Germans for *Graupeln*, or snow pellets, and for the semi-transparent mixture of snow and ice that in the dry weather of Central Europe nearly corresponds to the sleet of the coasts of England and America. △° small quantity of sleet; △² much sleet.

5. V SILVER FROST—(English, "silver thaw," French, *givre*, German, *Rauhfrost* or *duft-anhang*); this refers to an accumulation of snow and sleet on the limbs of trees, in which the snow is the main feature, so that the external appearance is silvery white and rough.

6. ⌒⌒ GLAZED FROST—(French, *verglas*, German, *Glatteis*); this refers to an accumulation of snow and ice on the trees, in which the ice is in excess and the external appearance is smooth and transparent. In using the symbols for "silver frost" and "glazed frost," the Munich Conference requests that these terms be considered as descriptive of the resulting phenomena, no matter how they are brought about, therefore the definitions avoid any statement as to the conditions attending the formation of the depositions. The same rule applies to the use of the symbol for "hoar frost."

7. ← ICE-NEEDLES—(Not yet well defined by international usage).

8. ⇸ DRIFTING SNOW—(German, *schneegestober*); this symbol indicates that strong winds are raising the snow from the ground, filling the air with it like dust, and transporting it horizontally; this may occur under a clear sky. The symbol does not refer to snow falling from the clouds, nor to the mere fact that the snow is lying in drifts on the ground. When the air is filled with blinding snow-dust, use the symbol ⇸², but for light winds and light snow-dust use ⇸°.

9. ⊠ SNOW-COVERING—Or quantity of snow lying on the ground; when more than half the soil in the neighborhood of any station is covered with snow this is indicated by ⊠, if the snow covering is thin, use ⊠°, but if it is considered deep for that station use ⊠².

10. ≡ FOG ≡ Ground fog not exceeding height of a man; ≡° thin fog or mist enveloping and above the observer; ≡² heavy fog or mist, such as the Scotch mist, drizzling down upon the observer. Fog symbols should not be used when an observer at a high station notices fog in the valley below him; this should be expressed by a note in the daily journal.

11. ∞ HIGH HAZE—Such as makes distant mountains appear hazy, or such as covers the sky in the case of Indian summer haze or prairie fires; German, *Moorrauch*. If clouds are also prevalent in connection with this haze, the additional cloud signal should be given. The intensity, or density, of the haze is expressed by ∞° for light haze and ∞² for dense haze. The symbol ∞ indicates merely the hazy condition, or the optical result, without considering whether the haze is caused by dust or moisture.

12. ⌒ DEW; ⌒° LIGHT DEW; ⌒² HEAVY DEW—As the formation of dew depends upon the nature and exposure of the horizontal surface on which dew is deposited, the observer should use the same horizontal object uniformly throughout the season.

13. ⊓ HOAR FROST; ⊓° LIGHT HOAR FROST; ⊓² HEAVY HOAR FROST, injurious to vegetation—The expression "frosty weather" refers to the low temperature as such; but the expression "hoar frost" to the crystalline ice deposited upon the surface of solids in the open air. Hoar frost is deposited on horizontal objects generally under a clear sky at night.

14. ⇉ STRONG WIND—An arrow with four feathers indicates a wind whose strength is 8, 9, 10, 11, or 12 on the Beaufort scale, or 8, 9, or 10, on the international scale, or anything in excess of 50 miles per hour or 20 metres per second in absolute measures; ⇉² a remarkably strong wind or one exceeding 11 on the Beaufort scale, or 80 miles per hour, or 35 metres per second.

15. ⚡ THUNDERSTORM—Namely thunder, whether with or without lightning, rain, hail, or wind.

16. ⚡ LIGHTNING—Distant lightning or any form of lightning that occurs without audible thunder, even when it occurs in the zenith, which is sometimes the case (this latter occurrence should be especially described in the journal of the observer); ⚡° infrequent lightning, or lightning that is confined to a small region of the sky; ⚡² lightning that occurs very frequently or extends over a large region of the sky. When distant lightning appears at a definite direction in the horizon, the observer should add the letters indicating the points of the compass, for instance, ⚡° NW. 10 p. indicates that occasional distant lightning occurred in the northwest at ten p. m.

17. ☉ SOLAR AUREOLA, CORONA, or GLORY—German, *Kranz Lichtkron*, "Corona," *Sonnenhof*. These are small circles of prismatic colors surrounding the sun, the radii of these circles are usually less than six degrees, but in the extreme case of Bishop's ring, its radius was fifteen degrees. Several concentric circles are sometimes visible; each circular band of prismatic colors has its red on the outside, and its blue, violet, or purple on the inside, with respect to the sun; such rings are generally formed when the sun shines through a thin cloud and may be seen if the sun is viewed through neutral-tinted glass or by reflection in water. Similar circles surrounding the shadow of the observer's head are called "anthelia," "aureolæ," "glories," or "fog-shadows," (German, *Gegensonne, Brockenspectra*).

18. ☽ LUNAR AUREOLA or CORONA—(German, *Mondhof*); circles surrounding the moon similar to the solar corona.

19. ⊕ SOLAR HALO—(German, *Sonnenring*); these are larger circles surrounding the sun whose sizes are quite definite, namely, about twenty-two degrees and about forty degrees radius from the sun; they are easily distinguishable from the coronæ by the fact that the colors are feebler and are so arranged that the red light is inside or nearest the sun and the blue light is outside; the greater part of the breadth of the halo is white. Complex combinations of halos, parhelia, horizontal circles, and vertical columns sometimes occur, all of which may be indicated in general by the symbol ⊕², where the figure ² indicates that the display is more brilliant than usual; a detailed statement of the radii or diameters of the rings and columns and of their arrangements should be given in the text.

20. ☡ LUNAR HALO—(German, *Mondring*); phenomena surrounding the moon similar to the solar halo.

21. ⌒ RAINBOW—Double rainbows and those with adjacent supernumerary bows may be indicated by ⌒².

22. ⌒ AURORAL LIGHTS—Namely, any display of the Aurora Borealis.

INDEX.

Abbe, C., work cited.................xxxiii, 1v
Absolute measure......................260, 261
    zero of temperature.......................259
Acceleration, dimensions of..................262
    of gravity....................................258
Activity, units of.........................260, 261
Air, coefficient of expansion of.............259
    density of, at different humidities,
        British..................liv–lv, 221–223
        Metric...................liv–lvi, 225–228
    density of, at different pressures,
        British.............liv, lv, lvi, 221–223
        Metric...................liv–lvi, 225–228
    density of, at different temperatures,
        British........................liv–lv, 220
        Metric...................liv–lvi, 224–225
    specific heat of dry........................259
    thermal conductivity of...............259
    weight of, dry..............................259
Ampere, dimensions of.......................262
Angle,..262
    conversion of days into............212–215
Angular velocity, dimensions of...........262
Angot, A., treatise cited.....................xxii
Aqueous vapor, decrease of pressure
    with altitude........................xliii, 146
    pressure of,
        British......xxxviii, xxxix, 134, 135
        Metric........................xli, 142, 143
    pressure of, at low temperature,
        xxxvi, 130, 131
    in saturated air,
        British.........xxxv, xxxvi, 122–127
        Metric..........xxxv, xxxvi, 128, 129
    weight of, British...............xxxvii, 132
    weight of, Metric...............xxxvii, 133
Arc, conversion into time.................li, 210
    of circle equal to its radius...............258
Area, dimensions of...........................262
    of surface of earth.........................258
    units of..................................260, 261
Astronomical constants......................259
Atmosphere, standard pressure of...259, 261
    weight of unit of volume...............259
Aureola, solar.................................266

Aureola, lunar................................266
Auroral lights.................................266
Avoirdupois, conversion into metric,
    lvi, 229–230, 260

Babinet, barometric formula of....xxxii, 118
Barometer, correction for average degree of humidity,
    xxviii, xxix, xxx, 108, 112, 113
    latitude and weight of mercury,
    xxviii, xxx, 106, 107, 114
    variation of gravity with altitude,
    xix, 109, 115
    determination of heights by,
        British measures,
        xxvi–xxix, xxxii, 100–109, 118
        Metric measures,
        xxix–xxxi, xxxii, 110–115, 118
    difference of height corresponding
        to .01 inch change...........xxxi, 116
        1 mm.   "   ....xxxii, 117
Barometric readings,
    reduction to standard gravity,
    xviii–xx, 58–59
    sea level, British...........xx–xxv, 60–77
        Metric................xx–xxvi, 78–98
    standard temperature,
        British...................xv–xvii, 14–33
        Metric................xvii–xviii, 34–56
    when below 0°C xviii
    pressures corresponding to temperature of boiling water,........xxxv, 119
Beaufort, Admiral, wind, scale xlvi, 160, 265
Belli, work cited............................xxxviii
Bessel,   "   " liii
Bishop's ring..................................266
Boiling point, of water..............xxxiii, 119
    corresponding barometric pressures,
    xxxv, 119
Brass, coefficient of linear expansion.....259
Broch, work cited.................xxxv, lv, 259
Brockenspectra................................266
Bunsen, work cited...........................259

Caloric............................259, 260, 261
Capacity, electromagnetic..................260
   magnetic inductive........................262
   measures of, British......................260
   "   " Metric........................261
   of an accumulator........................262
   specific inductive.........................262
   units of..................................260, 261
Centigrade, conversion into Fahrenheit,
                                   xii, 7-9
   when near the boiling point........xii, 9
   of differences into Fahrenheit....xiii, 9
C. G. S. unit of magnetic intensity...lvii, 231
Circle, arc of....................................258
   circumference of............................258
   diameter.......................................258
   circumference of earth..................258
          equator..................................258
Clarke, A. R., treatise cited........xxii, xlvii
Clarke's spheroid......................xlvii, 258
Clouds, names and abbreviations.........264
Coefficient of expansion of air..............259
   linear   "   " brass...........259
   cubical   "   " glass............259
   expansion of mercury.....................259
Conductivity, dimensions of, thermal...262
   specific, electromagnetic................262
   thermal, of air..............................259
Conductor, resistance of.....................262
Constants, astronomical......................259
   geodetical....................................258
   numerical.....................................258
   physical.......................................259
Continental measures of length and
   equivalents..................................208
Conversion of measures of time and
   angle......................................li, 209-218
   linear measures...............xlix, 180-208
   British and Metric units...........260-261
   thermometric scales..................xi, 2-9
Correction, for air temperature in determining heights by barometer, British....................xxvii-xxviii, 104-105
   for air temperature in determining
   heights by barometer, Metric,
                             xxix-xxx, 111
   for temperature of the Mercury in
   the thermometer stem...........xiv, 12
   for gravity, in determining heights
   by the barometer, British, xxviii, 109
   for gravity, in determining heights
   by the barometer, Metric...xxxi, 115
   for humidity in determining heights
   by the barometer, British, xxviii, 108

Correction (Continued).
   for humidity in determining heights
   by the barometer, Metric, xxx, 112-113
   for latitude, in determining heights
   by the barometer, British,
                             xxviii, 106-107
   for latitude, in determining heights
   by the barometer, Metric.....xxx, 114
Corona 266
Cosines, table of natural...........lix, 236-237
Cotangents, table of natural.....lix, 238-239
Coulomb..262
Current, intensity of...........................262

Days, conversion into decimals of year
   and angle....................lii, 212-215
Day decimals of, into hours, minutes
   and seconds........................liii, 216
   mean solar..................................259
   sidereal......................................259
Declination of sun..................xlix, 177
Degree, length of, of meridian and any
   parallel........................xlvii, 164, 165
   length of, of meridian and any
   parallel at different latitudes,
                             xlvii, 164, 165
Degrees, Centigrade into Fahrenheit
   and Reaumur...................xi, xii, 7, 8, 9
   Fahrenheit into Centigrade xi, xii, 3-6
   Reaumur into Fahrenheit and Centigrade............................xi, 2
Density, of air............liv-lvi, 220-228, 259
   of earth, mean 258
         surface...................................258
   dimensions of 262
         surface...................................262
   of mercury..................................259
Depth of rainfall, corresponding quantity of water.......................lviii, 232
Determination of heights by barometer,
   British measures.....xxvi-xxix, 100-109
   Metric measures.....xxix-xxxi, 110-115
Depression of dew-point................138-141
Dew..265
Dew-point ....xxxviii-xli
Difference of heights by barometer,
                       xxxi, xxxii, 116, 117
   of potential.................................262
Differences Fahrenheit to Centigrade, xiii, 9
   Centigrade to Fahrenheit..........xiii, 9
Dimensions, in electrostatic system......262
   electromagnetic system..................262
   of the earth.................................258
   physical quantities........................262

INDEX.   269

Distance, mean of earth from sun.........259
Division tables of, for 28, 29 and 31,
    lviii, lix, 233–235
Dove's pentades...........................lviii, 232
Dry air, weight of............................259
Drifting snow..................................265
Duft-anhang....................................265
Duration of sunshine..........xlviii, 166-177
Dyne..........................lvii, 231, 260, 261

Earth, area of surface of....................258
    density of..................................258
    dimensions of..............................258
    eccentricity of............................258
    ellipticity of..............................258
    equatorial semi-axis......................258
    flattening of..............................258
    mean distance from the sun............259
El, value of the................................208
Electricity quantity of........................262
Electric force or electro-motive intensity,
    262
Electrostatics, quantities in..................262
Electromagnetics "   "...................262
Energy, dimensions of.........................262
Equator, circumference of....................258
Equator, length of semi-axis.................258
Erg...260, 261
Espy, treatise cited........................xxxviii
Expansion, coefficient of, air................259
    brass..259
    glass..259
    mercury.......................................259

Fahrenheit, conversion into Centigrade
    and Reaumur......................xi, xii, 3–6
    differences into differences Centi-
    grade...xiii, 9
Farad, dimensions of............................262
Fathom, Swedish, value of...................208
Ferrel, Wm., treatise cited,
    xxii, xxxi, xxxix, xlix
Feet, conversion into metres........1, 200-201
    per second into miles per hour, xlv, 155
    metres per second..................260
    kilometres per hour................260
Flattening of the earth........................258
Fog, symbol for..................................265
Foot, value of, Austrian......................208
    old French.................................208
    Russian.....................................208
    Rhenish.....................................208
    Spanish.....................................208
    Swedish.....................................208

Foot-pound..................................260, 261
Foot-poundal................................260, 261
Force, dimensions of............................262
    units of..................................260, 261
    electric.......................................262
    magnetic.....................................262
    electromagnetic............................262
Force-de-cheval............................260, 261
Formula, Babinet's barometric....xxxii, 118
    Lambert's, wind direction, xliii, 148-153
Freezing point of mercury..................259
Frost, glazed, hoar, silver, symbols, 264, 265

Gallon (U. S.) and Imperial..........260, 261
Gaussian units.........................lvii, 231
Gegensonne, symbol for.....................266
Geodetical constants..........................258
    tables...................................xlvi, 161
Givre or silver frost...........................264
Glass, coefficient of cubical expansion...259
Glatteis, or glazed frost......................264
Glazed frost.....................................264
Glory or corona, symbol for...............266
Graetz, work cited.............................259
Grains, conversion into grammes...lvii, 230
Grammes, conversion into grains...lvii, 231
Gramme-degree or therm....................261
Grammes per square centimetre...........261
Graupeln...264
Gravitation measure, units in........260, 261
Gravity, acceleration of.......................258
    correction for variation of, with
    altitude,..............xix, xxxi, 109, 115
    correction for variation of, with
    latitude..............xix, xxx, 106, 114
    reduction of barometric readings to
    standard..............xviii-xxiv, 58–98
    relative acceleration of, in different
    latitudes....................xlvi, 162, 163
Guyot, A., treatise cited.....................xxii

Hailstones, description and symbol for..264
Halo, solar and lunar..........................266
Hann, J., treatise cited.......................xliii
Harkness, Wm., treatise cited,
    xix, xxii, xlvi, 258, 259
Haze, symbol for................................265
Hazen, H. A., treatise cited..xliv, xlv, lviii
Heat, dimensions of............................262
    latent, of liquefaction of ice...........259
    vaporization of water....................259
    mechanical equivalent of..............259

Heat (*Continued*).
   specific, of dry air..........................259
   ratio of the two, of air..............259
   units of..................................260, 261
Hectare.................................260, 261
Hectolitre...................................261
Heights, determinatian of, by barometer,
   British..............xxvi-xxix, 100-109
   Metric..............xxix-xxxi, 110-115
   thermometrical measurement of,
                                    xxxiii, 119
Hoar-frost, symbol for.........................265
Horse-power.................................260, 261
Hours, conversion into decimals of a
   day..lii, 216
   of minutes and seconds into decimals of..............................liii, 217
Humidity relative, British,
                     xxxviii-xlii, 138-141
   Metric,
                     xxxviii-xlii, 144-145
   term for....................................lvi, 225
Hygrometrical tables..............xxxv, 122-146
Hypsometry.................xxxiii, xxxiv, 119

Ice, latent heat of liquefaction of.........259
   needles, symbol for.......................264
Inches, conversion into millimetres,
                                    xlix, 180-186
Inductive capacity, magnetic..............262
   specific......................................262
Inertia, moment of.........................262
Intensity, electro-motive....................262
   of current.................................262
   of field....................................262
   of magnetization.........................262
Interconversion, of British and Metric
   units.....................................260, 261
   nautical and statute miles.........li, 208
   sidereal and solar time............liii, 218
International meteorological symbols...263

James, H., treatise cited.......................160
Joule, value of..............................259, 261

Kilogramme-degree...........................259
Kilogrammes, conversion into avoirdupois..............................lvi, 230, 260
Kilogram-metres.............................260, 261
Kilogramme, prototype......................lvi
Kilometres into miles...............li, 206, 207
   per hour into metres per second,
                                    xlvi, 159
   miles per hour.........................261

Klafter, Wiener, value in metres and
   feet...208
Kranz or corona..............................266

Lambert's formula, mean wind direction.................................xliii, 148-153
Laplace, formula of..............................xx
Latent heat of liquefaction of ice..........259
   vaporization of water.....259
Latitude, gravity correction for,
                           xix, xxx, 106, 114
Laughton, J. K. treatise cited...............160
Length of arc of meridian.......... xlvii, 164
                           parallel.............xlvii, 165
   dimensions of.............................262
   of equator of earth......................258
   meridian circumference of........258
   second's pendulum.....................258
   measures of, Continental with metric and British equivalents.......li, 208
   units of....................................260
Libby, Wm., work cited,
                           xxxviii, xlviii, lvii
Lichtkron or corona..........................266
Light, velocity of..............................259
Lightning, symbol for........................266
Line, old French, value of..................208
Linear measures....................xlix, 179-208
Litre, value of.............................260, 261
Logarithms, table of...............lix, 240-241
   Naperian base............................258
   Modulus of common.....................258
Lunar aureola, halo, corona. ....,......266

Magnetic force............................ 262
   inductive capacity........................262
   intensity, units of.........................lvii
      table for converting.........lvii, 231
   moment 262
   potential 262
Magnetism, quantity of.......................262
Magnetization, intensity of...................262
Marek, M., treatise cited....................xxi
Marvin, C. F., treatise cited............xxxvi
Mass, dimensions of.........................262
   units of........................260, 261
Mean density of the earth...................258
Mean distance of earth from sun..........259
Mean solar time, conversion into sidereal,
                                    liii, 218
Mean time at apparent noon.........liii, 217
Measures of angle.........................li, 209
   of length, continental, Metric and
      British..............................li, 208

INDEX.

Measures (*Continued*).
   time..li, 209
   tables for interconversion of....260, 261
Mechanical equivalent of heat..............259
Megalerg...261
Mercury, coefficient of expansion...xvi, 259
   density of.......................................259
   freezing point of.............................259
Mercurial column, one inch high, pressure of..259
Meridian, arcs of terrestrial..................xlvii
   length of a degree..................xlvii, 164
   ellipse, perimeter of........................258
Meteorological stations, list of...lix, 243-257
Metre...vi, xlix, 260, 261
Metres, conversion into feet, l, 202-203, 261
   per second into kilometres per hour, xlvi, 158, 159
      miles per hour....xlv, 156, 157, 261
Micron..261
Mile, Austrian post, value of................208
   Danish,   " 208
   German sea,   " 208
   Nautical,   "   ....li, 208
   Netherlands (migl), " 208
   Norwegian,   " 208
   Prussian,   " 208
   Swedish,   " 208
   Statute (British), " 208
Miles, conversion into kilometres, li, 204-205, 261
   nautical..li, 208
   statute..li, 208
   per hour into feet per second, xlv, 154, 155
      metres per second, xlv, 154, 157, 260
      kilometres per hour...xlv, 154, 260
Millimetres, conversion into inches, l, 187-199
Minutes of time, into arc....................lii, 211
   into decimals of a day..................lii, 216
   conversion of day into..................liii, 216
   conversion into decimals of an hour, liii, 217
Moment of inertia..................................262
Momentum, dimensions of....................262
Mondhof or lunar corona......................266
Mondring or lunar halo.......................266
Moorrauch or high haze......................265
Moritz, A., treatise cited......................xxxv
Munich Conference......................263, 264

Naperian base of logarithms.................258
Nautical mile, equivalent in statute, li, 208

Neumayer, G., treatise cited................160
Numerical constants............................258
Numbers, logarithms of..........lix, 240, 241

Ohm, dimensions of..............................262
Ounces, conversion into kilogrammes, lvi, 229
   kilogrammes into.......lvi, lvii, 230, 261
Palm, Netherlands, value of..................208
Parallel, length of a degree on.....xlvii, 165
Pendulum, length of second's..............258
Pentades, Dove's......................lviii, 232
Perimeter of meridian ellipse..............258
Physical constants..............................259
   quantities, dimensions of................262
Potential, difference of........................262
   in electro-magnetics......................262
   magnetic..262
Pound, avoirdupois, conversion into kilogramme.......................lvi, 229, 260
   imperial standard...........................lvi
Pounds, per square foot......................260
   inch..260
Poundal......................................260, 261
Pound-degree.......................259, 260, 261
Power or rate of working....................262
Pressure of aqueous vapor, British, xxxviii, 122, 128, 134, 135
Pressure of aqueous vapor, Metric, xli, 128, 129, 142
Pressure of aqueous vapor at low temperatures..................xxxvi, 130, 131
Pressure of standard atmosphere..........259
   decrease of vapor pressure with altitude........................xliii, 146
Prototype kilogramme...........................lvi
Psychrometer, whirled........xxxviii, xxxix
Psychrometric observations,
   reduction of, British.....xxxix, 134, 137
      Metric..........xli, 142-143

Quantity of electricity 262
      conveyed by current................262
      magnetism..................................262
Quantities physical, dimensions of.......262
Quantity of water corresponding to given depths of rainfall............lviii, 232

Rainbow, symbol for............................266
Rainfall, conversion of depth of, into gallons and tons...................lviii, 232
   symbol for.....................................264

Rate of decrease of vapor pressure with altitude..................xliii, 146
Rate of working, dimensions of............262
Ratio of specific heats of air.................259
  yard to metre vi
Rauhfrost, or silver frost......................264
Reaumur, conversion to centigrade.....xi, 2
  Fahrenheit............................xi, 2
Regnault, treatise cited...xxxiv, xxxv, 259
Recknagel, work cited.........................259
Reduction of barometer to sea level, xx–xxiv, 60–98
  standard temperature, xv–xix, 14–56
  gravity......................xxxi, 58, 59
  psychrometric observations, xli, 142, 143
  of snowfall measurement.........xlii, 146
Relative humidity, xxxviii, xlii, 138–141, 144–145
Relative intensity of solar radiation, xlix, 178
Resistance of a conductor.....................262
  specific..................................262
Rode, Danish, value of........................208
Röntgen, work cited............................259
Rowland, H. A., treatise cited..............259
Ruthe, Prussian, value of.....................208
  Norwegian, " 208

Sagene (Russian), value of...................208
Scales, comparison of thermometric.......xi
  Reaumur to Fahrenheit.................. 2
    Centigrade............................. 2
  Fahrenheit to Centigrade.............3–6
  Centigrade to Fahrenheit.............7–9
Schott, C. A., treatise cited..................160
Scott, R. H., treatise cited..... 160
Schneegestöber or drifting snow...........265
Sea-level, reduction of barometer to,
  British....................xx–xxv, 60–77
  Metric....................xx–xxvi, 78–98
Seconds, of time into arc.....................211
  decimals of a day...............lii, 216
  conversion of decimals of a day into, liii, 216
  of time into decimals of an hour, liii, 217
  pendulum, length of....................258
  reduction for, sidereal or solar time, liv, 218
  circumference of circle in..............258
  arc of circle in..........................258
Sidereal day and year..........................259
  time, conversion to mean solar, liii, liv, 218

Silver frost...264
Sines, table of natural..............lix, 236–237
Sleet, symbol for.................................264
Snowfall, symbol for...........................264
  weight of and corresponding depth of water.....................xlii–xliii, 146
Solar, day mean.................................259
  time, mean, conversion to sidereal, liii, liv, 218
  aureola, symbol for....................266
  corona   "   " 266
  halo   "   " 266
  radiation, relative intensity of, xlix, 178
Sonnenhof, symbol for........................266
Sonnenring "   " 266
Sound, velocity of..............................259
Specific heat of dry air........................259
  heats, ratio of, of air....................259
  conductivity.............................262
  inductive capacity......................262
  resistance 262
Spheroid, Clarke's.................. xlvii, 258
Standard atmosphere....................259, 261
Stations, International Polar...... 257
  list of meteorological.........lix, 243–257
  of first order...............................lix
  in Africa 256–257
    Asia, 254–255, Australasia, 256,
    Europe, 247–254, North America,
    244–246, South America, 246–247,
    Austro-Hungary, 247–248, Belgium, 248, British Isles, 248–249,
    Canada, 244, Central America, 244,
    Denmark, 249, France, 249, Germany, 250, Greece, 248, Greenland, 244, Holland, 248, Italy,
    251, Mexico, 244, Norway, 249,
    Portugal, 253, Roumania, 248,
    Russia, 251–253, Spain, 253,
    Sweden, 249, Switzerland, 253,
    254, Turkey, 248, United States,
    245–246, West Indies, 244.
Statute miles, conversion of............li, 208
Stere, value of..............................260, 261
Strength of field................................262
  pole, in magnetics......................262
Stress, dimensions of..........................262
  units of..............................260, 261
Sun, declination of....................xlix, 177
  mean distance from the earth........259
Sunshine, duration of, at different latitudes and declinations......xlviii, 166–177
Surface (area), units of.................260, 261

INDEX. 273

Surface (*Continued*).
    density of the earth..........................258
    in electro-statics.......................262
Symbols, International Meteorologic, 263–266
Synoptic conversion of British and Metric units..............................260–261

Table for conversion of arc into time li, 210
    linear measures........................180–208
    mean solar into sidereal.....liii, liv, 218
    measures of weight.................229–231
    sidereal into mean solar.....liii, liv, 218
    time into arc...........................lii, 211
    Centigrade readings into Fahrenheit and Reaumur.......xi, xii, 7, 8, 9
        near boiling point................xii, 9
    velocities...............................154–159
    differences F to differences C......xiii, 9
    " C " F......xiii, 9
    Fahrenheit readings into Centigrade.............................xi, xii, 3–6
    Reaumur readings into F and C...xi, 2
    determination of heights by barometer, British.....................100–109
    determination of heights by barometer, Metric...................110–115
    decrease of vapor pressure with altitude...................................146
    dividing by 28..............lviii, lix, 233
        29..........................lviii, lix, 234
        30..........................lviii, lix, 235
    density of air........................220, 228
    reduction of barometer to standard temperature.........................14–56
        gravity 58–59
        sea-level............................60–98
        psychrometric observations, 136–145
        snowfall measurements............146
        temperature to sea-level......10, 11
    of duration of sunshine............166–177
    intensity of solar radiation.............178
    lengths of degree....................164–165
    natural cosines and sines.........236–237
        cotangents and tangents....238–239
    pressure of aqueous vapor........122–131
    pressures and corresponding boiling points...........................119
    quantity of rainfall and corresponding depths........................232
    of relative acceleration of gravity, 162, 163

Tables of relative humidity, 138–141, 144–145
Table of weight of aqueous vapor...132–133
Tangents, table of natural......lix, 238, 239
Temperature, absolute zero of..............259
    of freezing point of mercury.........259
    decrease of, with altitude, xiii, xiv, 10, 11
    reduction to sea level, British, xiii, xiv, 10
    reduction to sea level, Metric, xiii, xiv, 11
Thermometer, hypsometric..............xxxiv
    stem, correction for temperature of mercury ....xiv, 12
Therm or gramme degree....................261
Thermal conductively of air.................259
Thermometric scales....................xi, 2–9
Thomson, W., treatise cited.................259
Thorpe, T. E., " xiv
Thunderstorm, symbol for...................265
Time, conversion into arc...............lii, 211
    of arc into................................li, 210
    dimensions of 262
    mean, at apparent noon...........liii, 217
    mean solar into sidereal......liii–liv, 218
    minutes of, into arc.......................211
    seconds of, " 211
    sidereal into mean solar.....liii–liv, 218
Toisè, old French...............................208
Ton...260
Tonne 261
Tropical year................................liii, 259

Units of magnetic intensity...........lvii, 231
    interconversion of British and Metric...................................260, 261

Vapor aqueous, pressure of, British, xxxv, 122–127
    Metric, xxxvi, 128–129
    at low temperature ......xxxvi, 130, 131
    decrease of pressure with altitude, xliii, 146
    weight of.......xxxvii, xxxviii, 132, 133
Vaporization, latent heat of, of water...259
Vara, Mexican, value of......................208
    Spanish, " 208
Velocity, dimensions of.......................262
    of light....................................259
    sound.....................................259
    units of..............................260, 261
Velocities, conversion of...xlv, xlvi, 154–159
Verglas or glazed frost, symbol for.......264

Versta or Werst (Russian)..................208
Volt, dimensions of...........................262
Volume, dimensions of.......................262
    units of................................260, 261
    of distilled water.........................259

Water, distilled, volume and weight of, 259
    latent heat of vaporization of.........259
    specific heat of, compared with air...259
Watt...260, 261
Weight of aqueous vapor,
            xxxvii, xxxviii, 132, 133
    distilled water..............................259
    dry air......................................259
    one grain in dynes.......................260
        pound in dynes.....................260
        dyne 261

Werst or versta, Russian.....................208
Wind, mean direction by Lambert's
    formula...........................xliii, 148-153
    scale, Beaufort, conversion.....xlvi, 160
    symbols for.................................265
    tables............................xliii, 148-160
Work, dimensions of..........................262
    units of, in absolute measure...260, 261
Working, rate of...............................262

Yard, ratio of to metre............................vi
Year, conversion of days into decimals
    of, and angle.........................lii, 212-215
    bissextile, days into decimals of,
                                    lii, 212-215
    length of tropical....................liii, 259
    sidereal 259

Zero, absolute, of temperature..............259

www.ingramcontent.com/pod-product-compliance
Lightning Source LLC
Chambersburg PA
CBHW021204230426
43667CB00006B/548